# The Receptors

Volume 28

*Series editor*

Giuseppe di Giovanni

More information about this series at http://www.springer.com/series/7668

Patrizio Blandina • Maria Beatrice Passani
Editors

# Histamine Receptors

## Preclinical and Clinical Aspects

Humana Press

*Editors*
Patrizio Blandina
Department of Neuroscience, Psychology
Drug Research and Child Health
University of Florence
Florence, Italy

Maria Beatrice Passani
Department of Health Science
University of Florence
Florence, Italy

The Receptors
ISBN 978-3-319-40306-9 ISBN 978-3-319-40308-3 (eBook)
DOI 10.1007/978-3-319-40308-3

Library of Congress Control Number: 2016952257



Printed on acid-free paper

This Humana Press imprint is published by Springer Nature
The registered company is Springer International Publishing AG
The registered company address is: Gewerbestrasse 11, 6330 Cham, Switzerland

*The understanding of the functions*
*of histamine owes much to the ideas*
*of Pier Francesco Mannaioni. This book*
*is dedicated to him, a mentor*
*and a good friend of both of us.*

Patrizio Blandina and
Maria Beatrice Passani

# Preface

Vasodilation and contraction of smooth muscles in the gut were the first biological properties of histamine. The name derives from the combination of *histo- + amine* (tissue amine), and it was first described more than a century ago by Henry H. Dale and Patrick P. Laidlaw. Histamine has long been known to stimulate gastric acid secretion and to cause constriction of the bronchi during anaphylactic shock. Histamine is involved in inflammatory responses and in the generation of pruritus. As part of an immune response to foreign pathogens, histamine released by basophils and mast cells increases the permeability of capillaries to white blood cells and proteins, to allow them to engage pathogens in the infected tissues. Despite early studies suggesting a function in the brain, histamine as a neurotransmitter gained general recognition only more recently, after the morphological characterization of histaminergic neurons. The sole source of brain histamine is the hypothalamic tuberomammillary nuclei, where the somata of neurons projecting axons to the whole brain are to be found. Histamine regulates homeostatic and other functions, including arousal, circadian rhythms, learning and memory, and feeding. The variety of peripheral and central effects mediated by histamine is the result of the activation of $H_1$, $H_2$, $H_3$, and $H_4$ receptor subtypes. $H_1$ and $H_2$ receptor antagonists have been very successful therapeutic agents in the treatment of allergic reactions and gastric ulcers, respectively. For the discovery of these drugs, Daniel Bovet and James Black were awarded the Nobel Prize in Physiology or Medicine in 1957 and 1988, respectively. In this book the current state of the art of these receptors is presented by a number of experts in the field. We wish that it will increase the interest of the scientific community for these receptors as targets for the development of novel drugs.

# Contents

# Contributors

**José-Antonio Arias-Montaño** Departamento de Fisiología, Biofísica y Neurociencias, Centro de Investigación y de Estudios Avanzados del IPN, Ciudad de México, Mexico

**C. Ballerini** Dipartimento di Neuroscienze, Psicologia, Area del Farmaco e Salute del Bambino-Universitá di Firenze, Florence, Italy

**L. Bernardino** Health Sciences Research Centre, Faculty of Health Sciences, University of Beira Interior, Covilhã, Portugal

**Patrizio Blandina** Department of Neuroscience, Psychology, Drug Research and Child Health, University of Florence, Florence, Italy

**R. Bosma** Amsterdam Institute for Molecules, Medicines and Systems, Division of Medicinal Chemistry, Faculty of Sciences, Vrije universiteit Amsterdam, Amsterdam, The Netherlands

**Yu-Chia Chen** Department of Anatomy and Neuroscience Center, University of Helsinki, Helsinki, Finland

**R. Ferreira** Health Sciences Research Centre, Faculty of Health Sciences, University of Beira Interior, Covilhã, Portugal

**Hiroyuki Fukui** Department of Molecular Studies for Incurable Diseases, Institute of Biomedical Sciences, Tokushima University Graduate School, Tokushima, Japan

**Cristiane R. G. Furini** Memory Center, Brain Institute of Rio Grande do Sul, Pontifical Catholic University of Rio Grande do Sul, Porto Alegre, RS, Brazil

**Ana-Maricela García-Gálvez** Departamento de Fisiología, Biofísica y Neurociencias, Centro de Investigación y de Estudios Avanzados del IPN, Ciudad de México, Mexico

**Patrizia Giannoni** Department of Neuroscience, Institute of Functional Genomics, CNRS-University of Montpellier (UMR 5203), Montpellier, France

**Helmut L. Haas** Faculty of Medicine, Department of Neuro- and Sensory Physiology, Heinrich-Heine-University Düsseldorf, Düsseldorf, Germany

**S. Hagenow** Institute of Pharmaceutical and Medicinal Chemistry, Heinrich Heine University, Düsseldorf, North Rhine-Westphalia, Germany

**Kotaro Hiraoka** Cyclotron and Radioisotope Center (CYRIC), Tohoku University, Sendai, Japan

**Ivan Izquierdo** Memory Center, Brain Institute of Rio Grande do Sul, Pontifical Catholic University of Rio Grande do Sul, Porto Alegre, RS, Brazil

**Anikó Kárpáti** Department of Pharmacology, Tohoku University School of Medicine, Sendai, Japan

**K. Kieć-Kononowicz** Department of Technology and Biotechnology of Drugs, Jagiellonian University Medical College, Kraków, Poland

**Yoshiaki Kitamura** Department of Otorhinolaryngology, Institute of Biomedical Sciences, Tokushima University Graduate School, Tokushima, Japan

**D. Łażewska** Department of Technology and Biotechnology of Drugs, Jagiellonian University Medical College, Kraków, Poland

**Rob Leurs** Amsterdam Institute for Molecules, Medicines and Systems, Division of Medicinal Chemistry, Faculty of Sciences, Vrije universiteit Amsterdam, Amsterdam, The Netherlands

**Ricardo Márquez-Gómez** Departamento de Fisiología, Biofísica y Neurociencias, Centro de Investigación y de Estudios Avanzados del IPN, Ciudad de México, Mexico

**Daniel A. McNaught-Flores** Amsterdam Institute for Molecules, Medicines and Systems, Division of Medicinal Chemistry, Faculty of Sciences, Vrije universiteit Amsterdam, Amsterdam, The Netherlands

**Stephany Micallef** School of Pharmacy and Pharmaceutical Sciences, Trinity Biomedical Sciences Institute, Trinity College Dublin, Dublin 2, Ireland

**Hiroyuki Mizuguchi** Department of Molecular Pharmacology, Institute of Biomedical Sciences, Tokushima University Graduate School, Tokushima, Japan

**T. A. M. Mocking** Amsterdam Institute for Molecules, Medicines and Systems, Division of Medicinal Chemistry, Faculty of Sciences, Vrije universiteit Amsterdam, Amsterdam, The Netherlands

**María Morales** Departamento de Biología Molecular y Bioquímica, Facultad de Ciencias, Universidad de Málaga, Malaga, Spain

**Guadalupe-Elide Morales-Figueroa** Departamento de Fisiología, Biofísica y Neurociencias, Centro de Investigación y de Estudios Avanzados del IPN, Ciudad de México, Mexico

**Jociane de Carvalho Myskiw** Memory Center, Brain Institute of Rio Grande do Sul, Pontifical Catholic University of Rio Grande do Sul, Porto Alegre, RS, Brazil

**Fumito Naganuma** Department of Pharmacology, Tohoku University School of Medicine, Sendai, Japan

Department of Pharmacology, Tohoku Medical and Pharmaceutical University, Sendai, Japan

**Tadaho Nakamura** Department of Pharmacology, Tohoku University School of Medicine, Sendai, Japan

Department of Pharmacology, Tohoku Medical and Pharmaceutical University, Sendai, Japan

**Saara Nuutinen** Division of Pharmacology and Pharmacotherapy, Faculty of Pharmacy, University of Helsinki, Helsinki, Finland

**Nobuyuki Okamura** Department of Pharmacology, Tohoku University School of Medicine, Sendai, Japan

Cyclotron and Radioisotope Center (CYRIC), Tohoku University, Sendai, Japan

Department of Pharmacology, Tohoku Medical and Pharmaceutical University, Sendai, Japan

**Pertti Panula** Department of Anatomy and Neuroscience Center, University of Helsinki, Helsinki, Finland

**Maria Beatrice Passani** Department of Health Science, University of Florence, Florence, Italy

**Gustavo Provensi** Department of Neuroscience, Psychology, Drug Research and Child Health, University of Florence, Florence, Italy

**Henri Puttonen** Department of Anatomy and Neuroscience Center, University of Helsinki, Helsinki, Finland

**S. N. Rahman** Amsterdam Institute for Molecules, Medicines and Systems, Division of Medicinal Chemistry, Faculty of Sciences, Vrije universiteit Amsterdam, Amsterdam, The Netherlands

**Rocío Rodríguez-López** Departamento de Biología Molecular y Bioquímica, Facultad de Ciencias, Universidad de Málaga, Malaga, Spain

Instituto de Biomedicina IBIMA, Malaga, Spain

**Francisca Sánchez-Jiménez** Departamento de Biología Molecular y Bioquímica, Facultad de Ciencias, Universidad de Málaga, Malaga, Spain

Unidad 741 CIBER de Enfermedades Raras, Malaga, Spain

Instituto de Biomedicina IBIMA, Malaga, Spain

**B. Sadek** Department of Pharmacology and Therapeutics, College of Medicine and Health Sciences, Al Ain, United Arab Emirates University, Al Ain, Abu Dhabi, United Arab Emirates

**Outi Salminen** Division of Pharmacology and Pharmacotherapy, Faculty of Pharmacy, University of Helsinki, Helsinki, Finland

**Astrid Sasse** School of Pharmacy and Pharmaceutical Sciences, Trinity Biomedical Sciences Institute, Trinity College Dublin, Dublin 2, Ireland

**Olga A. Sergeeva** Faculty of Medicine, Department of Neuro- and Sensory Physiology, Heinrich-Heine-University Düsseldorf, Düsseldorf, Germany

Department of Clinical Neuroscience and Medical Psychology, Heinrich-Heine-University Düsseldorf, Düsseldorf, Germany

**H. Stark** Institute of Pharmaceutical and Medicinal Chemistry, Heinrich Heine University, Düsseldorf, North Rhine-Westphalia, Germany

**Maria Sundvik** Department of Anatomy and Neuroscience Center, University of Helsinki, Helsinki, Finland

**Noriaki Takeda** Department of Otorhinolaryngology, Institute of Biomedical Sciences, Tokushima University Graduate School, Tokushima, Japan

**Manabu Tashiro** Cyclotron and Radioisotope Center (CYRIC), Tohoku University, Sendai, Japan

**Robin L. Thurmond** Department of Immunology, Janssen Research & Development, LLC, San Diego, CA, USA

**E. W. E. Verweij** Amsterdam Institute for Molecules, Medicines and Systems, Division of Medicinal Chemistry, Faculty of Sciences, Vrije universiteit Amsterdam, Amsterdam, The Netherlands

**Henry F. Vischer** Amsterdam Institute for Molecules, Medicines and Systems, Division of Medicinal Chemistry, Faculty of Sciences, Vrije universiteit Amsterdam, Amsterdam, The Netherlands

**Kazuhiko Yanai** Department of Pharmacology, Tohoku University School of Medicine, Sendai, Japan

Cyclotron and Radioisotope Center (CYRIC), Tohoku University, Sendai, Japan

**Takeo Yoshikawa** Department of Pharmacology, Tohoku University School of Medicine, Sendai, Japan

# Chapter 1
# Molecular Aspects of Histamine Receptors

**T.A.M. Mocking, R. Bosma, S.N. Rahman, E.W.E. Verweij, Daniel A. McNaught-Flores, Henry F. Vischer, and Rob Leurs**

**Abstract** Histamine mediates a multitude of physiological effects in the human body by activating four histamine receptor subtypes. Histamine receptors have proven to be promising drug targets in the treatment of a variety of diseases, including hay fever, gastric ulcers, and inflammatory and neuropathological diseases. In this chapter the molecular aspects of histamine receptors are described, including expression profile, intracellular signaling, and how histamine receptor activity can be modulated by ligands targeting the histamine receptor binding sites. Special attention is given to the possible effect of cellular contexts that might impact the signaling properties of these receptors, like receptor desensitization and internalization, formation of oligomers, and biased signaling. Moreover, insights on the structural effects of antihistamines on the binding affinity and the lifetime of the receptor-ligand compex are discussed.

**Keywords** Histamine receptors • Binding kinetics • Ligand-receptor interactions • G protein signaling • Biased signaling • Desensitization • Internalization

## 1.1 Histamine Receptors

Histamine is a secreted messenger involved in the local regulation of physiological processes. Histamine binds and activates four histamine receptor subtypes that are differentially expressed. The histamine $H_1$ receptor ($H_1R$) is expressed on the surface of a wide variety of cell types, including epithelial, smooth vascular, vascular endothelial, immune, glial, and neuronal cells [1]. $H_1R$ is therapeutically best known for its role in allergic and anaphylactic responses and nausea due to motion sickness [2]. The histamine $H_2$ receptor ($H_2R$) is found on neurons, epithelial, endothelial,

T.A.M. Mocking • R. Bosma • S.N. Rahman • E.W.E. Verweij • D.A. McNaught-Flores
H.F. Vischer (✉) • R. Leurs (✉)
Amsterdam Institute for Molecules, Medicines and Systems, Division of Medicinal Chemistry, Faculty of Sciences, Vrije universiteit Amsterdam, De Boelelaan 1083, 1108 HZ, Amsterdam, The Netherlands
e-mail: h.f.vischer@vu.nl; r.leurs@vu.nl



P. Blandina, M.B. Passani (eds.), *Histamine Receptors*, The Receptors 28,
DOI 10.1007/978-3-319-40308-3_1

immune, and gastric cells and is involved in the regulation of gastric acid secretion, heart rate, and blood pressure [3]. The histamine $H_3$ receptor ($H_3R$) is predominantly expressed in the central nervous system (CNS) and linked to various neurophysiological processes, including sleep-wake cycle, weight regulation, and homeostasis [4]. In addition, $H_3R$ is also expressed in the periphery, like e.g., the cardiovascular system and the gastrointestinal tract [5]. The histamine $H_4$ receptor ($H_4R$) is found to be widely expressed in tissues and cells that are involved in immune responses (e.g., the lung, spleen, lymph node, bone marrow, and leukocytes) [6, 7].

Because of their role in a wide variety of (patho)physiological processes, histamine receptors are important drug targets. Understanding the molecular aspects of histamine receptor subtypes will be helpful in understanding how these receptors can be (selectively) regulated by ligands, how the receptors regulate downstream signaling, and how the cellular background influences receptor-mediated responses. Together, this forms the fundaments for understanding how the receptor is involved in pathophysiology and how these receptors can be therapeutically exploited.

## 1.2 Genomic Organization of Histamine Receptors and Molecular Features

The open reading frames (ORF) of $H_1R$ and $H_2R$ are encoded by two intronless genes (Fig. 1.1a, b) [8–17]. The human $H_1R$ ($hH_1R$) and $H_2R$ ($hH_2R$) consist of 491 and 360 amino acids, respectively, and belong to the class A G protein-coupled receptor (GPCR) family. Members of this structural family are characterized by an extracellular N-terminus, seven transmembrane helices that are alternating connected by three intracellular loops (ICL) and three extracellular loops (ECL), and an intracellular C-terminus [18]. In the genomic organization of the $hH_1R$ gene, multiple transcription start sites (TSS) exist and alternative splicing events in 5′ exons (A-J) result in mRNA transcripts with distinct five prime untranslated region (5′ UTR) in various tissues (Fig. 1.1a) [19]. Importantly, these events do not result in alterations in the translated $H_1R$ protein, but the different mRNA transcripts might be important for tissue-specific mRNA processing. The $H_1R$ proteins have been highly conserved throughout evolution as $hH_1R$ displays high overall sequence

**Fig. 1.1** (continued) The ORF encodes the $hH_1R$ protein. (**b**) The human histamine $H_2$ receptor ($hH_2R$) gene is located at chromosome 5q32.2. The genomic organization of $hH_2R$ consists of three exons (A–C), of which exon C contains the ORF encoding the $hH_2R$ protein. Moreover, it contains multiple TSS that are important for the differential expression in cell and tissue types. (**c**) The human histamine $H_3$ receptor ($hH_3R$) gene is located at chromosome 20 q13.33. The ORF of $hH_3R$ consists of four exons (A, B, C, and D). The first exon encodes the 5′ UTR, transmembrane domain (TM) 1 (I), and 5′ end of TM2 (II), the second encodes the 3′ end of TM2 and TM3 (III), the third encodes TM4 (IV) and TM5 (V), and the fourth encodes TM6 (VI) and TM7 (VII) and the 3′ UTR. (**d**) The human histamine $H_4$ receptor ($hH_4R$) gene is located at chromosome 18q11.2 and its ORF consists of three exons (A, B, and C). The first exon encodes the 5′ UTR, TM 1, and 5′ end of TM2, the second encodes the 3′ end of TM2 and TM3, and the third encodes TM4-7 and the 3′ UTR

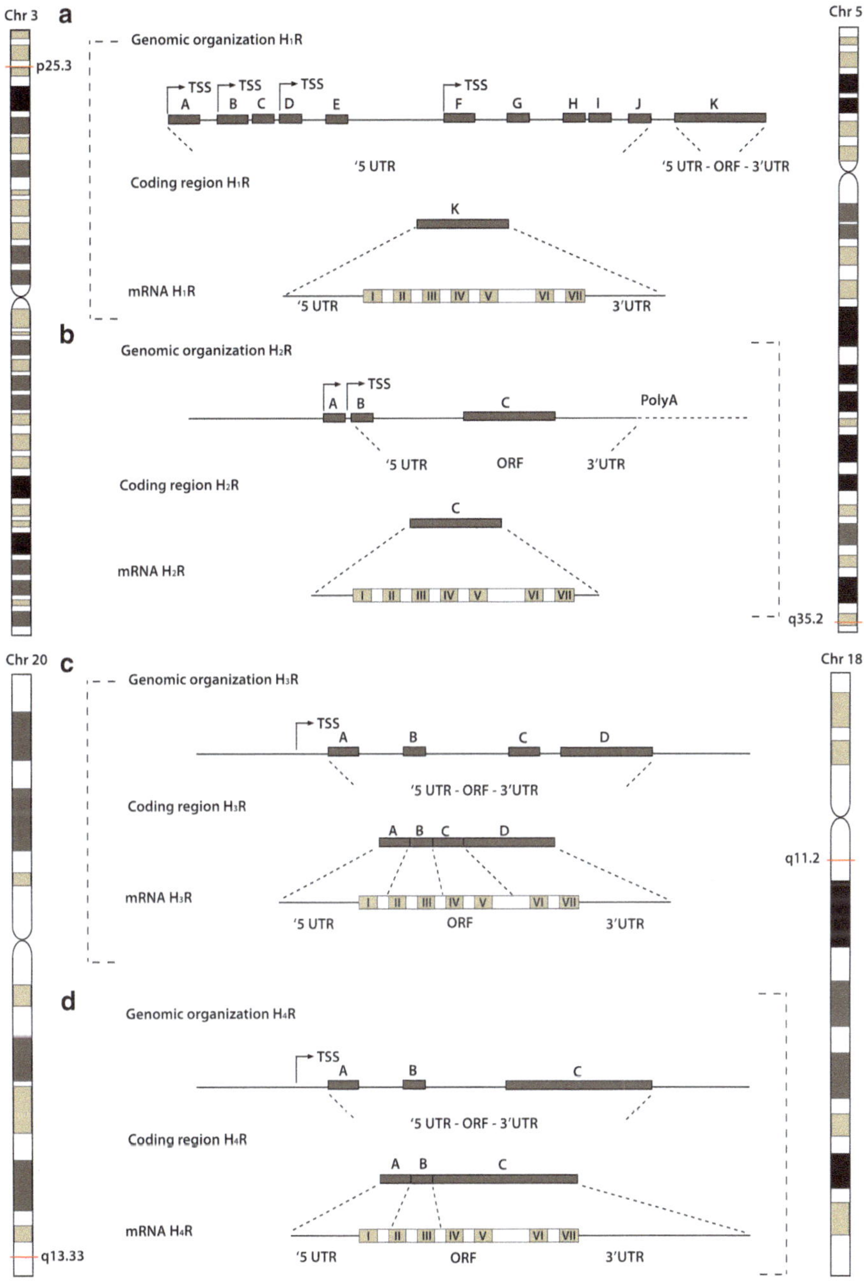


**Fig. 1.1** Schematic representation of genomic organization of genes encoding the human histamine receptors subtypes (based on [7, 19, 22, 24]): (**a**) the gene encoding the human histamine $H_1$ receptor ($hH_1R$) is located at chromosome 3p25.3 and consists of 11 exons (A–K) with multiple transcription start sites (TSS). $hH_1R$ is encoded by an intronless gene existing of a single exon (K), which consists of the open reading frame (ORF), and 5′ (5′UTR) and 3′ (3′UTR) untranslated regions.

identity with bovine (83 %), rat (79 %), mouse (76 %), and guinea pig (gp) (73 %) orthologs, with even more than 90 % sequence identity in the seven transmembrane domains (TMs) [11, 20]. The overall sequence identity of $H_1R$ orthologs from lower species, e.g., chick (46 %) and zebra fish (40 %), is less well conserved [21].

The human $H_2R$ gene has multiple transcription initiation sites located upstream mediating the differential expression of $H_2R$ in cell and tissue types (Fig. 1.1b) [17, 22]. Amino acid sequence alignments of $hH_2R$ and its orthologs in zebra fish (44 %), chick (56 %), mouse (85 %), guinea pig (86 %), rat (86 %), and canine (87 %) revealed moderate to high overall sequence identity [1, 13, 14, 16, 21, 23].

In contrast, the open reading frames of human $H_3R$ ($hH_3R$) and $H_4R$ ($hH_4R$) are encoded by four and three exons, respectively (Fig. 1.1c, d) [6, 24, 25]. The first identified $hH_3R$ and $hH_4R$ consist of 445 and 390 amino acids. However, isoforms have been described for both receptors as a consequence of alternative mRNA splicing. The genomic organization (i.e., exon-intron structure) of the $H_3R$ gene varies slightly in other species (e.g., rat and monkeys) as it consists of three exons and two introns [26, 27]. Nevertheless, from deduced amino acids (sequence identity), it is clear that $hH_3R$ and its zebra fish (50 %), chicken (59 %), rat (93 %), guinea pig (95 %), mouse (98 %), and monkey (98 %) orthologs have been highly conserved throughout evolution [4, 21, 28–30]. The $H_4R$ gene exhibits more interspecies differences. Identity of the amino acid sequence for rat, mouse, guinea pig, canine, porcine, and monkey ranged from 67 to 72 and 92 % for the evolutionary closest gene of monkeys [31–34].

Twenty $hH_3R$ isoforms have been described with deletions in different structural regions (Fig. 1.2) [35, 36]. Seven isoforms (i.e., $hH_3R$-453, -445, -415, -413, -373, -365, -329a) conserve the prototypical 7TM domain but are pharmacologically and functionally different as a consequence of differences in ICL3 and the C-tail (Fig. 1.2) [24, 37]. Thirteen isoforms (i.e., $hH_3R$-431, $hH_3R$-409, $hH_3R$-395, $hH_3R$-379, $hH_3R$-351, $hH_3R$-340, $hH_3R$-329b, $hH_3R$-293, $hH_3R$-326, $hH_3R$-290, $hH_3R$-309, $hH_3R$-301, and $hH_3R$-200) lack the typical 7TM domain and can act as dominant negatives by hampering cell surface expression of 7TM $H_3R$ isoforms [38]. The $hH_3R$ splice variants are found to be differentially expressed on presynaptic neurons found within the central nervous system (CNS) [24]. Similarly, non-7TM $hH_4R$ splice variants (i.e., $hH_4R$-67 and $hH_4R$-302) act as dominant negatives and retain full-length $hH_4R$ intracellularly through hetero-oligomerization [39]. Currently, no 7TM $H_4R$ isoforms have been described.

The four human histamine receptor subtypes share a moderate to low degree of amino acid sequence similarity (Table 1.1) but conserve multiple 7TM motifs that are characteristic for class A G protein-coupled receptors (GPCRs) [6, 40–43]. Phylogenetic analysis revealed that $H_1R$ is most closely related to the human muscarinic receptor family members, $hH_2R$ to serotonin receptors, whereas $hH_3R$ and $hH_4R$ are most similar to $\beta_2$-adrenergic receptors [43]. In general, histamine receptors contain several consensus sites important for glycosylation of the N-terminal tail, protein kinase A (PKA), and protein kinase C (PKC)-mediated phosphorylation of the intracellular loops and palmitoylation of the C-terminal tail.

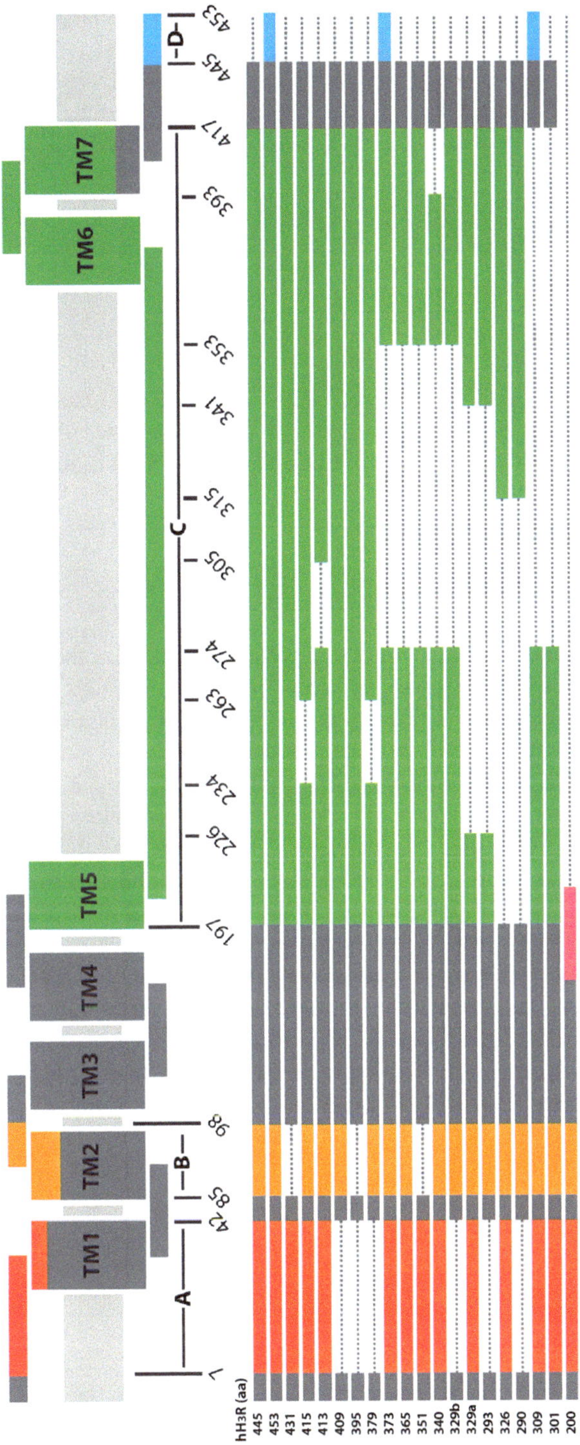


**Fig. 1.2** Schematic representation of human histamine $H_3$ receptor isoforms. In humans, 20 different $H_3$ receptor isoforms ($hH_3R$) exist. Alternative splicing events in four distinct regions have been reported (in *red* (7–42), *orange* (85–98), *green* (197–417), and *blue* (445–453)). The numbering is given according to the deduced amino acids (aa) of $hH_3R$-445. (modified from [36])

**Table 1.1** An overview of sequence identity at amino acid level shared between human histamine receptor ($hH_xR$) subtypes

| Sequence identity (%) | | | | | | | | |
|---|---|---|---|---|---|---|---|---|
| | $H_1R$ | | $H_2R$ | | $H_3R$ | | $H_4R$ | |
| | Overall | TMs | Overall | TMs | Overall | TMs | Overall | TMs |
| $H_1R$ | – | – | 21 [40] | 37 [287] | 22 [43] | 27 [43] | 19 [41] | 26 [287] |
| $H_2R$ | 21 [40] | 37 [287] | – | – | 20 [43] | 33 [43] | 19 [41] | 27 [287] |
| $H_3R$ | 22 [43] | 27 [43] | 20 [43] | 33 [43] | – | – | 37 [287] | 58 [287] |
| $H_4R$ | 19 [41] | 26 [287] | 19 [41] | 27 [287] | 37 [287] | 58 [287] | – | – |

Palmitoylation of cysteine is considered to be an important post-translational modification of many receptors, including $H_2R$, to anchor the C-terminal tail to the cell membrane [44]. Interestingly, a putative palmitoylation motif is absent in $hH_1R$.

## 1.3 Histamine Receptor Signaling

Histamine receptors activate heterotrimeric G proteins resulting in the dissociation of the α and βγ subunits. The histamine receptor subtypes display distinguished coupling preferences for the Gα subclasses and consequently regulate distinct intracellular signaling pathways.

The coupling of G proteins to GPCRs relies mainly on three domains on the receptor, i.e., ICL2, N-terminal, and C-terminal region of ICL3 [45]. The conserved DRY motif in TM3 is one of the key players in G protein coupling [46], with $R^{3.50}$ (Ballesteros-Weinstein numbering [47]) being crucial for G protein activation [48–52]. In addition, the crystal structures of rhodopsin and $D_3R$ showed a salt bridge between $R^{3.50}$ and $E^{6.30}$ in TM6 that stabilizes the receptor in the inactive state [53, 54], while in other crystallized GPCRs, the salt bridge was not present [55]. The residue at the 6.30 position for aminergic GPCRs is mostly negatively charged (D/E) highlighting the importance of this ionic lock. Although this motif is present in histamine receptors, this salt bridge was not observed in the $H_1R$ crystal structure [42].

Unique residues near ICL2 and ICL3 were exchanged between the $G_q$-coupled muscarinic $M_2$ receptor ($M_2R$) and $G_{i/o}$-coupled $M_3R$ to identify residues that confer $G\alpha_{q/11}$ protein coupling to the engineered $M_3R$ receptor. Interestingly, the identified combination of residues $S^{3.53}$, $R^{4.41}$, $Y^{5.58}$, $A^{6.33}$, $A^{6.34}$, and $L^{6.37}$ is conserved in the $G_q$-coupled $hH_1R$ [46, 56, 57]. Moreover, the $hH_2R$ only has the $R^{4.41}$, $Y^{5.58}$, $A^{6.33}$, and $L^{6.37}$ conserved and can, in addition to $G_s$, also couple $G_{q/11}$, while most of these residues are not present in the $hH_3R$ and $hH_4R$.

$G\alpha_{i/o}$, coupling to engineered $M_2R$, is dependent on residues located near the C-terminal end of ICL3, namely, $V^{6.33}$, $T^{6.34}$, $I^{6.37}$, and $L^{6.38}$ [56, 58]. From these residues $V^{6.33}$ in the $hH_3R$ is conserved, and both $hH_3R$ and $hH_4R$ have an $L^{6.37}$

instead of $I^{6.37}$ in wild-type $M_2R$ indicating that these positions might be mediating the coupling to the $G\alpha_{i/o}$ for the histamine receptors as well.

As mentioned before, the histamine receptors couple to one or more G proteins, which regulate different downstream signaling pathways. Signaling pathways of the respective histamine receptors are discussed below.

### 1.3.1 $H_1R$

The $H_1R$ couples to $G\alpha_{q/11}$ proteins resulting in the subsequent activation of phospholipase C-β (PLC-β), which catalyzes the hydrolysis of membrane phosphatidylinositol diphosphate ($PIP_2$) into membrane diacylglycerol (DAG) and the soluble second messenger inositol 1,4,5-trisphosphate ($IP_3$) (Fig. 1.3) [59, 60].

DAG-mediated activation of PKC results in a variety of physiological responses, including mitogen-activated protein kinases (MAPK)-induced cell proliferation and induction of interleukin 4 (IL-4) secretion (Table 1.2). Moreover, DAG activates the extracellular calcium influx via transient receptor potential cation channel 3 (TRPC3) and 6 in a PKC-independent fashion [61]. The second messenger $IP_3$ diffuses into the cytoplasm and binds to $IP_3$ receptors on the endoplasmic reticulum (ER) resulting in $Ca^{2+}$ mobilization. Intracellular calcium concentrations are therefore derived from both intracellular vesicles and the extracellular environment leading to various physiological responses, e.g., activation of specific PKC isoforms, smooth muscle contraction, nitric oxide synthase (NOS)-induced vasodilation, and stimulation of phospholipase $A_2$ (PLA2) to produce arachidonic acid (AA) from phospholipids (Table 1.2). Subsequently, AA is metabolized by cyclooxygenase (COX) into inflammatory mediators, and histamine release is induced [62–68]. In addition, $H_1R$-induced intracellular $Ca^{2+}$ binds to calmodulin (CaM) and activates CaM kinases (CaMK), resulting in the phosphorylation of cAMP response element-binding protein (CREB) and the subsequent proliferation of mouse cholangiocytes [65, 69]. CaM also binds to calcineurin (CaN) which dephosphorylates the nuclear factor of activated T cells (NFAT) and translocates it to the nucleus where it promotes, e.g., transcription of cytokines GM-CSF, IL-6, and IL-8 (CXCL8) in endothelial cells [70]. Another transcription factor activated by the $H_1R$ is the nuclear factor-κB (NF-κB) [71]. The activation of NF-κB is linked to the $Ca^{2+}$ signaling through CaM-CaN, PKC, and/or PI3K/Akt pathways [72], suggesting that these three pathways can activate the transcription factor when stimulated by $H_1R$.

### 1.3.2 $H_2R$

$H_2R$ predominantly couples to $G\alpha_s$ proteins in response to histamine, resulting in adenylyl cyclase (AC) activation and consequently increased 3′, 5′-cyclic adenosine monophosphate (cAMP) levels as observed in various tissues (Fig. 1.4) [73–77].

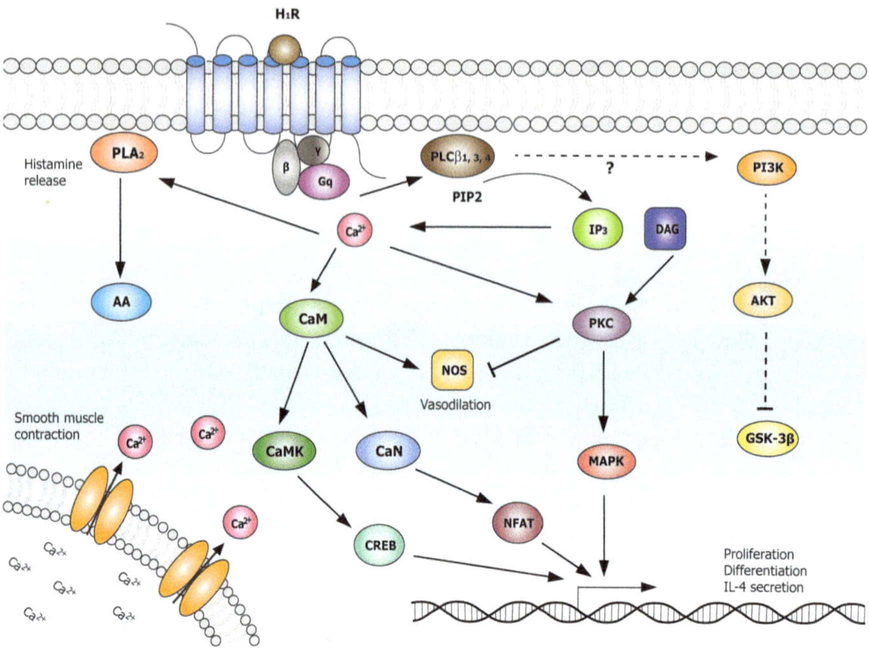


**Fig. 1.3** Signaling pathway of the $H_1R$. The $H_1R$ has been shown to modulate different signal transduction pathways including phospholipase C-β (PLC-β)-$IP_3$-diacylglycerol (DAG)-protein kinase (PKC) axis, mitogen-activated protein kinases (MAPK), nitric oxide synthase (NOS), calmodulin (CaM), calcineurin (CaN), phospholipase $A_2$ ($PLA_2$), and nuclear factor of activated T cells (NFAT). Dotted lines are suggested pathways; solid lines are canonical pathways

**Table 1.2** Overview of differential effects mediated by effectors after histamine receptor activation

| Receptor | Effector | Effect | Cell type | Reference |
|---|---|---|---|---|
| $H_1R$ | $Ca^{2+}$ | Increased NFAT-mediated cytokine production (GM-CSF/IL-6/IL-8) | HIUVEC cells | [70] |
| | | Increased NOS activity and cGMP production | Rat submandibular glands | [288] |
| | | (putatively by $Ca^{2+}$) NO/cGMP-induced blood vessel leakage | Hamster cheek pouch | [289] |
| | | Increased NO production | HUVEC cells | [65] |
| | | PKC-dependent cytokine production. (GM-CSF/IL-6/IL-8) | Human epidermal keratinocytes | [290] |
| | PKC | Induce IL-4 secretion | Mouse splenocytes | [291] |
| | | Differentiates pluripotent neural stem cell | Neural stem cells | [292] |
| | | Proliferation via MAPK | Astrocytoma U373 MG cells | [293] |

(continued)

**Table 1.2** (continued)

| Receptor | Effector | Effect | Cell type | Reference |
|---|---|---|---|---|
| $H_2R$ | PKA | Smooth muscle cells relaxation | Intestinal muscle cells | [63] |
| | | Reduce atrial natriuretic peptide (ANP) release | Heart atria | [294] |
| | | Induce melanogenesis | Melanocytes | [295] |
| | | Survival against NMDA toxicity | Cortical neurons | [296] |
| | | Cytokine release by STAT1 phosphorylation | Mouse splenocytes | [297] |
| | | Inhibit IL-12 production | Whole blood culture | [298] |
| | | Reduces cytotoxic response | Natural killer cells | [299] |
| | | Reduce potassium current | Inhibitory hypocampal interneurons | [300] |
| | | Increase calcium influx through phosphorylation of L-type $Ca^{2+}$ channels | Rat atrial myocytes | [79] |
| | | Gastric acid secretion | Parietal cells | [80] |
| | PKC | Inhibition of TRPV channels | HL-60 cells | [89] |
| $H_3R$ | Bγ-subunit | Neurotransmitter release | Guinea pig ileum | [99] |
| | PKA | Inhibit proliferation | Mouse pancreatic cells | [301] |
| | | Inhibit histamine synthesis | Histaminergic fibers | [302] |
| | | Inhibit L-DOPA accumulation | Nucleus accumbens | [96] |
| | | Reduces $Ca^{2+}$ conductances | Cardiac synaptosomes | [100] |
| | PI3K | Cell survival via Bcl-2 inhibition | Neural-derived cell types | [106, 107] |
| | $Na^+/H^+$ exchanger | Reduced norepinephrine release | Adrenergic nerve endings | [102–104] |
| $H_4R$ | PKA | Induces cell cycle arrest | Hematopoietic precursor cells | [303] |
| | | Inhibits TGF-β1 pathway | Lung cancer cells | [109] |
| | $Ca^{2+}$ | Induction of chemotaxis | Mast cells, eosinophils, dendritic cells, macrophages | [116–120] |
| | | production of inflammatory proteins | Mast cells, microglia, dendritic cells | [115, 120, 123, 125] |

The soluble second messenger cAMP activates PKA resulting in various physiological responses (Table 1.2) through activation of the transcription factor CREB [78]. Release of βγ-subunits by $G_{\alpha s}$-activated $H_2R$ inhibits potassium influx. Additionally, L-type $Ca^{2+}$ channels are phosphorylated via the cAMP-PKA pathway and increase $Ca^{2+}$ influx to the cell [79]. Furthermore, PKA activates the $H^+/K^+$ exchanger which leads to gastric acid secretion [80].

The $H_2R$ mediates phosphorylation of the extracellular signal-regulated kinases (ERK2) via the cAMP-PKA pathway in CA3 hypocampal neurons involved in fear memory consolidation [81]. However, a reduced level of $H_2R$-mediated ERK1/2 phosphorylation was observed in myelin-autoreactive T cells which showed a reduction in proliferation [82]. More recently, it was found that inhibition of the

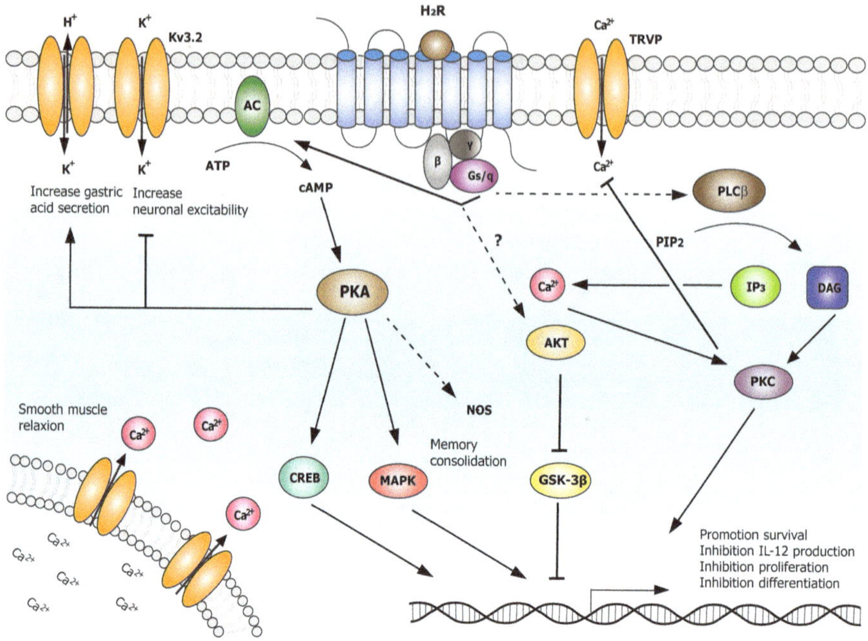


**Fig. 1.4** Signaling pathway of the $H_2R$. The $H_2R$ preferentially couples to $G\alpha_s$ proteins resulting in the activation of adenylyl cyclases (AC) to produce cAMP. Subsequent activation of PKA modulates CREB pathway, voltage-gated potassium channel (Kv3.2), and K+/H+ ATPase exchanger, MAPK, and Akt (PKB)-glycogen synthase kinase 3 beta (GSK3β). Coupling to $G\alpha_{q/11}$ proteins activates the PLC-β-$IP_3$/DAG-PKC pathway. Dotted lines are suggested pathways; solid lines are canonical pathways

MAPK pathway by PKA activation through $H_2R$ inhibited activator protein 1 (AP-1) translocation to the nucleus in human monocytoid THP-1 cells. This results in decreased production of pro-inflammatory tumor necrosis factors (TNF) [83, 84]. In addition, $H_2R$ signaling reduces Akt isoform 2 phosphorylation, decreasing the activity of glycogen synthase kinase (GSK)-3α/β. This was shown in the U937 promocytes cell line implicating the $H_2R$ as a potential target against acute myeloid leukemia [85].

In addition to $G\alpha_s$, the $H_2R$ can also couple $G\alpha_{q/11}$ proteins. $G\alpha_{q/11}$ coupling to the $H_2R$ led to an increase in cAMP and $IP_3$ levels as observed in COS-7 cells, CHO-$K_1$ cells, and guinea pig heart tissue [86–88]. Photoaffinity labeling of G proteins with [$\alpha^{32}$P]-GTP azidoaniline confirmed $H_2R$ coupling to both $G\alpha_s$ and $G\alpha_{q/11}$ proteins. PKC activity leads to the inhibition of the transient receptor potential cation channels (TRPC), modulating the differentiation of human promyelocytic leukemia HL-60 cells [89]. However, PLC activation by $H_2R$ is not unambiguously driven via the $G\alpha_q$ pathway, suggesting that coupling to the $G\alpha_q$ might be context dependent [87, 90].

### 1.3.3 $H_3R$

The $H_3R$ signals through pertussis toxin (PTx)-sensitive $G\alpha_{i/o}$ proteins resulting in AC inhibition (Fig. 1.5) [4, 91–93]. Consequently, cAMP formation, PKA activity, and CREB activation are reduced; the physiological consequences are summarized in Table 1.2 [94–96]. The released $G\beta\gamma$ subunit reduces the inward $Ca^{2+}$ conductance by voltage-gated calcium channel (VGCC) but activates G protein-gated inwardly rectifying potassium channels (GIRKs), resulting in reduced neuron excitability and neurotransmitter release [97–100]. In addition, $G\beta\gamma$ subunits are suggested to activate the PLC-$IP_3$/$Ca^{2+}$-PKC pathway although the mechanism remains to be elucidated [101]. The activation of this pathway depends on the type of $H_3R$ isoform, since histamine-induced $Ca^{2+}$ mobilization was observed with the $hH_3R$-445 but not $hH_3R$-365 [24]. Inhibition of the $Na^+/H^+$ exchanger (NHE) by the $H_3R$ results in reduced norepinephrine release in adrenergic nerve endings, which protect against myocardial ischemia [102–104].

$H_3R$ induces phosphorylation of ERK1/2 in native tissue and heterologous expression systems upon agonist binding [81, 105]. On the other hand, PKC trans-

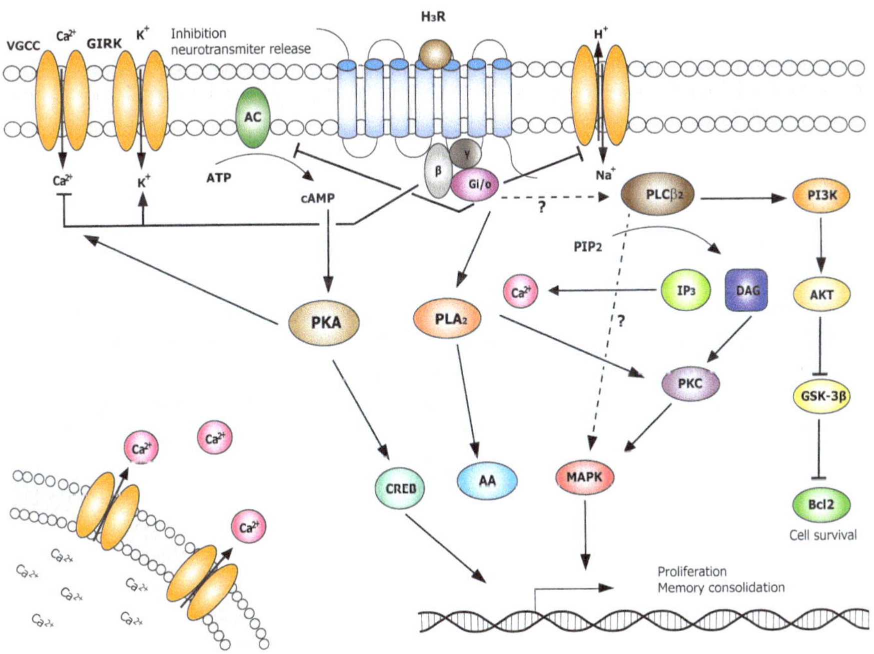


**Fig. 1.5** Signaling pathway of the $H_3R$. Upon activation of $H_3R$, it inhibits the cAMP-PKA-CREB pathway, activates MAPK and phosphoinositide 3-kinase (PI3K)-Akt-GSK3β pathway which inhibits Bcl-2 expression, activates the G protein inwardly rectifying potassium channels (GIRK), and inhibits the voltage-gated calcium channels (VGCC). *Dotted lines* are suggested pathways; *solid lines* are canonical pathways

location to the membrane in response to $H_3R$ activation results in ERK1/2 dephosphorylation and inhibition of cholangiocarcinoma cell proliferation [101]. The $H_3R$ activates Akt via PI3K, which inhibits GSK3β and Bcl-2 expression. This effect was observed in neural-derived cell types; activating this pathway promotes neuroprotection against serum deprivation and N-nitrosodimethylamine (NDMA)-induced cytotoxicity [106, 107].

Activation of $PLA_2$ via $H_3R$-induced $G\alpha_{i/o}$ activation results in the release of AA and COX activation to produce prostaglandin $E_2$ and AA metabolites, like 4-hydroxynonenal that is associated with neuronal apoptosis [108], which are associated with the progression of various CNS disorders like Alzheimer's disease [109, 110].

### *1.3.4 H₄R*

Similar to the $H_3R$, the $H_4R$ couples to PTx-sensitive $G\alpha_{i/o}$ proteins and consequently reduces PKA-mediated downstream signaling (Fig. 1.6 and Table 1.2) [31]. Additionally, induction of $Ca^{2+}$ release from intracellular stores is considered as an important cellular effect mediated by $H_4R$ activation [111–116]. $H_4R$-mediated $Ca^{2+}$ release was found to induce chemotaxis in mouse mast cells in a $G\alpha_{i/o}$-dependent manner [116]. Furthermore, $H_4R$-mediated chemotaxis was observed in dendritic cells, eosinophils, and macrophages [117–120]. Possibly, this effect is regulated by activation of the PLC-$IP_3$-$Ca^{2+}$ pathway initiated by Gβγ subunits upon receptor activation [116].

Actin polymerization, a prerequisite of cell migration, was observed upon $H_4R$ stimulation in human dendritic cells and eosinophils, and this effect was blocked in the presence of $H_4R$ antagonists JNJ7777120 or thioperamide [111, 121]. This process was demonstrated to be PTx sensitive, implicating the involvement of the $G\alpha_{i/o}$ protein [111]. Furthermore, $H_4R$-dependent Akt activation via PI3K was found to promote microglial, mast cells, eosinophils, and dendritic cells migration possibly as a consequence of actin polymerization [122, 123].

$H_4R$-mediated activation of various kinases, including ERK, PI3K, p38, and the AP-1 transcription factor has been reported [115, 120, 123–125]. In monocyte-derived dendritic cells, $H_4R$-mediated activation of AP-1 occurred independently of ERK1/2 activation [120]. An array of inflammatory proteins is regulated by $H_4R$ activation highlighting its involvement in inflammatory processes [115, 120, 123, 125].

### *1.3.5 Constitutive Activity of Histamine Receptors*

In addition to ligand-induced activation of GPCRs, receptor activation in the absence of an agonist can occur. This phenomenon is referred to as basal or constitutive activity and can be inhibited with an inverse agonist [126]. Also for histamine receptors, expressed in heterologous cell systems, constitutive activity is observed

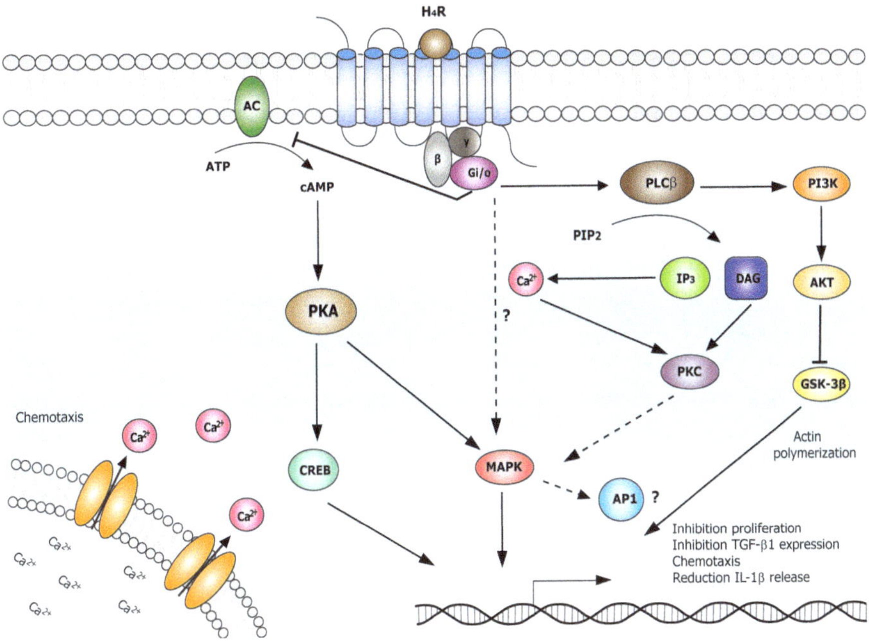


**Fig. 1.6** Signaling pathway of the $H_4R$. The $H_4R$ modulates the cAMP-PKA-CREB pathway and activates MAPK, PLC-β-$IP_3$/$Ca^{2+}$-DAG-PKC pathway and PI3K-Akt-GSK3β transduction. *Dotted lines* are suggested pathways; *solid lines* are canonical pathways

as revealed by inverse agonism of many of the known antihistamines [127–135]. In absence of a ligand an equilibrium exist in which the majority of the receptor population reside in the inactive state; while only a minor part of the receptor population is present in the active state; this is referred to as the two-state model [136]. Increased receptor and G protein expression levels are drivers for constitutive activity as increased number of these constituents will lead to a higher prevalence of the receptor in its active state [71, 137]. As heterologous cell systems often show receptor overexpression, constitutive activity is more readily observed in these cell systems compared to native tissue [138, 139].

To date, little is known about constitutive activity of histamine receptors in native tissue [131]. $H_2R$ constitutively inhibits serotonin release in isolated rat brain slices of the *substantia nigra*, which was not affected by the neutral $H_2R$ antagonist burimamide indicating that this $H_2R$ activity is histamine-independent. Importantly, selective $H_2R$ inverse agonists ranitidine and tiotidine attenuated the constitutive $H_2R$-mediated inhibition of serotonin release [131]. Additionally, G protein activation by native $H_3R$ in isolated mouse and rat cerebral membranes, as measured by [$^{35}$S]-GTPγS, was significantly reduced by the inverse agonists, FUB465, ciproxifan, and thioperamide, whereas neutral antagonist proxyfan did not affect $H_3R$-mediated [$^{35}$S]-GTPγS binding to these membranes confirming histamine-independent signaling [140]. $H_3R$ decreases histamine release in isolated mouse and rat cortical synaptosomes in response to agonist stimulation, whereas $H_3R$ inverse agonists increase histamine release. Moreover, $H_3R$ inverse agonists activated histaminergic neurons in vivo as revealed by elevated *tele*-methylhistamine levels in the mice brain [140]. Importantly, proxyfan did not affect the release of histamine and accumulation of *tele*-methylhistamine in vitro and in vivo, respectively, but antagonized both $H_3R$ agonists and inverse agonists [140, 141].

Mutational studies have revealed several motifs in GPCRs which might play a pivotal role in constitutive activity, e.g., the DRY motif, the sodium pocket, and the FF motif [142]. Probably these motifs are involved in stabilizing either the active or inactive state of the receptor, and attenuating these motifs therefore changes the balance between the two states. Also for the histamine receptors, these motifs were found to play a role [52, 143–146]. For example, mutating the FF motif in the ECL2 of the $hH_4R$ demonstrated a decrease in constitutive activity [134, 143]. However, one conserved motif which accounts for constitutive activity in all of the histamine receptors has not been identified.

### 1.3.6 Histamine Receptor Desensitization and Internalization

GPCR desensitization is an important mechanism to stop prolonged signaling in response to agonists. Desensitization is followed by internalization to downregulate cell responsiveness to continuous stimuli. Agonist-bound GPCRs are readily desensitized by phosphorylation at serine or threonine residues in their intracellular loops and C-terminus by G protein-coupled receptor kinases (GRKs) and/or second messenger-dependent kinases [147–149]. Subsequent recruitment of β-arrestin to phosphorylated GPCRs further prevents G protein coupling and targets GPCRs toward internalization via clathrin-coated pits [150]. Internalized GPCRs are either degraded in lysosomes or recycled to the cell membrane [151]. In addition, GPCRs can also internalize via caveolae and lipid rafts in a clathrin-independent manner [152].

$H_1R$ desensitization has been extensively described in ex vivo tissue and native cells. Prolonged stimulation of HeLa cells with histamine reduced $H_1R$-induced signaling, but their responsiveness is fully recovered 150 min after histamine removal [153]. Additionally, $H_1R$ desensitization was observed in bovine smooth tracheal muscle tissue upon extensive pretreatment with a PKC activator, as indicated by reduced contractility of the muscle and $IP_3$ accumulation in response to histamine [154]. Furthermore, synthetic peptides corresponding to the intracellular domain of $hH_1R$ are phosphorylated by PKA, PKC, protein kinase G (PKG), calcium-calmodulin-sensitive protein kinase II (CAM kinase II), and GRK2 [155]. Based on these synthetic peptides, $T140^{ICL2}$, $T^{4.40}$, $S396^{ICL3}$, $S398^{ICL3}$, and T478 were identified as potential phosphorylation sites [155]. Interestingly, Ala substitution of all five residues did not affect histamine-induced internalization in CHO cells as compared to wild-type $H_1R$. In contrast to wild-type $H_1R$, however, this mutant receptor was not downregulated upon prolonged stimulation (24h) with histamine [155]. Hence, these five residues are not involved in $H_1R$ internalization but rather trafficking of internalized receptors toward lysosomes, which leads to receptor degradation. Construction of $hH_1R$ mutants containing either single or multiple of the five residues substituted by alanine revealed that receptor down-regulation was mainly dependent on residue $T140^{ICL2}$ and $S398^{ICL3}$, while $T^{4.40}$ and $S396^{ICL3}$ demonstrated to have a modulatory role in this process [156]. $ZnCl_2$ and hypertonic conditions inhibit histamine-induced $hH_1R$ internalization in CHO cells, suggesting the involvement of GRKs and clathrin, respectively [157]. However, the involvement of β-arrestin in clathrin-dependent $H_1R$ internalization remains to be investigated. Caveolae/lipid raft inhibitors were unable to prevent $hH_1R$ internalization. In contrast, another study showed clathrin-independent internalization of $hH_1R$ via caveolae and lipid rafts in the same cell line [158].

$H_2R$ phosphorylation and desensitization is regulated by GRK2 and GRK3 [159–161] and involves β-arrestin, dynamin, and clathrin [162, 163]. Interestingly, cotransfection of a GRK2 phosphorylation inactive mutant revealed that $H_2R$ desensitization occurred in a phosphorylation-independent manner, while internalization and resensitization were precluded. The N-terminal RGS-homology domain (RH) of GRK2 was found to have an important role mediating desensitization of the $H_2R$. RHs were found to mostly modulate the G protein-mediated signal transduction mechanisms in a negative way [164]. Rapid desensitization of the $H_2R$ was observed in COS-7 cells upon agonist stimulation which could not be prevented by PKA or PKC inhibitors [159]. Site-directed mutagenesis identified T308 and T315, in the C-terminal tail, to be important for $H_2R$ desensitization and internalization, respectively [165]. Interestingly, agonist-induced signaling of the $H_2R$-T315A mutant was not affected, suggesting that desensitization and internalization is regulated by different mechanisms. The fate of $H_2R$ upon internalization can be differentially regulated by chemically distinctive ligands. For example, amthamine mainly induced recycling of receptors to the cell surface, whereas famotidine stimulation resulted in receptor downregulation [166]. Differences in receptor fate due to activation by chemically distinctive ligands are referred to as a biased signaling event.

Cells expressing the $H_1R$ and $H_2R$ demonstrated cross-desensitization between both receptors in a GRK2-dependent manner. Agonist-induced activation of one receptor resulted in the desensitization of the other receptor [167].

R-α-methylhistamine (RAMH) stimulation desensitized $H_3R$-mediated responses in the guinea pig ileum after 30 min, which could be partially reversed after 60 min [168]. RAMH induced $hH_3R$ internalization in CHO-K1 cells in a GRK2-dependent manner [93]. Additionally, $H_3R$ internalization was found to be clathrin-dependent, whereas caveolae inhibitors did not impair $H_3R$ internalization [93]. Interestingly, the non-imidazole compound ZEL-H16 acted as partial agonist $H_3R$-mediated G protein signaling to CREB-driven reporter gene expression and ERK1/2 phosphorylation but was fully effective in $H_3R$ internalization and delayed receptor recycling to the cell surface as compared to histamine when tested at saturating ligand concentrations (i.e., approximately $1000 \times K_i$) [169] (Fig. 1.7). The difference in efficacy between histamine and ZEL-H16 in G protein signaling on the one hand and receptor trafficking on the other hand might be related to differences in target residence time and/or biased signaling and requires further research.

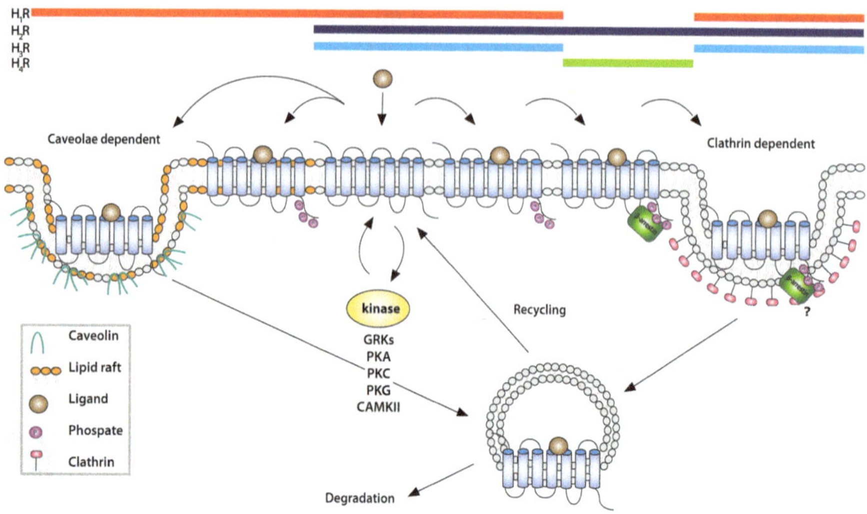


**Fig. 1.7** Overview of $H_xR$ subtype internalization. Upon ligand-induced receptor activation, kinases are recruited which phosphorylate the C-terminal tail of the receptor. Consequently, the receptor can be internalized either in a caveolae- or clathrin-dependent manner. The latter is dependent on phosphorylation motifs by GRKs and the recruitment of effector proteins, e.g., β-arrestin. After internalization the receptor can be degraded or recycled to the cell membrane. Area under the horizontal colored *bars*, presented at the *top* of the figure, indicates regulatory mechanisms known for the respective histamine receptor subtypes (*red*, $H_1R$; *dark blue*, $H_2R$; *cobalt blue*, $H_3R$; *green*, $H_4R$). For $H_1R$ and $H_3R$, only indirect evidence suggests that β-arrestin recruitment induces clathrin-dependent internalization, as indicated with a question mark

Hitherto, no studies have been reported on $H_4R$ desensitization or internalization. Yet, agonist-induced β-arrestin2 recruitment to the $hH_4R$ has been shown in $U_2OS$ and HEK293T cells suggesting that the receptor might desensitize in a phosphorylation-dependent manner [170, 171].

### 1.3.7 *Oligomerization of Histamine Receptors*

GPCRs can associate in multimeric complexes to modulate each other's function [172]. Homo-oligomerization has been demonstrated for most histamine receptors (i.e., $H_1R$, $H_2R$, and $H_4R$) [39, 173–179]. Intriguingly, homo-dimerization of ligand binding defective $H_1R$ mutants resulted in the reconstitution of $H_1R$ ligand binding [174]. Moreover, $H_2R$ and $H_4R$ homo-oligomers are found to be expressed at the cell surface, even at low receptor densities [173, 175]. Hetero-oligomerization of $H_3R$ with dopamine $D_1$ receptor ($D_1R$) shifted coupling of the latter from $G\alpha_s$ to $G\alpha_i$ [177], whereas $H_3R$-mediated MAPK activation required the presence of $D_1R$. Importantly, specific antagonists were able to cross-antagonize the agonist-induced signaling through the associated receptor, suggesting negative cooperativity within the oligomeric complex. Moreover, hetero-oligomerization of the $H_3R$ with the dopamine $D_2$ receptor ($D_2R$) decreased the ability of $D_2R$ to bind agonists (but not antagonists) upon stimulation of $H_3R$, implying negative cooperativity of $H_3R$ for this receptor as well [176]. Heteromerization of $H_1R$ with $H_4R$ synergistically effects MAPK signaling [175, 179]. In addition, co-expression of non-functional $H_1R$ mutant with the $\alpha_{1b}$-adrenoreceptor fused to a mutant $G\alpha_{11}$ resulted in heterodimers with reconstituted receptor signaling [180]. Intriguingly, truncated $H_3R$ and $H_4R$ splice variants decrease cell surface expression of full-length $H_3R$ and $H_4R$ by forming heteromeric complexes [38, 39]. The $H_1R$ and $H_2R$ can cross-desensitize each other and co-internalize, and heterodimerization is proposed to be a possible mechanism [167]. The physiological significance of cross talk and cross-regulation of the level of the histamine receptors requires in vivo and/or in vitro validation, for example, by experimental disruption of existing oligomeric complexes.

## 1.4 Ligand Binding to Receptors

Binding of ligands to their receptor is often quantified by their equilibrium dissociation constant ($K_d$) of ligand binding, which reflects the affinity in which the ligand binds the receptor at a steady state of the binding reaction. How fast this steady state, or equilibrium, is obtained depends on the rate by which the ligand associates and dissociates from the receptor.

In drug discovery, many lead compounds fail on their efficacy in clinical trials [181]. One reason for this might be that in vitro optimization criteria incorrectly reflect the efficacy of drugs in biological systems [182]. In particular, binding affinity is a commonly used criterion for lead development of antagonists. However, this parameter is only valid under equilibrium conditions, whereas in vivo, concentrations of endogenous ligands and drugs are changing continuously. Therefore, it was proposed that duration of antagonist binding on the target is a better denominator for in vivo efficacy. This is often denoted as the *residence time (RT)*, which is the reciprocal of the dissociation rate constant (i.e., $k_{off}$) [182–184]. Long-residence antagonists were found to have prolonged in vivo efficacy past the point that free ligand is still available in the blood (i.e., rate of clearance is higher than the rate of dissociation) [185, 186]. This prolonged inhibition after removing the antagonist has often been observed in organ bath experiments, in which a sustained suppression of the receptor response was measured after washout of the antagonist [184, 187, 188].

Hitherto, antihistamine therapeutics are designed to antagonize histamine-induced receptor activation. To this end, it is important to know how histamine binds to the four histamine receptor subtypes and design antihistamines that competitively block this binding site and show subtype selectivity.

### *1.4.1 Histamine Binding to Its Receptors*

Although histamine binds and subsequently activates four different histamine receptors, it does not do so with equal affinity. Histamine displays highest affinities for $hH_3R$ and $hH_4R$, i.e., $pK_i$ 8.0 and 7.8, respectively, whereas affinities for $hH_1R$ and $hH_2R$ are in the micromolar range, i.e., $pK_i$ 4.2 and 4.3, respectively [127, 189]. Similar to other aminergic GPCRs, the four histamine receptors conserve $D^{3.32}$ in TM3, which is a key ionic anchor point for ligands binding these receptors. Substitution of $D^{3.32}$ with Ala or Asn impaired binding of histamine to all four histamine receptors (Fig. 1.8) [128, 190–193]. In addition, binding of ligands that contain a basic amine moiety was found to be mediated via the $D^{3.32}$ as well, since upon Ala substitution of this residue, the binding affinity of these ligands was negatively affected [128, 191–195]. Moreover, this interaction was confirmed in the doxepin-bound crystal structure of $H_1R$ [196].

TM5 is the least conserved among the histamine receptor subtypes and might therefore shed some light on the higher affinity of histamine for $H_3R$ and $H_4R$ as compared to $H_1R$ and $H_2R$ (Fig. 1.8) [18]. Histamine interacts with residue 5.46 in TM5, which differs between histamine receptor subtypes. The $N^{5.46}$ and $T^{5.46}$ in $H_1R$ and $H_2R$, respectively, can make less strong ionic/hydrogen bond interaction with the $N^{\tau}$ of the imidazole ring compared to the negatively charged $E^{5.46}$ in $H_3R$ and $H_4R$, which explains the observed difference in affinity [18]. Nonetheless, mutating

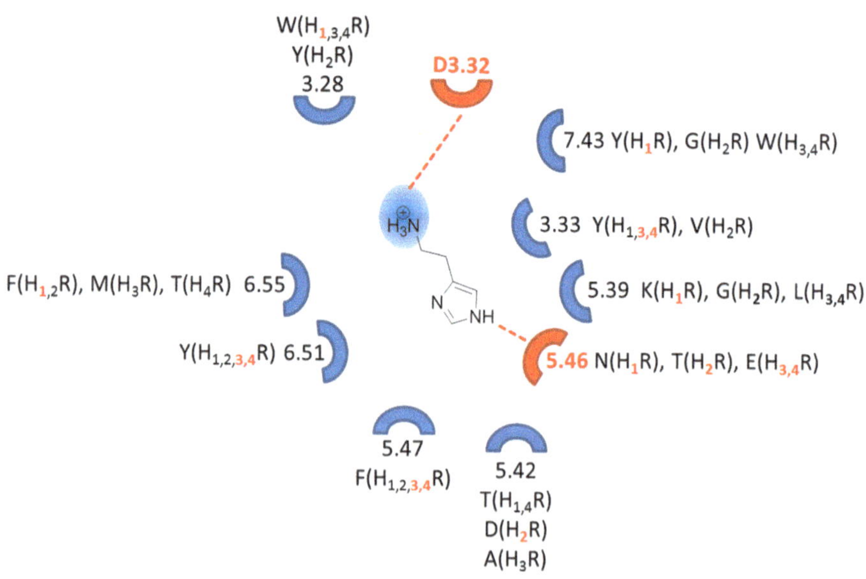


**Fig. 1.8** Histamine binding to the four histamine receptors. Conserved (known) interaction positions are denoted by *red* symbols. For each individual residue, the specific receptor for which residue was implicated to be important for histamine binding is denoted in *red*. Residues are represented in Ballesteros-Weinstein numbering [47]

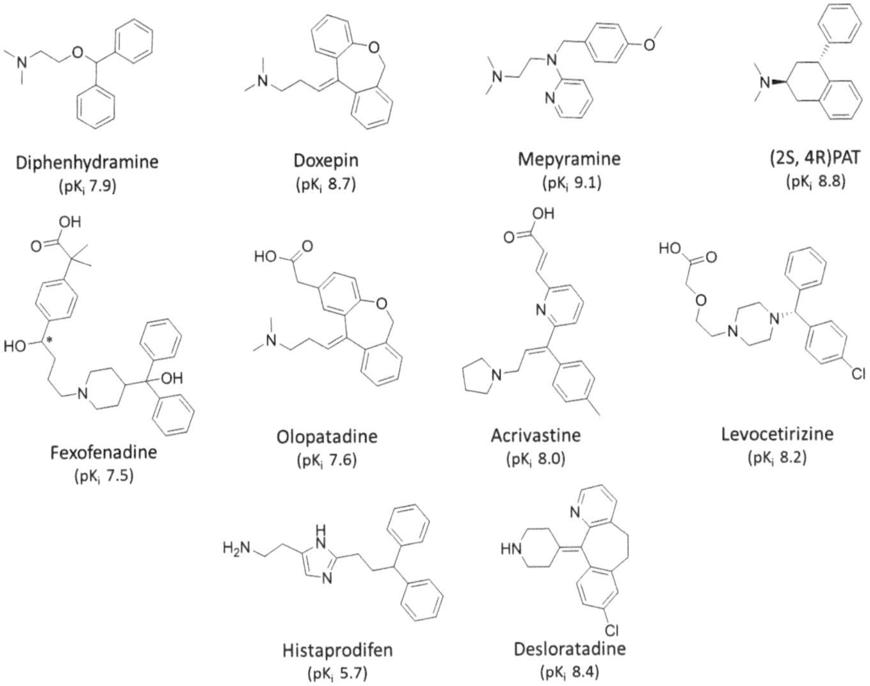


**Fig. 1.9** $H_1R$ ligand structures. Binding affinities were adapted from [42, 195, 200, 213]

residue 5.46 decreased histamine-binding affinity for all histamine receptor subtypes [190–193, 197, 198]. Interestingly, Ala substitution of $N^{5.46}$ did not impair affinity of histaprodifen-based agonists for the $H_1R$ (Fig. 1.9), but decreased their potency, suggesting that this interaction contributes to receptor activation [128].

Mutational analysis of the $H_1R$ binding pocket showed that histamine also interacts with $W^{3.28}$ [199], $K^{5.39}$ [200, 201], and $F^{6.55}$ [128, 195] (see Fig. 1.8). Ala substitution of $Y^{7.43}$ reduced binding affinity of histamine, which might be the consequence of the loss of a stabilizing interaction with $D^{3.32}$ upon histamine binding [195]. However, this was not confirmed in another study [202].

Similarly, Ala substitution of $S^{3.36}$ decreased histamine affinity for the mutant $H_1R$ [129]. Further mutational analysis revealed that $S^{3.36}$ undergoes a rotamer switch upon histamine binding, resulting in receptor activation via the interaction with $N^{7.45}$ and a consecutive H-bond network involving the N7.49 (NPxxY) and $D^{2.50}$ [129].

For the $H_3R$ and the $H_4R$, in silico studies suggested that histamine binds similarly in both receptors where hydrophobic residues $Y^{3.33}$, $F^{5.47}$, and $Y^{6.51}$ mediated histamine binding via non-bonding interactions [18, 190, 203].

In order for the body to rapidly attenuate histamine signaling, it is probably not beneficial to have prolonged histamine occupancy on the receptor after extracellular histamine levels have already been decreased. This is reflected by the relatively short residence time of histamine for the $H_1R$ (0.25 min) [204] and $H_4R$ (11 min).

Likewise, histamine levels have been reported to be often transiently elevated for similar timespans [205–208]. This might be in part mediated by the small molecular size of histamine, since this seems to be related to residence time of ligands for their receptor [209].

### 1.4.2 $H_1R$ Ligand Binding

Most $H_1R$ antihistamines are characterized by two aromatic rings that are coupled to a basic amine via a short linear or cyclic spacer (Figs. 1.9 and 1.10). As for histamine, the basic amine of antihistamines interacts with $D^{3.32}$, whereas their aromatic moiety interacts with aromatic residues in TM6. Indeed, the ring structures of doxepin interact with hydrophobic residues of TM6 ($F^{6.44}$, $w^{6.48}$, $Y^{6.51}$, $F^{6.52}$m and $F^{6.55}$ ) but also of TM3 ($Y^{3.33}$ and $I^{3.40}$) and TM4 ($W^{4.56}$) in the crystal structure of the doxepin-bound $H_1R$ [42]. Site-directed mutagenesis revealed similar interactions for [$^3$H]-mepyramine and [$^3$H]-(2S, 4R) PAT with $H_1R$ (Fig. 1.10) [195, 200]. In addition, $F^{5.47}$ was found to be crucial for [$^3$H]mepyramine binding to $H_1R$.

Doxepin and mepyramine are typical first-generation $H_1R$ antagonists for the treatment of allergic rhinitis, allergic conjunctivitis, and urticaria, but cause sedative side effects by their ability to cross the blood-brain barrier (BBB) [210]. Because of their effect on the sleep-wake cycle, first-generation antihistamines like diphenhydramine and doxylamine have been used in the treatment of insomnia [211]. Moreover, doxepin has recently been accepted for treatment of insomnia, and even further research of new $H_1R$ antihistamines with less side effects or a

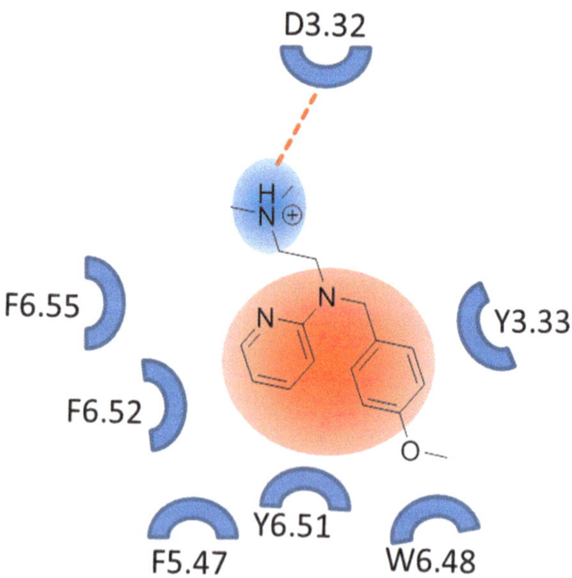


**Fig. 1.10** Binding interaction of mepyramine with the $H_1R$. Interactions are extrapolated from the doxepin-bound crystal structure and mutagenesis studies [42, 195]

polypharmacological profile (e.g., 5-$HT_{2A}$ receptor) for the treatment of insomnia is ongoing [211, 212]. On the other hand, second-generation antihistamines targeting the $H_1R$ were developed with reduced brain penetration [213]. One of the strategies was to make ligands more hydrophilic, and as such zwitterionic antihistamines harboring a carboxylic acid moiety were developed like levocetirizine and fexofenadine, which indeed are not able to penetrate the BBB [214]. An additional, or alternative, strategy was to produce antihistamines, which were actively transported out of the brain by targeting transporters like P-glycoprotein or ABCB1 [215, 216]. Therefore, some second-generation ligands still have comparable hydrophobicity to first-generation antihistamines (e.g., desloratadine).

Zwitterionic second-generation antihistamines still match the general pharmacophore of $H_1R$ antihistamines, but a carboxylic acid group is substituted from the hydrophobic ring (olopatadine, acrivastine) or from the basic amine (levocetirizine, fexofenadine). Especially fexofenadine has a large chemical group substituted from its basic amine. In general this seems to be permitted for $H_1R$ antihistamine, even for very large groups, and does not seem to hamper affinity. Carboxyl groups of acrivastine and cetirizine interact with the $K^{5.39}$ based on a decreased affinity upon mutation of this residue [200, 201]. Hence, this additional interaction of zwitterionic antihistamines with the receptor is important for high affinity. Residues that interact with histamine (Fig. 1.8) are not always important for binding of antihistamines. For instance, the $F^{6.55}A$ mutation decreased histamine-binding affinity 300-fold, whereas binding affinities of mepyramine and (2S, 4R) PAT were sixfold decreased and fourfold increased, respectively [128, 195]. Likewise, Ala substitution of $N^{5.46}$ resulted in a minor decrease of affinity for mepyramine, cetirizine, and olopatadine [18, 194, 197]. Interestingly, mutagenesis showed that residue 5.46 is important for binding of $H_2R$ and $H_4R$ antihistamines, and this was also supported by in silico data for $H_3R$ antihistamines [191, 192, 203].

Based on the interactions observed in the doxepin-bound $H_1R$ crystal structure and validated interactions for $H_1R$ antihistamines, an interaction fingerprint (IFP) was generated describing the key interaction points in the receptor. Using an energy-based scoring function combined with a filter selecting for ligands that satisfied the defined IFP, a confirmed hit rate of 73 % was obtained for a chemical diverse set of 26 ligands [217].

Successful therapeutics in the treatment of allergic rhinitis, allergic conjunctivitis, and urticaria like desloratadine and levocetirizine have significantly longer residence times on the $H_1R$ as compared to some of the first-generation antihistamines like mepyramine and diphenhydramine (Table 1.5) [218–220]. Levocetirizine, for example, had a residence time of 205 min compared to a residence time of 1.2 min of mepyramine, despite the similar binding affinities of 8.5 ($pK_i$) and 8.4 ($pK_d$), respectively (Table 1.1) [201]. The carboxyl group of levocetirizine is important for long-residence time on the $H_1R$, as revealed by 5- and 20-fold decreased residence times of analogs with alcohol or methyl ester substituents, respectively (Table 1.3) [201]. The long-residence time of levocetirizine is suggested to be mediated by the interaction of its carboxyl group with $K^{5.39}$ in $H_1R$, since levocetirizine has a tenfold decreased residence time on the $H_1R$-$K^{5.39}A$ mutant (Table 1.3). Moreover, the faster dissociation rates of structural analogs

**Table 1.3** Structure kinetic relationship $H_1R$ antihistamines

| Ligand structure | Name | $pK_i$ (WT) | *RT* (WT) (min) | *RT* (K191A) (min) | *RT* ratio (191A/WT) |
|---|---|---|---|---|---|
| 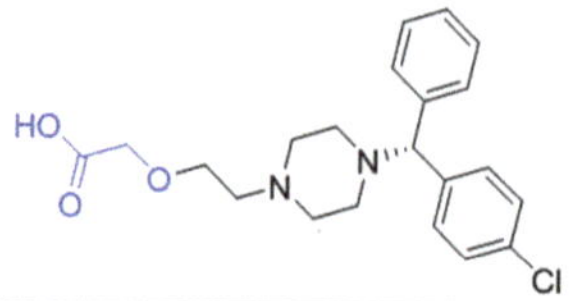 | levocetirizine | 8.5 | 200 | 19 | 10 |
| 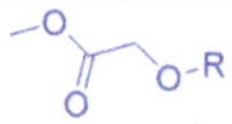 | (R)ucb29992 | 8.3 | 10 | 14 | 1 |
| 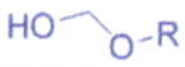 | (R)hydroxyzine | 8.7 | 45 | 25 | 2 |
| 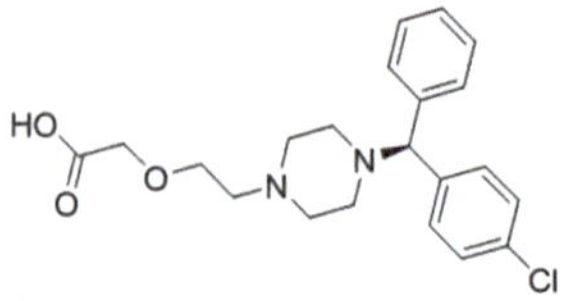 | dextrocetirizine | 7.1 | 8.7 | 1.4 | 6 |

Dissociation half-lives and binding affinities were determined using radioligand binding studies [201]. Residence times were calculated by dividing the dissociation half-life by the natural logarithm of two

without this carboxyl group were unaffected by this mutation. Also dextrocetirizine has strongly reduced residence time for the $H_1R$ as compared to its enantiomer levocetirizine. However, this decrease in residence time was not caused by the loss of interaction with $K^{5.39}$ since residence time decreased even more upon Ala substitution of $K^{5.39}$ (Table 1.3). Instead, dextrocetirizine seemed to be destabilized by $Y^{5.42}$, since Ala substitution of this residue largely restored affinity and residence time of this enantiomer for the $H_1R$, compared to levocetirizine [201]. In conclusion, for levocetirizine and dextrocetirizine, the interaction with $K^{5.39}$ increases the residence time of these antihistamines on the $H_1R$.

### 1.4.3 *$H_2R$ Ligand Binding*

The $H_2R$ antihistamine tiotidine shows considerable overlap with histamine binding to $H_2R$ by interacting with $D^{3.32}$, $D^{5.42}$, and $T^{5.46}$ based on mutagenesis studies (Fig. 1.11) [192]. Analogous to other $H_2R$ antihistamines (e.g., ranitidine and cimetidine), tiotidine has a guanidine group with reduced basicity due to substitution of the terminal amine with a cyanide group (Fig. 1.12). Guanidine groups are not uncommon in $H_2R$ agonists (e.g., impromidine) in which they are likely replacing the terminal amine of histamine, required to make an ionic interaction with the $D^{3.32}$. However, antihistamines containing a cyano-guanidine are unlikely to make an ionic interaction with the $D^{3.32}$ due to their reduced basicity, but hydrogen bonding can still be retained. In the context of cimetidine, this is probably the most logical binding mode,

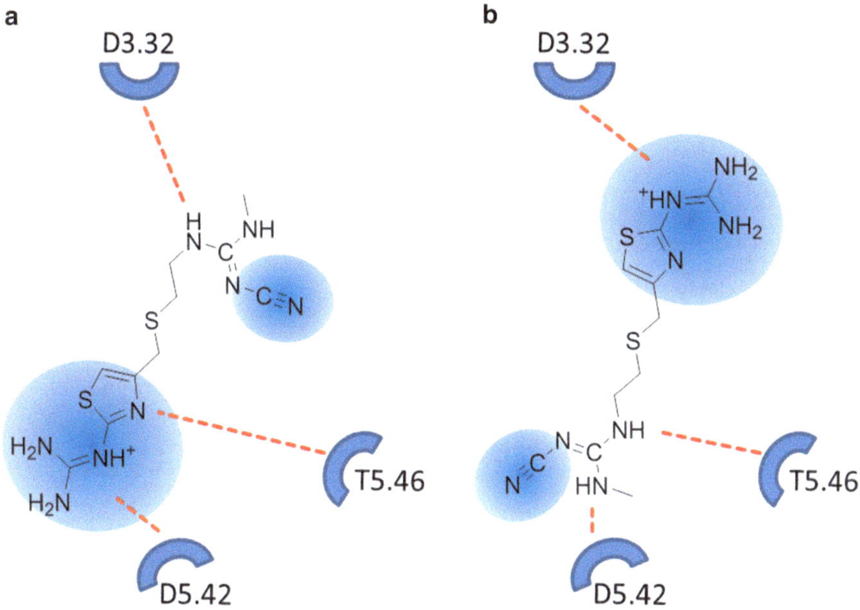


**Fig. 1.11** Binding interaction of tiotidine with the $H_2R$. Binding interactions are extrapolated from mutagenesis studies. Data could support two different binding modes of the ligand [192]. The *left graph* shows the hypothetical binding mode in which the cyanoguanidine group interacts with $D^{3.32}$. In contrast, the *right graph* shows a possible binding mode in which tiotidine is inverted and the basic guanidine groups interacts with the $D^{3.32}$

considering the resemblance with histamine (Fig. 1.12). However, unlike cimetidine, tiotidine has an additional guanidine group substituted from the thiazole (replacing the imidazole) group with much higher basicity. Therefore, it might well be that tiotidine is inverted in the binding pocket compared to histamine and cimetidine, making an interaction between the basic guanidine group and the $D^{3.32}$ (Fig. 1.11). Moreover, a basic amine in combination with a guanidine-like structure is a common motif in $H_2R$ antihistamines [213]. Therefore, for many antagonists it remains elusive how $H_2R$ binds in the pocket. Surprisingly, recent in silico analysis predicted that $D^{3.32}$ in fact makes an interaction with the imidazole group of cimetidine instead of the guanidine group, in contrast to the binding mode of histamine which imidazole ring is thought to bind with $D^{5.42}$ and $T^{5.46}$ [18, 221]. Moreover, the basic guanidine group of famotidine and the dimethylamine of ranitidine were also predicted to interact with $D^{3.32}$ [221]. This supports the inverted binding mode of tiotidine (Fig. 1.11b) but requires further experimental validation.

Potent, high-affinity agonists for the $H_2R$ were developed containing a $N^G$-acylated imidazolylpropylguanidines (e.g., UR-PG80). These agonists are due to their reduced basicity of the guanidine group more likely to cross the BBB and consequently can be used as a tool to study the $H_2R$ in the brain [222]. Drawback of these agonists is that they could also activate the $H_3R$ and $H_4R$. The imidazole moiety in

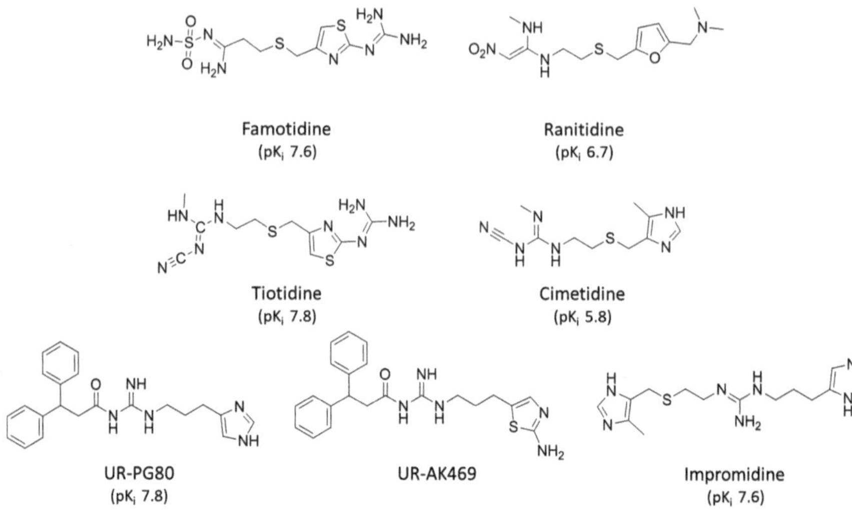


**Fig. 1.12** $H_2R$ ligand structures. Binding affinities were adapted from [213, 223]

these ligands is a chemical scaffold that binds all histamine receptors, which complicate the design of subtype-selective ligands. Recently some selective and potent $H_2R$ agonists were obtained by replacing the imidazole ring of $N^G$-acylated imidazolylpropylguanidines (e.g., UR-PG80) with an aminothiazole ring (e.g., UR-AK469) [223]. UR-PG80 has a good potency for $H_2R$, $H_3R$, and $H_4R$, while UR-AK469 had much reduced potency for the $H_3R$ and $H_4R$. It is therefore likely that aminothiazole rings are selective for the $H_2R$.

Long-residence time of $H_2R$ antihistamines was observed by means of insurmountable antagonism in functional experiments or by receptor recovery studies after preincubation with antagonist and subsequent removal of free antagonist followed by agonist stimulation [188, 224–228]. This suggested that the antihistamine iodoaminopotentidine and loxtidine had much longer residence times than antihistamine famotidine and ranitidine [188, 225, 226, 229, 230]. However, only for iodoaminopotentidine it was established what the residence time was by means of radioligand binding studies (Table 1.5). In vivo, it was shown that long-binding $H_2R$ antihistamines showed prolonged antagonism in time, indicating the therapeutic relevance of long-residence antihistamines [188, 231]. However, blocking $H_2R$ with long-binding $H_2R$ antihistamine loxtidine was associated with gastric cancers resulting from on-target toxicity [224].

IT-066 showed insurmountable $H_2R$ antagonism for positive chronotropic effect on the guinea pig atria in response to histamine, which could not be reversed by repeated wash steps [228]. Similar long-residence time was observed for the structural analog BMY25368, which has a shorter spacer and lacks the nitrogen on the phenyl ring (Table 1.4). Conversely, other changes in the spacer of IT-066 impaired long-residence binding (Table 1.4), while only marginally affecting $IC_{50}$ values

**Table 1.4** Structure kinetic relationship $H_2R$ antihistamines

| Ligand structure | name | $IC_{50}$ (μM)[a] | insurmountable |
|---|---|---|---|
| 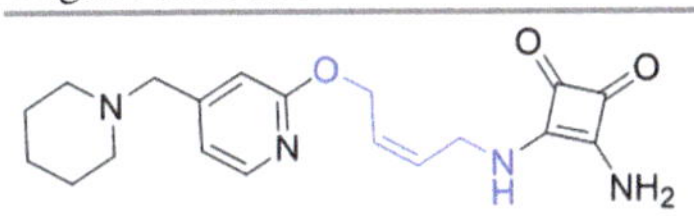 | IT-066 | 0.28 | **Yes.** ≥60 min post-washout |
| 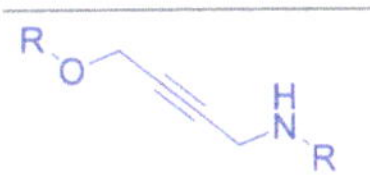 | – | 0.74 | Not after 20 min post-washout |
| 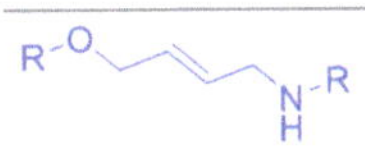 | – | 0.83 | Not after 20 min post-washout |
| 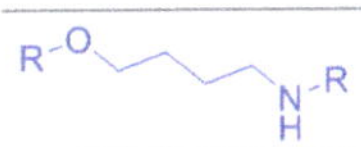 | – | 0.37 | Not after 20–40 min post-washout |
| 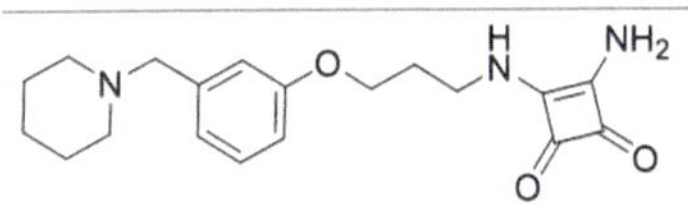 | BMY25368 | – | **Yes.** ≥60 min after removal of antagonist. |

The positive chronotropic response of histamine on the guinea pig atria, before and after preincubation with $H_2$ antihistamines, was investigated. It was observed that histamine response was blocked in an insurmountable way by the antihistamines. Histamine stimulation was repeated several times after 20 min washout steps. It could be observed that some antagonists remained bound to the receptor during these washout steps, since insurmountable antagonism remained apparent for the histamine response [228].

[a]$IC_{50}$ was determined after a 5 min preincubation of the antagonist followed by histamine stimulation.

[228]. IT-066 was also found to display insurmountable antagonism on the human $H_2R$ [227]. Both IT-066 and BMY25368 contain a 4-(piperidin-1-ylmethyl)pyridine group, which is also found in long-binding antihistamines loxtidine and iodoaminopotentidine (Table 1.5). This might be a chemical moiety imposing long residence on the $H_2R$. However, the structure (apparent)-binding kinetics relationship showed that this is at least partially mediated by the linker, bridging the amine-containing functional groups conserved for $H_2R$ antihistamines. In analogy to this observation, the linker of iodoaminopotentidine and loxtidine might be important as well.

Structural analogs were made from BMY25368 by substituting a spacer from the 4-amino group with a variable spacer length and a terminal propionamide [232]. At least for some of these compounds, insurmountable antagonism was reported. Moreover, one of the ligands ($n$=6 spacer) was radiolabeled allowing more in-depth analysis of its long-residence properties. This ligand called [$^3$H]UR-DE257 showed biphasic dissociation, indicating that there might be a more complicated binding mechanism than a simple one-step binding [232]. However, for guinea pig membranes, this biphasic dissociation was less apparent. Moreover, long-binding radioligand [$^{125}$I]iodoaminopotentidine did not show this biphasic dissociation [230], meaning that an aberrant binding mechanism is probably not conserved for long-binding $H_2$ antihistamines with a 4-(piperidin-1-ylmethyl)pyridine group. Aberrant dissociation might be related to rebinding of the ligand [232], and possibly the used concentration-unlabeled famotidine to displace [$^3$H]UR-DE257 was not sufficient to prevent rebinding.

**Table 1.5** Residence time of antihistamines

| | SHORT | MEDIUM | LONG |
|---|---|---|---|
| **$H_1R$** | 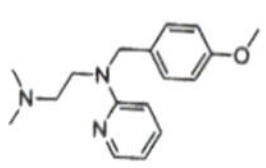<br>mepyramine<br>1.2 min[1] (37 °C)<br>1.4 min[2] (37 °C) | 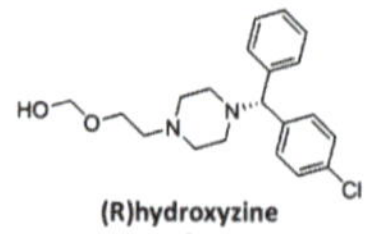<br>(R)hydroxyzine<br>45 min[1] (37 °C) | 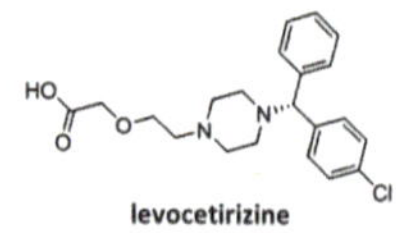<br>levocetirizine<br>200 min[1] (37 °C) |
| | 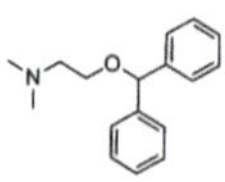<br>diphendyramine<br>0.43 min[2] (37 °C) | 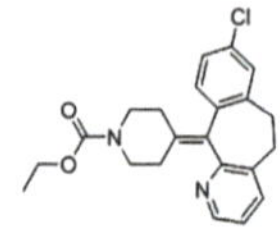<br>loratadine<br>7.2 min[1] (37 °C) | 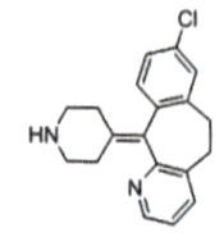<br>desloratadine<br>72* min[3] (37 °C) |
| **$H_2R$** | | | 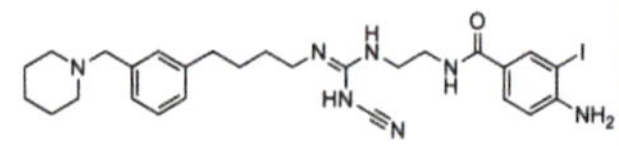<br>iodoaminopotentidine<br>77 min[4] (25 °C) |
| **$H_3R$** | 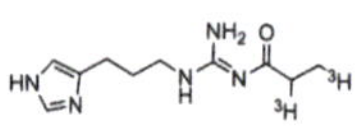<br>[3H]UR-PI294<br>3.3 min[5] (22 °C) | 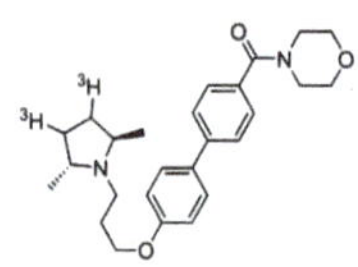<br>[3H]A-349821<br>14 min[6] (RT) | 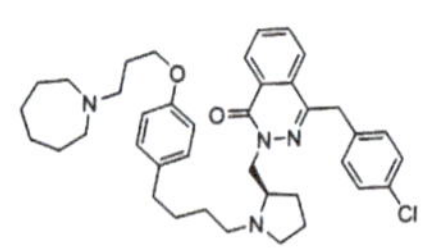<br>GSK1004723<br>128 min[7] (37 °C) |
| **$H_4R$** | 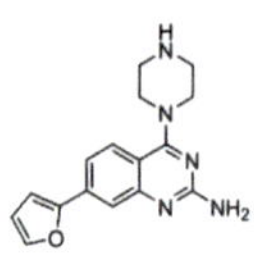<br>VUF11489<br>6 min[8] (25 °C) | 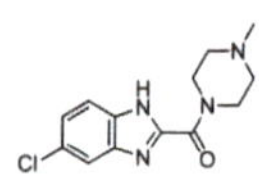<br>VUF6002<br>37 min[8] (25 °C) | 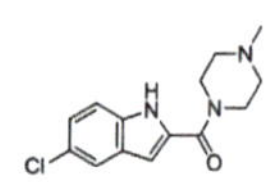<br>JNJ7777120<br>89 min[8] (25 °C) |

For each of histamine receptor subtypes, ligands were classified based on relative dissociation rate. Diphenhydramine and mepyramine (denoted with 2) were measured on rabbit arteries. For the $H_2R$ relative dissociation rate was determined for the guinea pig receptor. All other values were determined for the human isoforms of the respective histamine receptors. Residence times were calculated from the reported dissociation half-lives ($t_{1/2}$/ln2) and otherwise from the reported dissociation rate constants ($1/k_{off}$). References for the $H_1R$ [201, 219, 220], gp$H_2R$ [188, 225, 226, 229, 230], $H_3R$ [184, 222, 304], and $H_4R$ [270]. Specific conditions in which dissociation rates were measured are specified, since this can influence the measured values.

*In another publication, desloratadine had neglectable dissociation after 6 h indicating that the residence time might be much longer [218].

[1] 50 mM Tris–HCl, pH 7.4, 2 mM $MgCl_2$ at 37 °C

[2] Krebs-Henseleit solution pH 7,4 at 37 °C

[3] 50 mM sodium phosphate buffer pH 7.4 at 37 °C

[4] Phosphate buffer

[5] 12.5 mm $MgCl_2$, 1 mm EDTA, and 75 mm Tris–HCl, pH 7.4 at 22 °C

[6] 50 mM Tris, 5 mM EDTA, pH 7.4 at room temperature (RT)

[7] 50 mM HEPES, pH 7.4 at 37 °C

[8] 100 mM Tris–HCl, pH 7.4 at 25 °C

### 1.4.4 Dual-Acting Ligands

Several ligands display moderate to high affinity for two histamine receptor subtypes and are referred to as dual-acting ligands. Quinazolines that were originally designed for the $H_4R$ displayed similar affinities for the $H_1R$ (Fig. 1.13, VUF10499) [233]. In silico modeling of these ligands at the $H_4R$ revealed three pockets that are occupied by these ligands. As benzylquinoxalines also demonstrate the same trend with moderate $H_1R$ affinity, the three pocket models are well tolerated for both $H_1R$ and $H_4R$ [234]. Recently, 2,4-diaminopyrimidines (Fig. 1.13) were designed as dual-acting ligands based on the quinazolines where the rigidity of the tricyclic core was removed and side chains were swapped [235]. These ligands displayed moderate affinities for both receptors. Docking studies revealed that in both receptors, the piperazine-moiety interacts with $D^{3.32}$ and the aromatic side chain is located in a pocket between TM3, 5 and 7. For the $H_1R$ the 2,4-diaminopyrimidines bind in a hydrophobic pocket containing $Y^{3.33}$, $W^{6.48}$, $F^{5.47}$, $F^{6.52}$, and $F^{6.55}$. For $H_4R$ the diaminopyrimidine is located near $Y^{3.33}$, $E^{5.46}$, and $Y^{6.51}$ where the aromatic core makes an aromatic interaction with $F169^{ECL2}$. This study shows that despite the low sequence homology between these receptors and involvement of different residues (70 %) in ligand binding for both receptors, it's still feasible to design $H_1R/H_4R$ ligands [235].

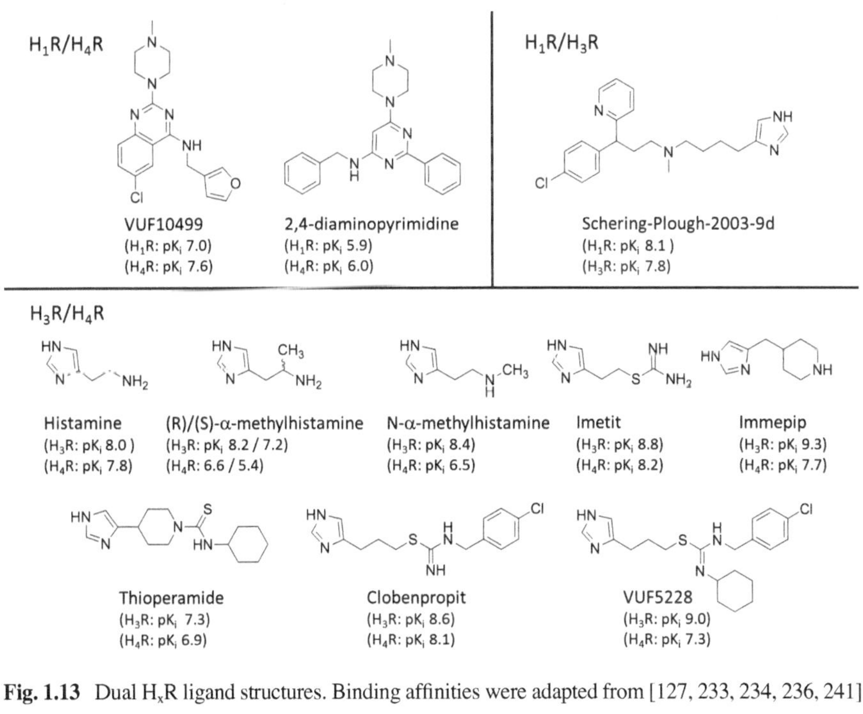


**Fig. 1.13** Dual $H_xR$ ligand structures. Binding affinities were adapted from [127, 233, 234, 236, 241]

Besides, some $H_1R/H_3R$ dual-acting ligands have been identified. Dual-acting $H_1R/H_3R$ ligands were designed by combining the $H_1R$ ligand chlorpheniramine with an alkylamine, a common moiety in $H_3R$ ligands, linked via the amine. This resulted in ligands with a good affinity for both receptor subtypes, $pK_i$ 8.1 and 7.8 for the $H_1R$ and $H_3R$, respectively (Fig. 1.13 Schering-Plough-2003-9d) [236]. Another type of dual $H_1R/H_3R$ ligand is the clinical candidate GSK100472 which also acts on both $H_1R$ and $H_3R$. However, the binding mode of these ligands, to the best of our knowledge, has not been studied.

Considering the high sequence similarity between the transmembrane domains (TMs) of the $hH_3R$ and the $hH_4R$, it might not be surprising that several dual-acting $H_3R/H_4R$ ligands have been identified. Histamine derivatives like N-α-methylhistamine (NAMH), R-α-methylhistamine (RAMH), S-α-methylhistamine (SAMH), imetit and immepip display moderate to high affinity for both $H_3R$ and $H_4R$ (Fig. 1.13). Similar to histamine, NAMH, RAMH, SAMH, and imetit bind by H-bonding of their imidazole nitrogen with $E^{5.46}$ in the $H_3R$ and $H_4R$. The $E^{5.46}Q$ substitution highly affects binding of these agonists as they directly interact with this residue by their imidazole nitrogen [190, 191]. In contrast to histamine, however, the agonists NAMH, RAMH, and imetit have 4–80-fold selectivity for $H_3R$ over $H_4R$ [127]. In addition, RAMH and SAMH display stereoselectivity on $H_3R$ and $H_4R$ as a consequence of the steric clash between the methyl group of SAMH and $Y^{3.33}$ in the binding pocket [127, 203]. Interestingly, in silico docking studies suggested that immepip binds with its imidazole toward $D^{3.32}$ due to steric hindrance with residues $L^{7.42}$ and $Y^{3.33}$ that prohibit binding in a histamine-like orientation [237].

Interestingly, the isothiourea-based $H_3R$ inverse agonists, thioperamide, clobenpropit, and iodophenpropit [238–240] (Fig. 1.13), act as inverse agonist, partial agonist, and antagonist on the $H_4R$, respectively [127]. In silico docking studies have indicated that clobenpropit binds to the $H_3R$ by H-bond interactions between the imidazole nitrogen and $E^{5.46}$ and the isothiourea group with $D^{3.32}$ and $Y^{6.51}$ [203]. In the $H_4R$, however, clobenpropit can adopt two possible binding modes where $E^{5.46}$ and $D^{3.32}$ can make interaction with either the imidazole nitrogen or the isothiourea-nitrogen [203, 241–243]. The residues $M^{6.55}$ and $L^{7.42}$ are located in two distinct binding pockets and confer a threefold higher affinity of clobenpropit for $H_3R$ as compared to $H_4R$. This higher affinity can be transferred to $H_4R$ by $T^{6.55}M$ and $Q^{7.42}L$ substitutions [241]. The more hydrophobic $M^{6.55}$ in $H_3R$ might be better able to stabilize clobenpropit as compared to $T^{6.55}$ in $H_4R$. The clobenpropit analog VUF5228 contains an additional cyclohexyl group and prefers binding to $H_3R$ over $H_4R$ (63-fold affinity difference) (Fig. 1.13). Interestingly, VUF5228 could only adopt one binding mode in the $H_4R$. The $H_4R$-$E^{5.46}Q$ displayed a sixfold-reduced affinity for both clobenpropit and VUF5228, suggesting that hydrophobic interactions could stabilize the loss of ionic interaction. In contrast to clobenpropit, VUF5228 was unaffected by $Q^{7.42}L$ and $T^{6.55}M$ mutation. In $H_3R$ $Y^{4.57}$ interacts with $E^{5.46}$ and restrains its side chain in an orientation not suitable for VUF5228 as its cyclohexyl group is too large to change binding position like what is observed for clobenpropit [241, 244]. Indeed, the $H_4R$-$N^{4.57}Y$ mutant has reduced affinity for VUF5228 but not clobenpropit. Thus, clobenpropit is affected by mutations in two distinct binding pockets (i.e., $Q^{7.42}L$ and $T^{6.55}M$), while VUF5228 is only affected by mutation $N^{4.57}Y$ supporting the two binding mode hypothesis for clobenpropit to the $H_4R$.

Imidazole-containing ligands have two major drawbacks as they block CYP450 enzymes and hardly pass the BBB [245, 246]. Therefore, the main focus for the design of selective ligands for the $H_3R$ and $H_4R$ was extended to ligands that lack this imidazole moiety to circumvent these adverse effects.

### 1.4.5 $H_3R$ Selective Ligands

$H_3R$ agonists are derived from chemical modification of histamine as the imidazole ring is required for receptor binding affinity. Addition of a methyl substituent to the nitrogen in the histamine side chain increased $H_3R$ selectivity of NAMH over $H_4R$. This methyl substituent disturbs the H-bonding network between $D^{3.32}$, $Q^{7.42}$, and $C^{3.36}$ in $H_4R$, resulting in decreased affinity for this receptor [191, 247]. Incorporation of cyclic amines, like piperidines, increases agonist affinity for $H_3R$ [18, 246, 248]. Furthermore, N-methylated immepip and replacement of the cyclic amine with a pyridine ring improved $H_3R$ selectivity of methimepip and immethridine (Fig. 1.14), respectively, as this was catastrophic for binding to other histamine receptor subtypes [249, 250]. To our best knowledge, only one non-imidazole-based $H_3R$ agonist, i.e., ZEL-H16, has so far been discovered (Fig. 1.14) [169].

Although $H_3R$ selective antagonists are structurally diverse [251], they all possess similar features. Non-imidazole $H_3R$ antagonists, such as ABT-239, require a basic "nitrogen moiety" (blue) and an aromatic moiety (red) to interact with $D^{3.32}$ and the aromatic residues in TM6, respectively (Fig. 1.15). In silico docking studies have shown that $Y^{3.33}$ and $W^{6.48}$ are important for interaction with this aromatic core [252]. This aromatic core is linked to the basic nitrogen moiety via a fixed alkyl linker (green) [109, 246, 253], which allows interactions with adjacent positioned

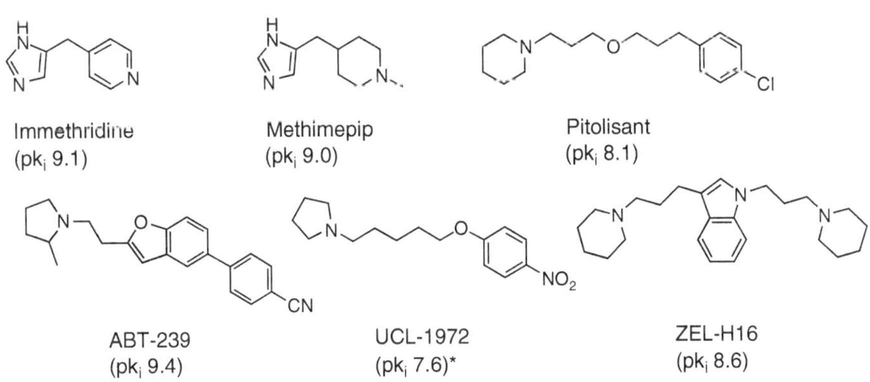


**Fig. 1.14** $H_3R$ selective ligand structures. Binding affinities were adapted from [127, 169, 213, 284]

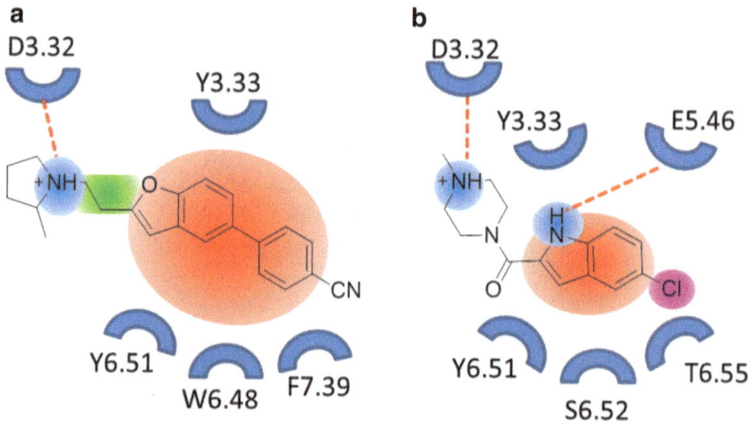


**Fig. 1.15** Pharmacophore features of the histamine $H_3$ receptor antagonist ABT-239 (**a**) and histamine $H_4$ receptor antagonist JNJ7777120 (**b**). Basic properties are shown in *blue*, aromatic properties are shown in *red*, halogens are depicted in *purple*, and linker is depicted in *green*. Interactions are extrapolated from mutagenesis data and docking studies [190, 191, 241, 262, 285]

$Y^{3.33}$ and $D^{3.32}$ in TM3, respectively [252]. In addition, this aromatic core may contain lipophilic or hydrophilic side chains that interact with residues in the lipophilic pocket amid TM 3-5-6. Furthermore, the hydrophobic subpocket between TM2-3-7 can be occupied by an aliphatic ring [203, 254].

Interestingly, rat $H_3R$ displays higher binding affinities for some ligands as compared to $hH_3R$, even though their sequence similarity is 93.5 %. Humanification of rat $H_3R$ by mutation of $V^{3.40}A$ and $A^{3.37}T$ confers human $H_3R$-like affinities to the mutant receptor [29, 255, 256]. Molecular modeling suggests that the tighter binding pocket formed by $V^{3.40}$ in rat $H_3R$ better stabilizes ligand binding via hydrophobic interaction [29].

The kinetic properties of several $H_3R$ radioligands have been investigated and range from fast to slow dissociating ligands (Table 1.5). [$^3$H]UR-PI294 dissociates rapidly (RT: 4.7 min) from the $H_3R$, where [$^3$H]A-349821 has a longer residence time of 14 min. Recently, the dual targeting $H_1R/H_3R$ antagonist GSK1004723 was developed which showed long-binding properties (RT: 111 and 143 min, respectively) for both receptors [184]. Several $H_3R$ antagonists have entered clinical trials for the treatment of CNS conditions like narcolepsy, Alzheimer's disease, and epilepsy. Thus far, only pitolisant (Wakix®) has recently been approved by the European Medicines Agency (EMA) for treatment of narcolepsy (Fig. 1.14) [257].

### 1.4.6 $H_4R$ Selective Ligands

The first discovered selective agonist for the $H_4R$ was 4-methylhistamine (Fig. 1.16) and is 150-fold selective for $H_4R$ compared to $H_2R$ [127, 247, 258]. Importantly, this methyl substitution of the imidazole ring was not tolerated by the $H_3R$. Interestingly,

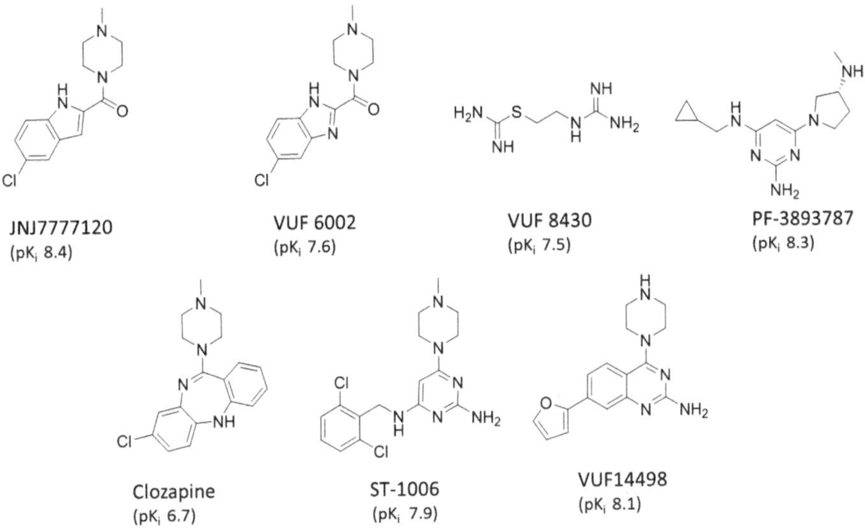


**Fig. 1.16** $H_4R$ selective ligands. Binding affinities were adapted from [127, 213, 270, 286]. (*asterisks*) affinity on $rH_3R$

in contrast to $H_3R$ agonists, the imidazole ring is not as important for binding of $H_4R$ agonists [259]. For example, agonists that lack the imidazole moiety are clozapine and dimaprit. Moreover, minor modifications to these ligands highly increased affinity and selectivity over $H_3R$, as demonstrated for VUF8430 and oxapine (VUF6884) (Fig. 1.16) [260]. However, clozapine was proposed to have a different binding mode as compared to histamine and VUF8430, since its potency was barely affected by $E^{5.46}Q$ mutation [191, 244, 261].

The main pharmacophore feature for $H_4R$ antagonists is the presence of an aliphatic ring containing a methylated basic nitrogen, like N-methylpiperazine, to interact with $D^{3.32}$ via ionic or H-bond interactions (Fig. 1.15 blue). Site-directed mutagenesis of $D^{3.32}$ to Ala or Asn totally abrogates binding of [$^3$H]JNJ7777120, while $H_4R$-$E^{5.46}Q$ mutant retains binding [191]. This illustrates that antagonists depend on only H-bond interaction, while histamine requires additional ionic interactions. Moreover, a second nitrogen atom is desired to form H-bond interaction with $E^{5.46}$ as the $E^{5.46}A$ mutation impaired [$^3$H]JNJ7777120 binding.

The second common feature for $H_4R$ antagonist is the presence of a central heteroaromatic system that can bind in the cavity between TM3 and TM6 as shown in silico where the aromatic moiety is forming π-π stacking with $Y^{3.33}$ and $Y^{6.51}$ (Fig. 1.15, red). In addition, this aromatic pocket also contains polar residues, i.e., $S^{6.52}$ and $T^{6.55}$ ($T^{6.52}$ and $M^{6.55}$ in the $H_3R$), where the heteroaromatic nitrogen of JNJ7777120, e.g., indole or aminopyrimidine, can interact via H-bond interactions with $E^{5.46}$ [18, 191, 262]. In contrast, only hydrophobic residues are present in this aromatic pocket of $H_1R$ and consequently can only bind aromatic systems.

The $H_4R$ contains two hydrophobic subpockets that are located between TM3, TM4, TM5, TM6 (subpocket A) and TM3, TM5, TM6, and ECL2 (subpocket B) [241, 261]. Ligands with hydrophobic substituents on their heteroaromatic systems are capable of binding in either of these subpockets. In silico studies have indicated that indolecarboxamides, like JNJ7777120, and 2-aminopyrimidines both occupy subpocket B with their aromatic moieties. Aminopyrimidines (e.g., ST-1006 [263]) and indolecarboxamides (e.g., JNJ7777120) (Fig. 1.16) make the described interactions with $D^{3.32}$ and $E^{5.46}$ (Fig. 1.15) [262], but do also interact with $L^{5.39}$. Substitution of $L^{5.39}$ with Val, as present in *M. fascicularis* monkey $H_4R$, affected the binding of large ligands like JNJ7777120 and aminopyrimidines, whereas smaller ligands like histamine and imetit were not affected by this mutation [264]. Moreover, chlorine-substituted aminopyrimidines and indolecarboxamides were more affected than their non-substituted analogs by this mutation. In silico a subtle orientational difference was found in the positioning of the chlorine substituent. Indolecarboxamides point their chlorine substituent toward ECL2 ($F168^{ECL2}$ and $F169^{ECL2}$) and the aromatic TM6 (T6.55 and L5.39) inducing a small binding cavity, while 2-aminopyrimidines direct their chlorine toward TM5 surrounding a larger binding cavity [262]. This indicates that different ligand classes, although binding similarly in subpocket B, orientate their chlorine substituent toward different binding cavities. A variety of these heteroaromatic scaffolds have been patented and include among others 2-aminopyrimidines, quinoxaline, benzofuropyrimidine, and benzimidazole [259].

In addition, recognition of interspecies differences has advanced our understanding of how small differences in amino acids residues can affect the binding mode of $H_4R$ ligands. Phenylalanine 169 in extracellular loop 2, which is part of the FF motif, was discovered as the residue responsible for the affinity change of histamine, VUF8430, and clozapine between the mouse ($V171^{ECL2}$) and $hH_4R$ ($F169^{EL2}$) [265]. It was suggested that mutation of F169 into Val disrupts the receptor stabilizing interactions of $F168^{ECL2}$ with hydrophobic residues buried deep inside the receptor. Moreover, it determines positioning of $F168^{ECL2}$ that is involved in ligand binding [266, 267]. In this sense, $F169^{ECL2}$ is a key player in determining the tertiary receptor structure and consequently influences ligand binding. This illustrates the important role of ECL2 in ligand binding [268].

Hitherto, the target residence time for only a few $H_4R$ ligands has been determined [269, 270]. The tested ligands showed big differences in their dissociation rates as observed in Table 1.5. Interestingly, VUF6002 only differed from JNJ7777120 by an additional amine in the indole group leading to a sixfold increase in residence time (18 and 111 min, respectively [270]. In contrast, aminopyrimidine VUF11489 and quinazolines displayed a residence time of approximately 5 min [269, 270]. Furthermore, $H_4R$ antagonist PF-3893787, which has reached phase II clinical trials, displays a residence time of 20 min [269].

## 1.5 New Opportunities in Histamine Receptor Therapeutics

GPCRs activate a variety of responses via different interaction partners. However, ligands do not always display similar efficacy toward all GPCR-mediated responses (Fig. 1.17). Therapeutical advantages of these so-called biased ligands have been

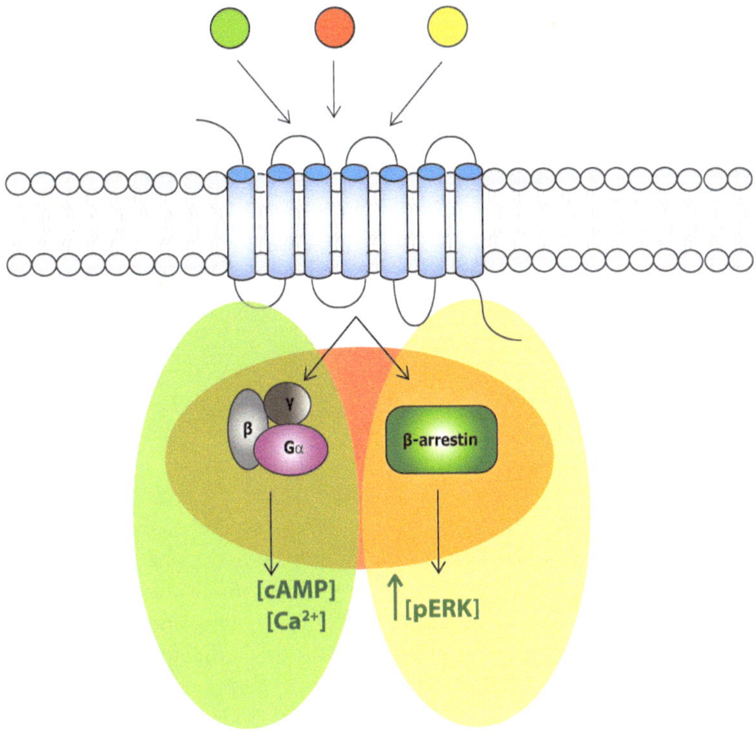


**Fig. 1.17** Principle of biased signaling. Chemically distinctive ligands can induce differential receptor conformations. As a result bias toward a certain signaling pathway, either G protein or β-arrestin, can be observed. Ligands that do not show preference toward one specific signaling pathway but stimulate both pathways to the same extent are known as non-biased ligands. *Green* ligand, G protein biased; *yellow* ligand, β-arrestin biased; *red* ligand, non-biased

demonstrated for the angiotensin II type 1 ($AT_1R$) receptor that effectively antagonizes angiotensin-induced G protein signaling while still activating β-arrestin-mediated responses in acute heart failure conditions [271–273].

Famotidine inhibits constitutive $hH_2R$-mediated cAMP production but stimulates $H_2R$ desensitization, internalization, and ERK1/2 phosphorylation [166]. These biased effects were not only observed in heterologous transfected cells, but also in native cells, highlighting that biased signaling might be physiologically relevant [166].

Biased β-arrestin2 recruitment to $H_4R$ followed by ERK1/2 phosphorylation was observed in response to JNJ7777120 in the absence of $G\alpha_i$ signaling [171]. Yet, JNJ7777120 is a widely used $H_4R$ antagonist in both in vitro and in vivo studies [274]. Reevaluation of representative $H_4R$ ligands from diverse chemical classes, including histamine analogs, triazoles, guanidines, isothioureas, dibenzodiazepines, aminopyrimidines, indolecarboxamides, quinoxalines, and quinazoline sulfonamides, identified additional β-arrestin-biased ligands. However, the relevance of β-arrestin recruitment to $H_4R$ remains to be investigated [275, 276] (Fig. 1.17).

Dual $H_3R/H_4R$ ligands might be useful in the clinic as these receptors might collectively induce the onset of pathophysiological condition like pain, itching, and breast cancer [277–279]. Dual-acting $H_1R/H_3R$ antihistamines [184, 280] antagonize the $H_3R$ and are hypothesized to prevent nasal congestion resulting from allergic rhinitis if $H_1R$ antagonists are inadequate [184]. However, in clinical trials both ligands were not able to reduce nasal congestion by $H_3R$ antagonism [281].

Recently, more attention is also given to dual-acting drug acting on non-histamine receptors. $H_3R$ antagonists that also display anticholinesterase activity were implicated for the treatment of Alzheimer's disease [282]. In addition, $H_1R$ antagonists with 5-$HT_{2A}$ activity were developed for the treatment of sleeping disorder [283].

## 1.6 Conclusion

Histamine receptors are considered as important drug targets in a variety of diseases. Indeed, antihistamines have proven to be successful in the treatment of $H_1R$ and $H_2R$-related conditions. This review provides a comprehensive overview of the current state of knowledge concerning histamine receptor ligand binding dynamics and signaling pathways. Increased complexity of histamine signaling might result from receptor oligomerization, desensitization of signaling, and constitutive activity, but are opportunities to explore novel ways to modulate histamine receptor activities using biased- and dual-acting ligands. In addition, increasing structural insights and appreciation of on-target antihistamine residence time could provide tools in developing new therapeutics with enhanced in vivo efficacy.

**Acknowledgments** The work was supported by the Innovative Medicines Initiative Joint Undertaking under grant agreement 115366, resources of which are composed of financial contribution from the European Union's Seventh Framework Program (FP7/2007–2013) and European Federation of Pharmaceutical Industries and Associations (EFPIA) companies' in-kind contribution. This work was additionally supported by the ECHO and TOPPUNT grant of the Netherlands Organization of Scientific Research. All authors participate in the European Cooperation in Science and Technology (COST) Action CM1207: GPCR-Ligand Interactions, Structures, and Transmembrane Signalling: a European Research Network (GLISTEN).

## References

1. Traiffort E, Leurs R, Arrang JM, Tardivel-Lacombe J, Diaz J, Schwartz J-C, Ruat M. Guinea pig histamine H1 receptor. I. Gene cloning, characterization, and tissue expression revealed by in situ hybridization. J Neurochem. 1994;62:507–18.
2. Simons FER, Simons KJ. Histamine and H1-antihistamines: celebrating a century of progress. J Allergy Clin Immunol. 2011;128:1139–50.e4.
3. Yellin TO, Sperow JW, Buck SH. Antagonism of tolazoline by histamine H2-receptor blockers. Nature. 1975;253:561–3.
4. Lovenberg TW, Roland BL, Wilson SJ, Jiang X, Pyati J, Huvar A, Jackson MR, Erlander MG. Cloning and functional expression of the human histamine H3 receptor. Mol Pharmacol. 1999;55:1101–7.

5. Malinowska B, Godlewski G, Schlicker E. Histamine H3 receptors—general characterization and their function in the cardiovascular system. J Physiol Pharmacol. 1998;49:191–211.
6. Nakamura T, Itadani H, Hidaka Y, Ohta M, Tanaka K. Molecular cloning and characterization of a new human histamine receptor, HH4R. Biochem Biophys Res Commun. 2000;279:615–20.
7. Cogé F, Guénin SP, Rique H, Boutin JA, Galizzi JP. Structure and expression of the human histamine H4-receptor gene. Biochem Biophys Res Commun. 2001;284:301–9.
8. Fujimoto K, Horio Y, Sugama K, Ito S, Liu YQ, Fukui H. Genomic cloning of the rat histamine H1 receptor. Biochem Biophys Res Commun. 1993;190:294–301.
9. Horio Y, Mori Y, Higuchi I, Fujimoto K, Ito S, Fukui H. Molecular cloning of the guinea-pig histamine H1 receptor gene. J Biochem. 1993;114:408–14.
10. Le Coniat M, Traiffort E, Ruat M, Arrang JM, Berger R. Chromosomal localization of the human histamine H1-receptor gene. Hum Genet. 1994;94:186–8.
11. Inoue I, Taniuchi I, Kitamura D, Jenkins N. Characteristics of the mouse genomic histamine H1 receptor gene. Genomics. 1996;36(1):178–81.
12. Fukui H, Fujimoto K, Mizuguchi H, Sakamoto K, Horio Y, Takai S, Yamada K, Ito S. Molecular cloning of the human histamine H1 receptor gene. Biochem Biophys Res Commun. 1994;201:894–901.
13. Gantz I, Schaeffer M, DelValle J, Logsdon C, Campbell V, Uhler M, Yamada T. Molecular cloning of a gene encoding the histamine H2 receptor. Proc Natl Acad Sci U S A. 1991;88:429–33.
14. Gantz I, Munzert G, Tashiro T, Schaffer M, Wang L, DelValle J, Yamada T. Molecular cloning of the human histamine H2 receptor. Biochem Biophys Res Commun. 1991;178:1386–92.
15. Ruat M, Traiffort E, Arrang J-M, Leurs R, Schwartz J-C. Cloning and tissue expression of a rat histamine H2-receptor gene. Biochem Biophys Res Commun. 1991;179:1470–8.
16. Kobayashi T, Inoue I, Jenkins NA, Gilbert DJ, Copeland NG, Watanabe T. Cloning, RNA expression, and chromosomal location of a mouse histamine H2 receptor gene. Genomics. 1996;37:390–4.
17. Traiffort E, Vizuete ML, Tardivellacombe J, Souil E, Schwartz JC, Ruat M. The guinea pig histamine H2 receptor: gene cloning, tissue expression and chromosomal localization of its human counterpart. Biochem Biophys Res Commun. 1995;211:570–7.
18. Kooistra AJ, Kuhne S, de Esch IJP, Leurs R, de Graaf C. A structural chemogenomics analysis of aminergic GPCRs: lessons for histamine receptor ligand design. Br J Pharmacol. 2013;170:101–26.
19. Swan C, Richards SA, Duroudier NP, Sayers I, Hall IP. Alternative promoter use and splice variation in the human histamine H1 receptor gene. Am J Respir Cell Mol Biol. 2006;35:118–26.
20. Moguilevsky N, Varsalona F, Noyer M, Gillard M, Guillaume J-P, Garcia L, Szpirer C, Szpirer J, Bollen A. Stable expression of human H1-histamine-receptor cDNA in Chinese hamster ovary cells. Eur J Biochem. 1994;224:489–96.
21. Peitsaro N, Sundvik M, Anichtchik OV, Kaslin J, Panula P. Identification of zebrafish histamine H1, H2 and H3 receptors and effects of histaminergic ligands on behavior. Biochem Pharmacol. 2007;73:1205–14.
22. Murakami H, Sun-Wada GH, Matsumoto M, Nishi T, Wada Y, Futai M. Human histamine H2 receptor gene: multiple transcription initiation and tissue-specific expression. FEBS Lett. 1999;451:327–31.
23. Karlstedt K, Senkas A, Ahman M, Panula P. Regional expression of the histamine H(2) receptor in adult and developing rat brain. Neuroscience. 2001;102:201–8.
24. Cogé F, Guénin SP, Audinot V, et al. Genomic organization and characterization of splice variants of the human histamine H3 receptor. Biochem J. 2001;355:279–88.
25. Wellendorph P, Goodman MW, Burstein ES, Nash NR, Brann MR, Weiner DM. Molecular cloning and pharmacology of functionally distinct isoforms of the human histamine H(3) receptor. Neuropharmacology. 2002;42:929–40.
26. Morisset S, Sasse A, Gbahou F, Héron A, Ligneau X, Tardivel-Lacombe J, Schwartz JC, Arrang JM. The rat H3 receptor: gene organization and multiple isoforms. Biochem Biophys Res Commun. 2001;280:75–80.

27. Strakhova MI, Fox GB, Carr TL, Witte DG, Vortherms TA, Manelli AM, Miller TR, Yao BB, Brioni JD, Esbenshade TA. Cloning and characterization of the monkey histamine H3 receptor isoforms. Eur J Pharmacol. 2008;601:8–15.
28. Chen JC, Liu CL, Lovenberg TW. Molecular and pharmacological characterization of the mouse histamine H-3 receptor. Eur J Pharmacol. 2003;467:57–65.
29. Yao BB, Hutchins CW, Carr TL, Cassar S, Masters JN, Bennani YL, Esbenshade TA, Hancock AA. Molecular modeling and pharmacological analysis of species-related histamine H3 receptor heterogeneity. Neuropharmacology. 2003;44:773–86.
30. Tardivel-Lacombe J, Rouleau A, Heron A, Morisset S, Pillot C, Cochois V, Schwartz JC, Arrang JM. Cloning and cerebral expression of the guinea pig histamine H3 receptor: evidence for two isoforms. Neuroreport. 2000;11:755–9.
31. Oda T, Morikawa N, Saito Y, Masuho Y, Matsumoto S. Molecular cloning and characterization of a novel type of histamine receptor preferentially expressed in leukocytes. J Biol Chem. 2000;275:36781–6.
32. Liu C, Wilson SJ, Kuei C, Lovenberg TW. Comparison of human, mouse, rat, and guinea pig histamine H4 receptors reveals substantial pharmacological species variation. J Pharmacol Exp Ther. 2001;299:121–30.
33. Oda T, Matsumoto S, Matsumoto M, Takasaki J, Kamohara M, Soga T, Hiyama H, Kobori M, Katoh M. Molecular cloning of monkey histamine H4 receptor. J Pharmacol Sci. 2005;98:319–22.
34. Jiang W, Lim HD, Zhang M, Desai P, Dai H, Colling PM, Leurs R, Thurmond RL. Cloning and pharmacological characterization of the dog histamine H4 receptor. Eur J Pharmacol. 2008;592:26–32.
35. Leurs R, Bakker RA, Timmerman H, de Esch IJP. The histamine H3 receptor: from gene cloning to H3 receptor drugs. Nat Rev Drug Discov. 2005;4:107–20.
36. Leurs R, Vischer HF, Wijtmans M, de Esch IJP. En route to new blockbuster anti-histamines: surveying the offspring of the expanding histamine receptor family. Trends Pharmacol Sci. 2011;32:250–7.
37. Bongers G, Krueger KM, Miller TR, et al. An 80-amino acid deletion in the third intracellular loop of a naturally occurring human histamine H3 isoform confers pharmacological differences and constitutive activity. J Pharmacol Exp Ther. 2007;323:888–98.
38. Bakker RA, Lozada AF, van Marle A, et al. Discovery of naturally occurring splice variants of the rat histamine H3 receptor that act as dominant-negative isoforms. Mol Pharmacol. 2006;69:1194–206.
39. van Rijn RM, van Marle A, Chazot PL, et al. Cloning and characterization of dominant negative splice variants of the human histamine H4 receptor. Biochem J. 2008;414:121–31.
40. Smit MJ, Hoffmann M, Timmerman H, Leurs R. Molecular properties and signalling pathways of the histamine H1 receptor. Clin Exp Allergy. 1999;29 Suppl 3:19–28.
41. Leurs R, Chazot PL, Shenton FC, Lim HD, de Esch IJP. Molecular and biochemical pharmacology of the histamine H4 receptor. Br J Pharmacol. 2009;157:14–23.
42. Shimamura T, Shiroishi M, Weyand S, et al. Structure of the human histamine H1 receptor complex with doxepin. Nature. 2011;475:65–70.
43. Leurs R, Hoffmann M, Wieland K, Timmerman H. H3 receptor gene is cloned at last. Trends Pharmacol Sci. 2000;21:11–2.
44. Fukushima Y, Saitoh T, Anai M, et al. Palmitoylation of the canine histamine H2 receptor occurs at Cys(305) and is important for cell surface targeting. Biochim Biophys Acta. 2001;1539:181–91.
45. Wess J. G-protein-coupled receptors: molecular mechanisms involved in receptor activation and selectivity of G-protein recognition. FASEB J. 1997;11:346–54.
46. Wess J. Molecular basis of receptor/G-protein-coupling selectivity. Pharmacol Ther. 1998;80:231–64.
47. Ballesteros JA, Weinstein H. Integrated methods for the construction of three-dimensional models and computational probing of structure-function relations in G protein-coupled receptors. Methods Neurosci. 1995;25:366–428.

48. Schneider EH, Schnell D, Strasser A, Dove S, Seifert R. Impact of the DRY motif and the missing "ionic lock" on constitutive activity and G-protein coupling of the human histamine H4 receptor. J Pharmacol Exp Ther. 2010;333:382–92.
49. Feng W, Song ZH. Effects of D3.49A, R3.50A, and A6.34E mutations on ligand binding and activation of the cannabinoid-2 (CB2) receptor. Biochem Pharmacol. 2003;65:1077–85.
50. Leschner J, Wennerberg G, Feierler J, Bermudez M, Welte B, Kalatskaya I, Wolber G, Faussner A. Interruption of the ionic lock in the bradykinin B2 receptor results in constitutive internalization and turns several antagonists into strong agonists. J Pharmacol Exp Ther. 2013;344:85–95.
51. Liu R, Groenewoud NJA, Peeters MC, Lenselink EB, IJzerman AP. A yeast screening method to decipher the interaction between the adenosine A2B receptor and the C-terminus of different G protein alpha-subunits. Purinergic Signal. 2014;10:441–53.
52. Alewijnse AE, Timmerman H, Jacobs EH, Smit MJ, Roovers E, Cotecchia S, Leurs R. The effect of mutations in the DRY motif on the constitutive activity and structural instability of the histamine H2 receptor. Mol Pharmacol. 2000;57:890–8.
53. Palczewski K, Kumasaka T, Hori T, et al. Crystal structure of rhodopsin: a G protein-coupled receptor. Science. 2000;289:739–45.
54. Chien EYT, Liu W, Zhao Q, et al. Structure of the human dopamine D3 receptor in complex with a D2/D3 selective antagonist. Science. 2010;330:1091–5.
55. Jaakola V-P, Griffith MT, Hanson MA, Cherezov V, Chien EY, Lane JR, Ijzerman AP, Stevens RC. The 2.6 angstrom crystal structure of a human A2a adenosine receptor bound to an antagonist. Science. 2008;322:1211–7.
56. Liu J, Conklin BR, Blin N, Yun J, Wess J. Identification of a receptor/G-protein contact site critical for signaling specificity and G-protein activation. Proc Natl Acad Sci. 1995;92:11642–6.
57. Blin N, Yun J, Wess J. Mapping of single amino acid residues required for selective activation of Gq/11 by the $m^3$ muscarinic acetylcholine receptor. J Biol Chem. 1995;270:17741–8.
58. Eason MG, Liggett SB. Chimeric mutagenesis of putative G-protein coupling domains of the alpha2A-adrenergic receptor. Localization of two redundant and fully competent gi coupling domains. J Biol Chem. 1996;271:12826–32.
59. Berridge MJ. Inositol trisphosphate and calcium signaling. Ann N Y Acad Sci. 1995;766:31–43.
60. Rhee SG. Regulation of phosphoinositide-specific phospholipase C. Annu Rev Biochem. 2001;70:281–312.
61. Hofmann T, Obukhov AG, Schaefer M, Harteneck C, Gudermann T, Schultz G. Direct activation of human TRPC6 and TRPC3 channels by diacylglycerol. Nature. 1999;397:259–63.
62. Choi SH, Sakamoto T, Fukutomi O, Inagaki N, Matsuura N, Nagai H, Koda A. Pharmacological study of phospholipase A2-induced histamine release from rat peritoneal mast cells. J Pharmacobiodyn. 1989;12:517–22.
63. Morini G, Kuemmerle JF, Impicciatore M, Grider JR, Makhlouf GM. Coexistence of histamine H1 and H2 receptors coupled to distinct signal transduction pathways in isolated intestinal muscle cells. J Pharmacol Exp Ther. 1993;264:598–603.
64. Galajda Z, Balla J, Szentmiklosi AJ, et al. Histamine and H1 -histamine receptors faster venous circulation. J Cell Mol Med. 2011;15:2614–23.
65. Lantoine F, Iouzalen L, Devynck MA, Millanvoye-Van Brussel E, David-Dufilho M. Nitric oxide production in human endothelial cells stimulated by histamine requires Ca2+ influx. Biochem J. 1998;330(Pt 2):695–9.
66. Stark H. Histamine H4 receptor: a novel drug target for immunoregulation and inflammation. Berlin: De Gruyter; 2013.
67. Morita Y, Aida N, Miyamoto T. Role of phospholipase A2 activation in histamine release from human basophils. Allergy. 1983;38:413–8.
68. Dennis EA. Diversity of group types, regulation, and function of phospholipase A2. J Biol Chem. 1994;269:13057–60.
69. Francis H, Glaser S, Demorrow S, et al. Small mouse cholangiocytes proliferate in response to H1 histamine receptor stimulation by activation of the IP3/CaMK I/CREB pathway. Am J Physiol Cell Physiol. 2008;295:C499–513.

70. Boss V, Wang XF, Koppelman LF, Xu KM, Murphy TJ. Histamine induces nuclear factor of activated T cell-mediated transcription and cyclosporin A-sensitive interleukin-8 mRNA expression in human umbilical vein endothelial cells. Mol Pharmacol. 1998;54:264–72.
71. Bakker RA, Schoonus SBJ, Smit MJ, Timmerman H, Leurs R. Histamine H-1-receptor activation of nuclear factor-kappa B: roles for G beta gamma- and G alpha(q/11)-subunits in constitutive and agonist-mediated signaling. Mol Pharmacol. 2001;60:1133–42.
72. Lilienbaum A, Israel A. From calcium to NF-kB signaling pathways in neurons. Mol Cell Biol. 2003;23(8):2680–98.
73. Houston C, Wenzel Seifert K, Bürckstümmer T, Seifert R. The human histamine H2-receptor couples more efficiently to Sf9 insect cell Gs-proteins than to insect cell Gq-proteins: limitations of Sf9 cells for the analysis of receptor/Gq-protein coupling. J Neurochem. 2002;80:678–96.
74. Al-Gadi M, Hill SJ. The role of calcium in the cyclic AMP response to histamine in rabbit cerebral cortical slices. Br J Pharmacol. 1987;91:213–22.
75. Batzri S, Gardner JD. Cellular cyclic AMP in dispersed mucosal cells from guinea pig stomach. Biochim Biophys Acta. 1978;541:181–9.
76. Chew CS, Hersey SJ, Sachs G, Berglindh T. Histamine responsiveness of isolated gastric glands. Am J Physiol. 1980;238:G312–20.
77. Grund VR, Goldberg ND, Hunninghake DB. Histamine receptors in adipose tissue: involvement of cyclic adenosine monophosphate and the H2-receptor in the lipolytic response to histamine in isolated canine fat cells. J Pharmacol Exp Ther. 1975;195:176–84.
78. Cheng X, Ji Z, Tsalkova T, Mei F. Epac and PKA: a tale of two intracellular cAMP receptors. Acta Biochim Biophys Sin (Shanghai). 2008;40:651–62.
79. Wellner-Kienitz MC, Bender K, Meyer T, Pott L. Coupling to Gs and G(q/11) of histamine H2 receptors heterologously expressed in adult rat atrial myocytes. Biochim Biophys Acta. 2003;1642:67–77.
80. Ahmad J, Misra M, Rizvi W, Kumar A. Histamine aspects in acid peptic diseases and cell proliferation BT—biomedical aspects of histamine. Dordrecht: Springer; 2010. p. 175–98.
81. Giovannini MG, Efoudebe M, Passani M-B, Baldi E, Bucherelli C, Giachi F, Corradetti R, Blandina P. Improvement in fear memory by histamine-elicited ERK2 activation in hippocampal CA3 cells. J Neurosci. 2003;23:9016–23.
82. Lapilla M, Gallo B, Martinello M, et al. Histamine regulates autoreactive T cell activation and adhesiveness in inflamed brain microcirculation. J Leukoc Biol. 2011;89:259–67.
83. Thomas CM, Hong T, van Pijkeren JP, Hemarajata P, Trinh DV, Hu W, Britton RA, Kalkum M, Versalovic J. Histamine derived from probiotic Lactobacillus reuteri suppresses TNF via modulation of PKA and ERK signaling. PLoS One. 2012;7, e31951.
84. Funaki C, Hodges RR, Dartt DA. Identification of the Raf-1 signaling pathway used by cAMP to inhibit p42/p44 MAPK in rat lacrimal gland acini: role in potentiation of protein secretion. Invest Ophthalmol Vis Sci. 2010;51:6321–8.
85. Werner K, Neumann D, Seifert R. High constitutive Akt2 activity in U937 promonocytes: effective reduction of Akt2 phosphorylation by the histamine H2-receptor and the β2-adrenergic receptor. Naunyn Schmiedebergs Arch Pharmacol. 2015;389(1):87–101.
86. Kühn B, Schmid A, Harteneck C, Gudermann T, Schultz G. G proteins of the Gq family couple the H2 histamine receptor to phospholipase C. Mol Endocrinol. 1996;10:1697–707.
87. Wang L, Gantz I, Delvalle J. Histamine H2 receptor activates adenylate cyclase and PLC via separate GTP-dependent pathways. Am J Physiol. 1996;271:G613–20.
88. Fukushima Y, Asano T, Katagiri H, et al. Interaction between the two signal transduction systems of the histamine H2 receptor: desensitizing and sensitizing effects of histamine stimulation on histamine-dependent cAMP production in Chinese hamster ovary cells. Biochem J. 1996;320:27–32.
89. Suh B-C, Lee H, Jun D-J, Chun J-S, Lee J-H, Kim K-T. Inhibition of H2 histamine receptor-mediated cation channel opening by protein kinase C in human promyelocytic cells. J Immunol. 2001;167:1663–71.
90. Mitsuhashi M, Mitsuhashi T, Payan DG. Multiple signaling pathways of histamine H2 receptors. Identification of an H2 receptor-dependent Ca2+ mobilization pathway in human HL-60 promyelocytic leukemia cells. J Biol Chem. 1989;264:18356–62.

91. Liu C, Ma X, Jiang X, et al. Cloning and pharmacological characterization of a fourth histamine receptor (H(4)) expressed in bone marrow. Mol Pharmacol. 2001;59:420–6.
92. Gbahou F, Rouleau A, Arrang JM. The histamine autoreceptor is a short isoform of the H3 receptor. Br J Pharmacol. 2012;166:1860–71.
93. Osorio-Espinoza A, Escamilla-Sánchez J, Aquino-Jarquin G, Arias-Montaño J-A. Homologous desensitization of human histamine H3 receptors expressed in CHO-K1 cells. Neuropharmacology. 2014;77:387–97.
94. Moreno-Delgado D, Gomez-Ramirez J, Torrent-Moreno A, González-Sepúlveda M, Blanco I, Ortiz J. Different role of cAMP dependent protein kinase and CaMKII in H3 receptor regulation of histamine synthesis and release. Neuroscience. 2009;164:1244–51.
95. Moreno-Delgado D, Torrent A, Gomez-Ramirez J, de Esch I, Blanco I, Ortiz J. Constitutive activity of H3 autoreceptors modulates histamine synthesis in rat brain through the cAMP/PKA pathway. Neuropharmacology. 2006;51:517–23.
96. Aquino-Miranda G, Escamilla-Sánchez J, González-Pantoja R, Bueno-Nava A, Arias-Montaño J-A. Histamine H3 receptor activation inhibits dopamine synthesis but not release or uptake in rat nucleus accumbens. Neuropharmacology. 2015.
97. Tedford HW, Zamponi GW. Direct G protein modulation of Cav2 calcium channels. Pharmacol Rev. 2006;58:837–62.
98. Sahlholm K, Nilsson J, Marcellino D, Fuxe K, Arhem P. The human histamine H3 receptor couples to GIRK channels in Xenopus oocytes. Eur J Pharmacol. 2007;567:206–10.
99. Blandizzi C, Tognetti M, Colucci R, Del Tacca M. Histamine H(3) receptors mediate inhibition of noradrenaline release from intestinal sympathetic nerves. Br J Pharmacol. 2000;129:1387–96.
100. Seyedi N, Mackins CJ, Machida T, Reid AC, Silver RB, Levi R. Histamine H3-receptor-induced attenuation of norepinephrine exocytosis: a decreased protein kinase a activity mediates a reduction in intracellular calcium. J Pharmacol Exp Ther. 2005;312:272–80.
101. Francis H, Onori P, Gaudio E, et al. H3 histamine receptor–mediated activation of protein kinase Cα Inhibits the growth of cholangiocarcinoma in vitro and in vivo. Mol Cancer Res. 2009;7:1704–13.
102. Silver RB, Mackins CJ, Smith NC, Koritchneva IL, Lefkowitz K, Lovenberg TW, Levi R. Coupling of histamine H3 receptors to neuronal Na+/H+ exchange: a novel protective mechanism in myocardial ischemia. Proc Natl Acad Sci U S A. 2001;98:2855–9.
103. Levi R, Smith NC. Histamine H(3)-receptors: a new frontier in myocardial ischemia. J Pharmacol Exp Ther. 2000;292:825–30.
104. van Willigen G, Nieuwland R, Nürnberg B, Gorter G, Akkerman JW. Negative regulation of the platelet Na+/H+ exchanger by trimeric G-proteins. Eur J Biochem. 2000;267:7102–8.
105. Flores-Clemente C, Osorio-Espinoza A, Escamilla-Sánchez J, Leurs R, Arias J-M, Arias-Montaño J-A. A single-point mutation (Ala280Val) in the third intracellular loop alters the signalling properties of the human histamine $H_3$ receptor stably expressed in CHO-K1 cells. Br J Pharmacol. 2013;170:127–35.
106. Bongers G, Sallmen T, Passani M-B, et al. The Akt/GSK-3β axis as a new signaling pathway of the histamine H3 receptor. J Neurochem. 2007;103:248–58.
107. Mariottini C, Scartabelli T, Bongers G, Arrigucci S, Nosi D, Leurs R, Chiarugi A, Blandina P, Pellegrini-Giampietro DE, Passani M-B. Activation of the histaminergic H3 receptor induces phosphorylation of the Akt/GSK-3 beta pathway in cultured cortical neurons and protects against neurotoxic insults. J Neurochem. 2009;110:1469–78.
108. Farooqui AA, Horrocks LA. Phospholipase A2-generated lipid mediators in the brain: the good, the bad, and the ugly. Neuroscientist. 2006;12:245–60.
109. Sander K, Kottke T, Stark H. Histamine H3 receptor antagonists go to clinics. Biol Pharm Bull. 2008;31:2163–81.
110. Levi R, Seyedi N, Schaefer U, Estephan R, Mackins CJ, Tyler E, Silver RB. Histamine H3-receptor signaling in cardiac sympathetic nerves: identification of a novel MAPK-PLA2-COX-PGE2-EP3R pathway. Biochem Pharmacol. 2007;73:1146–56.

111. Buckland KF, Williams TJ, Conroy DM. Histamine induces cytoskeletal changes in human eosinophils via the H(4) receptor. Br J Pharmacol. 2003;140:1117–27.
112. Thurmond RL, Desai PJ, Dunford PJ, et al. A potent and selective histamine H4 receptor antagonist with anti-inflammatory properties. J Pharmacol Exp Ther. 2004;309:404–13.
113. Raible DG, Lenahan T, Fayvilevich Y, Kosinski R, Schulman ES. Pharmacological characterization of a novel histamine-receptor on human eosinophils. Am J Respir Crit Care Med. 1994;149:1506–11.
114. Dijkstra D, Leurs R, Chazot P, Shenton FC, Stark H, Werfel T, Gutzmer R. Histamine downregulates monocyte CCL2 production through the histamine H4 receptor. J Allergy Clin Immunol. 2007;120:300–7.
115. Jemima EA, Prema A, Thangam EB. Functional characterization of histamine H4 receptor on human mast cells. Mol Immunol. 2014;62:19–28.
116. Hofstra CL, Desai PJ, Thurmond RL, Fung-Leung W-P. Histamine H4 receptor mediates chemotaxis and calcium mobilization of mast cells. J Pharmacol Exp Ther. 2003;305:1212–21.
117. Bäumer W, Wendorff S, Gutzmer R, Werfel T, Dijkstra D, Chazot P, Stark H, Kietzmann M. Histamine H4 receptors modulate dendritic cell migration through skin—immunomodulatory role of histamine. Allergy. 2008;63:1387–94.
118. Czerner CP, Klos A, Seifert R, Neumann D. Histamine induces chemotaxis and phagocytosis in murine bone marrow-derived macrophages and RAW 264.7 macrophage-like cells via histamine H4-receptor. Inflamm Res. 2014;63:239–47.
119. Ling P, Ngo K, Nguyen S, Thurmond RL, Edwards JP, Karlsson L, Fung-Leung W-P. Histamine H4 receptor mediates eosinophil chemotaxis with cell shape change and adhesion molecule upregulation. Br J Pharmacol. 2004;142:161–71.
120. Gutzmer R, Diestel C, Mommert S, Köther B, Stark H, Wittmann M, Werfel T. Histamine H4 receptor stimulation suppresses IL-12p70 production and mediates chemotaxis in human monocyte-derived dendritic cells. J Immunol. 2005;174:5224–32.
121. Gschwandtner M, Mommert S, Köther B, Werfel T, Gutzmer R. The histamine H4 receptor is highly expressed on plasmacytoid dendritic cells in psoriasis and histamine regulates their cytokine production and migration. J Invest Dermatol. 2011;131:1668–76.
122. Zampeli E, Tiligada E. The role of histamine H4 receptor in immune and inflammatory disorders. Br J Pharmacol. 2009;157:24–33.
123. Ferreira R, Santos T, Gonçalves J, Baltazar G, Ferreira L, Agasse F, Bernardino L. Histamine modulates microglia function. J Neuroinflammation. 2012;9:90.
124. Morse KL, Behan J, Laz TM, et al. Cloning and characterization of a novel human histamine receptor. J Pharmacol Exp Ther. 2001;296:1058–66.
125. Desai P, Thurmond RL. Histamine H 4 receptor activation enhances LPS-induced IL-6 production in mast cells via ERK and PI3K activation. Eur J Immunol. 2011;41:1764–73.
126. Costa T, Cotecchia S. Historical review: negative efficacy and the constitutive activity of G-protein-coupled receptors. Trends Pharmacol Sci. 2005;26:618–24.
127. Lim HD, van Rijn RM, Ling P, Bakker RA, Thurmond RL, Leurs R. Evaluation of histamine H1-, H2-, and H3-receptor ligands at the human histamine H4 receptor: identification of 4-methylhistamine as the first potent and selective H4 receptor agonist. J Pharmacol Exp Ther. 2005;314:1310–21.
128. Bruysters M, Pertz HH, Teunissen A, Bakker RA, Gillard M, Chatelain P, Schunack W, Timmerman H, Leurs R. Mutational analysis of the histamine H1-receptor binding pocket of histaprodifens. Eur J Pharmacol. 2004;487:55–63.
129. Jongejan A, Bruysters M, Ballesteros JA, Haaksma E, Bakker RA, Pardo L, Leurs R. Linking agonist binding to histamine H1 receptor activation. Nat Chem Biol. 2005;1:98–103.
130. Takahashi K, Tokita S, Kotani H. Generation and characterization of highly constitutive active histamine H3 receptors. J Pharmacol Exp Ther. 2003;307:213–8.
131. Threlfell S, Exley R, Cragg SJ, Greenfield SA. Constitutive histamine H2 receptor activity regulates serotonin release in the substantia nigra. J Neurochem. 2008;107:745–55.
132. Schneider EH, Schnell D, Papa D, Seifert R. High constitutive activity and a G-protein-independent high-affinity state of the human histamine H(4)-receptor. Biochemistry. 2009;48:1424–38.

133. Smit MJ, Leurs R, Alewijnse AE, Blauw J, Van Nieuw Amerongen GP, Van De Vrede Y, Roovers E, Timmerman H. Inverse agonism of histamine H2 antagonist accounts for upregulation of spontaneously active histamine H2 receptors. Proc Natl Acad Sci U S A. 1996;93:6802–7.
134. Wifling D, Löffel K, Nordemann U, Strasser A, Bernhardt G, Dove S, Seifert R, Buschauer A. Molecular determinants for the high constitutive activity of the human histamine H4 receptor: functional studies on orthologues and mutants. Br J Pharmacol. 2015;172:785–98.
135. Bakker RA, Nicholas MW, Smith TT, Burstein ES, Hacksell U, Timmerman H, Leurs R, Brann MR, Weiner DM. In vitro pharmacology of clinically used central nervous system-active drugs as inverse H1 receptor agonists. J Pharmacol Exp Ther. 2007;322:172–9.
136. Kenakin T. Ligand-selective receptor conformations revisited: the promise and the problem. Trends Pharmacol Sci. 2003;24:346–54.
137. Leurs R, Rodriguez Pena MS, Bakker RA, Alewijnse AE, Timmerman H. Constitutive activity of G protein coupled receptors and drug action. Pharm Acta Helv. 2000;74:327–31.
138. Seifert R, Wenzel Seifert K. Constitutive activity of G-protein-coupled receptors: cause of disease and common property of wild-type receptors. Naunyn Schmiedebergs Arch Pharmacol. 2002;366:381–416.
139. Giraldo J. How inverse can a neutral antagonist be? Strategic questions after the rimonabant issue. Drug Discov Today. 2010;15:411–5.
140. Morisset S, Rouleau A, Ligneau X, Gbahou F, Tardivel-Lacombe J, Stark H, Schunack W, Ganellin CR, Schwartz JC, Arrang JM. High constitutive activity of native H3 receptors regulates histamine neurons in brain. Nature. 2000;408:860–4.
141. Rouleau A, Ligneau X, Tardivel-Lacombe J, Morisset S, Gbahou F, Schwartz J-C, Arrang J-M. Histamine H3-receptor-mediated [35S]GTP gamma[S] binding: evidence for constitutive activity of the recombinant and native rat and human H3 receptors. Br J Pharmacol. 2002;135:383–92.
142. Smit MJ, Vischer HF, Bakker RA, Jongejan A, Timmerman H, Pardo L, Leurs R. Pharmacogenomic and structural analysis of constitutive g protein-coupled receptor activity. Annu Rev Pharmacol Toxicol. 2007;47:53–87.
143. Wifling D, Bernhardt G, Dove S, Buschauer A. The extracellular loop 2 (ECL2) of the human histamine H4 receptor substantially contributes to ligand binding and constitutive activity. PLoS One. 2015;10, e0117185.
144. Hancock AA, Esbenshade TA, Krueger KM, Yao BB. Genetic and pharmacological aspects of histamine H3 receptor heterogeneity. Life Sci. 2003;73:3043–72.
145. Wittmann H-J, Seifert R, Strasser A. Mathematical analysis of the sodium sensitivity of the human histamine H3 receptor. Silico Pharmacol. 2014;2:1.
146. Wittmann H-J, Seifert R, Strasser A. Sodium binding to hH3R and hH 4R—a molecular modeling study. J Mol Model. 2014;20:2394.
147. de Munnik SM, Kooistra AJ, van Offenbeek J, Nijmeijer S, de Graaf C, Smit MJ, Leurs R, Vischer HF. The viral G protein-coupled receptor ORF74 hijacks β-arrestins for endocytic trafficking in response to human Chemokines. PLoS One. 2015;10, e0124486.
148. Gimenez LE, Kook S, Vishnivetskiy SA, Ahmed MR, Gurevich E, Gurevich VV. Role of receptor-attached phosphates in binding of visual and non-visual arrestins to G protein-coupled receptors. J Biol Chem. 2012;287:9028–40.
149. Kelly E, Bailey CP, Henderson G. Agonist-selective mechanisms of GPCR desensitization. Br J Pharmacol. 2009;153:S379–88.
150. Moore CAC, Milano SK, Benovic JL. Regulation of receptor trafficking by GRKs and arrestins. Annu Rev Physiol. 2007;69:451–82.
151. Marchese A, Paing MM, Temple BRS, Trejo J. G protein-coupled receptor sorting to endosomes and lysosomes. Annu Rev Pharmacol Toxicol. 2008;48:601–29.
152. Kirkham M, Parton RG. Clathrin-independent endocytosis: new insights into caveolae and non-caveolae lipid raft carriers. Biochim Biophys Acta. 2005;1745:273–86.
153. Bristow DR, Zamani MR. Desensitization of histamine H1 receptor-mediated inositol phosphate production in HeLa cells. Br J Pharmacol. 1993;109:353–9.

154. Pype JL, Mak JC, Dupont LJ, Verleden GM, Barnes PJ. Desensitization of the histamine H1-receptor and transcriptional down-regulation of histamine H1-receptor gene expression in bovine tracheal smooth muscle. Br J Pharmacol. 1998;125:1477–84.
155. Horio S, Ogawa M, Kawakami N, Fujimoto K, Fukui H. Identification of amino acid residues responsible for agonist-induced down-regulation of histamine H(1) receptors. J Pharmacol Sci. 2004;94:410–9.
156. Horio S, Kato T, Ogawa M, Fujimoto K, Fukui H. Two threonine residues and two serine residues in the second and third intracellular loops are both involved in histamine H1 receptor downregulation. FEBS Lett. 2004;573:226–30.
157. Hishinuma S, Komazaki H, Fukui H, Shoji M. Ubiquitin/proteasome-dependent down-regulation following clathrin-mediated internalization of histamine H1-receptors in Chinese hamster ovary cells. J Neurochem. 2010;113:990–1001.
158. Self TJ, Oakley SM, Hill SJ. Clathrin-independent internalization of the human histamine H1-receptor in CHO-K1 cells. Br J Pharmacol. 2005;146:612–24.
159. Shayo C, Fernandez N, Legnazzi BL, Monczor F, Mladovan A, Baldi A, Davio C. Histamine H2 receptor desensitization: involvement of a select array of G protein-coupled receptor kinases. Mol Pharmacol. 2001;60:1049–56.
160. Fernández NN, Monczor FF, Lemos BB, Notcovich CC, Baldi AA, Davio CC, Shayo CC. Reduction of G protein-coupled receptor kinase 2 expression in U-937 cells attenuates H2 histamine receptor desensitization and induces cell maturation. Mol Pharmacol. 2002;62:1506–14.
161. Lemos Legnazzi B, Shayo C, Monczor F, Martin ME, Fernandez N, Brodsky A, Baldi A, Davio C. Rapid desensitization and slow recovery of the cyclic AMP response mediated by histamine H(2) receptors in the U937 cell line. Biochem Pharmacol. 2000;60:159–66.
162. Fernandez N, Monczor F, Baldi A, Davio C, Shayo C. Histamine H2 receptor trafficking: role of arrestin, dynamin, and clathrin in histamine H2 receptor internalization. Mol Pharmacol. 2008;74:1109–18.
163. Xu A-J, Kuramasu A, Maeda K, Kinoshita K, Takayanagi S, Fukushima Y, Watanabe T, Yanagisawa T, Sukegawa J, Yanai K. Agonist-induced internalization of histamine H2 receptor and activation of extracellular signal-regulated kinases are dynamin-dependent. J Neurochem. 2008;107:208–17.
164. Fernandez N, Gottardo FL, Alonso MN, Monczor F, Shayo C, Davio C. Roles of phosphorylation-dependent and -independent mechanisms in the regulation of histamine H2 receptor by G protein-coupled receptor kinase 2. J Biol Chem. 2011;286:28697–706.
165. Fukushima Y, Asano T, Takata K, et al. Role of the C terminus in histamine H2 receptor signaling, desensitization, and agonist-induced internalization. J Biol Chem. 1997;272:19464–70.
166. Alonso N, Zappia CD, Cabrera M, Davio CA, Shayo C, Monczor F, Fernández NC. Physiological implications of biased signaling at histamine H2 receptors. Front Pharmacol. 2015;6:1–9.
167. Alonso N, Fernandez N, Notcovich C, Monczor F, Simaan M, Baldi A, Gutkind JS, Davio C, Shayo C. Cross-desensitization and cointernalization of H1 and H2 histamine receptors reveal new insights into histamine signal integration. Mol Pharmacol. 2013;83:1087–98.
168. Perez-garcıa C, Morales L, Alguacil LF. Histamine H 3 receptor desensitization in the guinea-pig ileum. Eur J Pharmacol. 1998;341(2–3):253–6.
169. Shi Y, Hu Y, Xu Y, Zhong T, Chen X, Zhou N, Sheng R, Sun Y, Yang F, Yang D. Identification and characterization of ZEL-H16 as a novel agonist of the histamine H3 receptor. PLoS One. 2012;7, e42185.
170. Nijmeijer S, Engelhardt H, Schultes S, et al. Design and pharmacological characterization of VUF14480, a covalent partial agonist that interacts with cysteine 983.36 of the human histamine H4 receptor. Br J Pharmacol. 2013;170:89–100.
171. Rosethorne EM, Charlton SJ. Agonist-biased signaling at the histamine H4 receptor: JNJ7777120 recruits β-arrestin without activating G proteins. Mol Pharmacol. 2011;79:749–57.
172. Vischer HF, Castro M, Pin J. G protein-coupled receptor multimers: a question still open despite the use of novel approaches. Mol Pharmacol. 2015;88:561–71.

173. Fukushima Y, Asano T, Saitoh T, Anai M, Funaki M, Ogihara T, Katagiri H, Matsuhashi N, Yazaki Y, Sugano K. Oligomer formation of histamine H2 receptors expressed in Sf9 and COS7 cells 1. This research was supported in part by a grant from the Ministry of Education, Science and Culture, Japan. 1. FEBS Lett. 1997;409:283–6.
174. Bakker RA, Dees G, Carrillo JJ, Booth RG, Lo JF, Milligan G, Strange PG, Leurs R. Domain swapping in the human histamine H 1 receptor. J Pharmacol Exp Ther. 2004;311:131–8.
175. van Rijn RM, Chazot PL, Shenton FC, Sansuk K, Bakker RA, Leurs R. Oligomerization of recombinant and endogenously expressed human histamine H(4) receptors. Mol Pharmacol. 2006;70:604–15.
176. Ferrada C, Ferré S, Casadó V, et al. Interactions between histamine H3 and dopamine D2 receptors and the implications for striatal function. Neuropharmacology. 2008;55:190–7.
177. Ferrada C, Moreno E, Casadó V, et al. Marked changes in signal transduction upon heteromerization of dopamine D1 and histamine H3 receptors. Br J Pharmacol. 2009;157:64–75.
178. Nijmeijer S, Leurs R, Smit MJ, Vischer HF. The epstein-barr virus-encoded G protein-coupled receptor BILF1 hetero-oligomerizes with human CXCR4, scavenges $G\alpha_i$ proteins, and constitutively impairs CXCR4 functioning. J Biol Chem. 2010;285:29632–41.
179. Beermann S, Bernhardt G, Seifert R, Buschauer A, Neumann D. Histamine H1- and H4-receptor signaling cooperatively regulate MAPK activation. Biochem Pharmacol. 2015;98:432–9.
180. Carrillo JJ, Pediani J, Milligan G. Dimers of class A G protein-coupled receptors function via agonist-mediated trans-activation of associated G proteins. J Biol Chem. 2003;278:42578–87.
181. Waring MJ, Arrowsmith J, Leach AR, et al. An analysis of the attrition of drug candidates from four major pharmaceutical companies. Nat Rev Drug Discov. 2015;14:475–86.
182. Swinney DC. Biochemical mechanisms of drug action: what does it take for success? Nat Rev Drug Discov. 2004;3:801–8.
183. Copeland RA, Pompliano DL, Meek TD. Drug-target residence time and its implications for lead optimization. Nat Rev Drug Discov. 2006;5:730–9.
184. Slack RJ, Russell LJ, Hall DA, Luttmann MA, Ford AJ, Saunders KA, Hodgson ST, Connor HE, Browning C, Clark KL. Pharmacological characterization of GSK1004723, a novel, long-acting antagonist at histamine H 1 and H 3 receptors. Br J Pharmacol. 2011;164:1627–41.
185. Ramsey SJ, Attkins NJ, Fish R, van der Graaf PH. Quantitative pharmacological analysis of antagonist binding kinetics at CRF1 receptors in vitro and in vivo. Br J Pharmacol. 2011;164:992–1007.
186. Bradshaw JM, McFarland JM, Paavilainen VO, et al. Prolonged and tunable residence time using reversible covalent kinase inhibitors. Nat Chem Biol. 2015;11(7):525–31. doi:10.1038/nchembio.1817.
187. Christophe B, Carlier B, Gillard M, Chatelain P, Peck M, Massingham R. Histamine H1 receptor antagonism by cetirizine in isolated guinea pig tissues: Influence of receptor reserve and dissociation kinetics. Eur J Pharmacol. 2003;470:87–94.
188. Arai H, Nakagawa M, Tanada K, Yamaguchi H, Hirai S. Histamine H2-receptor antagonism of T-593: studies on positive chronotropic responses in guinea pig atria. Jpn J Pharmacol. 1994;64:251–6.
189. Leurs R, Smit MJ, Timmerman H. Molecular pharmacological aspects of histamine receptors. Pharmacol Ther. 1995;66:413–63.
190. Uveges AJ, Kowal D, Zhang Y, Spangler TB, Dunlop J, Semus S, Jones PG. The role of transmembrane helix 5 in agonist binding to the human H3 receptor. J Pharmacol Exp Ther. 2002;301:451–8.
191. Jongejan A, Lim HD, Smits RA, De Esch IJP, Haaksma E, Leurs R. Delineation of agonist binding to the human histamine H4 receptor using mutational analysis, homology modeling, and ab initio calculations. J Chem Inf Model. 2008;48:1455–63.
192. Gantz I, DelValle J, Wang LD, Tashiro T, Munzert G, Guo YJ, Konda Y, Yamada T. Molecular basis for the interaction of histamine with the histamine H2 receptor. J Biol Chem. 1992;267:20840–3.

193. Ohta K, Hayashi H, Mizuguchi H, Kagamiyama H, Fujimoto K, Fukui H. Site-directed mutagenesis of the histamine H1 receptor: roles of aspartic acid107, asparagine198 and threonine194. Biochem Biophys Res Commun. 1994;203:1096–101.
194. Nonaka H, Otaki S, Ohshima E, Kono M, Kase H, Ohta K, Fukui H, Ichimura M. Unique binding pocket for KW-4679 in the histamine H1 receptor. Eur J Pharmacol. 1998;345:111–7.
195. Cordova-Sintjago TC, Fang L, Bruysters M, Leurs R, Booth RG. Molecular determinants of ligand binding at the human histamine H1 receptor: site-directed mutagenesis results analyzed with ligand docking and molecular dynamics studies at H1 homology and crystal structure models. J Chem Pharm Res. 2012;4:2937–51.
196. Shiroishi M, Kobayashi T, Ogasawara S, Tsujimoto H, Ikeda-Suno C, Iwata S, Shimamura T. Production of the stable human histamine H1 receptor in Pichia pastoris for structural determination. Methods. 2011;55:281–6.
197. Leurs R, Smit MJ, Tensen CP, Terlaak AM, Timmerman H. Site-directed mutagenesis of the histamine H1-receptor reveals a selective interaction of asparagine207 with subclasses of H1-receptor agonists. Biochem Biophys Res Commun. 1994;201:295–301.
198. Shin N, Coates E, Murgolo NJ, Morse KL, Bayne M, Strader CD, Monsma FJ. Molecular modeling and site-specific mutagenesis of the histamine-binding site of the histamine H4 receptor. Mol Pharmacol. 2002;62:38–47.
199. Bakker RA, Weiner DM, ter Laak T, Beuming T, Zuiderveld OP, Edelbroek M, Hacksell U, Timmerman H, Brann MR, Leurs R. 8R-lisuride is a potent stereospecific histamine H1-receptor partial agonist. Mol Pharmacol. 2004;65:538–49.
200. Wieland K, Laak AM, Smit MJ, Kühne R, Timmerman H, Leurs R. Mutational analysis of the antagonist-binding site of the histamine H(1) receptor. J Biol Chem. 1999;274:29994–30000.
201. Gillard M, Van Der Perren C, Moguilevsky N, Massingham R, Chatelain P. Binding characteristics of cetirizine and levocetirizine to human H(1) histamine receptors: contribution of Lys(191) and Thr(194). Mol Pharmacol. 2002;61:391–9.
202. Bruysters M, Jongejan A, Gillard M, Van De Manakker F, Bakker RA, Chatelain P, Leurs R. Pharmacological differences between human and guinea pig histamine H 1 receptors: Asn 84 (2. 61) as key residue within an additional binding pocket in the H 1. Receptor. 2005;67:1045–52.
203. Kim S-K, Fristrup P, Abrol R, Goddard WA. Structure-based prediction of subtype selectivity of histamine H3 receptor selective antagonists in clinical trials. J Chem Inf Model. 2011;51:3262–74.
204. Wittmann H-J, Seifert R, Strasser A. Influence of the N-terminus and the E2-loop onto the binding kinetics of the antagonist mepyramine and the partial agonist phenoprodifen to H(1) R. Biochem Pharmacol. 2011;82:1910–8.
205. Pollard H, Bischoff S, Schwartz JC. Turnover of histamine in rat brain and its decrease under barbiturate anesthesia. J Pharmacol Exp Ther. 1974;190:88–99.
206. Dismukes K, Snyder SH. Histamine turnover in rat brain. Brain Res. 1974;78:467–81.
207. Hough LB, Khandelwal JK, Green JP. Histamine turnover in regions of rat brain. Brain Res. 1984;291:103–9.
208. Petersen LJ, Mosbech H, Skov PS. Allergen-induced histamine release in intact human skin in vivo assessed by skin microdialysis technique: Characterization of factors influencing histamine releasability. J Allergy Clin Immunol. 1996;97:672–9.
209. Pan AC, Borhani DW, Dror RO, Shaw DE. Molecular determinants of drug-receptor binding kinetics. Drug Discov Today. 2013;18:667–73.
210. Kay GG. The effects of antihistamines on cognition and performance. J Allergy Clin Immunol. 2000;105:S622–7.
211. Moree WJ, Li B-F, Jovic F, et al. Characterization of novel selective H1-antihistamines for clinical evaluation in the treatment of insomnia. J Med Chem. 2009;52:5307–10.
212. Sullivan S. Update on emerging drugs for insomnia. Expert Opin Emerg Drugs. 2012;17:295–8.
213. Panula P, Chazot PL, Cowart M, Gutzmer R, Leurs R, Liu WLS, Stark H, Thurmond RL, Haas HL. International Union of Basic and Clinical Pharmacology. XCVIII. Histamine receptors. Pharmacol Rev. 2015;67:601–55.

214. Hiraoka K, Tashiro M, Grobosch T, Maurer M, Oda K, Toyohara J, Ishii K, Ishiwata K, Yanai K. Brain histamine $H_1$ receptor occupancy measured by PET after oral administration of levocetirizine, a non-sedating antihistamine. Expert Opin Drug Saf. 2015;14:199–206.
215. Broccatelli F, Carosati E, Cruciani G, Oprea TI. Transporter-mediated efflux influences CNS side effects: ABCB1, from antitarget to target. Mol Inform. 2010;29:16–26.
216. Obradovic T, Dobson GG, Shingaki T, Kungu T, Hidalgo IJ. Assessment of the first and second generation antihistamines brain penetration and role of P-glycoprotein. Pharm Res. 2006;24:318–27.
217. de Graaf C, Kooistra AJ, Vischer HF, et al. Crystal structure-based virtual screening for fragment-like ligands of the human histamine H(1) receptor. J Med Chem. 2011;54:8195–206.
218. Anthes JC, Gilchrest H, Richard C, et al. Biochemical characterization of desloratadine, a potent antagonist of the human histamine H(1) receptor. Eur J Pharmacol. 2002;449:229–37.
219. Gillard M, Chatelain P. Changes in pH differently affect the binding properties of histamine H1 receptor antagonists. Eur J Pharmacol. 2006;530:205–14.
220. Onaran HO, Bökesoy TA. Kinetics of antagonism at histamine-H1 receptors in isolated rabbit arteries. Naunyn Schmiedebergs Arch Pharmacol. 1990;341:316–23.
221. Saxena M, Bhunia SS, Saxena AK. Molecular modelling studies on 2-substituted octahydropyrazinopyridoindoles for histamine H2 receptor antagonism. SAR QSAR Environ Res. 2015;26:739–55.
222. Ghorai P, Kraus A, Keller M, et al. Acylguanidines as bioisosteres of guanidines: NG-acylated imidazolylpropylguanidines, a new class of histamine H2 receptor agonists. J Med Chem. 2008;51:7193–204.
223. Kraus A, Ghorai P, Birnkammer T, Schnell D, Elz S, Seifert R, Dove S, Bernhardt G, Buschauer A. N(G)-acylated aminothiazolylpropylguanidines as potent and selective histamine H(2) receptor agonists. ChemMedChem. 2009;4:232–40.
224. Brittain RT, Jack D, Reeves JJ, Stables R. Pharmacological basis for the induction of gastric carcinoid tumours in the rat by loxtidine, an insurmountable histamine H2-receptor blocking drug. Br J Pharmacol. 1985;85:843–7.
225. Bertaccini G, Coruzzi G, Poli E, Adami M. Pharmacology of the novel H2 antagonist famotidine: in vitro studies. Agents Actions. 1986;19:180–7.
226. Coruzzi G, Adami M, Pozzoli C, Giorgi F, Bertaccini G. Cardiac and gastric effects of histamine H2 receptor antagonists: no evidence for a correlation between lipophilicity and receptor affinity. Br J Pharmacol. 1996;118:1813–21.
227. Otsuka H, Fukushima Y, Tamai M, et al. Long-lasting binding of IT-066 to human histamine H2 receptor. Dig Dis Sci. 2000;45:796–801.
228. Kijima H, Isobe Y, Muramatsu M, Yokomori S, Suzuki M, Higuchi S. Structure-activity characterization of an H2-receptor antagonist, 3-amino-4-[4-[4-(1-piperidinomethyl)-2-pyridyloxyl]-cis-2butenylamino[-3-cyclobutene-1,2-dione hydrochloride (IT-066), involved in the insurmountable antagonism against histamine-induced p. Biochem Pharmacol. 1998;55:151–7.
229. Coruzzi G, Pozzoli C, Poli E, Coppelli G, Bertaccini G. Effects of histamine H2 receptor agonists and antagonists on the isolated guinea pig gallbladder. Fundam Clin Pharmacol. 1999;13:84–90.
230. Ruat M, Traiffort E, Bouthenet ML, Schwartz JC, Hirschfeld J, Buschauer A, Schunack W. Reversible and irreversible labeling and autoradiographic localization of the cerebral histamine H2 receptor using [125I]iodinated probes. Proc Natl Acad Sci U S A. 1990;87:1658–62.
231. Brittain RT, Jack D. Histamine H2-antagonists—past, present and future. J Clin Gastroenterol. 1983;5 Suppl 1:71–9.
232. Baumeister P, Erdmann D, Biselli S, Kagermeier N, Elz S, Bernhardt G, Buschauer A. [3 H] UR-DE257: development of a tritium-labeled squaramide-type selective histamine H 2 receptor antagonist. ChemMedChem. 2015;10:83–93.
233. Smits RA, de Esch IJP, Zuiderveld OP, Broeker J, Sansuk K, Guaita E, Coruzzi G, Adami M, Haaksma E, Leurs R. Discovery of quinazolines as histamine H 4 receptor inverse agonists using a scaffold hopping approach. J Med Chem. 2008;51:7855–65.

234. Smits RA, Lim HD, Hanzer A, Zuiderveld OP, Guaita E, Adami M, Coruzzi G, Leurs R, de Esch IJP. Fragment based design of new H 4 receptor—ligands with anti-inflammatory properties in vivo. J Med Chem. 2008;51:2457–67.
235. Hammer SG, Gobleder S, Naporra F, Wittmann H-J, Elz S, Heinrich MR, Strasser A. 2,4-Diaminopyrimidines as dual ligands at the histamine H1 and H4 receptor—H1/H4-receptor selectivity. Bioorg Med Chem Lett. 2016;26:292–300.
236. Aslanian R, Mutahi MWA, Shih N-Y, Piwinski JJ, West R, Williams SM, She S, Wu R-L, Hey JA. Identification of a dual histamine H1/H3 receptor ligand based on the H1 antagonist chlorpheniramine. Bioorg Med Chem Lett. 2003;13:1959–61.
237. Ishikawa M, Watanabe T, Kudo T, Yokoyama F, Yamauchi M, Kato K, Kakui N, Sato Y. Investigation of the histamine H3 receptor binding site. Design and synthesis of hybrid agonists with a lipophilic side chain. J Med Chem. 2010;53:6445–56.
238. Arrang JM, Garbarg M, Lancelot JC, Lecomte JM, Pollard H, Robba M, Schunack W, Schwartz J-C. Highly potent and selective ligands for histamine H3-receptors. Nature. 1987;327:117–23.
239. van der Goot H, Schepers M, Sterk G, Timmerman H. Isothiourea analogues of histamine as potent agonists or antagonists of the histamine H3-receptor. Eur J Med Chem. 1992;27:511–7.
240. Wieland K, Bongers G, Yamamoto Y, Hashimoto T, Yamatodani A, Menge WM, Timmerman H, Lovenberg TW, Leurs R. Constitutive activity of histamine h(3) receptors stably expressed in SK-N-MC cells: display of agonism and inverse agonism by H(3) antagonists. J Pharmacol Exp Ther. 2001;299:908–14.
241. Istyastono EP, Nijmeijer S, Lim HD, van de Stolpe A, Roumen L, Kooistra AJ, Vischer HF, de Esch IJP, Leurs R, de Graaf C. Molecular determinants of ligand binding modes in the histamine H(4) receptor: linking ligand-based three-dimensional quantitative structure-activity relationship (3D-QSAR) models to in silico guided receptor mutagenesis studies. J Med Chem. 2011;54:8136–47.
242. Kiss R, Noszál B, Rácz Á, Falus A, Eros D, Keseru GM. Binding mode analysis and enrichment studies on homology models of the human histamine H4 receptor. Eur J Med Chem. 2008;43:1059–70.
243. Lorenzi S, Mor M, Bordi F, Rivara S, Rivara M, Morini G, Bertoni S, Ballabeni V, Barocelli E, Plazzi PV. Validation of a histamine H3 receptor model through structure-activity relationships for classical H3 antagonists. Bioorg Med Chem. 2005;13:5647–57.
244. Istyastono EP, de Graaf C, de Esch IJP, Leurs R. Molecular determinants of selective agonist and antagonist binding to the histamine H4 receptor. Curr Top Med Chem. 2011;11:661–79.
245. Lin JH, Lu AY. Inhibition and induction of cytochrome P450 and the clinical implications. Clin Pharmacokinet. 1998;35:361–90.
246. Celanire S, Wijtmans M, Talaga P, Leurs R, de Esch IJP. Histamine H-3 receptor antagonists reach out for the clinic. Drug Discov Today. 2005;10:1613–27.
247. Gbahou F, Vincent L, Humbert-Claude M, Tardivel-Lacombe J, Chabret C, Arrang J-M. Compared pharmacology of human histamine H3 and H4 receptors: structure-activity relationships of histamine derivatives. Br J Pharmacol. 2006;147:744–54.
248. Kuhne S, Wijtmans M, Lim HD, Leurs R, de Esch IJ. Several down, a few to go: histamine H3 receptor ligands making the final push towards the market? Expert Opin Investig Drugs. 2011;20:1629–48.
249. Kitbunnadaj R, Hashimoto T, Poli E, et al. N-substituted piperidinyl alkyl imidazoles: discovery of methimepip as a potent and selective histamine H3 receptor agonist. J Med Chem. 2005;48:2100–7.
250. Kitbunnadaj R, Zuiderveld OP, Christophe B, et al. Identification of 4-(1H-imidazol-4(5)-ylmethyl)pyridine (immethridine) as a novel, potent, and highly selective histamine H(3) receptor agonist. J Med Chem. 2004;47:2414–7.
251. Lazewska D, Kiec-Kononowicz K. Recent advances in histamine H3 receptor antagonists/inverse agonists. Expert Opin Ther Pat. 2010;20:1147–69.
252. Axe FU, Bembenek SD, Szalma S. Three-dimensional models of histamine H3 receptor antagonist complexes and their pharmacophore. J Mol Graph Model. 2006;24:456–64.

253. Hancock AA, Diehl MS, Faghih R, et al. In vitro optimization of structure activity relationships of analogues of A-331440 combining radioligand receptor binding assays and micronucleus assays of potential antiobesity histamine H3 receptor antagonists. Basic Clin Pharmacol Toxicol. 2004;95:144–52.
254. Sander K, von Coburg Y, Camelin J-C, Ligneau X, Rau O, Schubert-Zsilavecz M, Schwartz J-C, Stark H. Acidic elements in histamine H(3) receptor antagonists. Bioorg Med Chem Lett. 2010;20:1581–4.
255. Hancock AA. The challenge of drug discovery of a GPCR target: analysis of preclinical pharmacology of histamine H3 antagonists/inverse agonists. Biochem Pharmacol. 2006;71:1103–13.
256. Ligneau X, Morisset S, Tardivel-Lacombe J, Gbahou F, Ganellin CR, Stark H, Schunack W, Schwartz JC, Arrang JM. Distinct pharmacology of rat and human histamine H(3) receptors: role of two amino acids in the third transmembrane domain. Br J Pharmacol. 2000;131:1247–50.
257. Wakix pitolisant. In: Summ. Opin. (initial Auth. pitolisant). 2015. http://www.ema.europa.eu/docs/en_GB/document_library/Summary_of_opinion_-_Initial_authorisation/human/002616/WC500196742.pdf. Accessed 22 Nov 2015
258. Durant GJ, Ganellin CR, Parsons ME. Chemical differentiation of histamine H1- and H2-receptor agonists. J Med Chem. 1975;18:905–9.
259. Smits RA, Leurs R, de Esch IJP. Major advances in the development of histamine H4 receptor ligands. Drug Discov Today. 2009;14:745–53.
260. Lim HD, Adami M, Guaita E, Werfel T, Smits RA, de Esch IJP, Bakker RA, Gutzmer R, Coruzzi G, Leurs R. Pharmacological characterization of the new histamine H4 receptor agonist VUF 8430. Br J Pharmacol. 2009;157:34–43.
261. Nijmeijer S, de Graaf C, Leurs R, Vischer HF. Molecular pharmacology of histamine H4 receptors. Front Biosci (Landmark Ed). 2012;17:2089–106.
262. Schultes S, Nijmeijer S, Engelhardt H, Kooistra AJ, Vischer HF, de Esch IJP, Haaksma EEJ, Leurs R, de Graaf C. Mapping histamine H4 receptor–ligand binding modes. Med Chem Commun. 2012;39:193–204.
263. Gschwandtner M, Koether B, Werfel T, Stark H, Gutzmer R. Profiling of histamine H4 receptor agonists in native human monocytes. Br J Pharmacol. 2013;170:136–43.
264. Lim HD, De Graaf C, Jiang W, Sadek P, Mcgovern PM, Istyastono EP, Bakker RA, Esch IJP D, Thurmond RL, Leurs R. Molecular determinants of ligand binding to H 4 R species. Mol Pharmacol. 2010;77:734–43.
265. Lim HD, Jongejan A, Bakker RA, Haaksma E, de Esch IJP, Leurs R. Phenylalanine 169 in the second extracellular loop of the human histamine H4 receptor is responsible for the difference in agonist binding between human and mouse H4 receptors. J Pharmacol Exp Ther. 2008;327:88–96.
266. Cherezov V, Rosenbaum DM, Hanson MA, et al. High-resolution crystal structure of an engineered human beta2-adrenergic G protein-coupled receptor. Science. 2007;318:1258–65.
267. Lebon G, Warne T, Edwards PC, Bennett K, Langmead CJ, Leslie AGW, Tate CG. Agonist-bound adenosine A2A receptor structures reveal common features of GPCR activation. Nature. 2011;474:521–5.
268. Wheatley M, Wootten D, Conner MT, Simms J, Kendrick R, Logan RT, Poyner DR, Barwell J. Lifting the lid on GPCRs: the role of extracellular loops. Br J Pharmacol. 2012;165:1688–703.
269. Andaloussi M, Lim HD, Van Der Meer T, Sijm M, Poulie CBM, De Esch IJP, Leurs R, Smits RA. A novel series of histamine H4 receptor antagonists based on the pyrido[3,2-d]pyrimidine scaffold: comparison of hERG binding and target residence time with PF-3893787. Bioorg Med Chem Lett. 2013;23:2663–70.
270. Smits RA, Lim HD, van der Meer T, Kuhne S, Bessembinder K, Zuiderveld OP, Wijtmans M, de Esch IJP, Leurs R. Ligand based design of novel histamine H4 receptor antagonists; fragment optimization and analysis of binding kinetics. Bioorg Med Chem Lett. 2012;22:461–7.
271. Violin JD, Soergel DG, Boerrigter G, Burnett JC, Lark MW. GPCR biased ligands as novel heart failure therapeutics. Trends Cardiovasc Med. 2013;23:242–9.
272. Boerrigter G, Lark MW, Whalen EJ, Soergel DG, Violin JD, Burnett JC. Cardiorenal actions of TRV120027, a novel ß-arrestin-biased ligand at the angiotensin II type I receptor, in

healthy and heart failure canines: a novel therapeutic strategy for acute heart failure. Circ Heart Fail. 2011;4:770–8.
273. Kingwell K. Pioneering biased ligand offers efficacy with reduced on-target toxicity. Nat Rev Drug Discov. 2015;14:809–10.
274. Thurmond RL. The histamine H4 receptor: from orphan to the clinic. Front Pharmacol. 2015;6:1–11.
275. Nijmeijer S, Vischer HF, Rosethorne EM, Charlton SJ, Leurs R. Analysis of multiple histamine H4 receptor compound classes uncovers G i protein- and -arrestin2-biased ligands. Mol Pharmacol. 2012;82:1174–82.
276. Nijmeijer S, Vischer HF, Sirci F, Schultes S, Engelhardt H, de Graaf C, Rosethorne EM, Charlton SJ, Leurs R. Detailed analysis of biased histamine $H_4$ receptor signalling by JNJ 7777120 analogues. Br J Pharmacol. 2013;170:78–88.
277. Kottke T, Sander K, Weizel L, Schneider EH, Seifert R, Stark H. Receptor-specific functional efficacies of alkyl imidazoles as dual histamine H3/H4 receptor ligands. Eur J Pharmacol. 2011;654:200–8.
278. Davenas E, Rouleau A, Morisset S, Arrang JM. Autoregulation of McA-RH7777 hepatoma cell proliferation by histamine H3 receptors. J Pharmacol Exp Ther. 2008;326:406–13.
279. Medina V, Croci M, Crescenti E, et al. The role of histamine in human mammary carcinogenesis: H3 and H4 receptors as potential therapeutic targets for breast cancer treatment. Cancer Biol Ther. 2008;7:28–35.
280. Norman P. New H1/H3 antagonists for treating allergic rhinitis: WO2010094643. Expert Opin Ther Pat. 2011;21:425–9.
281. Daley-Yates P, Ambery C, Sweeney L, Watson J, Oliver A, McQuade B. The efficacy and tolerability of two novel $H_1/H_3$ receptor antagonists in seasonal allergic rhinitis. Int Arch Allergy Immunol. 2012;158:84–98.
282. Incerti M, Flammini L, Saccani F, Morini G, Comini M, Coruzzi M, Barocelli E, Ballabeni V, Bertoni S. Dual-acting drugs: an in vitro study of nonimidazole histamine H 3 receptor antagonists combining anticholinesterase activity. ChemMedChem. 2010;5:1143–9.
283. Gianotti M, Botta M, Brough S, et al. Novel spirotetracyclic zwitterionic dual H1/5-HT2A receptor antagonists for the treatment of sleep disorders. J Med Chem. 2010;53:7778–95.
284. Esbenshade TA, Fox GB, Krueger KM, et al. Pharmacological properties of ABT-239 [4-(2-{2-[(2R)-2-Methylpyrrolidinyl]ethyl}-benzofuran-5-yl)benzonitrile]: I. Potent and selective histamine H3 receptor antagonist with drug-like properties. J Pharmacol Exp Ther. 2005;313:165–75.
285. Smits RA, Lim HD, Stegink B, Bakker RA, De Esch IJP, Leurs R. Characterization of the histamine H4 receptor binding site. Part 1. Synthesis and pharmacological evaluation of dibenzodiazepine derivatives. J Med Chem. 2006;49:4512–6.
286. Lim HD, Smits RA, Bakker RA, van Dam CME, de Esch IJP, Leurs R. Discovery of S -(2-Guanidylethyl)-isothiourea (VUF 8430) as a potent nonimidazole histamine H 4 receptor agonist. J Med Chem. 2006;49:6650–1.
287. Nguyen T, Shapiro DA, George SR, et al. Discovery of a novel member of the histamine receptor family. Mol Pharmacol. 2001;59:427–33.
288. Borda E, Stranieri G, Sterin-Borda L. H(1)-Receptor activation triggers the endogenous nitric oxide signalling system in the rat submandibular gland. Mediators Inflamm. 2002;11:337–43.
289. Mayhan WG. Nitric oxide accounts for histamine-induced increases in macromolecular extravasation. Am J Physiol. 1994;266:H2369–73.
290. Matsubara M, Tamura T, Ohmori K, Hasegawa K. Histamine H1 receptor antagonist blocks histamine-induced proinflammatory cytokine production through inhibition of Ca2+-dependent protein kinase C, Raf/MEK/ERK and IKK/IkB/NF-kB signal cascades. Biochem Pharmacol. 2005;69:433–49.
291. Kharmate G, Liu Z, Patterson E, Khan MM. Histamine affects STAT6 phosphorylation via its effects on IL-4 secretion: Role of H1 receptors in the regulation of IL-4 production. Int Immunopharmacol. 2007;7:277–86.

292. Molina Hernández A, Velasco I. Histamine induces neural stem cell proliferation and neuronal differentiation by activation of distinct histamine receptors. J Neurochem. 2008;106:706–17.
293. Hernández-Angeles A, Soria-Jasso LE, Ortega A, Arias-Montaño JA. Histamine H1 receptor activation stimulates mitogenesis in human astrocytoma U373 MG cells. J Neurooncol. 2001;55:81–9.
294. Li D, Wen JF, Jin JY, Jin H, Ann HS, Kim SZ, Kim SH, Lee HS, Cho KW. Histamine inhibits atrial myocytic ANP release via H2 receptor-cAMP-protein kinase signaling. Am J Physiol Regul Integr Comp Physiol. 2003;285:R380–93.
295. Yoshida M, Takahashi Y, Inoue S. Histamine induces melanogenesis and morphologic changes by protein kinase A activation via H2 receptors in human normal melanocytes. J Invest Dermatol. 2000;114:334–42.
296. Dai H, Zhang Z, Zhu Y, Shen Y, Hu W, Huang Y, Luo J, Timmerman H, Leurs R, Chen Z. Histamine protects against NMDA-induced necrosis in cultured cortical neurons through H2 receptor/cyclic AMP/protein kinase A and H3 receptor/GABA release pathways. J Neurochem. 2006;96:1390–400.
297. Sakhalkar SP, Patterson EB, Khan MM. Involvement of histamine H1 and H2 receptors in the regulation of STAT-1 phosphorylation: Inverse agonism exhibited by the receptor antagonists. Int Immunopharmacol. 2005;5:1299–309.
298. van der Pouw Kraan TC, Snijders A, Boeije LC, de Groot ER, Alewijnse AE, Leurs R, Aarden LA. Histamine inhibits the production of interleukin-12 through interaction with H2 receptors. J Clin Invest. 1998;102:1866.
299. Truta-Feles K, Lagadari M, Lehmann K, et al. Histamine modulates γδ-T lymphocyte migration and cytotoxicity, via Gi and Gs protein-coupled signalling pathways. Br J Pharmacol. 2010;161:1291–300.
300. Atzori M, Lau D, Tansey EP, Chow A, Ozaita A, Rudy B, McBain CJ. H2 histamine receptor-phosphorylation of Kv3.2 modulates interneuron fast spiking. Nat Neurosci. 2000;3:791–8.
301. Nakamura T, Yoshikawa T, Noguchi N, Sugawara A, Kasajima A, Sasano H, Yanai K. The expression and function of histamine H3 receptors in pancreatic beta cells. Br J Pharmacol. 2013;171:171–85.
302. Gomez-Ramirez J, Ortiz J, Blanco I. Presynaptic H3 autoreceptors modulate histamine synthesis through cAMP pathway. Mol Pharmacol. 2002;61:239–45.
303. Petit-Bertron A-F, Machavoine F, Defresne MP, Gillard M, Chatelain P, Mistry P, Schneider E, Dy M. H4 histamine receptors mediate cell cycle arrest in growth factor-induced murine and human hematopoietic progenitor cells. PLoS One. 2009;4, e6504.
304. Witte DG, Yao BB, Miller TR, et al. Detection of multiple H3 receptor affinity states utilizing [3H]A-349821, a novel, selective, non-imidazole histamine H3 receptor inverse agonist radioligand. Br J Pharmacol. 2006;148:657–70.

# Chapter 2
# Genetic Polymorphisms in the Histamine Receptor Family

**Stephany Micallef and Astrid Sasse**

**Abstract** Histamine is a biogenic amine which has an inherent biological importance in many physiological functions, both in the central nervous system and in the periphery. With the new genomic era we are facing, personalized care and treatment is becoming one of the major focal points in research. This chapter focuses on the genetic variations and polymorphisms localized on genes encoding for human histamine receptors (HRHs) where it provides an up-to-date collection of polymorphisms found on genes encoding HRHs and their association to diseases. There is a clear need to highlight the specific implications polymorphisms have on this family of G-protein-coupled receptors. This book chapter collates recent and other important publications related to polymorphisms and genetic linkage of histamine receptors. New association studies have been published for the gene encoding the HRH4, linking SNPs to asthma, cancer, and atopic dermatitis. For example, rs17187619, rs527790, rs487202, rs1421125, and rs615283 have been associated with infection induced in asthma patients. Other SNPs found to harbor a link in breast cancer include rs623590, rs11662595, and rs1421125. With the increasing interest in cancer research, a polymorphism (rs2607474) discovered on the gene encoding for HRH2 was also found to have an association to gastric atrophy leading to gastric cancer. Looking into pharmacogenetics, a linkage was found between risperidone treatment and histamine receptor 3 (HRH3), where rs3787430 could be a potential biological marker for treatment. With these significant genetic variations recently discovered and their potential contribution to the common diseases, this chapter gathers the knowledge to date for SNPs identified on the human histamine receptors.

**Keywords** Histamine receptors • Genetic polymorphisms • Single-nucleotide polymorphisms • Association studies • Linkage analysis • Human histamine 1 receptor • Human histamine 2 receptor • Human histamine 3 receptor • Human histamine 4 receptor

S. Micallef, Ph.D. • A. Sasse, Ph.D. (✉)
School of Pharmacy and Pharmaceutical Sciences, Trinity Biomedical Sciences Institute, Trinity College Dublin, Dublin 2, Ireland
e-mail: micalles@tcd.ie; sassea@tcd.ie

 
P. Blandina, M.B. Passani (eds.), *Histamine Receptors*, The Receptors 28,
DOI 10.1007/978-3-319-40308-3_2

Single-nucleotide polymorphisms can show an effect at a phenotypic level. In humans, these SNPs occur in at least 1 % of the population with at a frequency of 1 every 100–300 bases along the three billion-base human genome [1]. Understanding the genetic basis of receptors generates better comprehension of underlying diseases leading to novel strategies for improved therapeutic agents. Hence, considering the biological significance of histamine, polymorphisms of histamine-related genes represent potential genetic factors that are likely to influence a disease or a therapy. In this chapter, we look into the genetic composition of the histamine receptor family and identify genetic variants associated with disorders and how they impact on gene function.

The oldest family members of the amine subtype are histamine $H_1$ and histamine $H_2$ receptors. They were the first histamine GPCRs discovered, and they were first cloned in 1991 [2]. The existence of a third histamine receptor was known pharmacologically, but the histamine $H_3$ receptor was first cloned only in 1999, with the discovery that cells transfected with GPCR97 were able to inhibit adenylate cyclase in response to histamine [3]. Homology and phylogenetic analysis of the $H_3$ receptor showed it to be markedly different from the previously cloned $H_1$ receptor and $H_2$ receptors, which suggests that these histamine receptors evolved from different ancestor sequences [2]. This discovery also led to the identification of a fourth receptor, the human histamine $H_4$ receptor in the year 2000/2001 by several research groups [4–10].

All four histamine receptors belong to the rhodopsin-like family of the G-protein-coupled receptors [11]. Even though they form part of the same family, they exhibit different genomic and amino acid organization. Using the bioinformatics tool GenomeNet and multiple sequence alignment CLUSTALW, homology for the genes can be estimated. Comparing the DNA sequence alignment between all HRH genes, the homology seems to be in the same range (18–20 %) between human histamine receptors (HRHs) (Table 2.1) [12].

However, the translated genomic sequences for the HRH family into amino acids give a different picture. An amino acid sequence resemblance can be seen between two GPCR family members, namely, the $H_3$ receptor and the $H_4$ receptor. The histamine $H_3$ receptor displays a 37 % homology to the $H_4$ receptor, but when compared to $H_1$ and $H_2$ receptors, the similarity drops to 23 % and 22 %, respectively (Table 2.1) [12]. Looking at the transmembrane domains, the similarity between the $H_3$ receptor and $H_4$ receptor increases up to 54 %, and if the physiochemical properties of the amino acids is taken into account, this similarity increases even more to 68 % [13]. This similarity between the $H_3$ receptor and $H_4$

**Table 2.1** Gene and protein homology for the histamine receptor subtypes

| | | Gene % | | | |
|---|---|---|---|---|---|
| Homology for HRHs | | HRH1 | HRH2 | HRH3 | HRH4 |
| Protein % | HRH1 | | 18.60 | 20.20 | 20.60 |
| | HRH2 | 21.50 | | 20.70 | 18.95 |
| | HRH3 | 25.70 | 19.50 | | 18.30 |
| | HRH4 | 23.00 | 22.00 | 37.00 | |

receptor also led to the development of imidazole-containing ligands that have a dual action on histamine $H_3$ and histamine $H_4$ receptors [14]. Similarly, ligands targeting $H_1$ and $H_2$ receptors were also investigated using a series of cyanoguanidine compounds that were synthesized by linking mepyramine-type $H_1$ receptor antagonist substructures with roxatidine-, tiotidine-, or ranitidine-type $H_2$ receptor antagonist moieties [15]. However, these need to be further improved to provide an enhanced affinity binding profile. The combination of $H_1$ and $H_4$ receptor antagonists was also studied, where benefit was observed by inhibiting both receptors in histamine-induced scratching. This study suggested that combined $H_1$ and $H_4$ antagonists provided to be more effective than monotherapy [16]. The dual antagonistic benefit was also identified when the $H_1$ antagonist, olopatadine, and $H_4$ antagonist, JNJ17777120, were used in allergic dermatitis mice models. The combined effect proved to be as effective as prednisolone. Glucocorticoids such as prednisolone are known for atrophy of the skin and systemic side effects. The $H_4$ antagonist is currently in phase II clinical study and is proving to have a safe drug profile [17]. These results indicate a promising future for the treatment of these allergic conditions.

Important cellular processes such as DNA binding, protein interactions, and enzyme activity are highly impacted by conserved domains in proteins [18]. Alignment of the amino acid sequences of the GPCR family provides further evidence of the similarities and conserved domains between these receptors. Figure 2.1 represents the alignment of $H_1$ and $H_2$ receptors on the amino acid level. The dark-gray region highlights conserved amino acids between the two receptors, while the light-gray regions represent strong similarities between conserved domains. The yellow highlight on the other hand represents the seven transmembrane domains of the GPCRs. Even though the homology is significant between these proteins, the $H_1$ receptor is characterized by a large intracellular loop 3 (ICL3) unlike the receptor $H_2$ receptor subtype, with a small ICL3 and a longer C-terminal. The asterisks (*) showed in Fig. 2.1 represent amino acids which are singly fully conserved residue.

Figure 2.2 shows a similar representation for the newer histamine receptors, histamine $H_3$ and $H_4$ receptors; noticeably, both show a large intracellular loop 3 (ICL3), comparable to the ICL in histamine $H_1$ receptor. GPCRs are known to interact with a wide variety of protein domains through the intracellular loops, transmembrane and C-terminal domains. The most abundant interaction is with the C-terminus of the GPCR [19]. Both histamine $H_3$ and $H_4$ receptors show a longer C-terminus which is comparable to the histamine $H_1$ receptor.

Studies on the crystalline structures of GPCRs identified common similarities between class A receptors. The extracellular loops (ECL) are considered as peptide linkers that hold together functionally important transmembrane domains. This link also keeps these helices in a stable position in the cell membrane. The ECL has also the important role of receptor activation and ligand binding [20]. The first extracellular loop (ECL1) is generally small and made up of only a few amino acids. This is also represented in all four histamine receptors. Despite this, it was identified in different publications that the ECL1 affects the binding pockets, as

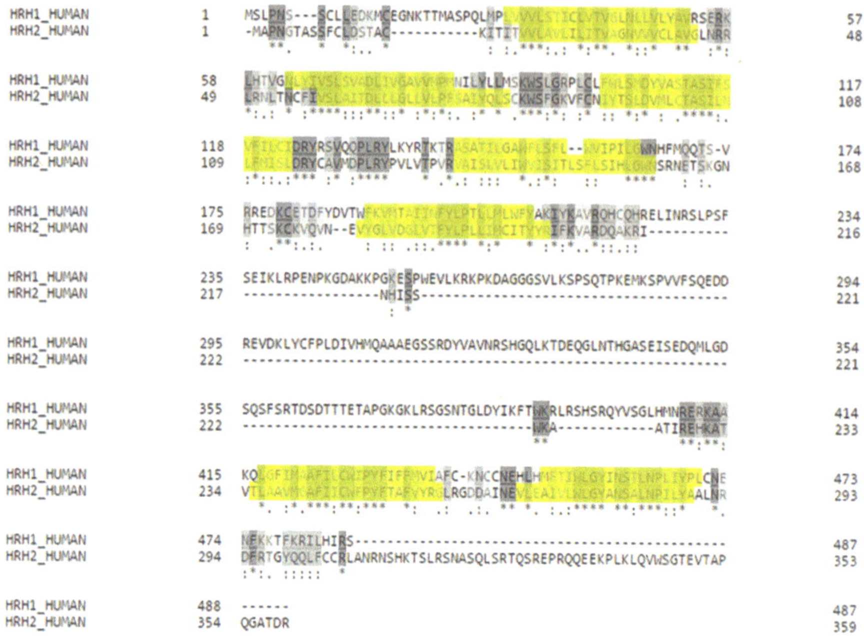

**Fig. 2.1** *Protein sequences alignment between $H_1$ receptor and $H_2$ receptor. Yellow* highlighted sequences refer to the seven transmembrane domains. The *dark gray* indicates the conserved domains, and the *light gray* refers to similarities between amino acids between the two GPCRs. *Asterisks* indicate positions which have a single, fully conserved residue. *Colon* indicates conservation between groups of strongly similar properties. *Dot* indicates conservation between groups of weakly similar properties [85]

with the help of other ECLs this loop can provide rigidity and structure essential for receptor activation [20].

Histamine $H_1$ receptor and $H_2$ subtypes have proven to be excellent drug targets, and the more recently discovered histamine $H_3$ and $H_4$ receptors remain under investigation as future drug targets. First- and second-generation antihistamines are widely used to relief symptoms of allergic rhinitis, urticarial, and other allergies and can cost between \$8 and \$200 per month per patient [21]; hence, the search for histamine ligands is appealing for the pharmaceutical industry. Ligands for histamine $H_3$ receptor are currently in clinical studies. Antagonists for this receptor are currently in Phase 1 and Phase 2 clinical trials under review that indicate there is a potential treatment for Alzheimer's disease, ADHD, schizophrenia, epilepsy, narcolepsy, obesity, neuropathic pain, and allergic rhinitis [22]. The histamine $H_3$ receptor antagonist pitolisant is in Phase 3 clinical trial for treatment of narcolepsy and excessive daytime sleepiness associated with Parkinson's disease. The recently discovered $H_4$ receptor is still undergoing intense preclinical research, with some novel $H_4$ receptor antagonists having entered clinical trials, such as UR-63325 which recently completed Phase 2 clinical trial for allergic rhinitis [23],

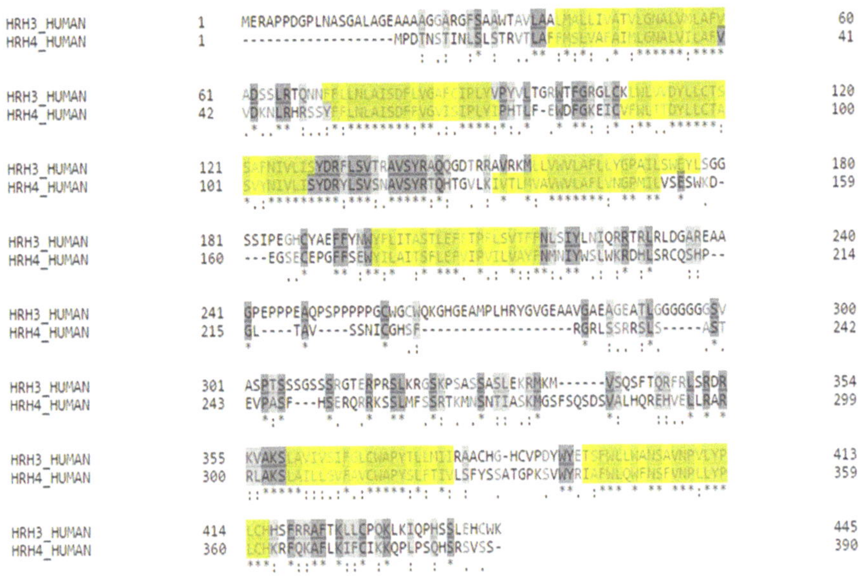


**Fig. 2.2** *Protein sequences alignment between $H_3$ receptor and $H_4$ receptor. Yellow* highlighted sequences refer to the seven transmembrane domains. The *dark gray* indicates the conserved domains, and the *light gray* refers to similarities between amino acids between the two GPCRs. *Asterisks* indicate positions which have a single, fully conserved residue. *Colon* indicates conservation between groups of strongly similar properties. *Dot* indicates conservation between groups of weakly similar properties [85]

JNJ-39758979 completing Phase 2 clinical trial for patients suffering from asthma, and PF-3893787 which is in Phase 1 clinical trial for asthmatic patients subjected to allergen change [24].

## 2.1 Human Histamine $H_1$ Receptor

The human histamine $H_1$ receptor is the largest protein in the histamine receptor family [2] and is expressed primarily in smooth muscle, endothelial cells, the adrenal medulla, the heart, and also in the central nervous system (CNS). The $H_1$ receptor is predominantly involved in smooth muscle contraction and increased vascular permeability [7]. The main therapeutic target for $H_1$ receptor antagonists are allergic conditions and sleep disorders [4].

Histamine $H_1$ gene (HRH1) is located on chromosome 3; 3p25. Figure 2.3 shows the complete organization of this receptor. In (A), the genomic organization is seen where it is characterized by a large intron between the UTR region at the 5′-end and has only one exon [12]. The amino acid sequence of this receptor (B) in turn shows a large intracellular loop (ICL3) similar to the $H_3$ and $H_4$ receptor. The snake plot for the amino acid sequence is represented in (C) which illustrates the structural formation of the receptor [25].

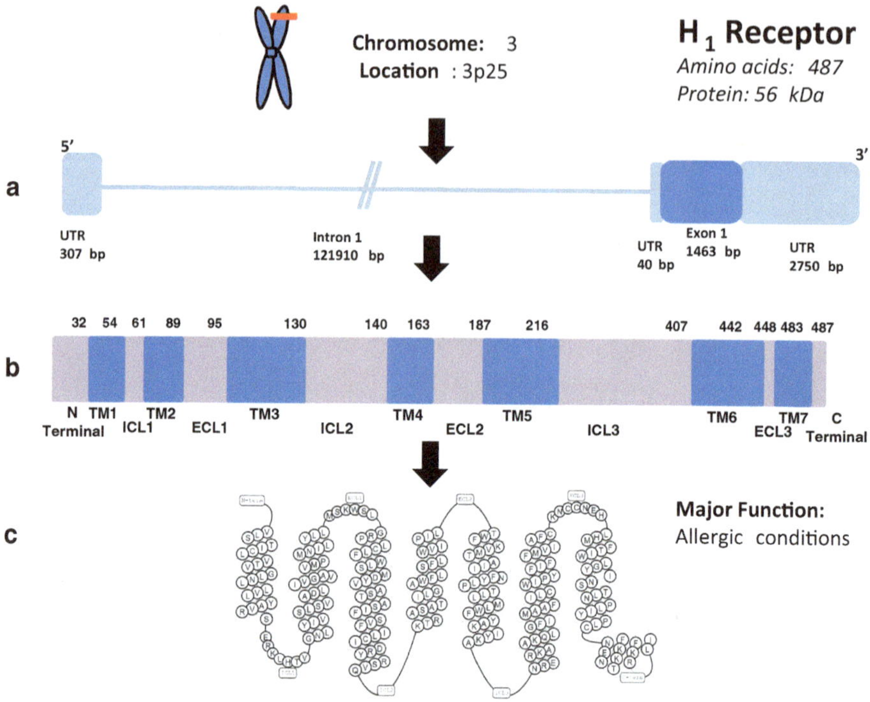


**Fig. 2.3** *Histamine $H_1$ receptor molecular organization.* (**a**) Genomic organization of HRH1. (**b**) Amino acid organization, where TM is transmembrane domain, ICL is intracellular loop, and ECL is extracellular loop. (**c**) GPCR snake plot

## *2.1.1 Histamine $H_1$ Receptor Polymorphisms in CNS Disorders*

The histaminergic system has proven to be one of the aminergic systems present in the mammalian CNS. The discovery that histamine is one of the major neurotransmitters in the CNS encouraged studies that proved it is involved in a wide range of physiological functions such as cognition, emotion, feeding behavior, and sleep-wake cycle [26]. Studies on schizophrenia observed that the levels of N-tele-methylhistamine, a major brain histamine metabolite, were elevated in the cerebrospinal fluid of schizophrenia patients, and $H_1$ receptor binding sites were decreased [27–29]. The role of this receptor in psychiatric conditions was further evidenced by showing that antidepressants and antipsychotics exhibit a high binding affinity to $H_1$ receptor with a very pronounced antagonistic effect [30]. High binding affinity to the H1 receptor in antidepressants has been linked to a higher incidence of metabolic syndrome in patients with bipolar disorder [31]. The histaminergic system may also be a target for the prevention of obesity and metabolic syndrome through histamine $H_1$ and $H_3$ receptors with combined activity compounds displaying agonism at $H_1$ receptors and antagonism at $H_3$ receptors [32].

Genetic variations in histamine $H_1$ receptors were studied due to its interesting role in CNS disorders. Single-nucleotide polymorphisms (SNPs) were found to play a role in depression in bipolar disorder, where the SNPs in the dopamine $D_3$ receptor and histamine $H_1$ receptor genes indicated a significant association to the improvement of the condition following olanzapine and fluoxetine treatment [33]. Studying SNP variations can also be of interest in predicting side effects of medications and developing personalized medicines. Antipsychotic medication such as clozapine and olanzapine are associated with an increased risk of obesity. Clozapine, the atypical antipsychotic, has a diverse binding profile to all four aminergic systems (serotonergic, dopaminergic, histaminergic, and noradrenergic) [34]. A study on antipsychotic affinity for the histamine $H_1$ receptor and the muscarinic $M_3$ receptor has been linked to side effects such as weight gain. A significant association was also identified between patients using antipsychotics with high affinity for the $H_1$ receptor (clozapine, olanzapine, and quetiapine) in obesity and BMI, where haplotype for rs346074–rs346070 A-T associated with obesity $P=0.025$ and haplotype A-T associated with BMI $P=0.005$ [35]. In another study, however, looking into a polymorphism on exon 1 which leads to a change in Glu349Asp on this receptor (which lies on ICL3) found no association for antipsychotic induced weight gain in a Chinese population diagnosed with schizophrenia [36].

Variations found on the promoter region of histamine H1 receptor were investigated for association for psychiatric disorders. A significant linkage disequilibrium (LD) ($P<10^{-6}$) was found in the excess of C allele, but no strong association with schizophrenia was identified in multiple testing. However, the polymorphism for G/C allele may represent a marker for more functional polymorphisms located further upstream [37].

The histamine receptors are also of interest in Parkinson's disease patients (PD) as histamine is involved in neuronal degeneration and neurotoxicity. The expression and density of histamine receptors were observed to change in PD patients [38], and histamine receptor antagonists were noticed to improve motor symptoms. However, a study that looked into the a C/T polymorphism in exon 1 which results in a Leu449Ser amino acid substitution was not able to link this SNP to PD [39].

### 2.1.2 Histamine $H_1$ Receptor Polymorphisms in Inflammation

Histamine levels have been studied in various inflammatory diseases suggesting an association of these conditions to histamine receptors. Histamine levels are found to be elevated in bronchoalveolar fluids extracted from asthma patients, in the skin and plasma of patients with atopic dermatitis, in chronic urticaria biopsies, in both plasma and synovial fluid of patients with rheumatoid arthritis, and in the plasma of patients with psoriatic arthritis [40]. The major receptors shown to be linked to inflammation and immune response are $H_1$, $H_2$, and $H_4$ receptors [41]. Activation of $H_1$ receptors increases histamine release and the release of other mediators while also augmenting the pro-inflammatory activity of the immune system by promoting migration to the area of inflammation.

To date, the $H_1$ receptor has been the major target in the treatment of allergic conditions. Histamine $H_1$ receptor antagonists have been used for decades to treat allergies, and although they provide symptomatic relief for most allergic conditions, they are not as effective for allergic asthma [42]. A study in a Korean population looked into associations between aspirin-induced urticarial/angioedema to polymorphisms on $H_1$ receptor. The polymorphisms studied were a C/T change in the promoter region and a G/A change in exon 1 that leads to D349A amino acid change. However, no association was found between these genetic variations and the allergic response [43].

### 2.1.3 Histamine $H_1$ Receptor Polymorphisms in Cancer

The HRH1 gene was investigated for a potential link in various types of cancer. In a study conducted by Wang et al., 88 SNPs were identified that may cause an impact of histamine on $H_1$ receptor. Out of these 88 SNPs, 84 led to a missense mutation and 4 alleles that disrupt an exon splicing enhancer. From 153 tests carried out on the HRH1 gene, 23 tests proved an association between microarray expression in the HRH1 gene and cancer prognosis (including bladder, blood, brain, breast, colorectal, eye, lung, ovarian, soft tissue cancers) with 5 % level of significance. It is also worth noting that the level between HRH1 expression and the different types of cancer prognosis varied, suggesting that the function of HRH1 gene varies between cancers [44].

In a large population study to determine genetic variability, polymorphisms were selected to identify opioid efficacy in cancer patients. Experience has shown that opioid doses and efficacy vary in cancer patients, and studies in small population samples were not found to be reliable. This study conducted by Klepstad et al. included polymorphisms from $H_1$ receptor (rs2606731, rs346076, rs901865, rs346070). However, opioid efficacy in cancer patients for pain is not genetically associated with these SNPs [45].

A summary of the above mentioned polymorphisms and their respective SNP location is shown in Table 2.2 below.

**Table 2.2** HRH1 polymorphisms in diseases

| rs number | SNP | Location on gene | Condition | PubMed ID |
|---|---|---|---|---|
| NA[a] | G/C | Promoter (intron) | Schizophrenia | 12429384 [37] |
| rs346074 | A/G | Intron 1 | Schizophrenia | 21336576 [35] |
| rs346070 | C/T | 3′ UTR | | |
| rs2067467 | G/T | Exon 1 | Schizophrenia | 21937795 [36] |
| rs2067470 | C/T | Exon 1 | Parkinson's disease | 18366640 [39] |
| NA[b] | C/T | Promoter | Aspirin-induced urticaria/angioedema | 15953854 [43] |
| NA[b] | G/A | Exon 1 | | |
| rs2606731 | G/T | Intron 1 | Opioid use in cancer pain | 21398039 [45] |
| rs346076 | A/G | Intron 1 | | 21570824 [46] |
| rs346070 | C/T | 3′UTR | | |
| rs901865 | G/A | 5′UTR | | |

NA[a] denotes rs number that could not be identified

NA[b] denotes rs number that was not assigned

## 2.2 Human Histamine $H_2$ Receptor

The human histamine $H_2$ receptor is mostly expressed in gastric parietal cells, vascular smooth muscle, suppressor T cells, neutrophils, the CNS, and the heart. Its main involvement is in stimulation of gastric acid secretion and hence an important therapeutic target for gastric ulcers [7]. The histamine $H_2$ receptor is located in chromosome 5 and contains two introns and two exons. Similar to the $H_1$ receptor, HRH2 gene has intron 1 located in the 5′-UTR region; however, this region is much smaller in length with 24,561 base pairs compared to 121,910 base pairs in the $H_1$ receptor gene [12]. Looking into the translated amino acid sequence, this member displays a different structure when compared to the other family members (Fig. 2.4). The receptor is characterized by a small ICL3 and a larger C-terminal with around 54 amino acids, compared to 4, 18, and 20 amino acids for $H_1$, $H_3$, and $H_4$ receptors, respectively.

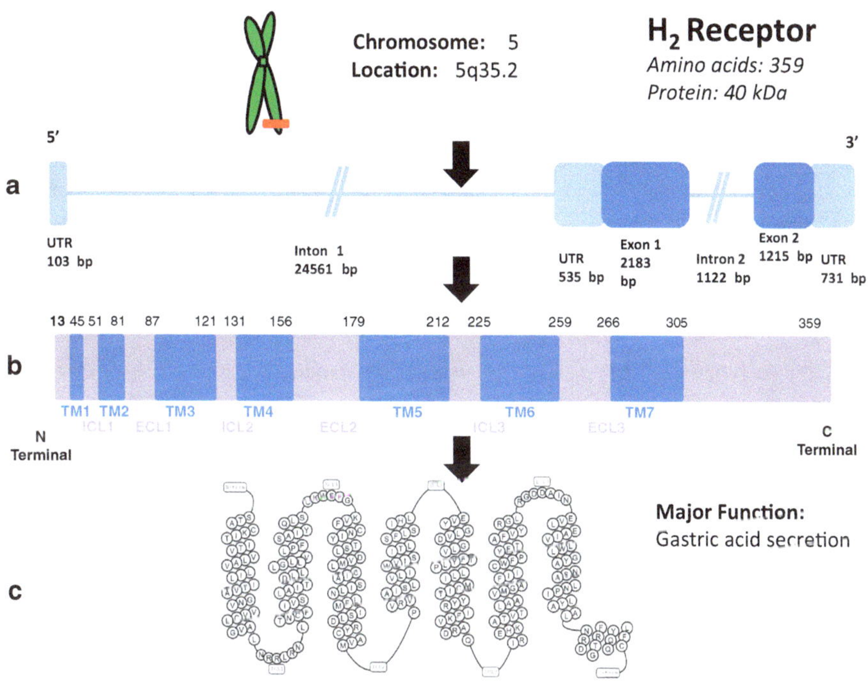


**Fig. 2.4** *Histamine $H_2$ receptor molecular organization.* (**a**) Genomic organization of HRH2. (**b**) Amino acid organization, where TM is transmembrane domain, ICL is intracellular loop, and ECL is extracellular loop. (**c**) GPCR snake plot

### 2.2.1 Histamine $H_2$ Polymorphisms in CNS Disorders

Human histamine 2 receptor has also been localized in the neocortex, hippocampus, caudate, and putamen regions of the brain [37]. However, studies focusing on the $H_2$ receptor and schizophrenia so far have been controversial. The amino acid exchange of Asn217Asp was suggested to be more common in schizophrenia patients [47, 48]; however, this could not be replicated in other studies [49]. To date no strong association has been linked between the $H_2$ receptor and the disease, yet famotidine, a selective $H_2$ receptor antagonist, has been successfully used to treat the negative symptoms of schizophrenia [50]. Low brain distribution of available $H_2$ receptor antagonists/inverse agonists most probably contribute to the missing significance of data. The interactions between antipsychotics and the four histamine receptor subtypes generate an interesting perspective for the involvement of HRHs and the disease.

A study for the $H_2$ polymorphism −1018-G/A was carried out on 164 patients treated with clozapine and diagnosed with schizophrenia showed a strong linkage disequilibrium (LD) between −1018-G/A and −592-A/G (LD $P < 10^{-6}$). This relationship was not observed in multiple testing possibly due to small patient sample and due to the differences in diagnostics methods employed to assess drug response. However, due to limitations in the study, there could still be a potential role in patient outcome to clozapine treatment upon further studies [37]. This same polymorphism (−1018-G/A) was also looked into for association with Parkinson's disease (PD), since this genetic variation lies in the promoter region of the receptor. Variations in the promoter region may contribute to changes in expression of the receptor. Nevertheless, no major risk factor was observed for the risk of PD to this SNP [39]. This promoter polymorphism was included in multiple studies which are summarized in Table 2.3.

### 2.2.2 Histamine $H_2$ Polymorphisms in Inflammatory Disorders

The histamine $H_2$ receptor works by suppressing inflammation by decreasing eosinophil and neutrophil chemotaxis, decreasing IL-12 by dendritic cells, increasing IL-10, and inducing the development of Th2 or tolerance-inducing dendritic cells

**Table 2.3** HRH2 polymorphisms in diseases

| rs number | SNP | Location on gene | Condition | PubMed ID |
|---|---|---|---|---|
| rs2067474 | G/A | Promoter | Schizophrenia | 12429384 [37] |
| rs2067474 | G/A | Promoter | Parkinson's disease | 18366640 [39] |
| NA[a] | G/A | Exon 1 | Aspirin-induced urticaria/angioedema | 15953854 [43] |
| NA[a] | C/T | Exon 1 | | |
| rs2607474 | G/A | Promoter | Gastric mucosal atrophy | 22720301 [53] |
| rs1800689 | G/A | Exon 1 | Non-melanoma skin cancer | 21760883 [60] |
| rs2607474 | G/A | Promoter | Gastric cancer | 22615049 [58] |

NA[a] denotes rs number was not assigned

[51]. This receptor also became pharmacologically important in gastric acid production. Studies observed that histamine-evoked gastric acid secretion were not blocked with classical antihistamines; hence, this led to the conclusion that histamine $H_2$ receptors were involved in gastric acid secretion [52].

The same polymorphism mentioned earlier, −1018-G/A, located in the promoter region of the $H_2$ receptor was identified to be associated with gastric mucosal atrophy [53], which is interesting considering that $H_2$ receptor blockers are the major therapeutic strategy used against gastric acid disorders. Atrophic gastritis is a chronic inflammation in the mucosal lining and, together with *H. pylori* infections, can even result in gastric cancer if left untreated [54, 55]. The GG genotype of rs2607474 proved to give a significant increase in the risk of gastric mucosal atrophy, especially in *H. pylori*-infected patients. The minor allele frequency for the GA genotype in gastric atrophy patients compared to controls was 8.02 % and 13.3 %, respectively ($P=0.057$), with the GG genotype being significantly increased in the patient group ($P=0.055$) [53]. As a result, this polymorphism in particular seems to provide a biological marker for this condition and can be used as a tool to identify the risk and severity of gastric mucosal atrophy in combination with *H. pylori* infection.

Genetic variation on the exon of histamine H2 receptor was also studied in relation to aspirin-induced urticaria/angioedema. The polymorphisms located on exon 1, G543A, and C826T, however, did not show any significant difference in allele and genotype frequencies of their SNPs, which indicated that these genetic variations are not related with the development of aspirin-induced urticaria/angioedema phenotype [43].

### 2.2.3 Histamine $H_2$ Polymorphisms in Cancer

Histamine seems to play a role in cell proliferation in malignant cells, where high histamine biosynthesis was reported in different human neoplasias [56, 57]. It has been shown that $H_1$ and $H_2$ receptors are expressed in normal and malignant cells lines, and that $H_3$ and $H_4$ receptors are expressed in cell lines of human mammary gland [57]. Analysis carried out on the $H_2$ receptor in gastric carcinoma identified promoter polymorphism rs2607474 (−1018-G/A) to be associated with the disease. The homozygous GG genotype was found to be associated with gastric mucosal atrophy ($P=0.0052$) with a subsequent development of gastric cancer (mostly intestinal, $P=0.0047$) especially at an advanced age [58]. This polymorphism was also identified by the same group to be associated with gastric mucosal atrophy as mentioned in the previous section. And this association with gastric cancer was even higher at an advanced age where the atrophy and metaplasia scores were found to be higher [58]. Due to the location of this SNP in an important region for gene regulation, it seems like it has an important implication in gene transcription, but further analyses are needed to prove this biological significance. Another recent study also investigated the potential impact this polymorphism has in breast cancer. Two hundred and one Chinese Han women were genotyped for rs2607474; however, there is no significant difference in the frequencies of genotypes ($P=0.174$) or alleles ($P=0.054$) between breast cancer patients and health controls [59]. Hence, while this SNP indicates a potential role in gastric cancer, it might not be a risk factor in breast cancer. However, further investigations need to be conducted to confirm this is not ethnic or population related.

The polymorphisms rs1800689 which is located on exon 1 (G/A; codon 181) was included in a study in UV-susceptibility in non-melanoma skin cancer. However it was concluded that there was no association between this SNP with UV-induced immunosuppression and risk of non-melanoma skin cancer [60].

The table below (Table 2.3) shows all polymorphisms related to histamine $H_2$ receptor and associated diseases.

## 2.3 Human Histamine $H_3$ Receptor

The histamine $H_3$ receptor, first cloned in 1999 by J&J [3], is a presynaptic autoreceptor on histamine neurons and heteroreceptor on various neurotransmitter neurons and is expressed in the CNS and peripheral nerves [7]. The genomic organization of this receptor is slightly different from the previous two receptors. It is similar to histamine $H_2$ where both have two introns. Though, intron two is the largest with 1549 base pairs, unlike histamine $H_2$ where intron one was the largest intron with 24561 base pairs. The histamine $H_3$ gene also has three exons with exon 3 being the largest with 941 base pairs [12]. These exons are translated into amino acids forming the seven transmembrane domains with a large ICL3, similar to histamine $H_1$ and $H_4$ receptors. Exon 1 translates the N-terminal to the mid of TM2, exon 2 translates to the mid of TM2 to the beginning of ICL2, while exon 3 of histamine $H_3$ receptor translates into ICL2 to the C-terminal. Figure 2.5 shows these levels of structural organization leading to the GPCR.

### *2.3.1 Histamine $H_3$ Polymorphisms in CNS Disorders*

Migraine is a neurovascular disorder, and histamine is thought to participate in the pathophysiology of the disease, where the frequency of migraine is more pronounced in patients with allergic diseases and an increase in histamine levels was also observed in patients susceptible to migraine attacks [61, 62]. A polymorphism located in the third intracellular loop of histamine H3 receptor (A280V), which leads to the amino acid exchange from alanine to valine, was found to be a risk factor for migraine [62]. The V allele for this SNP was found to be higher in patients when compared to controls, and especially the genotype VV or VA was found to be 12.92 % when compared to controls with 3.22 % ($P=0.001$). This same polymorphism was also reported to influence the signaling of the histamine $H_3$ receptor without affecting the binding characteristics in the Shy-Drager syndrome study.

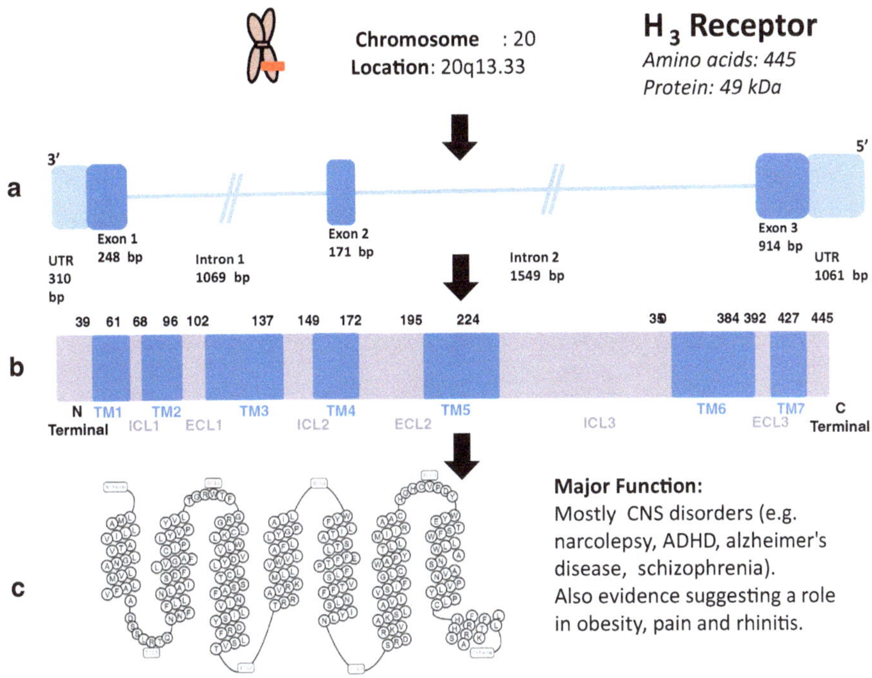


**Fig. 2.5** *Histamine $H_3$ receptor molecular organization.* (**a**) Genomic organization of HRH3. (**b**) Amino acid organization, where TM is transmembrane domain, ICL is intracellular loop, and ECL is extracellular loop. (**c**) GPCR snake plot

**Table 2.4** HRH3 polymorphisms in diseases

| rs number | SNP | Location on gene | Condition | PubMed ID |
|---|---|---|---|---|
| rs3787430 | C/T | Exon 3 | Schizophrenia | 21652606 [66] |
| rs3787429 | C/T | Exon 3 | | |
| rs752380770 | C/T | Exon 3 | Migraine | 21376262 [62] |
| rs752380770 | C/T | Exon 3 | Shy-Drager syndrome | 23713487 [63, 64] |
| rs6062144 | C/G | Upstream 5′UTR | Asthma | 19824886 [67] |

This SNP was initially noticed in one patient when HRH3 was originally cloned (shown in Table 2.4) [63]. This syndrome is marked with multiple system atrophy with orthostatic hypotension. In this publication, a functional implication of the amino acid exchange of alanine to valine was suggested it reduces signaling efficacy in $H_3$ receptors when expressed in CHO-K1 cell line [64]. This was suggested to be of relevance in the pathophysiology of conditions linked to this A/V mutation such as migraine and multiple system atrophy with orthostatic hypotension; however, further research is needed to confirm this hypothesis [64].

This third member of the histamine family was also studied for its impact in schizophrenia. A publication by Southam et al. identified a possible role for the histamine

H3 receptor in schizophrenia. A histamine H3 receptor antagonist, GSK207040, was found to exhibit a behavioral and neurochemical profile on this receptor and beneficial to treat the cognitive and sensory gating deficits of schizophrenia [65]. The polymorphism found in exon 3, giving a C/T allele exchange, was investigated in risperidone response in Chinese Han population. Strong association was identified between rs3787430 ($P$=0.024, 0.010) and a positive response to risperidone treatment in a Chinese Han population after 4 and 8 weeks, respectively. This polymorphism could be a potential genetic marker for treatment. Association for rs3787429 was only noted after 4 weeks ($P$=0.013) [66]. The histaminergic system has been linked to the pathophysiology of schizophrenia and with histamine $H_3$ receptor, being mostly located in the central nervous system, will continue to encourage research in this area.

### *2.3.2 Histamine $H_3$ Polymorphisms in Inflammatory Disorders*

Allergic asthma is an inheritable condition, and many genetic variations were investigated in order to establish a link to the disease and its treatment. The histamine $H_3$ receptor was not excluded from these investigations. The study by Ferreira et al. aimed to establish a link between candidate genes and the disorder, including eight SNPs on HRH3. The SNP rs6062144, located 8 kbp upstream to the gene, was the most associated ($P$=0.036), but there was no overall evidence for association [67]. The evidence for a strong link between asthma and histamine $H_3$ receptor is still far from being confirmed, and as seen from Table 2.4, the majority of association studies between this third receptor and diseases are mostly related to CNS disorders.

The role of histamine and histamine receptors in experimental autoimmune encephalitis has been looked into. This third histamine receptor and its polymorphisms were investigated for a potential role in multiple sclerosis pathogenesis. The functional polymorphism G293D located in the third intracellular loop of the GPCR was studied by expressing the two alleles in mice models and comparing the pharmacological properties of H3 receptor ligands. Overall, no significant difference in affinity was observed; however, this polymorphism seems to have an impact in the HRH3 isoform for this receptor. This study indicates the HRH3 polymorphism isoform has a role in regulating neurogenic control in experimental autoimmune encephalitis and T cell response in mice [68]. With further research into this role, we might be a step closer into identifying pharmacological targets in preventing the development of new lesions in multiple sclerosis.

### *2.3.3 Histamine $H_3$ Polymorphisms in Cancer*

Unlike the other family members of this GPCR family, the histamine $H_3$ receptor is less investigated in its impact in cancer. The SNPs located on this gene were genotyped in a Chinese Han population together with histamine decarboxylase (HDC), and histamine N-methytransferase (HNMT) was genotyped for an association in breast cancer. Two hundred and one patient samples were analyzed for a potential role in this condition. The SNPs rs3787429 and rs3787430 were identified as the only two-tag SNPs for HRH3 gene. However, even though, as mentioned earlier, these two polymorphisms were identified to contribute to risperidone treatment, no significant association was identified in the risk of breast cancer in this group [69]. Further to this finding, a larger population study would be required to omit the role of these SNPs in breast cancer.

## 2.4 Human Histamine $H_4$ Receptor

Several research groups cloned and identified the new member of the histamine receptor family, the human histamine receptor 4 [2, 4, 7, 70]. To date, only two isoforms have been identified for the HRH4. However when compared to the other member HRH3, 20 naturally occurring isoforms have been identified so far, where some display differential expression patterns in various brain areas and functionalities compared with its respective full-length isoform [22]. Investigations of potential therapeutic use for this receptor are strongly related to its localization and expression. The HRH4 is preferentially expressed in various cells of the immune system and mast cells. It also induces chemotaxis of eosinophils, mast cells, and monocyte-derived dendritic cells. Other involvements of HRH4 were also reported in the control of interleukin (IL)-16 release from human lymphocytes, and it has been speculated that H4 selective antagonists might be useful in the treatment of asthma [26, 70]. Various research groups suggest that this receptor can also be a potential therapeutic target for other inflammatory diseases, such as chronic allergy, atopic dermatitis, and inflammatory bowel disease. Other possible roles for HRH4 are rheumatoid arthritis and colorectal cancer since this receptor has been detected in the primary synovial fluid of these tissues [70].

The HRH4 gene is located in chromosome 18 in section q11.2 close to the centromere. On comparing the four receptors, the $H_3$ receptor and $H_4$ receptor genes (HRH3 and HRH4, respectively) share a similar genomic organization, consisting of three exons and two introns, and in both receptors, intron 2 is the largest with 7804 base pairs for HRH4 [12]. Similar to $H_1$ and $H_4$ receptors, the intracellular loop 3 is large and is the largest for this family member. The gene, amino acid, and GPCR organization are shown in Fig. 2.6, respectively.

### 2.4.1 *Histamine $H_4$ Polymorphisms in CNS Disorders*

The histamine $H_4$ receptor was also studied for associations with CNS disorders. Immunological and pharmacological studies have suggested that HRH4 is expressed on fibers emanating from brain regions, most likely the hippocampus or thalamus

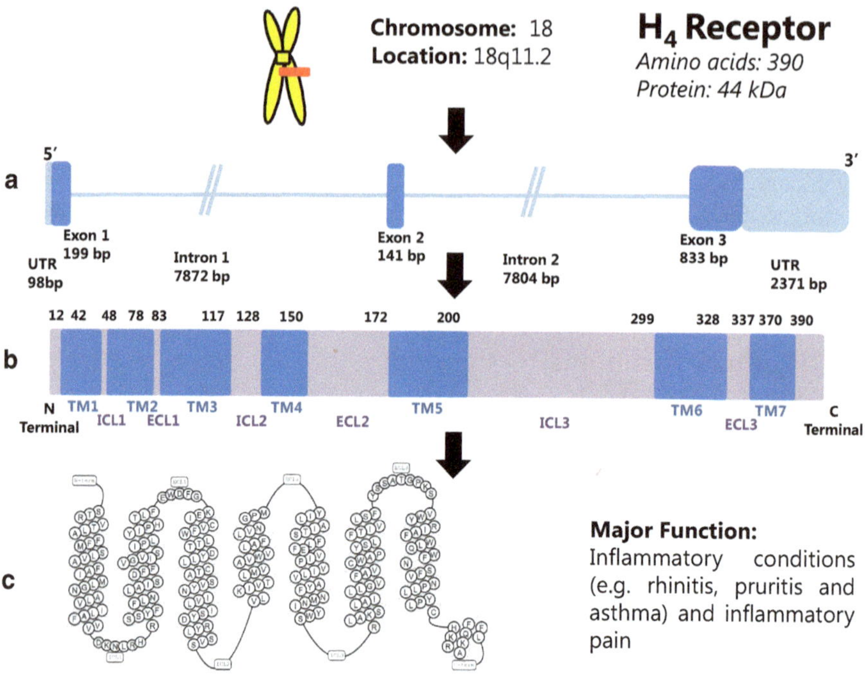


**Fig. 2.6** *Histamine $H_4$ receptor molecular organization.* (**a**) Genomic organization of HRH4. (**b**) Amino acid organization, where TM is transmembrane domain, ICL is intracellular loop, and ECL is extracellular loop. (**c**) GPCR snake plot

[71]. It was also suggested that it is expressed in subpopulations of neurons within the sensory dorsal root ganglia and dorsal horn of the spinal cord and in restricted cortical thalamic areas of the brain [72]. So far, the expression of HRH4 in the brain is still controversial. Several groups could not detect mRNA of this receptor in the CNS [4, 6], while other research groups reported the presence in various parts of the CNS, including the amygdala, cerebellum, hippocampus, caudate nucleus, substantia nigra, thalamus, and hypothalamus [9, 10].

Evidence was put forward for the role in the schizophrenia, and this was also seen by the different binding affinity that therapeutic drugs have for this condition. Clozapine and amitriptyline were found to bind to the $H_4$ receptor in the low micromolar range where clozapine showed partial agonism for the $H_4$ receptor and a high binding affinity to the $H_2$ receptor, both in a clinically relevant concentration [30]. An association study was conducted to identify a link between a polymorphism (rs4483927) with histamine $H_4$ and the efficacy of risperidone, an antipsychotic drug used in schizophrenia. A significant association for risperidone efficacy where the TT genotype predicts poor response to treatment both on the positive, negative, and general subscales and on the total scale of PANSS scores ($P$=0.017, 0.019, 0.021, 0.002, respectively). The SNP could be used as a biological marker for therapeutic response in personalized medicine [73].

### 2.4.2 Histamine $H_4$ Polymorphisms in Inflammatory Disorders

Histamine $H_1$, $H_2$, and $H_4$ receptors are the three main receptors shown to be linked to inflammation and immune response [41]. The $H_4$ receptor promotes the accumulation of inflammatory cells at sites of allergic inflammation by increasing calcium flux in human eosinophils, increasing eosinophil chemotaxis, and increasing IL-16 production [51, 74].

The main focus for this new receptor was with inflammatory disorders. Only few genetic variations on this gene have been associated so far with asthma and underlying conditions, as shown in Table 2.5. These SNPs (rs17187619, rs527790, rs487202, rs1421125, rs615283) do not give a missense mutation since they are not located in an exon leading to an amino acid exchange, yet they still seem to contribute to the condition.

Other polymorphisms found on the $H_4$ receptor have also been linked to infections in asthma patients. The recent SNPs described in the study by Simon et al. identified an association to conjunctivitis in asthma patients for the polymorphism rs615283 in the GG genotype ($P=0.008$). An association was also identified in polymorphisms rs17187619, rs487202, rs527790 in patients with or without infection-induced asthma ($P=0.002$, 0.0002, 0.00007 respectively). Haplotypes with a CC allele combination from polymorphisms rs487202 and rs574913 were found to be associated with infection-induced asthma ($P=0.0009$) [75]. The CA haplotype for these polymorphisms was also more common in infection-induced asthma patients ($P=0.0006$). SNPs located downstream to the HRH4 gene, rs1421125, rs487202, and rs527790, have shown to be linked to asthmatics with

**Table 2.5** HRH4 polymorphisms in diseases

| rs number | SNP | Location on gene | Condition | PubMed ID |
|---|---|---|---|---|
| rs4483927 | G/T | Intron 1 | Schizophrenia | 23422377 [73] |
| rs17187619 | T/C | Intron | Asthma | 22653292 [75] |
| rs527790 | G/A | After H4 | | |
| rs487202 | C/G | After H4 | | |
| rs1421125 | C/A | 3′UTR | | |
| rs615283 | G/A | promoter | | |
| rs77485247 | A/T | 5′UTR | Atopic dermatitis | 20199554 [41] |
| rs74604924 | A/T | Exon 3 | | |
| rs77041280 | A/T | Exon 3 | | |
| rs77485247 | A/T | 5′UTR | Lupus erythematosus | 20618322 [78] |
| rs74604924 | A/T | Exon 3 | | |
| rs77041280 | A/T | Exon 3 | | |
| rs623590 | A/C/T | Intron 1 | Breast cancer | 23481304 [83] |
| rs16940762 | A/G | Intron 1 | | |
| rs11662595 | A/G | Exon 3 | | |
| rs1421125 | A/C | 3′UTR | | |

or without allergic rhinitis ($P=0.005$, 0.006, 0.006, respectively) [75], which enhances the concept that this receptor could play a role in this condition.

The histamine $H_4$ receptor was also associated to allergic atopic dermatitis (AAD). This receptor has been found to be expressed in dendritic epithelial cells with a role in regulating histamine levels. It was suggested that $H_4$ receptor agonists can be beneficial in downregulating inflammation in AAD [76]. One of the SNPs identified in this linkage study was rs77485247, located 51 bp upstream to the gene in the 5′UTR [41], representing the potential promoter region of the gene. A variation in this part of the gene could be responsible for gene expression [77], but further investigations need to be done to confirm. Two other polymorphisms located in exon 3 were also investigated in systemic lupus erythematosus and atopic dermatitis. The nonsense mutation of SNP rs74604924 leads to a stop codon, whereas rs77041280 results in a missense mutation leading to an amino acid exchange. These SNPs suggest a genetic linkage to atopic dermatitis where rs74604924 gives a functional amino acid exchange Lys376Ter[*] and rs77041280 gives an amino acid exchange Ile376Lys (rs77485247, rs74604924, rs77041280 where $P=0.002$, $P<0.001$, $P<0.001$, respectively) [41]. However, these SNPs were found to show no significant association between rs77485247, rs74604924, and rs77041280 in lupus erythematosus ($P<0.694$, 0.439, 0.561, respectively) [78]; however copy number variants found on the $H_4$ receptor were identified as a significant link to the disease [78]. These copy number variations (CNVs) were recently found to be associated with allergic atopic dermatitis. Variations included in the study were both deletions (CNVs less than 2) and amplifications (CNVs more than 4). Amplification on the HRH4 gene were found to be significantly associated to AAD ($P<0.05$), whereas deletions on the HRH4 were not found to be linked to the condition ($P=0.3$). The HRH4 mRNA levels were also found to be higher in CNV amplifications, suggesting that HRH4 expression is regulated by inflammatory stimuli from this autoimmune disease [79].

This potential linkage between SNPs and AAD encourages studies to identify ligands that can treat or prevent this condition. The $H_4$ receptor antagonists, JNJ7777120 and JNJ28307474, were investigated for their potential use when compared to $H_1$ receptor antagonists, cetirizine and hydroxyzine, but it was shown that there was no impact on the skin lesion prevention for AAD [80]. Interactions between $H_4$ receptor and these antagonists, JNJ3975879, are to date the most studied and published as they are a potent selective $H_4$ antagonist that proved effective in preclinical models of pruritus, dermatitis, asthma, and arthritis [81]. A recent study found that co-administration of histamine $H_1$ and $H_4$ receptor antagonists had potent inhibitory effects that were equal to those of steroids in a chronic allergic dermatitis mouse model, and JNJ39758979 exhibited antipruritic effects in moderate AD patients [82]. This encourages further investigations for $H_4$ receptor antagonists in these allergic inflammatory conditions.

### 2.4.3 Histamine $H_4$ Polymorphisms in Cancer

This new member of the histamine receptor family has been recently associated with malignancy of breast cancer. Evidence was shown in the analysis of genotypes and haplotypes of a Chinese Han population. This study showed that SNPs rs623590, rs11662595, and rs1421125 had a significant association with the risk and malignancy of breast cancer. The T allele in rs623590 had a decreased risk of breast cancer ($P=0.012$), while A allele of rs1421125 had an increased risk of cancer ($P=0.008$) [83]. Furthermore, the haplotype C-A-A of rs623590- rs11662595-rs1421125 was found to be more common in breast cancer patients ($P=0.003$) [83]. More investigations need to be performed to understand if these results are replicable in other populations and if this association is ethnicity based or not.

The potential link between histamine H4 receptor and gastric carcinoma was also studied. However, no significant difference was found between copy number variations on $H_4$ receptor genes and patients with gastric carcinoma, but there was a substantial correlation between copy number deletions and $H_4$ receptor downregulation. This suggests that abnormalities in $H_4$ receptor have a potential role in histamine-mediated regulation of tumor growth in gastric cancer [84].

The summary of the polymorphisms in histamine H4 receptor and associated diseases are presented in Table 2.5 below.

Pharmacogenetics is progressively becoming important aspect for patient's health, where the drug targets more efficiently and effectively the disease and provides personalized care. The biogenic amine histamine plays an essential role in controlling many physiological functions, both in the central nervous system (CNS) and the peripheral nervous system (PNS) [22, 70]; hence, genetic variation in the four receptor subtypes becomes an important focal point. The knowledge to date on polymorphisms for the histamine receptor family encourages further interest in identifying novel genetic variations that contribute to a disease, especially with the importance histamine has in biological processes. The studies mentioned in this book chapter encourage the continued quest for personalized care to patients. However, more genomic research in the context of biogenic amines, here histamine, is needed to establish a direct association between diseases, biological markers, and drug treatments in this new era of personalized medicine.

## References

1. Consortium IH. The International HapMap Project. Nature. 2003;426(6968):789–96. doi:10.1038/nature02168.
2. Zhu Y, Michalovich D, Wu H, Tan KB, Dytko GM, Mannan IJ, et al. Cloning, expression, and pharmacological characterization of a novel human histamine receptor. Mol Pharmacol. 2001;59(3):434–41.

3. Lovenberg TW, Roland BL, Wilson SJ, Jiang X, Pyati J, Huvar A, et al. Cloning and functional expression of the human histamine H3 receptor. Mol Pharmacol. 1999;55(6):1101–7.
4. Oda T, Morikawa N, Saito Y, Masuho Y, Matsumoto S. Molecular cloning and characterization of a novel type of histamine receptor preferentially expressed in leukocytes. J Biol Chem. 2000;275(47):36781–6. doi:10.1074/jbc.M006480200.
5. Leurs R, Chazot PL, Shenton FC, Lim HD, de Esch IJ. Molecular and biochemical pharmacology of the histamine H4 receptor. Br J Pharmacol. 2009;157(1):14–23. doi:10.1111/j.1476-5381.2009.00250.x.
6. Nguyen T, Shapiro DA, George SR, Setola V, Lee DK, Cheng R, et al. Discovery of a novel member of the histamine receptor family. Mol Pharmacol. 2001;59(3):427–33.
7. Nakamura T, Itadani H, Hidaka Y, Ohta M, Tanaka K. Molecular cloning and characterization of a new human histamine receptor, HH4R. Biochem Biophys Res Commun. 2000;279(2):615–20. doi:10.1006/bbrc.2000.4008.
8. Morse KL, Behan J, Laz TM, West Jr RE, Greenfeder SA, Anthes JC, et al. Cloning and characterization of a novel human histamine receptor. J Pharmacol Exp Ther. 2001;296(3):1058–66.
9. Coge F, Guenin SP, Rique H, Boutin JA, Galizzi JP. Structure and expression of the human histamine H4-receptor gene. Biochem Biophys Res Commun. 2001;284(2):301–9. doi:10.1006/bbrc.2001.4976.
10. Liu C, Ma X, Jiang X, Wilson SJ, Hofstra CL, Blevitt J, et al. Cloning and pharmacological characterization of a fourth histamine receptor (H(4)) expressed in bone marrow. Mol Pharmacol. 2001;59(3):420–6.
11. Garcia-Martin E, Ayuso P, Martinez C, Blanca M, Agundez JA. Histamine pharmacogenomics. Pharmacogenomics. 2009;10(5):867–83. doi:10.2217/pgs.09.26.
12. Micallef S, Stark H, Sasse A. Polymorphisms and genetic linkage of histamine receptors. Life Sci. 2013;93(15):487–94. doi:10.1016/j.lfs.2013.08.012.
13. de Esch IJ, Thurmond RL, Jongejan A, Leurs R. The histamine H4 receptor as a new therapeutic target for inflammation. Trends Pharmacol Sci. 2005;26(9):462–9. doi:10.1016/j.tips.2005.07.002.
14. Kottke T, Sander K, Weizel L, Schneider EH, Seifert R, Stark H. Receptor-specific functional efficacies of alkyl imidazoles as dual histamine H3/H4 receptor ligands. Eur J Pharmacol. 2011;654(3):200–8. doi:10.1016/j.ejphar.2010.12.033.
15. Sadek B, Alisch R, Buschauer A, Elz S. Synthesis and dual histamine H(1) and H(2) receptor antagonist activity of cyanoguanidine derivatives. Molecules. 2013;18(11):14186–202. doi:10.3390/molecules181114186.
16. Thurmond RL, Gelfand EW, Dunford PJ. The role of histamine H1 and H4 receptors in allergic inflammation: the search for new antihistamines. Nat Rev Drug Discov. 2008;7(1):41–53.
17. Ohsawa Y, Hirasawa N. The antagonism of histamine H1 and H4 receptors ameliorates chronic allergic dermatitis via anti-pruritic and anti-inflammatory effects in NC/Nga mice. Allergy. 2012;67(8):1014–22. doi:10.1111/j.1398-9995.2012.02854.x.
18. Venter JC, Adams MD, Myers EW, Li PW, Mural RJ, Sutton GG, et al. The sequence of the human genome. Science. 2001;291(5507):1304–51. doi:10.1126/science.1058040.
19. Bockaert J, Marin P, Dumuis A, Fagni L. The 'magic tail' of G protein-coupled receptors: an anchorage for functional protein networks. FEBS Lett. 2003;546(1):65–72. doi:10.1016/S0014-5793(03)00453-8.
20. Peeters MC, van Westen GJ, Li Q, IJzerman AP. Importance of the extracellular loops in G protein-coupled receptors for ligand recognition and receptor activation. Trends Pharmacol Sci. 2011;32(1):35–42. doi:10.1016/j.tips.2010.10.001.
21. ConsumerReports.org. Using the antihistamines to treat: Allergies, Hay Fever, & Hives. Consumer reports. 2013. https://www.consumerreports.org/health/resources/pdf/best-buy-drugs/Antihistamines.pdf. Accessed 12 Oct 2015.
22. Leurs R, Vischer HF, Wijtmans M, de Esch IJ. En route to new blockbuster anti-histamines: surveying the offspring of the expanding histamine receptor family. Trends Pharmacol Sci. 2011;32(4):250–7. doi:10.1016/j.tips.2011.02.004.
23. Proof of activity study of UR-63325 in allergic rhinitis induced by nasal challenge [database on the Internet]. 2011. http://clinicaltrials.gov/ct2/show/NCT01260753?term=UR-63325&rank=1. Accessed 12 Oct 2015.

24. Salcedo C, Pontes C, Merlos M. Is the H4 receptor a new drug target for allergies and asthma? Front Biosci (Elite Ed). 2013;5:178–87.
25. Isberg V, Vroling B, van der Kant R, Li K, Vriend G, Gloriam D. GPCRDB: an information system for G protein-coupled receptors. Nucleic Acids Res. 2014;42(Database issue):D422–5. doi:10.1093/nar/gkt1255.
26. Parsons ME, Ganellin CR. Histamine and its receptors. Br J Pharmacol. 2006;147 Suppl 1:S127–35. doi:10.1038/sj.bjp.0706440.
27. Ito C. The role of the central histaminergic system on schizophrenia. Drug News Perspect. 2004;17(6):383–7.
28. Nakai T, Kitamura N, Hashimoto T, Kajimoto Y, Nishino N, Mita T, et al. Decreased histamine H1 receptors in the frontal cortex of brains from patients with chronic schizophrenia. Biol Psychiatry. 1991;30(4):349–56.
29. Prell GD, Green JP, Kaufmann CA, Khandelwal JK, Morrishow AM, Kirch DG, et al. Histamine metabolites in cerebrospinal fluid of patients with chronic schizophrenia: their relationships to levels of other aminergic transmitters and ratings of symptoms. Schizophr Res. 1995;14(2):93–104.
30. Appl H, Holzammer T, Dove S, Haen E, Strasser A, Seifert R. Interactions of recombinant human histamine H1R, H2R, H3R, and H4R receptors with 34 antidepressants and antipsychotics. Naunyn Schmiedebergs Arch Pharmacol. 2012;385(2):145–70. doi:10.1007/s00210-011-0704-0.
31. Salvi V, Barone-Adesi F, D'Ambrosio V, Albert U, Maina G. High H1-affinity antidepressants and risk of metabolic syndrome in bipolar disorder. Psychopharmacology (Berl). 2016;233(1):49–56. doi:10.1007/s00213-015-4085-9.
32. Provensi G, Blandina P, Passani MB. The histaminergic system as a target for the prevention of obesity and metabolic syndrome. Neuropharmacology. 2015;106:3–12. doi:10.1016/j.neuropharm.2015.07.002.
33. Perlis RH, Adams DH, Fijal B, Sutton VK, Farmen M, Breier A, et al. Genetic association study of treatment response with olanzapine/fluoxetine combination or lamotrigine in bipolar I depression. J Clin Psychiatry. 2010;71(5):599–605. doi:10.4088/JCP.08m04632gre.
34. Basile VS, Masellis M, Potkin SG, Kennedy JL. Pharmacogenomics in schizophrenia: the quest for individualized therapy. Hum Mol Genet. 2002;11(20):2517–30.
35. Vehof J, Risselada AJ, Al Hadithy AF, Burger H, Snieder H, Wilffert B, et al. Association of genetic variants of the histamine H1 and muscarinic M3 receptors with BMI and HbA1c values in patients on antipsychotic medication. Psychopharmacol (Berl). 2011;216(2):257–65. doi:10.1007/s00213-011-2211-x.
36. Wu R, Zhao J, Shao P, Ou J, Chang M. Genetic predictors of antipsychotic-induced weight gain: a case-matched multi-gene study. Zhong Nan Da Xue Xue Bao Yi Xue Ban. 2011;36(8):720–3. doi:10.3969/j.issn.1672-7347.2011.08.003.
37. Mancama D, Arranz MJ, Munro J, Osborne S, Makoff A, Collier D, et al. Investigation of promoter variants of the histamine 1 and 2 receptors in schizophrenia and clozapine response. Neurosci Lett. 2002;333(3):207–11.
38. Anichtchik OV, Rinne JO, Kalimo H, Panula P. An altered histaminergic innervation of the sub stantia nigra in Parkinson's disease. Exp Neurol. 2000;163(1):20–30. doi:10.1006/exnr.2000.7362.
39. García-Martín E, Ayuso P, Luengo A, Martínez C, Agúndez JA. Genetic variability of histamine receptors in patients with Parkinson's disease. BMC Med Genet. 2008;9:15. doi:10.1186/1471-2350-9-15.
40. Pino-Ángeles A, Reyes-Palomares A, Melgarejo E, Sánchez-Jiménez F. Histamine: an undercover agent in multiple rare diseases? J Cell Mol Med. 2012;16(9):1947–60. doi:10.1111/j.1582-4934.2012.01566.x.
41. Yu B, Shao Y, Zhang J, Dong XL, Liu WL, Yang H, et al. Polymorphisms in human histamine receptor H4 gene are associated with atopic dermatitis. Br J Dermatol. 2010;162(5):1038–43. doi:10.1111/j.1365-2133.2010.09675.x.
42. Zampeli E, Tiligada E. The role of histamine H4 receptor in immune and inflammatory disorders. Br J Pharmacol. 2009;157(1):24–33. doi:10.1111/j.1476-5381.2009.00151.x.

43. Choi JH, Kim SH, Suh CH, Nahm DH, Park HS. Polymorphisms of high-affinity IgE receptor and histamine-related genes in patients with ASA-induced urticaria/angioedema. J Korean Med Sci. 2005;20(3):367–72.
44. Wang M, Wei X, Shi L, Chen B, Zhao G, Yang H. Integrative genomic analyses of the histamine H1 receptor and its role in cancer prediction. Int J Mol Med. 2014;33(4):1019–26. doi:10.3892/ijmm.2014.1649.
45. Klepstad P, Fladvad T, Skorpen F, Bjordal K, Caraceni A, Dale O, et al. Influence from genetic variability on opioid use for cancer pain: a European genetic association study of 2294 cancer pain patients. Pain. 2011;152(5):1139–45. doi:10.1016/j.pain.2011.01.040.
46. Laugsand EA, Fladvad T, Skorpen F, Maltoni M, Kaasa S, Fayers P, et al. Clinical and genetic factors associated with nausea and vomiting in cancer patients receiving opioids. Eur J Cancer. 2011;47(11):1682–91. doi:10.1016/j.ejca.2011.04.014.
47. Orange PR, Heath PR, Wright SR, Pearson RC. Allelic variations of the human histamine H2 receptor gene. Neuroreport. 1996;7(7):1293–6.
48. Orange PR, Heath PR, Wright SR, Ramchand CN, Kolkeiwicz L, Pearson RC. Individuals with schizophrenia have an increased incidence of the H2R649G allele for the histamine H2 receptor gene. Mol Psychiatry. 1996;1(6):466–9.
49. Ito C, Morisset S, Krebs MO, Olie JP, Loo H, Poirier MF, et al. Histamine H2 receptor gene variants: lack of association with schizophrenia. Mol Psychiatry. 2000;5(2):159–64.
50. Rosse RB, Kendrick K, Tsui LC, Fay-McCarthy M, Collins JP, Rosenberg P, et al. Famotidine adjunctive pharmacotherapy of schizophrenia: a case report. Clin Neuropharmacol. 1995;18(4):369–74.
51. Akdis CA, Simons FE. Histamine receptors are hot in immunopharmacology. Eur J Pharmacol. 2006;533(1–3):69–76.
52. Alonso N, Zappia CD, Cabrera M, Davio CA, Shayo C, Monczor F, et al. Physiological implications of biased signaling at histamine H2 receptors. Front pharmacol. 2015;6:45. doi:10.3389/fphar.2015.00045.
53. Yamada H, Tahara T, Shiroeda H, Hayashi R, Saito T, Nakamura M, et al. Effects of -1018G>A polymorphism of HRH2 (rs2607474) on the severity of gastric mucosal atrophy. J Gastrointestin Liver Dis. 2012;21(2):139–43.
54. Sande N, Nikulin M, Nilsson I, Wadstrom T, Laxen F, Harkonen M, et al. Increased risk of developing atrophic gastritis in patients infected with CagA+ Helicobacter pylori. Scand J Gastroenterol. 2001;36(9):928–33.
55. Ebule I, Longdoh A, Paloheimo I. Helicobacter pylori infection and atrophic gastritis. Afr Health Sci. 2013;13(1):112–7. doi:10.4314/ahs.v13i1.15.
56. Rivera ES, Cricco GP, Engel NI, Fitzsimons CP, Martin GA, Bergoc RM. Histamine as an autocrine growth factor: an unusual role for a widespread mediator. Semin Cancer Biol. 2000;10(1):15–23.
57. Medina VA, Brenzoni PG, Lamas DJ, Massari N, Mondillo C, Nunez MA, et al. Role of histamine H4 receptor in breast cancer cell proliferation. Front Biosci (Elite Ed). 2011;3:1042–60.
58. Arisawa T, Tahara T, Ozaki K, Matsue Y, Minato T, Yamada H, et al. Association between common genetic variant of HRH2 and gastric cancer risk. Int J Oncol. 2012;41(2):497–503. doi:10.3892/ijo.2012.1482.
59. Cai WK, Zhang JB, Wang NM, Wang YL, Zhao CH, Lu J, et al. Lack of association between rs2067474 polymorphism in histamine receptor H2 gene and breast cancer in Chinese Han population. ScientificWorldJournal. 2015;2015:545292. doi:10.1155/2015/545292.
60. Welsh MM, Karagas MR, Kuriger JK, Houseman A, Spencer SK, Perry AE, et al. Genetic determinants of UV-susceptibility in non-melanoma skin cancer. PLoS One. 2011;6(7), e20019. doi:10.1371/journal.pone.0020019.
61. Gupta S, Nahas SJ, Peterlin BL. Chemical mediators of migraine: preclinical and clinical observations. Headache. 2011;51(6):1029–45. doi:10.1111/j.1526-4610.2011.01929.x.
62. Millán-Guerrero RO, Baltazar-Rodríguez LM, Cárdenas-Rojas MI, Ramírez-Flores M, Isais-Millán S, Delgado-Enciso I, et al. A280V polymorphism in the histamine H3 receptor as a risk factor for migraine. Arch Med Res. 2011;42(1):44–7. doi:10.1016/j.arcmed.2011.01.009.

63. Wiedemann P, Bönisch H, Oerters F, Brüss M. Structure of the human histamine H3 receptor gene (HRH3) and identification of naturally occurring variations. J Neural Transm. 2002;109(4): 443–53. doi:10.1007/s007020200036.
64. Flores-Clemente C, Osorio-Espinoza A, Escamilla-Sanchez J, Leurs R, Arias JM, Arias-Montano JA. A single-point mutation (Ala280Val) in the third intracellular loop alters the signalling properties of the human histamine H receptor stably expressed in CHO-K1 cells. Br J Pharmacol. 2013;170(1):127–35. doi:10.1111/bph.12257.
65. Southam E, Cilia J, Gartlon JE, Woolley ML, Lacroix LP, Jennings CA, et al. Preclinical investigations into the antipsychotic potential of the novel histamine H3 receptor antagonist GSK207040. Psychopharmacology (Berl). 2009;201(4):483–94. doi:10.1007/s00213-008-1310-9.
66. Wei Z, Wang L, Zhang M, Xuan J, Wang Y, Liu B, et al. A pharmacogenetic study of risperidone on histamine H3 receptor gene (HRH3) in Chinese Han schizophrenia patients. J Psychopharmacol. 2012;26(6):813–8. doi:10.1177/0269881111405358.
67. Ferreira MA, Zhao ZZ, Thomsen SF, James M, Evans DM, Postmus PE, et al. Association and interaction analyses of eight genes under asthma linkage peaks. Allergy. 2009;64(11):1623–8. doi:10.1111/j.1398-9995.2009.02091.x.
68. Krementsov DN, Wall EH, Martin RA, Subramanian M, Noubade R, Del Rio R, et al. Histamine H(3) receptor integrates peripheral inflammatory signals in the neurogenic control of immune responses and autoimmune disease susceptibility. PLoS One. 2013;8(7), e62743. doi:10.1371/journal.pone.0062743.
69. He GH, Lin JJ, Cai WK, Xu WM, Yu ZP, Yin SJ, et al. Associations of polymorphisms in histidine decarboxylase, histamine N-methyltransferase and histamine receptor H3 genes with breast cancer. PLoS One. 2014;9(5), e97728. doi:10.1371/journal.pone.0097728.
70. Lim HD, Adami M, Guaita E, Werfel T, Smits RA, de Esch IJ, et al. Pharmacological characterization of the new histamine H4 receptor agonist VUF 8430. Br J Pharmacol. 2009;157(1): 34–43. doi:10.1111/j.1476-5381.2009.00200.x.
71. Connelly WM, Shenton FC, Lethbridge N, Leurs R, Waldvogel HJ, Faull RL, et al. The histamine H4 receptor is functionally expressed on neurons in the mammalian CNS. Br J Pharmacol. 2009;157(1):55–63. doi:10.1111/j.1476-5381.2009.00227.x.
72. Strakhova MI, Nikkel AL, Manelli AM, Hsieh GC, Esbenshade TA, Brioni JD, et al. Localization of histamine H4 receptors in the central nervous system of human and rat. Brain Res. 2009;1250:41–8. doi:10.1016/j.brainres.2008.11.018.
73. Wei Z, Wang L, Yu T, Wang Y, Sun L, Wang T, et al. Histamine H4 receptor polymorphism: a potential predictor of risperidone efficacy. J Clin Psychopharmacol. 2013;33(2):221–5. doi:10.1097/JCP.0b013e318283963b.
74. Jutel M, Watanabe T, Akdis M, Blaser K, Akdis CA. Immune regulation by histamine. Curr Opin Immunol. 2002;14(6):735–40.
75. Simon T, Semsei AF, Ungvári I, Hadadi E, Virág V, Nagy A, et al. Asthma endophenotypes and polymorphisms in the histamine receptor HRH4 gene. Int Arch Allergy Immunol. 2012;159(2):109–20. doi:10.1159/000335919.
76. Dijkstra D, Stark H, Chazot PL, Shenton FC, Leurs R, Werfel T, et al. Human inflammatory dendritic epidermal cells express a functional histamine H4 receptor. J Invest Dermatol. 2008;128(7):1696–703. doi:10.1038/sj.jid.5701250.
77. Brouwer JR, Willemsen R, Oostra BA. Microsatellite repeat instability and neurological disease. Bioessays. 2009;31(1):71–83. doi:10.1002/bies.080122.
78. Yu B, Shao Y, Li P, Zhang J, Zhong Q, Yang H, et al. Copy number variations of the human histamine H4 receptor gene are associated with systemic lupus erythematosus. Br J Dermatol. 2010;163(5):935–40. doi:10.1111/j.1365-2133.2010.09928.x.
79. Chen B, Ye T, Shao Y, Zhang J, Zhong Q, Hu X, et al. Association between copy-number variations of the human histamine H4 receptor gene and atopic dermatitis in a Chinese population. Clin Exp Dermatol. 2013;38(3):295–300. doi:10.1111/ced.12117. quiz-1.
80. Baumer W, Stahl J, Sander K, Petersen LJ, Paps J, Stark H, et al. Lack of preventing effect of systemically and topically administered histamine H(1) or H(4) receptor antagonists in a dog model of acute atopic dermatitis. Exp Dermatol. 2011;20(7):577–81. doi:10.1111/j.1600-0625.2011.01268.x.

81. Thurmond RL. The histamine H4 receptor: from orphan to the clinic. Front Pharmacol. 2015;6:65. doi:10.3389/fphar.2015.00065.
82. Ohsawa Y, Hirasawa N. The role of histamine H1 and H4 receptors in atopic dermatitis: from basic research to clinical study. Allergol Int. 2014;63(4):533–42. doi:10.2332/allergolint.13-RA-0675.
83. He GH, Lu J, Shi PP, Xia W, Yin SJ, Jin TB, et al. Polymorphisms of human histamine receptor H4 gene are associated with breast cancer in Chinese Han population. Gene. 2013;519(2):260–5. doi:10.1016/j.gene.2013.02.020.
84. Zhang C, Xiong Y, Li J, Yang Y, Liu L, Wang W, et al. Deletion and down-regulation of HRH4 gene in gastric carcinomas: a potential correlation with tumor progression. PLoS One. 2012;7(2), e31207. doi:10.1371/journal.pone.0031207.
85. McWilliam H, Li W, Uludag M, Squizzato S, Park YM, Buso N, et al. Analysis tool web services from the EMBL-EBI. Nucleic Acids Res. 2013;41(Web Server issue):W597–600. doi:10.1093/nar/gkt376.

# Chapter 3
# The Role of Histamine $H_4$ Receptor in Asthma and Atopic Dermatitis

**Robin L. Thurmond**

**Abstract** Histamine acts through four different receptors and ligands for two of these receptors have been used in clinical practice for many decades. Efficacy of drugs that target the histamine $H_1$ receptor in allergic disease such as urticaria, allergic rhinitis and allergic conjunctivitis has provided a rationale for the association of histamine with allergy. However, these drugs are not efficacious in other diseases thought to be allergic in nature such as asthma and atopic dermatitis. Recent preclinical and clinical data has shown a role for histamine $H_4$ receptor antagonists in these diseases. Histamine $H_4$ receptor antagonists are efficacious in preclinical models of asthma, atopic dermatitis and pruritus. One histamine $H_4$ receptor antagonists, JNJ 39758979, has shown clinical efficacy in reducing histamine-induced pruritus and in improving the skin lesions and pruritus in patients with atopic dermatitis. Therefore, the histamine $H_4$ receptor may play a role in diseases such as asthma and atopic dermatitis, which have long since been believed not to be mediated by histamine due to lack of efficacy of traditional antihistamines.

**Keywords** Allergy • Eosinophils • Dendritic cells • Pruritus • Antihistamine • JNJ7777120 • JNJ39758979

## 3.1 Introduction

Histamine mediates physiological functions via activation of one of four G-protein-coupled receptors: the histamine $H_1$ receptor ($H_1R$), the histamine $H_2$ receptor ($H_2R$), the histamine $H_3$ receptor ($H_3R$), and the histamine $H_4R$ receptor ($H_4R$) [1]. Drugs that target the $H_1R$ and $H_2R$ have been used for decades for the treatment of allergic reactions, insomnia, gastric ulcers and gastroesophageal reflux disease. A drug targeting the $H_3R$, pitolisant, has recently been approved for the treatment of narcolepsy [2, 3]. There are no currently available drugs that target the $H_4R$, but

R.L. Thurmond, Ph.D. (✉)
Department of Immunology, Janssen Research & Development, LLC,
3210 Merryfield Row, San Diego, CA 92107, USA
e-mail: rthurmon@its.jnj.com



P. Blandina, M.B. Passani (eds.), *Histamine Receptors*, The Receptors 28,
DOI 10.1007/978-3-319-40308-3_3

ligands for this receptor are potentially useful for the treatment of immune-mediated diseases [4].

One of the first actions identified for histamine was in modulating smooth muscle contraction, and this was shown to be blocked even by the very first antihistamines [5, 6]. This action on smooth muscle cells leads to histamine-induced bronchoconstriction when histamine is administered to humans, and this effect can be blocked by antagonists of the $H_1R$ [7–10]. In addition to this, it is known that mast cell degranulation occurs during an asthma attack, and there is an increase in the number of mast cells [11–13]. Consistent with this, asthma attacks lead to a rapid increase in airway histamine levels, and this correlates with asthma severity [11, 13–16]. But even early on, it was noted that one of the first antihistamines, diphenhydramine, was not very effective in asthma, although it did show a clear benefit in hay fever [17–20]. Some studies have shown effects of antihistamines in blocking decreases in lung function upon allergen challenge in humans especially on the early phase response, but others have shown no effect, particularly on the late phase response [8–10, 21–24]. Overall, it is not thought that antihistamines that target the $H_1R$ are efficacious for asthma in humans. A meta-analysis of 19 studies of antihistamines in asthma with approximately 580 total subjects on an antihistamine and 560 subjects on placebo concluded that the antihistamines conferred no benefit with respect to airway function compared to placebo [25]. Thus, antihistamines are not thought to be useful for the treatment of asthma [26].

A similar story exists for atopic dermatitis, which is an inflammatory skin disease characterized by eczematous plaques and pruritus [27]. The disease is thought to be driven, like asthma, by Th2 cell responses. Indeed novel drugs that target these responses such as dupilumab, which targets the IL-4 receptor α-subunit and thus blocks IL-4 and IL-13 signaling, have shown efficacy in the disease [28]. Histamine has been shown to be higher in the plasma and skin lesions of atopic dermatitis patients compared to health subjects [29–31]. Most of the attention for antihistamines in atopic dermatitis has been directed toward treating pruritus, since histamine is known to induce pruritus in humans and drugs that target the $H_1R$ have been shown to be effective in a number of other pruritic diseases [32]. However, it has been shown that while the $H_1R$ antagonists terfenadine and clemastine could inhibit histamine-induced itch in patients with atopic dermatitis, there was no effect on the pruritus associated with the disease [33]. Many clinical trials have assessed the efficacy of a variety of antihistamines on pruritus in patients with atopic dermatitis, but the general consensus is that there is limited evidence for efficacy [34–36]. Therefore, it is generally believed that histamine does not play a large role in pruritus in atopic dermatitis.

Thus, there is quite a bit of historical data linking histamine to both asthma and atopic dermatitis. Despite this, clinically available antihistamine drugs have not been proven to be efficacious in these indications, and this has led to the conclusion that histamine is not an important mediator of the pathophysiology of asthma or atopic dermatitis. All of the clinically available compounds are ligands for either the $H_1R$ or $H_2R$ and have little, if any, affinity for the $H_3R$ and $H_4R$ [32]. One possibility for the lack of efficacy observed to date with antihistamines in asthma and atopic dermatitis is that the histamine-mediated functions in these diseases are driven by activation of a different histamine receptor. Therefore, when the $H_4R$ receptor was discovered, it was natural to explore if this receptor was the missing link between histamine and these diseases.

## 3.2 The Histamine $H_4$ Receptor

The $H_4R$ was discovered based on its sequence homology to the $H_3R$. The $H_4R$ and $H_3R$ also share similar pharmacology as many of the older $H_3R$ ligands, such as thioperamide, also having affinity for the $H_4R$. In addition to this, both the $H_3R$ and $H_4R$ receptors are high-affinity histamine receptors in contrast to the $H_1R$ and $H_2R$ receptors. Histamine binds to the $H_4R$ with $K_i$ of 5 nM, whereas the affinity ($K_i$) of histamine for the $H_1R$ and $H_2R$ receptor is 40 μM and 8 μM, respectively [32]. This may reflect the sources of histamine needed to trigger responses via the various receptors. For example, the $H_1R$ and $H_2R$ receptors are activated by histamine release in high concentrations from cells that store histamine such as mast cells, enterochromaffin-like cells and neurons. This, however, may oversimplify the situation as the concentration of ligand needed to activate any receptor is dependent on the agonist affinity, the receptor density on the cell surface, and the fraction of receptors that need to be activated in order to drive a downstream process. The high affinity of histamine for the $H_4R$ suggests that this receptor may be tuned to respond to low-level secretion of histamine from cells that synthesize it de novo such as many cells involved in immune responses. As evidence of this, $H_4R$-mediated effects in some asthma and pruritus models have been shown to be mast cell independent [37–39]. The role of de novo-synthesized histamine can at times be confusing for in vitro pharmacology experiments. For example, Dunford et al. [38] showed that dendritic cells treated with an $H_4R$ antagonist or those from $H_4R$-deficient mice produced lower levels of chemokines and cytokines when stimulated. The fact that the antagonist mimics the effects observed in the absence of the receptor ($H_4R$-deficient cells) proves that this effect is mediated by the $H_4R$. However, traditional pharmacology would dictate that additional proof of receptor activity should be provided by showing that an agonist of the receptor can induce a result that can be reversed by an antagonist. In the dendritic cell case, no effect could be produced by adding histamine, thus potentially confounding the conclusion that the $H_4R$ was involved. This was resolved by showing that simulation of these dendritic cells induced the production of histamine and this was enough to trigger the $H_4R$. Only after the synthesis of histamine was blocked by inhibiting histidine decarboxylase could the addition of exogenous histamine produce an effect that was then blocked by an $H_4R$ antagonist [38].

The high affinity of histamine for the $H_4R$ is important in driving chemotaxis of cells. Many cells migrate toward a histamine concentration gradient via activation of the $H_4R$, and since histamine has such a high affinity for the receptor, these cells may be able to respond to weak histamine signals. The difference in affinity between the different receptors is also important in this respect. Low concentrations of histamine induce chemotaxis via the $H_4R$, but at higher histamine levels activation of the $H_2R$

occurs and chemotaxis is inhibited. This allows cells to migrate along a histamine concentration gradient, but as the source of histamine is approached, the cells stop.

The $H_4R$ is expressed on many cells involved in the immune response; however, a complete understanding of this is hampered by lack of selective antibodies [40, 41]. PCR can be used to detect RNA expression, but this is not suitable for primary tissues where multiple cell types are present, is sensitive to the purity of cells when isolated, and does not necessarily correlate with the levels of functional receptor expression on the cell surface. The best proof of the existence of the receptor on any particular cell type should rely on both antibody and RNA detection as well as characterization of function using multiple ligands. In addition to this, the receptor expression may vary depending on the inflammatory state of the cell [42, 43]. With these caveats in mind, expression and/or function of the receptor has been reported on mast cells; eosinophils; neutrophils; dendritic cells; Langerhans cells; NK cells; monocytes; T cells including γδ T cells, Th1, Th2, Th17 cells, and $CD8^+$ T cells; monocytes; keratinocytes; inflammatory dendritic epidermal cells; epithelial cells, basophils; and fibroblasts [42–67].

The $H_4R$ mediates proinflammatory functions in a number of cell types involved in allergic responses. Chemotaxis of mast cells, eosinophils, dendritic cells, T cells, and fibroblasts can be induced by activation of the $H_4R$ [44, 47, 56–58, 68–70]. For mast cells, histamine-induced chemotaxis is reduced in cells derived from $H_4R$-deficient mice [58] and can be blocked by antagonists of the receptor [58, 71, 72]. This histamine-induced migration of mast cells may be an important driver for increases in tissue mast cell numbers that occur in allergic diseases [73–77]. Reductions in tissue mast cell numbers have been reported after treatment in vivo with $H_4R$ antagonists in models of dermal inflammation and arthritis [37, 45, 78–81]. In vivo $H_4R$ effects on dendritic cell migration have also been observed in an allergic skin model upon injection of an $H_4R$ agonist in vivo [37, 56]. In addition, histamine or an $H_4R$ agonist induced the migration of human Langerhans cells from the epidermis in an ex vivo assay, and these effects could be inhibited by an $H_4R$ antagonist [56]. In eosinophils, histamine-induced migration (or the shape change that precedes this) was blocked by $H_4R$ antagonists [61, 82–88]. Changes in actin polymerization which precedes migration can also be induced by histamine in an $H_4R$-dependent fashion. One way to detect this change is by measuring the change in forward scattering of eosinophils by flow cytometry. In this case, histamine causes an increase in forward scattering by the eosinophils, and this is blocked by $H_4R$ antagonists [61, 82]. This assay can be used in the clinical setting. It was shown that after oral dosing of the $H_4R$ antagonist JNJ-39758979, the histamine-induced increase in forward scattering was reduced in a dose-dependent fashion [87].

In addition to chemotaxis, the $H_4R$ can mediate other functions of mast cells, dendritic cells, and T cells. Histamine or 4-methylhistamine has been shown to induce degranulation in human mast cell lines and cord blood-derived mast cells, and the effects of 4-methylhistamine were reversed by treatment with an $H_4R$ antagonist [60]. In mouse bone marrow-derived mast cells histamine added after IgE sensitization, but before addition of antigen, did not impact antigen-induced degranulation nor was there any difference in mast cells derived from $H_4R$-deficient mice [58, 63]. However, an $H_4R$ antagonist was shown to reduce degranulation indirectly when added prior to IgE sensitization by inhibiting the IgE-mediated upregulation of the FcεRI [63]. The production of inflammatory mediators by mast cells can also be modulated by the $H_4R$. Histamine or $H_4R$ agonists can induce IL-6 secretion from mouse bone marrow-derived mast cells, and this was blocked with $H_4R$ antagonists or in cells derived from $H_4R$-deficient mice [89]. The production of $LTB_4$ in human cord blood-derived mast cells was induced by histamine or 4-methylhistamine, and in both cases, this effect was completely inhibited by an $H_4R$ antagonist [60]. $LTB_4$ produced by antigen stimulation of mouse bone marrow mast cells was not modulated by the $H_4R$ [58].

The $H_4R$ also modulates cytokine and chemokine production from other inflammatory cells. In human NK cells, an $H_4R$ agonist induced the production of CCL3, and this effect was blocked with an $H_4R$ antagonist [90]. In vivo injection of an NKT cell agonist induced the production of IL-4 and IFNγ, and this was reduced in $H_4R$-deficient mice indicating that the $H_4R$ may be involved in NKT cell function as well [91]. In addition, this cytokine production was inhibited in mice that cannot produce histamine, restored when histamine was injected along with the NKT cell agonists but blocked when the mice were pretreated with an $H_4R$ antagonist [91]. The spontaneous production of CCL2 from human monocytes was inhibited by histamine or $H_4R$ agonists, and this effect was reversed by an $H_4R$ antagonist [42]. Several $H_4R$ agonists were able to inhibit the production of IL-12 upon stimulation of human monocytes with IFNγ and LPS, but not all of the effects were reversed by an $H_4R$ antagonists suggesting that other receptors, in addition to the $H_4R$, may be involved [54]. Both stimulated IL-27 and IP-10 production by human monocytes appeared to be suppressed by activation of either the $H_2R$ or $H_4R$, although the activity at the $H_2R$ appeared to be more significant [53, 90].

Some of the same effects of the $H_4R$ in monocytes can also be observed in dendritic cells. Stimulated production of IP-10 in human monocyte-derived dendritic cells also appeared to be inhibited by activation of $H_2R$ or $H_4R$, although, in contrast to the effects on monocytes, the activity at the $H_4R$ appeared to be more significant [90]. Similarly for IL-27 production, bone marrow-derived dendritic cells from $H_4R$-deficient mice produced higher amounts of IL-27 upon stimulation compared to those from wild-type mice [53]. IL-12 production appears to be downregulated by activation of the $H_4R$ upon stimulation of human monocyte-derived dendritic cells [57]. The activity of other specific dendritic cell types or subsets particularly relevant for inflammatory skin diseases is also regulated by the $H_4R$. These include plasmacytoid dendritic cells, Langerhans cells, inflammatory dendritic epidermal cells, and the dendritic cell subset that express 6-sulfo LacNAc where levels of IL-12, TNF, CCL2, IFNα, and CXCL8 can be modulated [49, 55, 56, 69]. Mouse $CD11c^+$ dendritic cells from either $H_4R$-deficient mice or treated with an $H_4R$ antagonist showed reduced production of IL-6, KC, MIP-1α, and IP-10 upon stimulation [38]. This translated into a reduced potential to polarize Th2 T cells in vitro [38]. Dendritic cell-mediated, antigen-specific proliferation of human T cells was reduced when human monocyte-derived dendritic cells were treated with an $H_4R$ antagonist, and this correlated with a reduction in cell surface activation markers [92]. There may also be direct effects on T-cell function. It has been shown that the $H_4R$ is important for IL-16 production from $CD8^+$ T cells, and IL-13 expression has been shown to be induced by $H_4R$ agonists in Th2 cells [43, 50]. Differentiation of human Th17 cells in vitro can be inhibited by $H_4R$ antagonists as evidenced by a reduction in the production of IL-17 [45]. In addition, the secretion of IL-17 upon stimulation of Th17 cells with staphylococcal enterotoxin B was augmented by histamine, and this could be inhibited by an $H_4R$ antagonist [64]. In vivo reduction in Th17 cell differentiation in $H_4R$-deficient or $H_4R$ antagonist-treated mice has been observed as well as a reduction in IL-17 levels in mouse asthma, dermatitis, and arthritis models [37, 38, 45]. One intriguing finding is that the interaction of basophils with human memory T cells can amplify IL-17 release and lead to an increase in the number of IL-$17^+$ cells in a process that partially relies on activation of the $H_2R$ and $H_4R$ [93]. These effects observed on IL-17 production and Th17 cells could implicate a role for $H_4R$ antagonists for the treatment of IL-17-driven diseases such as psoriasis.

It is potentially interesting to note that many of the functions of the $H_4R$ are revealed upon stimulation of cells via toll-like receptors (TLRs). For example, much of the work on dendritic cells shows effects of $H_4R$ ligands upon cytokine and chemokine production induced by stimulation of cells with TLR agonists such as LPS, polyinosinic-polycytidylic acid or CpG [38, 49, 53, 55, 57]. A potential interaction between TLR and $H_4R$ activation was explored in mast cells [89]. Although both TLR and $H_4R$ activation on their own were shown to increase IL-6 production from mouse bone marrow-derived mast cells, the combination led to a synergistic increase in IL-6 production. This was explained by a sustained activation of the downstream kinases ERK and phosphatidylinositide 3-kinase that was greater when both receptors were activated [89]. This interaction may also occur in vivo where administration of LPS leads to an increase in plasma TNF levels. This increase was reduced in $H_4R$-deficient mice or mice treated with $H_4R$ antagonists [94]. Furthermore, treatment with an $H_4R$ antagonist reduced liver damage in an LPS-induced liver injury model and reduced tissue TNF in a colitis model thought to be driven by TLR activation [94, 95]. Finally, it was shown that LPS is required in a mouse asthma model in order to invoke $H_4R$ sensitivity [94]. These results suggest that $H_4R$ may be important in modulating other inflammatory responses.

## 3.3 Preclinical Data on the Role of the $H_4R$ in Asthma

Several groups have reported efficacy of $H_4R$ antagonists in models of asthma. In an acute mouse T-cell-dependent asthma model, three compounds have been shown to reduce the number of total cells, eosinophils, and lymphocytes in the bronchoalveolar fluid (BALF) when administered during the allergen challenge phase of the model [38, 87, 96]. Histological analysis showed that there was also a reduction in airway inflammation upon treatment with an $H_4R$ antagonist [96]. In addition, there was a decrease in IL-5 and IL-17 upon restimulation of lymphocytes [38]. Treatment with JNJ 7777120 also led to an improvement in lung function as measured by reductions in central airway resistance, tissue stiffness, and tissue damping [97]. $H_4R$-deficient mice were protected in this disease model and showed a reduction in the number of BALF total cells, macrophages, eosinophils, and lymphocytes, as well as a reduction in IL-4, IL-5, IL-6, IL-13, and IL-17 upon restimulation of both lymphocytes and splenocytes compared to wild-type mice [38]. Furthermore, there was a reduction in antigen (ovalbumin)-specific IgE and $IgG_1$ levels in the $H_4R$-deficient mice compared to wild-type mice [38]. In a more chronic asthma model, the $H_4R$-antagonist JNJ 7777120 was able to reduce BALF eosinophils and reduce inflammation when dosed even after disease onset has occurred [97]. This also led to an inhibition of IL-4, IL-5, and IL-13 levels upon lymphocyte restimulation and, for IL-5 and IL-13, when measured in lung tissue. In addition to this, treatment with the $H_4R$ antagonist reduced the number of T cells in the lung, and there was a reduction in the levels of T-cell chemokines [97]. This data, coupled with the effects on Th2 cell cytokines in vivo and in vitro, points to a role for the $H_4R$ in Th2 cell development and/or activation. Consistent with this, administering an $H_4R$ antagonist during the T-cell priming phase of the acute mouse model also led to reduction in BALF total cell and eosinophils, as well as reductions in airway inflammation, cytokines, and antibody levels [38, 96]. To further explore this, Hartwig [98] used an adoptive transfer approach to determine the role of the $H_4R$ on T cells and dendritic cells in a mouse asthma model. In this model, mice were sensitized to ovalbumin by injections of ovalbumin-specific T cells that had been polarized to a Th2 phenotype in vitro. The in vitro polarization utilized either wild-type or $H_4R$-deficient $CD11c^+$ cells in combination with either wild-type or $H_4R$-deficient transgenic T cells. Transfer of wild-type T cells polarized with wild-type CD11c+ cells resulted in an increase in BALF eosinophils, airway inflammation, and a reduction in lung function. If the T cells were polarized in the presence of $H_4R$-deficient CD11c+ cells, then these changes were not observed. However, there was no impact if only the T cells lacked the $H_4R$. The importance of dendritic cells in vivo is also suggested by the fact that transfer of wild-type dendritic cells into $H_4R$-deficient mice can restore susceptibility of these mice to a collagen antibody-induced arthritis model and that the $H_4R$ is required on host cells in an adoptive transfer Th17 differentiation model [45]. All of these results are completely consistent with the in vitro findings that $H_4R$-deficient CD11c+ cells or those treated with an $H_4R$ antagonist were impaired in their ability to activate T cells [38, 98]. Interestingly, host mice deficient in the $H_4R$ were protected from eosinophilia, lung inflammation, and loss of airway function when primed with wild-type T cells polarized with wild-type CD11c+ cells [98]. This may also implicate the $H_4R$ as being important on antigen presenting cells even during the challenge or effector phase of the model. Consistent with this hypothesis, Amaral [99] found that bone marrow-derived dendritic cells pulsed with thioperamide (a dual $H_4R/H_3R$ antagonist) and ovalbumin inhibited BALF eosinophils and ovalbumin-specific IgE levels when delivered intratracheally in a mouse asthma model compared to dendritic cells without thioperamide. These treated dendritic cells were also able to increase the number of lung regulatory T cells ($CD4^+/CD25^+/FoxP3^+$), IL-10 levels, and the inhibitory capacity of lung dendritic cells ($CD11c^+$). The bone-marrow-derived dendritic cells only expressed low levels of the $H_3R$, and therefore the activity of thioperamide was ascribed to the $H_4R$; however, it should be noted that thioperamide does have activity at other receptors and does not completely mimic the effects of other $H_4R$ antagonists [100, 101]. The use of a more specific ligand would be needed to confirm that these effects are indeed mediated via the $H_4R$. The same is true for the finding that intratracheal administration of 4-methylhistamine inhibits lung inflammation and function in a mouse asthma model [102]. In this case too, it appeared that the compound increased the number of regulatory T cells. 4-Methylhistamine has been described as an agonist of both the $H_4R$ and the $H_2R$ as well as having cross reactivity versus other receptors [103]. Studies with more selective ligands and in $H_4R$-deficient mice would be needed to define the receptor involved in these effects.

An $H_4R$ antagonist was also shown to reduce BALF eosinophils and improve lung function in a rat asthma model [104]. In a guinea pig model, JNJ 7777120 was shown to decrease airflow resistance, improve lung inflammation, and reduce the number of lung eosinophils and degranulated mast cells [105]. Furthermore, there was a reduction in lung tissue levels of TNF, IL-4, leukotriene $B_4$, and prostaglandin $D_2$ in this model.

## 3.4 The Role of the $H_4R$ in Atopic Dermatitis

The effect of $H_4R$ antagonists in mouse asthma models clearly points to a role for the receptor in mediating Th2 responses. Another disease thought to be mediated by Th2 cytokines is atopic dermatitis. A proof of the involvement of these cytokines in the pathophysiology of atopic dermatitis is shown by the effectiveness of dupilumab, an antibody that blocks the IL-4 receptor α-subunit and thus inhibits IL-4 and IL-13 signaling [28]. The $H_4R$ can modulate Th2 responses in preclinical dermatology models. In a FITC-induced dermatitis model that has the characteristics of a Th2-driven inflammation (i.e., increases in Th2 cytokines and infiltration of eosinophils), treatment with an $H_4R$ antagonist reduced the tissue levels of IL-4 [37]. In addition, restimulation of T cells from $H_4R$ antagonist-treated mice led to lower levels of IL-4 and IL-5 secretion compared to untreated mice [37]. Similar changes were seen in a chronic allergic dermatitis model induced by 2,4,6-trinitro-1-chlorobenzene, where the $H_4R$ antagonist JNJ 7777120 was able to reduce IL-4, IL-5, and IL-6 levels in the lesions [78, 80]. The dual $H_2R/H_4R$ agonist 4-methylhistamine increased the levels of IL-4 and IL-5 [80]. Interestingly this correlated with an increase in IL-12 for JNJ 7777120 and decrease in IL-12 for 4-methylhistamine [80]. Reductions in IL-4 levels were also observed in a chronic model using picryl chloride after treatment with JNJ 7777120 [79]. Using an adoptive transfer model where antigen-specific T cells were given to mice prior to antigen challenge in the skin, JNJ 7777120 was shown to reduce the number of antigen-specific T cells in the skin [106]. In this case, there was no effect on Th2 cytokines except when JNJ 7777120 was given in combination with an $H_1R$ antagonist. This combination also had a more dramatic effect on the number of antigen-specific T cells in the skin compared to either agent alone. Other cell types are also impacted by the $H_4R$ in these models. Reduction in eosinophil and mast cell numbers has been observed after treatment with $H_4R$ antagonists in several of these models [37, 78–81]. However, in some dermatology models, the $H_4R$ does not appear to play a role as $H_4R$ antagonists are not effective in some acute hapten models [80, 107] and in mouse and dog models where *Dermatophagoides farinae* was used as the allergen [108, 109]. Of note, JNJ 7777120 was able to inhibit dermal inflammation induced by the application of croton oil in CD-1 mice but not in other strains suggesting that models are not equivalent in different strains and underscores the potential difficulty in translating preclinical data to human efficacy [110].

In addition to possible effects on inflammation in atopic dermatitis, the $H_4R$ may also directly impact pruritus. Histamine is one of the best characterized pruritogens both in humans and in mice. Injection of histamine into the skin causes dermal reactions such as wheal and flare but also induces pruritus. In mice, $H_4R$ agonists mimic the pruritic effects of histamine and induce scratching [39, 88, 111]. Scratching induced by histamine or $H_4R$ agonists can be inhibited by $H_4R$ antagonists and is reduced in mice deficient in $H_4R$ [39, 67, 71, 85, 88, 111–114]. The $H_4R$-mediated pruritus may be mediated by direct effects on neurons. Stimulation of afferent fibers by intradermal injection of compound 48/80 induced scratching in mice that was reduced by $H_4R$ antagonists and in $H_4R$-deficient mice [39]. Compound 48/80 also degranulates mast cells, but the pruritic effects were noted even in mast cell-deficient mice. Similarly, histamine-induced scratching also occurred in mast cell-deficient mice, and this could still be inhibited by an $H_4R$ antagonist [39]. Finally, reconstituting $H_4R$-deficient mice with wild-type bone marrow so that now all of the hematopoietic cells express the $H_4R$ did not restore histamine-induced itch. In support of a neuron effect, skin-specific sensory neurons in mice have been shown to express the $H_4R$, and activation of the receptor led to an increase in calcium [115].

Preclinically it has been shown that $H_4R$ antagonists not only reduce histamine-induced itch, but they can also block itch driven by other pruritogens such as substance P or $H_3R$ antagonists [67, 115]. Itch observed in dermatology models driven by haptens can also be reduced by $H_4R$ antagonists. JNJ 7777120 was able to reduce the scratching observed immediately upon application of fluorescein isothiocyanate to the ears of sensitized mice [37]. Itch generated by the acute application of either toluene-2,4-disiocyanate or 2,4-dinitrochlorobenzene was also blocked by JNJ 7777120 [107]. Chronic application of 2,4,6-trinitrochlorobenzene three times a week over 99 days induced scratching that could be inhibited by JNJ 7777120, as could the scratching observed when NC/Nga mice where sensitized and then challenged once a week for 10 weeks with picryl chloride [79, 81]. This may not be a general finding as $H_4R$ antagonists had no effect on scratching when a protein allergen was used to induce dermatitis in NC/Nga mice [109] or when croton oil was used in CD-1 mice [110].

The role of the $H_4R$ in histamine-induced itch has been validated in humans in a clinical study [116]. In this study, subjects were given a single dose of either an $H_4R$ antagonist, JNJ 39758979; an $H_1R$ antagonist, cetirizine; or placebo. JNJ 39758979 is a potent antagonist of the $H_4R$ with no affinity for the $H_1R$ [71]. Conversely, cetirizine is a potent $H_1R$ antagonist with no $H_4R$ affinity [103]. Therefore, these two compounds allow for the definitive characterization of the role of the $H_1R$ and $H_4R$ in mediating histamine-induced itch in humans. Itch was induced by intradermal injection of histamine into the forearm of the subjects and the itch sensation assessed on a 0–10 scale. This assessment was carried out the day before dosing (baseline) as well as 2 and 6 h after dosing. The primary endpoint for the study was the change in the AUC for pruritus score assessed over a 10-min period after histamine injection.

The itch AUC was similar for all groups at baseline. Placebo had no impact on the sensation of itch at either 2 or 6 h after dosing. As expected, the positive control, cetirizine, reduced the itch when assess 6 h after dosing (there was no effect at the 2 h time point which was consistent with the known onset of action of cetirizine). The $H_4R$ antagonist, JNJ 39758979, reduced the itch when assessed at both 2 and 6 h after dosing. Of note the histamine-induced wheal and flare response was also assessed in this study. The $H_1R$ antagonist, cetirizine, completely blocked this response as expected since it is known that these effects are mediated via the $H_1R$. However, JNJ 39758979 had no effect on wheal and flare. This is consistent with animal data [72] and provides proof that the effects seen on the itch parameter were not the results of cross reactivity at the $H_1R$. These data then provide conclusive evidence that histamine-induced pruritus in humans is mediated in part by the $H_4R$, consistent with what has been observed in mouse models.

Effects on histamine-induced pruritus are interesting, but it does not provide evidence that $H_4R$ antagonists will be effective in pruritic diseases where drugs that target the $H_1R$ are not effective. The most obvious one of these is atopic dermatitis. In this case, the rationale for testing $H_4R$ antagonists comes both from the preclinical and clinical antipruritic effects as well as the postulated anti-inflammatory effects. To establish whether $H_4R$ antagonists are efficacious in atopic dermatitis, a clinical study was conducted in patients with active moderate atopic dermatitis [117]. It was planned for the study to enroll approximately 105 patients across three arms—placebo, 100 mg JNJ 39758979, and 300 mg JNJ 39758979. Patients took the medication once a day for up to 6 weeks. Unfortunately, a serious safety issue arose that resulted in the early termination of the study (see [117] for details). This resulted in only a limited number of patients (50) reaching the 6-week endpoint thereby reducing the statistical power and leading to all of the analysis to be conducted post hoc. With these caveats in mind, it still appeared that the patients in the two JNJ 39758979 arms had numerical improvements in the Eczema Area and Severity Index (EASI) [118] compared to placebo. The reductions were on the order of 20–30 % and did exceed the threshold considered clinically meaningful [119]. However, the changes were not statistically significant perhaps due to the limited number of patients. Furthermore, assessment of the time course for the change in EASI suggests that improvements were generally seen by week 4 and patients may continue to improve with longer treatment (Fig. 3.1). Also consistent with a positive effect on the disease, less patients in the JNJ 39758979 arms needed to use rescue topical steroid medication [117].

Itch is not a component of the EASI assessment even though it is the most troubling symptom for most patients [120]. Therefore, the effect of JNJ 39758979 on itch was assessed by several methods. One of these was a global impression of change, where at each visit patients were asked to rate the severity of itch and amount of time itching compared to the start of the study. At week 6, more patients in the JNJ 39758979 groups reported improvements in these parameters compared to the patients in the placebo group. In general approximately 75–90 % of the patients in the JNJ 39758979 groups rated their itch severity and duration as better, whereas only 30–35 % of the placebo subject did [117]. These results reached nominal statistical significance. Another method used to assess the itch was daily diaries where patients were asked to rate the itch severity and duration every morning and

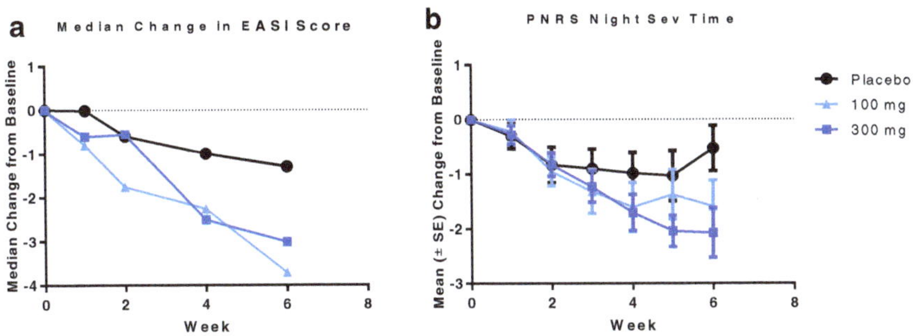


**Fig. 3.1** The time course for effects on skin lesions and pruritus from a clinical study with JNJ 39758979 in atopic dermatitis. (**a**) Changes from baseline to week 6 in EASI scores, shown as median changes. (**b**) Mean changes from baseline over time to week 6 in Pruritus Numeric Rating Scale, PNRS, Nighttime Pruritus Severity. Adapted from Murata et al. [117]

evening. At week 6, patients in the JNJ 39758979 groups had a numerical reduction (improvement) in itch severity and duration both at daytime and nighttime when compared to placebo (Fig. 3.1 and [117]). In this case, only the results from the 300 mg group reached nominal statistical significance.

One interesting note is the time course for the improvement in itch. The effects in the histamine-induced itch clinical study were acute. A single dose of JNJ 39758979 was able to reduce the itch sensation when histamine challenge was given 2 or 6 h after dosing [116]. It has been hypothesized that the mechanism for the antipruritic effect of $H_4R$ antagonists is due to blocking $H_4R$ signaling in neurons [121]. If this is the case, then there should be an immediate reduction in the itch sensation after compound dosing since it would block the transmission of the pruritic signal to the brain. However, the time course for the antipruritic effect in the atopic dermatitis study show that there is a slow onset of action and relief is not observed until after 3–4 weeks of dosing (Fig. 3.1 and [117]). This suggests that in atopic dermatitis, the antipruritic effects of the $H_4R$ antagonist is not due to blocking neuronal signaling and that it is unlikely that histamine itself is the pruritogen. More likely is that the $H_4R$ antagonist is acting as an anti-inflammatory agent and then indirectly reducing itch or that it is directly blocking the production of another pruritogen. One possible candidate is IL-31 that has been shown to be overexpressed in AD skin [122], and its expression can be induced with an $H_4R$ agonist in Th2 cells or in peripheral blood mononuclear cells from AD patients [43].

## 3.5 Conclusions

$H_4R$ antagonists are still early in clinical development. However, some recent clinical data has validated preclinical evidence for a role of the receptor in atopic dermatitis and pruritus. The results from the clinical study in atopic dermatitis indicate that such antagonists may have utility in improving the skin lesions and in providing relief to the pruritus associated with the disease. The results showing that an $H_4R$ antagonist can reduce histamine-induced itch in humans suggest utility in pruritus known to be driven by the release of histamine where $H_1R$ antagonists already have efficacy. This overlap between the two receptors may indicate that the combination of $H_1R$ and $H_4R$ antagonists may have additional benefits than either alone. As to date, no clinical data have been reported for asthma, but preclinical models indicate that $H_4R$ antagonists may also be useful for the treatment of allergic asthma. Therefore, the $H_4R$ may play a role in diseases such as asthma and atopic dermatitis, which have long since been believed not to be mediated by histamine due to lack of efficacy of $H_1R$ antagonists.

## References

1. Panula P, Chazot PL, Cowart M, Gutzmer R, Leurs R, Liu WLS, Stark H, Thurmond RL, Haas HL. International union of basic and clinical pharmacology. XCVIII. Histamine receptors. Pharmacol Rev. 2015;67:601–55.
2. Dauvilliers Y, Bassetti C, Lammers GJ, Arnulf I, Mayer G, Rodenbeck A, Lehert P, Ding C-L, Lecomte J-M, Schwartz J-C. Pitolisant versus placebo or modafinil in patients with narcolepsy: a double-blind, randomised trial. Lancet Neurol. 2013;12:1068–75.
3. Inocente C, Arnulf I, Bastuji H, Thibault-Stoll A, Raoux A, Reimao R, Lin J-S, Franco P. Pitolisant, an inverse agonist of the histamine H3 receptor: an alternative stimulant for narcolepsy-cataplexy in teenagers with refractory sleepiness. Clin Neuropharmacol. 2012;35:55–60.
4. Yu F, Bonaventure P, Thurmond Robin L. The future antihistamines: histamine H3 and H4 receptor ligands. Adv Exp Med Biol. 2010;709:125–40.
5. Bovet D, Walthert F. Synthetic antihistamine agents. Chemical structure and pharmacodynamic activities. Ann Pharm Fr. 1944;2:43.
6. Salva JA. Antihistamine RP 3277. Arch Inst Farmacol Exp Madrid. 1949;1:155–60.
7. Curry JJ. The action of histamine on the respiratory tract in normal and asthmatic subjects. J Clin Invest. 1946;25:785–91.
8. Rafferty P, Beasley R, Holgate ST. The contribution of histamine to immediate bronchoconstriction provoked by inhaled allergen and adenosine 5′ monophosphate in atopic asthma. Am Rev Respir Dis. 1987;136:369–73.
9. Rafferty P, Holgate ST. The inhibitory effect of azelastine hydrochloride on histamine- and allergen-induced bronchoconstriction in atopic asthma. Clin Exp Allergy. 1989;19:315–20.
10. Town GI, Holgate ST. Comparison of the effect of loratadine on the airway and skin responses to histamine, methacholine, and allergen in subjects with asthma. J Allergy Clin Immunol. 1990;86:886–93.
11. Broide DH, Gleich GJ, Cuomo AJ, Coburn DA, Federman EC, Schwartz LB, Wasserman SI. Evidence of ongoing mast cell and eosinophil degranulation in symptomatic asthma airway. J Allergy Clin Immunol. 1991;88:637–48.
12. Wardlaw AJ, Dunnette S, Gleich GJ, Collins JV, Kay AB. Eosinophils and mast cells in bronchoalveolar lavage in subjects with mild asthma. Relationship to bronchial hyperreactivity. Am Rev Respir Dis. 1988;137:62–9.
13. Wenzel SE, Fowler 3rd AA, Schwartz LB. Activation of pulmonary mast cells by bronchoalveolar allergen challenge. In vivo release of histamine and tryptase in atopic subjects with and without asthma. Am Rev Respir Dis. 1988;137:1002–8.
14. Casale TB, Wood D, Richerson HB, Trapp S, Metzger WJ, Zavala D, Hunninghake GW. Elevated bronchoalveolar lavage fluid histamine levels in allergic asthmatics are associated with methacholine bronchial hyperresponsiveness. J Clin Invest. 1987;79:1197–203.

15. Jarjour N, Calhoun W, Schwartz L, Busse W. Elevated bronchoalveolar lavage fluid histamine levels in allergic asthmatics are associated with increased airway obstruction. Am Rev Respir Dis. 1991;144:83–7.
16. Liu MC, Bleecker ER, Lichtenstein LM, Kagey-Sobotka A, Niv Y, McLemore TL, Permutt S, Proud D, Hubbard WC. Evidence for elevated levels of histamine, prostaglandin D2, and other bronchoconstricting prostaglandins in the airways of subjects with mild asthma. Am Rev Respir Dis. 1990;142:126–32.
17. Feinberg SM, Friedlaender S. Relief of dermographism and other urticarias of histamine origin by a synthetic benzhydryl alkamine ether. J Allergy. 1945;16:296–8.
18. McElin TW, Horton BT. Clinical observations on the use of benadryl; a new antihistamine substance. Proc Annu Meet Cent Soc Clin Res U S. 1945;18:45.
19. McGavack TH, Elias H, Boyd LJ. Some pharmacological and clinical experiences with dimethylaminoethyl benzhydryl ether hydrochloride (benadryl). Bull N Y Acad Med. 1946;22:481.
20. Taub SJ. The use of a new anti-histaminic drug (benadryl) in allergic diseases. Chic Med Sch Q. 1945;7:6.
21. Bentley AM, Walker S, Hanotte F, De Vos C, Durham SR. A comparison of the effects of oral cetirizine and inhaled beclomethasone on early and late asthmatic responses to allergen and the associated increase in airways hyperresponsiveness. Clin Exp Allergy. 1996;26:909–17.
22. Cockcroft DW, Keshmiri M, Murdock KY, Gore BC. Allergen-induced increase in airway responsiveness is not inhibited by acute treatment with ketotifen or clemastine. Ann Allergy. 1992;68:245–50.
23. Motojima S, Ohashi Y, Otsuka T, Fukuda T, Makino S. Effects of azelastine on allergen- and exercise-induced asthma. Asian Pac J Allergy Immunol. 1985;3:174–8.
24. Rafferty P, Ng WH, Phillips G, Clough J, Church MK, Aurich R, Ollier S, Holgate ST. The inhibitory actions of azelastine hydrochloride on the early and late bronchoconstrictor responses to inhaled allergen in atopic asthma. J Allergy Clin Immunol. 1989;84:649–57.
25. Van Ganse E, Kaufman L, Derde MP, Yernault JC, Delaunois L, Vincken W. Effects of antihistamines in adult asthma: a meta-analysis of clinical trials. Eur Respir J. 1997;10:2216–24.
26. Simons FER. Is antihistamine (H1-receptor antagonist) therapy useful in clinical asthma? Clin Exp Allergy. 1999;29:98–104.
27. Bieber T. Atopic dermatitis. N Engl J Med. 2008;358:1483–94.
28. Thaci D, Simpson EL, Beck LA, Bieber T, Blauvelt A, Papp K, Soong W, Worm M, Szepietowski JC, Sofen H, Kawashima M, Wu R, Weinstein SP, Graham NMH, Pirozzi G, Teper A, Sutherland ER, Mastey V, Stahl N, Yancopoulos GD, Ardeleanu M. Efficacy and safety of dupilumab in adults with moderate-to-severe atopic dermatitis inadequately controlled by topical treatments: a randomised, placebo-controlled, dose-ranging phase 2b trial. Lancet. 2016;387(10013):40–52.
29. Imaizumi A, Kawakami T, Murakami F, Soma Y, Mizoguchi M. Effective treatment of pruritus in atopic dermatitis using H1 antihistamines (second-generation antihistamines): changes in blood histamine and tryptase levels. J Dermatol Sci. 2003;33:23–9.
30. Johnson Jr HH, DeOreo GA, Lascheid WP, Mitchell F. Skin histamine levels in chronic atopic dermatitis. J Invest Dermatol. 1960;34:237–8.
31. Juhlin L. Localization and content of histamine in normal and diseased skin. Acta Derm Venereol. 1967;47:383–91.
32. Thurmond RL, Kazerouni K, Chaplan SR, Greenspan AJ. Peripheral neuronal mechanism of itch: histamine and itch. Boca Raton: CRC Press; 2014. p. 143–92.
33. Wahlgren CF, Hagermark O, Bergstrom R. The antipruritic effect of a sedative and a non-sedative antihistamine in atopic dermatitis. Br J Dermatol. 1990;122:545–51.
34. Akdis CA, Akdis M, Bieber T, Bindslev-Jensen C, Boguniewicz M, Eigenmann P, Hamid Q, Kapp A, Leung Donald YM, Lipozencic J, Luger Thomas A, Muraro A, Novak N, Platts-Mills Thomas AE, Rosenwasser L, Scheynius A, Simons FER, Spergel J, Turjanmaa K, Wahn U, Weidinger S, Werfel T, Zuberbier T. Diagnosis and treatment of atopic dermatitis in children and adults: European Academy of Allergology and Clinical Immunology/American Academy of Allergy, Asthma and Immunology/PRACTALL Consensus Report. J Allergy Clin Immunol. 2006;118:152–69.

35. Klein PA, Clark RA. An evidence-based review of the efficacy of antihistamines in relieving pruritus in atopic dermatitis. Arch Dermatol. 1999;135:1522–5.
36. Saeki H, Furue M, Furukawa F, Hide M, Ohtsuki M, Katayama I, Sasaki R, Suto H, Takehara K. Guidelines for management of atopic dermatitis. J Dermatol. 2009;36:563–77.
37. Cowden JM, Zhang M, Dunford PJ, Thurmond RL. The histamine $H_4$ receptor mediates inflammation and pruritus in Th2-dependent dermal inflammation. J Invest Dermatol. 2010;130:1023–33.
38. Dunford PJ, O'Donnell N, Riley JP, Williams KN, Karlsson L, Thurmond RL. The histamine $H_4$ receptor mediates allergic airway inflammation by regulating the activation of CD4+ T cells. J Immunol. 2006;176:7062–70.
39. Dunford PJ, Williams KN, Desai PJ, Karlsson L, McQueen D, Thurmond RL. Histamine $H_4$ receptor antagonists are superior to traditional antihistamines in the attenuation of experimental pruritus. J Allergy Clin Immunol. 2007;119:176–83.
40. Beermann S, Seifert R, Neumann D. Commercially available antibodies against human and murine histamine H4-receptor lack specificity. Naunyn Schmiedebergs Arch Pharmacol. 2012;385:125–35.
41. Gutzmer R, Werfel T, Baeumer W, Kietzmann M, Chazot PL, Leurs R. Well characterized anti-histamine 4 receptor antibodies contribute to current knowledge of the expression and biology of the human and murine histamine 4 receptor. Naunyn Schmiedebergs Arch Pharmacol. 2012;385:853–4.
42. Dijkstra D, Leurs R, Chazot P, Shenton FC, Stark H, Werfel T, Gutzmer R. Histamine down-regulates monocyte CCL2 production through the histamine $H_4$ receptor. J Allergy Clin Immunol. 2007;120:300–7.
43. Gutzmer R, Mommert S, Gschwandtner M, Zwingmann K, Stark H, Werfel T. The histamine H4 receptor is functionally expressed on TH2 cells. J Allergy Clin Immunol. 2009;123:619–25.
44. Bäumer W, Wendorff S, Gutzmer R, Werfel T, Dijkstra D, Chazot P, Stark H, Kietzmann M. Histamine H4 receptors modulate dendritic cell migration through skin—immunomodulatory role of histamine. Allergy. 2008;63:1387–94.
45. Cowden JM, Yu F, Banie H, Farahani M, Ling P, Nguyen S, Riley JP, Zhang M, Zhu J, Dunford PJ, Thurmond RL. The histamine H4 receptor mediates inflammation and Th17 responses in preclinical models of arthritis. Ann Rheum Dis. 2014;73:600–8.
46. Czerner CP, Klos A, Seifert R, Neumann D. Histamine induces chemotaxis and phagocytosis in murine bone marrow-derived macrophages and RAW 264.7 macrophage-like cells via histamine H4-receptor. Inflamma Res. 2014;63:239–47.
47. Damaj BB, Becerra CB, Esber HJ, Wen Y, Maghazachi AA. Functional expression of $H_4$ histamine receptor in human natural killer cells, monocytes, and dendritic cells. J Immunol. 2007;179:7907–15.
48. Dib K, Perecko T, Jenei V, McFarlane C, Comer D, Brown V, Katebe M, Scheithauer T, Thurmond RL, Chazot PL, Ennis M. The histamine H4 receptor is a potent inhibitor of adhesion-dependent degranulation in human neutrophils. J Leukoc Biol. 2014;96:411–8.
49. Dijkstra D, Stark H, Chazot PL, Shenton FC, Leurs R, Werfel T, Gutzmer R. Human inflammatory dendritic epidermal cells express a functional histamine $H_4$ receptor. J Invest Dermatol. 2008;128:1696–703.
50. Gantner F, Sakai K, Tusche MW, Cruikshank WW, Center DM, Bacon KB. Histamine $H_4$ and $H_2$ receptors control histamine-induced interleukin-16 release from human CD8+ T cells. J Pharmacol Exp Ther. 2002;303:300–7.
51. Glatzer F, Gschwandtner M, Ehling S, Rossbach K, Janik K, Klos A, Baeumer W, Kietzmann M, Werfel T, Gutzmer R. Histamine induces proliferation in keratinocytes from patients with atopic dermatitis through the histamine 4 receptor. J Allergy Clin Immunol. 2013;132:1358–67.
52. Godot V, Arock M, Garcia G, Capel F, Flys C, Dy M, Emilie D, Humbert M. H4 histamine receptor mediates optimal migration of mast cell precursors to CXCL12. J Allergy Clin Immunol. 2007;120:827–34.
53. Gschwandtner M, Bunk H, Koether B, Thurmond RL, Kietzmann M, Werfel T, Baeumer W, Gutzmer R. Histamine down-regulates IL-27 production in antigen-presenting cells. J Leukoc Biol. 2012;92:21–9.

54. Gschwandtner M, Koether B, Werfel T, Stark H, Gutzmer R. Profiling of histamine H4 receptor agonists in native human monocytes. Br J Pharmacol. 2013;170:136–43.
55. Gschwandtner M, Mommert S, Koether B, Werfel T, Gutzmer R. The histamine H4 receptor is highly expressed on plasmacytoid dendritic cells in psoriasis and histamine regulates their cytokine production and migration. J Invest Dermatol. 2011;131:1668–76.
56. Gschwandtner M, Rossbach K, Dijkstra D, Baeumer W, Kietzmann M, Stark H, Werfel T, Gutzmer R. Murine and human Langerhans cells express a functional histamine H4 receptor: modulation of cell migration and function. Allergy. 2010;65:840–9.
57. Gutzmer R, Diestel C, Mommert S, Koether B, Stark H, Wittmann M, Werfel T. Histamine $H_4$ receptor stimulation suppresses IL-12p70 production and mediates chemotaxis in human monocyte-derived dendritic cells. J Immunol. 2005;174:5224–32.
58. Hofstra CL, Desai PJ, Thurmond RL, Fung-Leung W-P. Histamine $H_4$ receptor mediates chemotaxis and calcium mobilization of mast cells. J Pharmacol Exp Ther. 2003;305:1212–21.
59. Ikawa Y, Shiba K, Ohki E, Mutoh N, Suzuki M, Sato H, Ueno K. Comparative study of histamine H4 receptor expression in human dermal fibroblasts. J Toxicol Sci. 2008;33:503–8.
60. Jemima EA, Prema A, Thangam EB. Functional characterization of histamine H4 receptor on human mast cells. Mol Immunol. 2014;62:19–28.
61. Ling P, Ngo K, Nguyen S, Thurmond RL, Edwards JP, Karlsson L, Fung-Leung W-P. Histamine $H_4$ receptor mediates eosinophil chemotaxis with cell shape change and adhesion molecule upregulation. Br J Pharmacol. 2004;142:161–71.
62. Lippert U, Artuc M, Grutzkau A, Babina M, Guhl S, Haase I, Blaschke V, Zachmann K, Knosalla M, Middel P, Kruger-Krasagakis S, Henz BM. Human skin mast cells express H2 and H4, but not H3 receptors. J Invest Dermatol. 2004;123:116–23.
63. Mirzahosseini A, Dalmadi B, Csutora P. Histamine receptor H4 regulates mast cell degranulation and IgE induced FcεRI upregulation in murine bone marrow-derived mast cells. Cell Immunol. 2013;283:38–44.
64. Mommert S, Gschwandtner M, Koether B, Gutzmer R, Werfel T. Human memory Th17 cells express a functional histamine H4 receptor. Am J Pathol. 2012;180:177–85.
65. Sander LE, Lorentz A, Sellge G, Coeffier M, Neipp M, Veres T, Frieling T, Meier PN, Manns MP, Bischoff SC. Selective expression of histamine receptors H1R, H2R, and H4R, but not H3R, in the human intestinal tract. Gut. 2006;55:498–504.
66. Truta-Feles K, Lagadari M, Lehmann K, Berod L, Cubillos S, Piehler S, Herouy Y, Barz D, Kamradt T, Maghazachi A, Norgauer J. Histamine modulates γδ-T lymphocyte migration and cytotoxicity, via Gi and Gs protein-coupled signalling pathways. Br J Pharmacol. 2010;161:1291–300.
67. Yamaura K, Oda M, Suwa E, Suzuki M, Sato H, Ueno K. Expression of histamine H4 receptor in human epidermal tissues and attenuation of experimental pruritus using H4 receptor antagonist. J Toxicol Sci. 2009;34:427–31.
68. Bryce PJ, Mathias CB, Harrison KL, Watanabe T, Geha RS, Oettgen HC. The H1 histamine receptor regulates allergic lung responses. J Clin Invest. 2006;116:1624–32.
69. Gschwandtner M, Schaekel K, Werfel T, Gutzmer R. Histamine H4 receptor activation on human slan-dendritic cells down-regulates their pro-inflammatory capacity. Immunology. 2011;132:49–56.
70. Kohyama T, Yamauchi Y, Takizawa H, Kamitani S, Kawasaki S, Nagase T. Histamine stimulates human lung fibroblast migration. Mol Cell Biochem. 2010;337:77–81.
71. Savall BM, Chavez F, Tays K, Dunford P, Cowden J, Hack MD, Wolin R, Thurmond RL, Edwards JP. Discovery and SAR of 6-Alkyl-2,4-diaminopyrimidines as H4 receptor antagonists. J Med Chem. 2014;57:2429–39.
72. Thurmond RL, Desai PJ, Dunford PJ, Fung-Leung W-P, Hofstra CL, Jiang W, Nguyen S, Riley JP, Sun S, Williams KN, Edwards JP, Karlsson L. A potent and selective histamine $H_4$ receptor antagonist with anti-inflammatory properties. J Pharmacol Exp Ther. 2004;309:404–13.
73. Amin K, Ludviksdottir D, Janson C, Nettelbladt O, Bjornsson E, Roomans GM, Boman G, Seveus L, Venge P. Inflammation and structural changes in the airways of patients with atopic and nonatopic asthma. BHR Group. Am J Respir Crit Care Med. 2000;162:2295–301.

74. Crimi E, Chiaramondia M, Milanese M, Rossi GA, Brusasco V. Increased numbers of mast cells in bronchial mucosa after the late-phase asthmatic response to allergen. Am Rev Respir Dis. 1991;144:1282–6.
75. Gauvreau GM, Lee JM, Watson RM, Irani AM, Schwartz LB, O'Byrne PM. Increased numbers of both airway basophils and mast cells in sputum after allergen inhalation challenge of atopic asthmatics. Am J Respir Crit Care Med. 2000;161:1473–8.
76. Kassel O, De Blay F, Duvernelle C, Olgart C, Israel-Biet D, Krieger P, Moreau L, Muller C, Pauli G, Frossard N. Local increase in the number of mast cells and expression of nerve growth factor in the bronchus of asthmatic patients after repeated inhalation of allergen at low-dose. Clin Exp Allergy. 2001;31:1432–40.
77. Kirby JG, Hargreave FE, Gleich GJ, O'Byrne PM. Bronchoalveolar cell profiles of asthmatic and nonasthmatic subjects. Am Rev Respir Dis. 1987;136:379–83.
78. Matsushita A, Seike M, Okawa H, Kadawaki Y, Ohtsu H. Advantages of histamine H4 receptor antagonist usage with H1 receptor antagonist for the treatment of murine allergic contact dermatitis. Exp Dermatol. 2012;21:714–5.
79. Ohsawa Y, Hirasawa N. The antagonism of histamine H1 and H4 receptors ameliorates chronic allergic dermatitis via anti-pruritic and anti-inflammatory effects in NC/Nga mice. Allergy. 2012;67:1014–22.
80. Seike M, Furuya K, Omura M, Hamada-Watanabe K, Matsushita A, Ohtsu H. Histamine H4 receptor antagonist ameliorates chronic allergic contact dermatitis induced by repeated challenge. Allergy. 2010;65:319–26.
81. Suwa E, Yamaura K, Oda M, Namiki T, Ueno K. Histamine H4 receptor antagonist reduces dermal inflammation and pruritus in a hapten-induced experimental model. Eur J Pharmacol. 2011;667:383–8.
82. Buckland KF, Williams TJ, Conroy DM. Histamine induces cytoskeletal changes in human eosinophils via the $H_4$ receptor. Br J Pharmacol. 2003;140:1117–27.
83. O'Reilly M, Alpert R, Jenkinson S, Gladue RP, Foo S, Trim S, Peter B, Trevethick M, Fidock M. Identification of a histamine $H_4$ receptor on human eosinophils-role in eosinophil chemotaxis. J Recept Signal Transduct Res. 2002;22:431–48.
84. Reher TM, Neumann D, Buschauer A, Seifert R. Incomplete activation of human eosinophils via the histamine H4-receptor: Evidence for ligand-specific receptor conformations. Biochem Pharmacol. 2012;84:192–203.
85. Shin N, Covington M, Bian D, Zhuo J, Bowman K, Li Y, Soloviev M, Qian D-Q, Feldman P, Leffet L, He X, Wang KH, Krug K, Bell D, Czerniak P, Hu Z, Zhao H, Zhang J, Yeleswaram S, Yao W, Newton R, Scherle P. INCB38579, a novel and potent histamine H4 receptor small molecule antagonist with anti-inflammatory pain and anti-pruritic functions. Eur J Pharmacol. 2012;675:47–56.
86. Strakhova MI, Cuff CA, Manelli AM, Carr TL, Witte DG, Baranowski JL, Vortherms TA, Miller TR, Rundell L, McPherson MJ, Adair RM, Brito AA, Bettencourt BM, Yao BB, Wetter JM, Marsh KC, Liu H, Cowart MD, Brioni JD, Esbenshade TA. In vitro and in vivo characterization of A-940894: a potent histamine H4 receptor antagonist with anti-inflammatory properties. Br J Pharmacol. 2009;157:44–54.
87. Thurmond RL, Chen B, Dunford PJ, Greenspan AJ, Karlsson L, La D, Ward P, Xu XL. Clinical and preclinical characterization of the histamine H4 receptor antagonist JNJ-39758979. J Pharmacol Exp Ther. 2014;349:176–84.
88. Yu F, Wolin RL, Wei J, Desai PJ, McGovern PM, Dunford PJ, Karlsson L, Thurmond RL. Pharmacological characterization of oxime agonists of the histamine H4 receptor. J Receptor Ligand Channel Res. 2010;3:37–49.
89. Desai P, Thurmond RL. Histamine H4 receptor activation enhances LPS-induced IL-6 production in mast cells via ERK and PI3K activation. Eur J Immunol. 2011;41:1764–73.
90. Glatzer F, Mommert S, Koether B, Gschwandtner M, Stark H, Werfel T, Gutzmer R. Histamine downregulates the Th1-associated chemokine IP-10 in monocytes and myeloid dendritic cells. Int Arch Allergy Immunol. 2014;163:11–9.

91. Leite-de-Moraes MC, Diem S, Michel M-L, Ohtsu H, Thurmond RL, Schneider E, Dy M. Cutting edge: histamine receptor H4 activation positively regulates in vivo IL-4 and IFN-gamma production by invariant NKT cells. J Immunol. 2009;182:1233–6.
92. Lundberg K, Broos S, Greiff L, Borrebaeck CAK, Lindstedt M. Histamine H4 receptor antagonism inhibits allergen-specific T-cell responses mediated by human dendritic cells. Eur J Pharmacol. 2011;651:197–204.
93. Wakahara K, Baba N, Van VQ, Begin P, Rubio M, Ferraro P, Panzini B, Wassef R, Lahaie R, Caussignac Y, Tamaz R, Richard C, Soucy G, Delespesse G, Sarfati M. Human basophils interact with memory T cells to augment Th17 responses. Blood. 2012;120:4761–71.
94. Cowden JM, Yu F, Challapalli M, Huang J-F, Kim S, Fung-Leung W-P, Ma JY, Riley JP, Zhang M, Dunford PJ, Thurmond RL. Antagonism of the histamine H4 receptor reduces LPS-induced TNF production in vivo. Inflamm Res. 2013;62:599–607.
95. Varga C, Horvath K, Berko A, Thurmond RL, Dunford PJ, Whittle BJR. Inhibitory effects of histamine $H_4$ receptor antagonists on experimental colitis in the rat. Eur J Pharmacol. 2005;522:130–8.
96. Beermann S, Glage S, Jonigk D, Seifert R, Neumann D. Opposite effects of mepyramine on JNJ 7777120-induced amelioration of experimentally induced asthma in mice in sensitization and provocation. PLoS One. 2012;7, e30285.
97. Cowden JM, Riley JP, Ma JY, Thurmond RL, Dunford PJ. Histamine H4 receptor antagonism diminishes existing airway inflammation and dysfunction via modulation of Th2 cytokines. Respir Res. 2010;11:86.
98. Hartwig C, Munder A, Glage S, Wedekind D, Schenk H, Seifert R, Neumann D. The histamine H4-receptor (H4R) regulates eosinophilic inflammation in ovalbumin-induced experimental allergic asthma in mice. Eur J Immunol. 2015;45(4):1129–40.
99. Amaral MM, Davio C, Ceballos A, Salamone G, Canones C, Geffner J, Vermeulen M. Histamine improves antigen uptake and cross-presentation by dendritic cells. J Immunol. 2007;179:3425–33.
100. Leurs R, Tulp MTM, Menge WMBP, Adolfs MJP, Zuiderveld OP, Timmerman H. Evaluation of the receptor selectivity of the H3 receptor antagonists, iodophenpropit and thioperamide: an interaction with the 5-HT3 receptor revealed. Br J Pharmacol. 1995;116:2315–21.
101. Neumann D, Beermann S, Burhenne H, Glage S, Hartwig C, Seifert R. The dual H3/4R antagonist thioperamide does not fully mimic the effects of the 'standard' H4R antagonist JNJ 7777120 in experimental murine asthma. Naunyn Schmiedebergs Arch Pharmacol. 2013;386:983–90.
102. Morgan RK, McAllister B, Cross L, Green DS, Kornfeld H, Center DM, Cruikshank WW. Histamine 4 receptor activation induces recruitment of FoxP3+ T cells and inhibits allergic asthma in a murine model. J Immunol. 2007;178:8081–9.
103. Lim HD, van Rijn RM, Ling P, Bakker RA, Thurmond RL, Leurs R. Evaluation of histamine H1-, H2-, and H3-receptor ligands at the human histamine H4 receptor: Identification of 4-methylhistamine as the first potent and selective H4 receptor agonist. J Pharmacol Exp Ther. 2005;314:1310–21.
104. Alfon J, Ardanaz N, Gil-Torregrosa B, Fernandez A, Balsa D, Carceller E, Gomez L, Merlos M, Cortijo J, Morcillo E, Bartroli X. UR-60427, a novel H4 receptor-inverse agonist that shows good efficacy in a rat asthma model. Inflamm Res. 2010;59 Suppl 2:S199–200.
105. Somma T, Cinci L, Formicola G, Pini A, Thurmond R, Ennis M, Bani D, Masini E. A selective antagonist of histamine H4 receptors prevents antigen-induced airway inflammation and bronchoconstriction in guinea pigs: involvement of lipocortin-1. Br J Pharmacol. 2013;170:200–13.
106. Mahapatra S, Albrecht M, Behrens B, Dittrich A-M, Jirmo A, Behrens G, Hartwig C, Neumann D, Bahre H, Raap U, Herrick C. Delineating the role of histamine-1- and -4-receptors in a mouse model of th2-dependent antigen-specific skin inflammation. PLoS One. 2014;9, e87296.
107. Rossbach K, Wendorff S, Sander K, Stark H, Gutzmer R, Werfel T, Kietzmann M, Baeumer W. Histamine H4 receptor antagonism reduces hapten-induced scratching behaviour but not inflammation. Exp Dermatol. 2009;18:57–63.

108. Baeumer W, Stahl J, Sander K, Petersen LJ, Paps J, Stark H, Kietzmann M, Olivry T. Lack of preventing effect of systemically and topically administered histamine H1 or H4 receptor antagonists in a dog model of acute atopic dermatitis. Exp Dermatol. 2011;20:577–81.
109. Kamo A, Negi O, Tengara S, Kamata Y, Noguchi A, Ogawa H, Tominaga M, Takamori K. Histamine H4 receptor antagonists ineffective against itch and skin inflammation in atopic dermatitis mouse model. J Invest Dermatol. 2014;134:546–8.
110. Coruzzi G, Pozzoli C, Adami M, Grandi D, Guido N, Smits R, de Esch I, Leurs R. Strain-dependent effects of the histamine H4 receptor antagonist JNJ7777120 in a murine model of acute skin inflammation. Exp Dermatol. 2012;21:32–7.
111. Bell JK, McQueen DS, Rees JL. Involvement of histamine $H_4$ and $H_1$ receptors in scratching induced by histamine receptor agonists in BalbC mice. Br J Pharmacol. 2004;142:374–80.
112. Cowart MD, Altenbach RJ, Liu H, Hsieh GC, Drizin I, Milicic I, Miller TR, Witte DG, Wishart N, Fix-Stenzel SR, McPherson MJ, Adair RM, Wetter JM, Bettencourt BM, Marsh KC, Sullivan JP, Honore P, Esbenshade TA, Brioni JD. Rotationally constrained 2,4-diamino-5,6-disubstituted pyrimidines: a new class of histamine H4 receptor antagonists with improved druglikeness and in vivo efficacy in pain and inflammation models. J Med Chem. 2008;51:6547–57.
113. Koenig JR, Liu H, Drizin I, Witte DG, Carr TL, Manelli AM, Milicic I, Strakhova MI, Miller TR, Esbenshade TA, Brioni JD, Cowart M. Rigidified 2-aminopyrimidines as histamine H4 receptor antagonists: effects of substitution about the rigidifying ring. Bioorg Med Chem Lett. 2010;20:1900–4.
114. Liu H, Altenbach RJ, Carr TL, Chandran P, Hsieh GC, Lewis LGR, Manelli AM, Milicic I, Marsh KC, Miller TR, Strakhova MI, Vortherms TA, Wakefield BD, Wetter JM, Witte DG, Honore P, Esbenshade TA, Brioni JD, Cowart MD. cis-4-(Piperazin-1-yl)-5,6,7a,8,9,10,11,11a-octahydrobenzofuro[2,3-h]quinazolin-2-amine (A-987306), a new histamine H4R antagonist that blocks pain responses against carrageenan-induced hyperalgesia. J Med Chem. 2008;51:7094–8.
115. Rossbach K, Nassenstein C, Gschwandtner M, Schnell D, Sander K, Seifert R, Stark H, Kietzmann M, Baeumer W. Histamine H1, H3 and H4 receptors are involved in pruritus. Neuroscience. 2011;190:89–102.
116. Kollmeier A, Francke K, Chen B, Dunford PJ, Greenspan AJ, Xia Y, Xu XL, Zhou B, Thurmond R. The H4 receptor antagonist, JNJ 39758979, is effective in reducing histamine-induced pruritus in a randomized clinical study in healthy subjects. J Pharmacol Exp Ther. 2014;350:181–7.
117. Murata Y, Song M, Kikuchi H, Hisamichi K, Xu XL, Greenspan A, Kato M, Chiou C-F, Kato T, Guzzo C, Thurmond RL, Ohtsuki M, Furue M. Phase 2a, randomized, double-blind, placebo-controlled, multicenter, parallel-group study of a H4R-antagonist (JNJ-39758979) in Japanese adults with moderate atopic dermatitis. J Dermatol. 2015;42:129–39.
118. Hanifin JM, Thurston M, Omoto M, Cherill R, Tofte SJ, Graeber M. The eczema area and severity index (EASI): assessment of reliability in atopic dermatitis. EASI Evaluator Group. Exp Dermatol. 2001;10:11–8.
119. Schram ME, Spuls PI, Leeflang MMG, Lindeboom R, Bos JD, Schmitt J. EASI, (objective) SCORAD and POEM for atopic eczema: responsiveness and minimal clinically important difference. Allergy. 2012;67:99–106.
120. Williams HC. Clinical practice. Atopic dermatitis. N Engl J Med. 2005;352:2314–24.
121. Thurmond RL, Gelfand EW, Dunford PJ. The role of histamine $H_1$ and $H_4$ receptors in allergic inflammation: the search for new antihistamines. Nat Rev Drug Discov. 2008;7:41–53.
122. Sonkoly E, Muller A, Lauerma AI, Pivarcsi A, Soto H, Kemeny L, Alenius H, Dieu-Nosjean MC, Meller S, Rieker J, Steinhoff M, Hoffmann TK, Ruzicka T, Zlotnik A, Homey B. IL-31: a new link between T cells and pruritus in atopic skin inflammation. J Allergy Clin Immunol. 2006;117:411–7.

# Chapter 4
# Identification and Roles of Zebrafish Histamine Receptors

**Maria Sundvik, Yu-Chia Chen, Henri Puttonen, and Pertti Panula**

**Abstract** Zebrafish, a small vertebrate model organism, has become a widely used tool in neuroscience. We describe the methods for genome modification and translation inhibition to produce new models. The brain histaminergic system comprises neurons in the hypothalamus and histamine receptors hrh1-hrh3, which are concentrated in specific brain regions. Their roles can be studied with several quantitative individual and social behavioral methods.

**Keywords** Model organism • Morpholino oligonucleotide • ZFN • TALEN • CRISPR/Cas9 • Behavioral analysis

## 4.1 Zebrafish as a Model Organism in Studies of Histaminergic Neurotransmitter System

### *4.1.1 Zebrafish as an Experimental Tool*

From the 1980s, zebrafish has been used in research laboratories to address fundamental questions related to vertebrate development, gene functions, mechanisms of diseases, and toxicology. Within neuroscience, the zebrafish, *Danio rerio*, has emerged as a versatile tool to understand normal neuronal development [1–3], neural basis of behavior [4–15], sensory processing [16–20], neurotransmitter pathways [15, 21–25], and disease mechanisms [17, 26–30].

The central nervous system of zebrafish shares the general characteristics of other teleostean brains [31] and is structurally very close to the vertebrate brain archetype [32]. All major neurotransmitter systems found in humans, including the dopaminergic, noradrenergic, serotonergic, histaminergic, cholinergic, glutamatergic, and GABAergic systems, are also found in zebrafish. The systems are very similar as several comparative studies have shown [33]. The basic anatomical core

M. Sundvik, Ph.D. • Y.-C. Chen, Ph.D. • H. Puttonen, M.D. • P. Panula, M.D., Ph.D. (✉)
Department of Anatomy and Neuroscience Center, University of Helsinki,
POB 63 (Haartmaninkatu 8), Helsinki 00014, Finland
e-mail: maria.sundvik@helsinki.fi; yu-chia.chen@helsinki.fi; henri.puttonen@helsinki.fi; pertti.panula@helsinki.fi



P. Blandina, M.B. Passani (eds.), *Histamine Receptors*, The Receptors 28,
DOI 10.1007/978-3-319-40308-3_4

structures in the human brain and zebrafish brain are similar, and they execute similar functions. For example, the hypothalamus and brain stem motor systems regulate hormone balance and movement, respectively. The proportional anatomical differences between the human and zebrafish are large due to the relative size of the telencephalon and mesencephalon. A major advantage of zebrafish in neuroscience studies is the fact that the intact brain can be investigated and imaged in a behavioral context at single cell level [4, 6, 7]. This allows a comprehensive analysis that has not yet been available in other vertebrate species in an organism for which there is a good selection of available genetic methods.

The histaminergic system in the zebrafish comprises the histamine-producing histidine decarboxylase-positive neurons in the caudal recess of the posterior hypothalamus and their widespread projections throughout the brain [21, 34, 35]. The dorsal telencephalon receives the densest afferent fiber innervation from the histaminergic neurons. Tract-tracing experiments showed that dorsal telencephalic target neurons project back to innervate the histaminergic neurons in the hypothalamus and thus form a telencephalic-hypothalamic loop [35]. The factors that control histamine neuron development are currently poorly understood. To date, the only known factor regulating the development of histaminergic neurons is the gene encoding the catalytic subunit of the γ-secretase, *presenilin 1* [29]. Mutations in this gene resulting in an increased production of the long form of amyloid β protein are strongly associated with the familial forms of Alzheimer's disease in humans [36, 37]. The histamine system in zebrafish has not been studied as extensively as the other aminergic systems, like the dopaminergic and 5-HT systems [24, 27, 28, 30, 38–43]. The basic features of the anatomy and function of the histaminergic system in zebrafish larvae and adults and the G protein-coupled histamine receptors have been described and will be presented in this chapter.

### 4.1.2 Molecular Tools for Zebrafish Genetic Manipulation

Morpholino oligonucleotides are short synthetic forms of DNA where the deoxyribose ring is modified to a morpholine [44]. Transient knockout of gene expression by injecting antisense morpholino oligonucleotides (AMOs) into one-to-four cell stage embryos is commonly used in the zebrafish. AMOs specifically bind to the target sequence and perturb translation or splicing of a specific mRNA resulting in reduction of the respective gene product [45]. This approach is rather useful for studying gene functions when mutants are unavailable or a knockout mutation produces a lethal effect. Kok et al. [46] reported that the majority of AMO-induced phenotypes rarely match the corresponding mutant phenotypes. Discrepancies may result from the different expression levels between knockdown and knockout or off-target effects in either morphants or mutants. It is also possible that genetic compensation occurs in null mutants that cannot be observed in knockdown effects [47]. Although the specificity of MOs is remaining questionable, off-targeting effects such as upregulation of the apoptosis factor p53 induced cell death which may mask the actual phenotypes [48]; this unexpected phenotype can be compensated by co-injections of p53 MO with gene-specific AMOs. However, AMOs may still be used if proper controls are conducted, such as rescue experiments with corresponding mRNAs or several different MOs targeting the same gene [49]. The use of AMOs is particularly useful in cases where specific tools are available to verify lack of active protein, such as an antibody against the resulting protein. In such cases, not only is it possible to verify that the target protein is lost like in the case of the neuropeptide galanin or tyrosine hydroxylase [50, 51], but the duration of active translation inhibition can also be verified. When the histamine system has been studied using AMOs for histidine decarboxylase, the antibody against histamine has been used to verify that a nearly complete lack of histamine is evident at 5 days post fertilization (dpf) and a noticeable effect at 11 dpf following a single AMO injection at 1–4 cell stages [35]. It is more difficult to apply AMOs for histamine receptors, because no specific antibodies are available for zebrafish receptor proteins.

Genome editing methods using zinc finger nucleases (ZFNs) [52], transcription activator-like effector nucleases (TALENs) [53], and clustered regularly interspaced short palindromic repeat (CRISPR) [54] systems have been successfully applied in zebrafish. ZFNs consist of two domains: a specific zinc finger DNA-binding domain containing small protein motifs, each binding to a specific DNA triplet, and a nonspecific DNA cleavage domain from FokI nuclease. When two ZFNs bind to adjacent target sites, the FokI activity is induced and leads to a double-stranded break (DSB) followed by nonhomologous end joining (NHEJ) DNA damage response that introduces insertions or deletions (Indel) in DNA causing frameshift mutations. Similar to the ZFN mechanism, TALENs are composed of the FokI endonuclease and a site-specific target domain. Most recently, the CRISPR system modified from the bacterial immune system, in which RNA strands guide CRISPR-associated 9 (Cas9) nuclease to cleave foreign DNA, has been actively used in zebrafish. It simply requires co-injections of the RNA encoding Cas9 endonuclease and a 20-nt guide RNA matching the genomic sequence of interest to introduce inherent mutations and germline transmission stability [55]. The bound single guide RNA guides the Cas9 nuclease to the desired target sites leading to the DSB and NHEJ causing Indel mutations. It is more efficient than TALENs and ZFINs. To avoid potential off-target effects of the CRISPR/Cas9 system, the target sequence must be chosen as unique as possible in the whole genome since the binding specificity is directed by only 20-bp sequence of RNA. Alternatively, the Cas9 seems to improve the specificity of cleavages [56]. Moreover, by constructing the promoters and combing the Tol2 system with the CRISPR/Cas9 system to generate heritable and tissue-specific knockouts, new tools are also available in zebrafish [57].

### 4.1.3 Behavioral Methods for Zebrafish

Behavioral methods for zebrafish now comprise, e.g., detailed analyses of components of sensorimotor reflexes and motor regulation including touch response and prepulse inhibition [13, 58–61], all essential aspects of swimming patterns, social behavior and shoaling [62–64], cognitive behavior [65, 66], olfactory behavior [67], and sleep [68, 69]. Initially, locomotion and particularly its visual regulation have been extensively studied due to availability of methods and the obvious importance of vision in fish behavior [4, 7, 70, 71]. Until now, receptor function using specific receptor ligands has been assessed rather seldom for individual receptors using systematic analysis of different ligands for individual receptors. Sets of antipsychotic and psychoactive drugs have been used to study locomotor activity and memory [72, 73], and a number of receptor ligands have been included in behavioral screens [74, 75].

In a standard locomotor assay, adult or larval fish are tracked in arenas or wells of a multiwell plate (different sized wells can be used successfully), and their movements are video recorded most commonly with video camera and automated computerized software [76]. The recorded coordinates can then be analyzed quantitatively to allow assessment of parameters such as total distance moved (cm), turn angle (degrees), angular velocity (degrees/s), and meander (degrees/cm). If the arenas are divided into zones or subregions, the time spent in different zones can be analyzed to allow the assessment of place preference within the arena.

To study the dark-induced flash response, infrared light can be used to illuminate the arena, and visible light is used as a cue. Turning off the lights for a short period of time (e.g., 2 min) elicits increased activity in wild-type fish during the initial seconds of sudden darkness [35]. This is likely to be similar to the O-bend [8, 9] that the fish exhibit when they navigate away from the darkness toward the perceived light source.

## 4.2 Structure and Function of G Protein-Coupled Receptors Including Histamine Receptors in Zebrafish

### *4.2.1 G Protein-Coupled Receptors in Zebrafish*

G protein-coupled receptors (GPCRs) in vertebrates can in general be divided into five subfamilies, glutamate (Class C), rhodopsin (Class A), adhesion, frizzled, and secretin (Class B) families, which are further divided in subfamilies. A decade ago, the total number of *rhodopsin* GPCRs in zebrafish was reported to be 591 and to constitute the majority of GPCRs in the zebrafish [77]. The rhodopsin GPCRs have been classified into 13 branches [78], and of these nearly 600 predicted GPCRs in zebrafish, 122 belong to the AMIN group, encoding the GPCRs binding biogenic amines such as serotonin, dopamine, histamine, trace amines, adrenaline, noradrenaline, and acetylcholine; 35 encode receptors for phospholipids, melanocortin, cannabinoids, and somatostatin receptors together with 3 orphan GPCRs; 6 encode melatonin and orphan receptor GPR50; 31 encode opsin/putative opsin receptors and orphan receptors GPR21 and GPR52; 22 prostaglandin receptors and orphan receptors SREB 1-3, GPR26, GPR62, and GPR78; 67 receptors for NPY, tachykinins, neurotensin, hypocretin/orexin, neuromedin, NPFF, PrRP, GnRH, CCK, etc.; 77 bradykinin receptors and receptors/putative receptors for chemokines; 4

receptors for melanocyte-concentrating hormone; 34 somatostatin, opsin, and galanin receptors; 8 orphan LGR receptors and receptors for relaxin, FSH, TSH, and LH; none for MRG and MAS receptors; 31 for olfactory receptors [79]; 80 purine/putative purine receptors, formyl-peptide receptors, retinoic acid receptors, and orphan GPCRs; and finally 74 unclassified receptors [77].

A closer look at the AMIN branch in zebrafish shows that there are five alpha2-adrenoceptors that mediate the effects of adrenaline and noradrenaline [80] compared with three found in mammals. In mammals, at least five dopamine receptors are acknowledged [81], of which D1–D4 have also been identified and described in zebrafish [82–84]. Of the 14 serotonin receptors in humans [85], fewer than a handful have been studied in zebrafish [86, 87].

### 4.2.2 Zebrafish Genome Duplication and the Histaminergic System

The quest to identify histamine receptors in zebrafish was initiated at the turn of the century to find out if all four G protein-coupled receptors are found in zebrafish. The focus was initially on the histamine receptor 3 (*hrh3*) [88], which had been identified as a key regulator of the histaminergic and several other transmitter systems in the brain [89]. A more comprehensive study of the other histamine receptor subtypes was then undertaken [76]. At the time of these studies, the entire zebrafish genome had not yet been sequenced. This caveat has now been overcome, which allows a genome-wide analysis of all receptors. The latest reference assembly of the zebrafish genome is the Genome Reference Consortium z10 (GRCz10) [90]. Similarities between the zebrafish and human genomes are significant: a comparison between the two genomes shows that about 70 % of the human genes have at least one zebrafish orthologue [90]. The zebrafish genome has undergone an additional, teleost fish-specific genome duplication event [91–95], and for some genes, there are two (or more) orthologues that correspond to one single human gene. Examples of this phenomenon are duplicated tyrosine hydroxylase (*th*) [51, 96] and adrenergic alpha2Da and alpha2Db receptors, which share the distribution patterns of zebrafish and mammalian alpha2A receptor [80]. The two *th* genes encoding tyrosine hydroxylases are expressed in a complementary manner in zebrafish brain and share the functions of the single mammalian gene [51, 97]. In contrast to zebrafish th, the histamine-synthesizing enzyme L-histidine decarboxylase is not duplicated [34, 35]. It is expressed in the zebrafish brain in the posterior recess of the hypothalamus, which corresponds to the location of histaminergic neurons in all vertebrates studied so far [34, 98–101], and the projections in the brain are also very similar in zebrafish and mammals [21, 100]. Three of the four vertebrate histamine GPCRs (*hrh1–hrh3*) have so far been identified in zebrafish [76, 88], whereas *hrh4* has not been reported on in any publication (Fig. 4.1). No detailed signaling or receptor radioligand binding data has yet been published for any of these receptors expressed in cell lines.

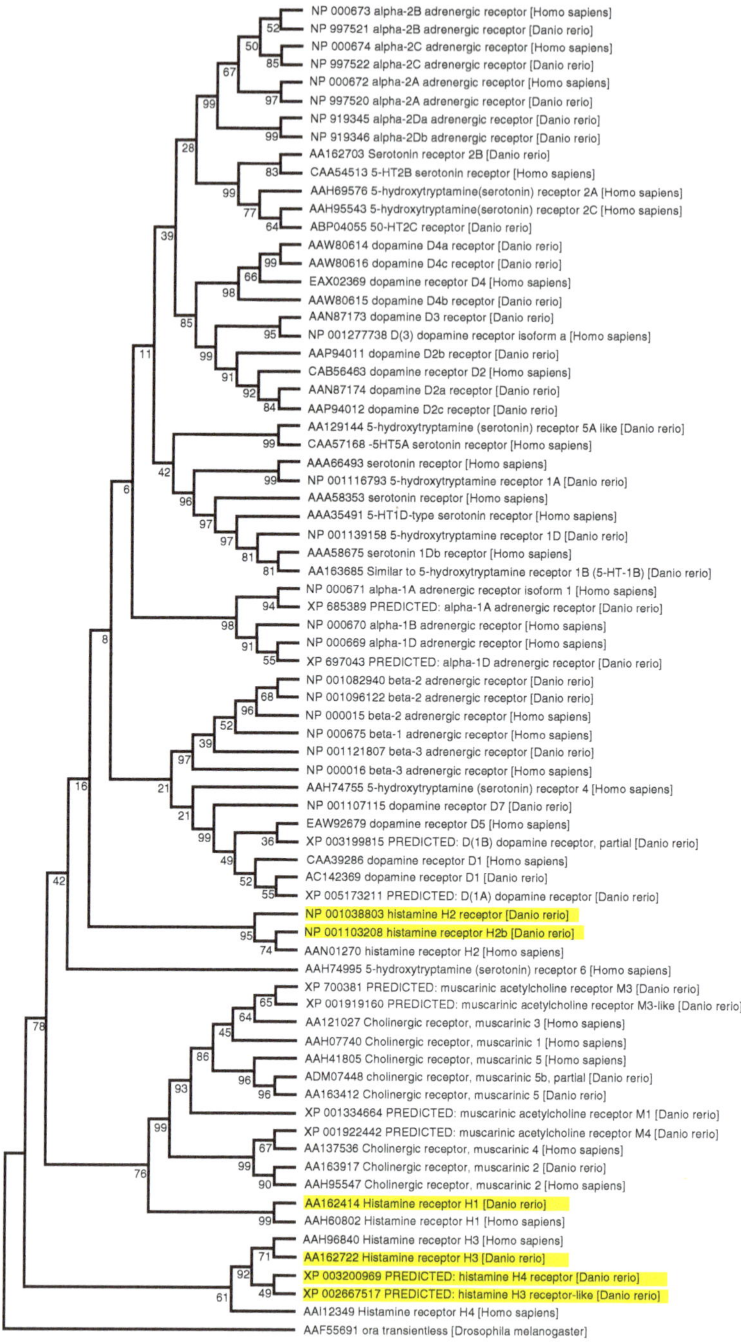


**Fig. 4.1** A phylogenetic tree of human and zebrafish biogenic amine receptor groups including adrenoceptors, serotonin, muscarinic cholinergic, dopamine, and histamine receptors. Zebrafish histamine receptors are highlighted in *yellow*. The tree was reconstructed using the neighbor-joining method with the program MEGA 6 based on the protein sequence alignment by the program Clustal X. Bootstrap values are shown on the internal nodes. Protein sequences were retrieved from GenBank (http://www.ncbi.nlm.nih.gov/genbank/, last accessed on November 30, 2015). The fruit fly ora transientless, a histamine-gated chloride channel, is used as an out-group

#### 4.2.2.1 Histamine Receptor 1, hrh1

Histamine receptor 1, *hrh1*, was identified and cloned from the zebrafish brain about a decade ago [76]. The zebrafish *hrh1* gene is an intronless gene encoding a 534-amino acid-long protein on chromosome 8. The human *hrh1* receptor gene is 57 amino acids shorter, and the canine *hrh1* receptor gene is 2 amino acids longer than the zebrafish *hrh*1 gene. Compared to the human *hrh1* gene, the zebrafish *hrh1* has a 20-amino acid extension in the 5′ end and amino acid inserts in the third intracellular loop (IC) compared to human *hrh1*. Interestingly, the zebrafish hrh1 protein has (or the gene that encodes for it) a second methionine 17 amino acids downstream from the initial methionine, which we suggested could be the real translation site. Sequence alignment analysis shows a 40–46 % identity between the mammalian and zebrafish *hrh1* receptor.

The *hrh1* transcript can be detected by RT-PCR already at 3 h post fertilization (hpf) in the zebrafish whole embryo, and in adulthood, *hrh1* transcript is detected in the brain, intestine, liver, and spleen [76]. The *hrh1* receptor mRNA is expressed most prominently in the dorsal telencephalon and anterior hypothalamus of larval zebrafish [35]. In contrast to mammals, there is no known common source of histamine in peripheral organs of the zebrafish [34]. Mast cells in zebrafish do not express histidine decarboxylase or contain histamine according to current knowledge. In mice, the *hrh1* mRNA can be detected mainly in the isocortex and, to some extent in the septum, hypothalamus and pons [102]. In humans, the *hrh1* receptor mRNA can be detected in a laminar manner in the cortex layers V and VI and receptor binding in layers III and IV [103] (Fig. 4.2).

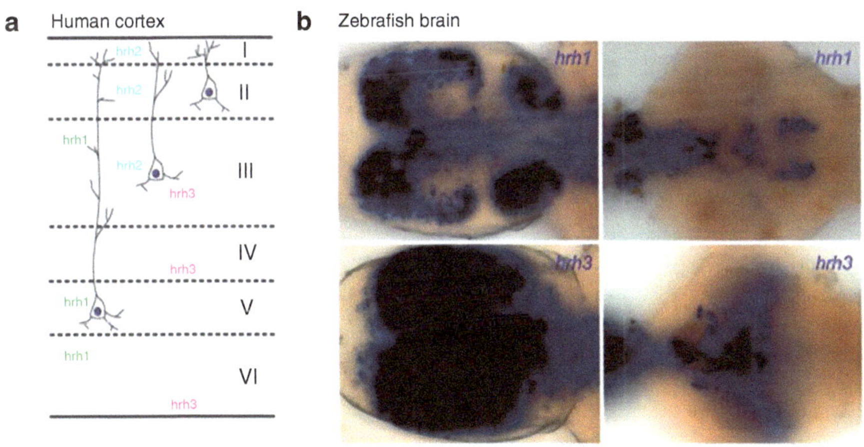


**Fig. 4.2** The distribution of histaminergic receptors in human and zebrafish brain. (**a**) The histamine receptors are distributed in a laminar manner in the human cerebral cortex. Here, both the receptor binding and mRNA distribution of each receptor are depicted as follows: hrh1 in layers III, V, and VI, hrh2 in layers I–III, and finally hrh3 in layers III, IV, and VI. Pyramidal neurons are located in layers II, III, and V. (**b**) In the zebrafish brain, the *hrh1* and *hrh3* mRNAs are expressed in the dorsal telencephalon; additionally, *hrh1* can be detected in the zebrafish habenula, anterior diencephalon, and locus coeruleus and *hrh3* in the histaminergic posterior hypothalamus, optic tectum, and anterior diencephalon

Histamine receptor 1 is the histamine receptor studied most widely in zebrafish. Pharmacological antagonism of the receptor has a sedative effect on the developing zebrafish larvae when compared to nontreated wild-type animals, as it affects locomotion dose dependently by reducing total distance moved and movement [76], increases rest state in the sleep-wake cycle [74], and impairs the response to sudden changes in the environment [35]. Inhibition of the hrh1 activity by pyrilamine resulted in a lower number of neuropeptidergic hypocretin/orexin-positive neurons in the anterior hypothalamus [35]. In humans, the functions of hrh1 have been studied in more detail, and inhibition of its function is mainly antiallergic and sedative [89]. Antagonism of hrh1 has also been shown to induce weight gain [104]. In rodents, similar features are reported; additionally, the effect of hrh1 knockout has been linked to the functioning of the hypocretin/orexin system [105] and impairments in spatial and fear-associated learning and memory [106, 107].

#### 4.2.2.2 Histamine Receptor 2, hrh2

The zebrafish histamine receptor 2, *hrh2*, was identified at the same time as *hrh1* [76]. This receptor is also encoded by an intronless gene, but located on chromosome 15. Sequence analysis shows 43–47 % identity on the peptide level to the corresponding canine, murine, and human hrh2 proteins. The zebrafish hrh2 is 51 amino acids longer than the human hrh2 receptor, with a 27-amino acid-longer 5′ end, a 6-amino acid-longer third IC, and a 13-amino acid-longer 3′ end. In the latest reference assembly, there are two annotations for *hrh2*: (1) *hrh2a* identical with the published *hrh2* and (2) a recently discovered *hrh2b* with 53 % protein similarity with human *hrh2* isoform 2 (Fig. 4.1).

The *hrh2* transcript can be detected by RT-PCR from 3 hpf onward in the zebrafish embryo and in adulthood in tissues such as the brain, gills, heart, and spleen [76]. The distribution of hrh2 has been analyzed by binding of $^{[125I]}$ aminopotentidine and was detected in most parts of the zebrafish brain. Highest densities of binding were found in the optic tectum, hypothalamus, locus coeruleus, and the superior reticular formation [76]. To date, *hrh2* in situ hybridization has been unsuccessful in zebrafish, most likely because of its even distribution within the brain and generally low expression level

In mice, the *hrh2* mRNA can be detected mainly in the olfactory areas and to some very minor extent in the cortex. Other parts of the brain seem to be devoid of *hrh2* receptor expression [102]. In humans, the *hrh2* receptor mRNA can be detected in a laminar manner in the cortex layer II, and higher-density receptor binding sites have been seen in the superficial layers I–III [108] (Fig. 4.2).

Histamine receptor 2 has been implicated in aggressive behaviors in adult zebrafish, since it is overexpressed in the hypothalamus of dominant male and female zebrafish [109]. Interestingly, the hrh2 also plays another role, tentatively in modulation of the fear response, as it is overexpressed in the telencephalon that harbors the amygdala, in subordinate male zebrafish [109]. Pharmacological inhibition of hrh2 has a sedative effect on zebrafish larvae, as it reduces locomotion of the treated fish when compared with nontreated siblings [76]. The hrh2 knockout mice exhibit severe cognitive impairment in tasks for fear-associated learning and memory, showing that also in mice the hrh2 mediates fear-associated behaviors [106]. In humans, hrh2 is best known for its role in gastric acid secretion, which made its inverse agonists blockbuster drugs in the 1970s [110], and antagonism of the receptor has also been shown to improve the negative symptoms of schizophrenic patients [111].

#### 4.2.2.3 Histamine Receptor 3, hrh3

The zebrafish histamine receptor 3, *hrh3*, was identified already at the turn of the century [88]. Histamine receptor 3 was initially identified as an autoreceptor regulating histamine synthesis and release in different brain areas [112], and later on it was shown to regulate the release of several neurotransmitters apart from histamine, such as the serotonin, noradrenaline, dopamine, and acetylcholine [89] in mammals. The gene encoding zebrafish *hrh3* is located on chromosome 7 and differs significantly from *hrhr1* and *hrh2*, as it contains three exons. The gene consists of 1422 nucleotides and has two possible translation start sites, of which the latter (nucleotides [101–103]) is most likely the active one. The initial exon covers sequence from the start site and until the middle part of transmembrane domain 2, and the second exon covers the sequence until intercellular loop 2. The third and final exon contains the rest of the coding sequence. Sequence comparison analysis shows a 42–50 % identity similarity between the zebrafish, mouse, rat, dog, and human sequences.

The expression of *hrh3* transcript can be detected as early as 11 hpf in the developing zebrafish embryo, and in adult tissues, the highest expression is found in the brain, with lower levels in both the heart and spleen [76]. mRNA expression detection by in situ hybridization of *hrh3* is strongest in the dorsal telencephalon of zebrafish larvae, and a very faint expression of the receptor mRNA is also detected in the histaminergic neurons of the posterior hypothalamus [35] (Fig. 4.2). In mice, the *hrh3* mRNA is widely expressed and can be found in high intensities in the isocortex, olfactory areas, hippocampal formation, cortical subplate, striatum, pallidum, thalamus, hypothalamus, midbrain, and cerebellum [102]. In humans, the *hrh3* receptor mRNA can be detected in a laminar manner in the cortex layer V, and higher densities of receptor binding sites were seen in the middle layers III and IV, similarly as was for *hrh1* receptor binding [108] (Fig. 4.2).

In zebrafish, pharmacological treatment by both hrh3 agonist and antagonist has the same sedative effect on locomotion [76], the reason why the properties of this receptor should be studied in more detail. No other published studies have addressed the role of this receptor in zebrafish. In rodents, hrh3 regulates alcohol drinking in at least three different drinking paradigms, and antagonists inhibit the rewarding effects of alcohol [113, 114]. Dopamine and histamine receptor dimerization [115, 116] is thought to play an essential role in mediating this behavior [117]. Similar studies have not been conducted in zebrafish. Hrh3 is the first histamine receptor that has been studied primarily for treatment of a central nervous system-derived disease, narcolepsy [118].

Moreover, there are additional transcripts for the *hrh3* receptor that have not been characterized in detail and reported on yet (Fig. 4.1). In mammals, the *hrh3* receptor has many splice variants [119, 120], and these additional transcripts in zebrafish might correspond to the reported splice variants in mammals.

#### 4.2.2.4 Histamine Receptor 4, hrh4

Histamine receptor 4, *hrh4*, is annotated in the most recent version of the published zebrafish genome assembly. However, the identity of protein is more similar with the human hrh3 (76 %) than human hrh4 (42 %). Therefore, it is likely that there is no true *hrh4* in the zebrafish (Fig. 4.1). Signaling and ligand profiling studies are needed to clarify the functions of these hrh3-like receptors.

## 4.3 Future Directions

For further studies, the zebrafish histamine receptors need to be expressed in cell lines to establish the ligand binding properties and signaling characteristics of all receptors. Signaling properties of, e.g., hrh3 are highly variable in different species, and agonist/antagonist properties of ligands also differ. When these studies are available, zebrafish can be used to pharmacologically assess the behavioral roles of histamine receptors. Using the CRISPR/Cas9 system for production of gene-modified zebrafish is now straightforward and widely applied, and new models lacking histamine receptors and other components of the histaminergic system are likely to emerge in the near future.

## 4.4 Concluding Remarks

Many aspects of histamine receptor function can be studied using zebrafish as a tool. The similarities of histamine receptors with those of mammals have rendered these studies relevant to assess the general properties of the central histaminergic system, because the mast cells of zebrafish do not contain histamine. New genetic tools are currently making development of new zebrafish models possible and cost-effective. The possibility to carry out large-scale screening experiments will open important new approaches in drug development and assessment of the role of histamine in brain functions.

## References

1. Haffter P, et al. The identification of genes with unique and essential functions in the development of the zebrafish, Danio rerio. Development. 1996;123:1.
2. Fetcho JR, Liu KS. Zebrafish as a model system for studying neuronal circuits and behavior. Ann N Y Acad Sci. 1998;860:333.
3. Machluf Y, Gutnick A, Levkowitz G. Development of the zebrafish hypothalamus. Ann N Y Acad Sci. 2011;1220:93.
4. Ahrens MB, et al. Brain-wide neuronal dynamics during motor adaptation in zebrafish. Nature. 2012;485:471.

5. Douglass AD, Kraves S, Deisseroth K, Schier AF, Engert F. Escape behavior elicited by single, channelrhodopsin-2-evoked spikes in zebrafish somatosensory neurons. Curr Biol. 2008;18:1133.
6. Naumann EA, Kampff AR, Prober DA, Schier AF, Engert F. Monitoring neural activity with bioluminescence during natural behavior. Nat Neurosci. 2010;13:513.
7. Orger MB, Kampff AR, Severi KE, Bollmann JH, Engert F. Control of visually guided behavior by distinct populations of spinal projection neurons. Nat Neurosci. 2008;11:327.
8. Burgess HA, Granato M. Modulation of locomotor activity in larval zebrafish during light adaptation. J Exp Biol. 2007;210:2526.
9. Burgess HA, Granato M. Sensorimotor gating in larval zebrafish. J Neurosci. 2007;27:4984.
10. Burgess HA, Johnson SL, Granato M. Unidirectional startle responses and disrupted left-right co-ordination of motor behaviors in robo3 mutant zebrafish. Genes Brain Behav. 2009;8(500).
11. Burgess HA, Schoch H, Granato M. Distinct retinal pathways drive spatial orientation behaviors in zebrafish navigation. Curr Biol. 2010;20:381.
12. Budick SA, O'Malley DM. Locomotor repertoire of the larval zebrafish: swimming, turning and prey capture. J Exp Biol. 2000;203:2565.
13. Fetcho JR, O'Malley DM. Visualization of active neural circuitry in the spinal cord of intact zebrafish. J Neurophysiol. 1995;73:399.
14. McElligott MB, O'Malley DM. Prey tracking by larval zebrafish: axial kinematics and visual control. Brain Behav Evol. 2005;66(177).
15. O'Malley DM, Kao YH, Fetcho JR. Imaging the functional organization of zebrafish hindbrain segments during escape behaviors. Neuron. 1996;17:1145.
16. Baier H, Rotter S, Korsching S. Connectional topography in the zebrafish olfactory system: random positions but regular spacing of sensory neurons projecting to an individual glomerulus. Proc Natl Acad Sci U S A. 1994;91:11646.
17. Bandmann O, Burton EA. Genetic zebrafish models of neurodegenerative diseases. Neurobiol Dis. 2010;40:58.
18. Friedrich RW. Calcium imaging in the intact olfactory system of zebrafish and mouse. Cold Spring Harb Protoc. 2014;2014:310.
19. Friedrich RW, Korsching SI. Combinatorial and chemotopic odorant coding in the zebrafish olfactory bulb visualized by optical imaging. Neuron. 1997;18:737.
20. Gabriel JP, Trivedi CA, Maurer CM, Ryu S, Bollmann JH. Layer-specific targeting of direction-selective neurons in the zebrafish optic tectum. Neuron. 2012;76:1147.
21. Kaslin J, Panula P. Comparative anatomy of the histaminergic and other aminergic systems in zebrafish (Danio rerio). J Comp Neurol. 2001;440:342.
22. Panula P, et al. The comparative neuroanatomy and neurochemistry of zebrafish CNS systems of relevance to human neuropsychiatric diseases. Neurobiol Dis. 2010;40:46.
23. McLean DL, Fetcho JR. Using imaging and genetics in zebrafish to study developing spinal circuits in vivo. Dev Neurobiol. 2008;68:817.
24. McLean DL, Fetcho JR. Relationship of tyrosine hydroxylase and serotonin immunoreactivity to sensorimotor circuitry in larval zebrafish. J Comp Neurol. 2004;480:57.
25. Ono F, Higashijima S, Shcherbatko A, Fetcho JR, Brehm P. Paralytic zebrafish lacking acetylcholine receptors fail to localize rapsyn clusters to the synapse. J Neurosci. 2001;21:5439.
26. Leimer U, et al. Zebrafish (Danio rerio) presenilin promotes aberrant amyloid beta-peptide production and requires a critical aspartate residue for its function in amyloidogenesis. Biochemistry. 1999;38:13602.
27. Sallinen V, et al. Hyperserotonergic phenotype after monoamine oxidase inhibition in larval zebrafish. J Neurochem. 2009;109:403.
28. Sallinen V, et al. MPTP and MPP+ target specific aminergic cell populations in larval zebrafish. J Neurochem. 2009;108:719.
29. Sundvik M, Chen YC, Panula P. Presenilin1 regulates histamine neuron development and behavior in zebrafish, danio rerio. J Neurosci. 2013;33:1589.
30. Sallinen V, et al. Dopaminergic cell damage and vulnerability to MPTP in Pink1 knockdown zebrafish. Neurobiol Dis. 2010;40:93.

31. Nieuwenhuys R, ten Donkelar HJ, Nicholson C. The central nervous system of vertebrates. Berlin: Springer; 1998.
32. Striedter GF. Principles of brain evolution. Sunderland: Sinauer Associates; 2005.
33. Panula P, et al. Modulatory neurotransmitter systems and behavior: towards zebrafish models of neurodegenerative diseases. Zebrafish. 2006;3:235.
34. Eriksson KS, Peitsaro N, Karlstedt K, Kaslin J, Panula P. Development of the histaminergic neurons and expression of histidine decarboxylase mRNA in the zebrafish brain in the absence of all peripheral histaminergic systems. Eur J Neurosci. 1998;10:3799.
35. Sundvik M, et al. The histaminergic system regulates wakefulness and orexin/hypocretin neuron development via histamine receptor H1 in zebrafish. FASEB J. 2011;25(12):4338–47.
36. Campion D, et al. Mutations of the presenilin I gene in families with early-onset Alzheimer's disease. Hum Mol Genet. 1995;4:2373.
37. Borchelt DR, et al. Familial Alzheimer's disease-linked presenilin 1 variants elevate Abeta1-42/1-40 ratio in vitro and in vivo. Neuron. 1996;17:1005.
38. McLean DL, Fetcho JR. Ontogeny and innervation patterns of dopaminergic, noradrenergic, and serotonergic neurons in larval zebrafish. J Comp Neurol. 2004;480:38.
39. Rink E, Wullimann MF. Connections of the ventral telencephalon (subpallium) in the zebrafish (Danio rerio). Brain Res. 2004;1011:206.
40. Rink E, Wullimann MF. The teleostean (zebrafish) dopaminergic system ascending to the subpallium (striatum) is located in the basal diencephalon (posterior tuberculum). Brain Res. 2001;889:316.
41. Ryu S, et al. Orthopedia homeodomain protein is essential for diencephalic dopaminergic neuron development. Curr Biol. 2007;17:873.
42. Scharer YP, Shum J, Moressis A, Friedrich RW. Dopaminergic modulation of synaptic transmission and neuronal activity patterns in the zebrafish homolog of olfactory cortex. Front Neural Circuits. 2012;6:76.
43. Thirumalai V, Cline HT. Endogenous dopamine suppresses initiation of swimming in prefeeding zebrafish larvae. J Neurophysiol. 2008;100:1635.
44. Summerton J, Weller D. Morpholino antisense oligomers: design, preparation, and properties. Antisense Nucleic Acid Drug Dev. 1997;7:187.
45. Bill BR, Petzold AM, Clark KJ, Schimmenti LA, Ekker SC. A primer for morpholino use in zebrafish. Zebrafish. 2009;6:69.
46. Kok FO, et al. Reverse genetic screening reveals poor correlation between morpholino-induced and mutant phenotypes in zebrafish. Dev Cell. 2015;32:97.
47. Rossi A, et al. Genetic compensation induced by deleterious mutations but not gene knockdowns. Nature. 2015;524:230.
48. Robu ME, et al. p53 activation by knockdown technologies. PLoS Genet. 2007;3, e78.
49. Eisen JS, Smith JC. Controlling morpholino experiments: don't stop making antisense. Development. 2008;135:1735.
50. Podlasz P, et al. Galanin gene expression and effects of its knock-down on the development of the nervous system in larval zebrafish. J Comp Neurol. 2012;520:3846.
51. Chen YC, Priyadarshini M, Panula P. Complementary developmental expression of the two tyrosine hydroxylase transcripts in zebrafish. Histochem Cell Biol. 2009;132:375.
52. Beerli RR, Barbas 3rd CF. Engineering polydactyl zinc-finger transcription factors. Nat Biotechnol. 2002;20:135.
53. Boch J, et al. Breaking the code of DNA binding specificity of TAL-type III effectors. Science. 2009;326:1509.
54. Hwang WY, et al. Efficient genome editing in zebrafish using a CRISPR-Cas system. Nat Biotechnol. 2013;31:227.
55. Gonzales AP, Yeh JR. Cas9-based genome editing in zebrafish. Methods Enzymol. 2014;546:377.
56. Cho SW, et al. Analysis of off-target effects of CRISPR/Cas-derived RNA-guided endonucleases and nickases. Genome Res. 2014;24:132.
57. Ablain J, Durand EM, Yang S, Zhou Y, Zon LI. A CRISPR/Cas9 vector system for tissue-specific gene disruption in zebrafish. Dev Cell. 2015;32:756.

58. Eaton RC, Farley RD, Kimmel CB, Schabtach E. Functional development in the Mauthner cell system of embryos and larvae of the zebra fish. J Neurobiol. 1977;8:151.
59. Saint-Amant L, Drapeau P. Synchronization of an embryonic network of identified spinal interneurons solely by electrical coupling. Neuron. 2001;31:1035.
60. Saint-Amant L, Drapeau P. Motoneuron activity patterns related to the earliest behavior of the zebrafish embryo. J Neurosci. 2000;20:3964.
61. Saint-Amant L, Drapeau P. Time course of the development of motor behaviors in the zebrafish embryo. J Neurobiol. 1998;37:622.
62. Larson ET, O'Malley DM, Melloni Jr RH. Aggression and vasotocin are associated with dominant-subordinate relationships in zebrafish. Behav Brain Res. 2006;167:94.
63. Perez-Escudero A, Vicente-Page J, Hinz RC, Arganda S, de Polavieja GG. idTracker: tracking individuals in a group by automatic identification of unmarked animals. Nat Methods. 2014;11:743.
64. Pavlidis M, Sundvik M, Chen YC, Panula P. Adaptive changes in zebrafish brain in dominant-subordinate behavioral context. Behav Brain Res. 2011;225(2):529–37.
65. Levin ED, Cerutti DT. Behavioral neuroscience of zebrafish. In: Buccafusco JJ, editor. Cerutti methods of behavior analysis in neuroscience. Boca Raton: CRC Press/Taylor & Francis; 2009.
66. Peitsaro N, Kaslin J, Anichtchik OV, Panula P. Modulation of the histaminergic system and behaviour by alpha-fluoromethylhistidine in zebrafish. J Neurochem. 2003;86:432.
67. Vitebsky A, Reyes R, Sanderson MJ, Michel WC, Whitlock KE. Isolation and characterization of the laure olfactory behavioral mutant in the zebrafish, Danio rerio. Dev Dyn. 2005;234:229.
68. Zhdanova IV. Sleep in zebrafish. Zebrafish. 2006;3:215.
69. Prober DA, Rihel J, Onah AA, Sung RJ, Schier AF. Hypocretin/orexin overexpression induces an insomnia-like phenotype in zebrafish. J Neurosci. 2006;26:13400.
70. Brockerhoff SE, et al. A behavioral screen for isolating zebrafish mutants with visual system defects. Proc Natl Acad Sci U S A. 1995;92:10545.
71. Neuhauss SC, et al. Genetic disorders of vision revealed by a behavioral screen of 400 essential loci in zebrafish. J Neurosci. 1999;19:8603.
72. Darland T, Dowling JE. Behavioral screening for cocaine sensitivity in mutagenized zebrafish. Proc Natl Acad Sci U S A. 2001;98:11691.
73. Giacomini NJ, Rose B, Kobayashi K, Guo S. Antipsychotics produce locomotor impairment in larval zebrafish. Neurotoxicol Teratol. 2006;28:245.
74. Rihel J, et al. Zebrafish behavioral profiling links drugs to biological targets and rest/wake regulation. Science. 2010;327:348.
75. MacRae CA, Peterson RT. Zebrafish as tools for drug discovery. Nat Rev Drug Discov. 2015;14:721.
76. Peitsaro N, Sundvik M, Anichtchik OV, Kaslin J, Panula P. Identification of zebrafish histamine H1, H2 and H3 receptors and effects of histaminergic ligands on behavior. Biochem Pharmacol. 2007;73:1205.
77. Fredriksson R, Schioth HB. The repertoire of G-protein-coupled receptors in fully sequenced genomes. Mol Pharmacol. 2005;67:1414.
78. Fredriksson R, Lagerstrom MC, Lundin LG, Schioth HB. The G-protein-coupled receptors in the human genome form five main families. Phylogenetic analysis, paralogon groups, and fingerprints. Mol Pharmacol. 2003;63:1256.
79. Zozulya S, Echeverri F, Nguyen T. The human olfactory receptor repertoire. Genome Biol. 2001;2:RESEARCH0018.
80. Ruuskanen JO, Peitsaro N, Kaslin JV, Panula P, Scheinin M. Expression and function of alpha-adrenoceptors in zebrafish: drug effects, mRNA and receptor distributions. J Neurochem. 2005;94:1559.
81. Sokoloff P, Schwartz JC. Novel dopamine receptors half a decade later. Trends Pharmacol Sci. 1995;16:270.
82. Yamamoto K, et al. Evolution of dopamine receptor genes of the D1 class in vertebrates. Mol Biol Evol. 2013;30:833.

83. Boehmler W, et al. D4 Dopamine receptor genes of zebrafish and effects of the antipsychotic clozapine on larval swimming behaviour. Genes Brain Behav. 2007;6:155.
84. Boehmler W, et al. Evolution and expression of D2 and D3 dopamine receptor genes in zebrafish. Dev Dyn. 2004;230:481.
85. McCorvy JD, Roth BL. Structure and function of serotonin G protein-coupled receptors. Pharmacol Ther. 2015;150:129.
86. Norton WH, Folchert A, Bally-Cuif L. Comparative analysis of serotonin receptor (HTR1A/HTR1B families) and transporter (slc6a4a/b) gene expression in the zebrafish brain. J Comp Neurol. 2008;511:521.
87. Schneider H, et al. Cloning and expression of a zebrafish 5-HT(2C) receptor gene. Gene. 2012;502:108.
88. Peitsaro N, Anichtchik OV, Panula P. Identification of a histamine H(3)-like receptor in the zebrafish (Danio rerio) brain. J Neurochem. 2000;75:718.
89. Panula P, et al. International union of basic and clinical pharmacology. XCVIII histamine receptors. Pharmacol Rev. 2015;67:601.
90. Howe K, et al. The zebrafish reference genome sequence and its relationship to the human genome. Nature. 2013;496:498.
91. Postlethwait J, Amores A, Force A, Yan YL. The zebrafish genome. Methods Cell Biol. 1999;60:149.
92. Woods IG, et al. A comparative map of the zebrafish genome. Genome Res. 2000;10:1903.
93. Catchen JM, Braasch I, Postlethwait JH. Conserved synteny and the zebrafish genome. Methods Cell Biol. 2011;104:259.
94. Cresko WA, et al. Genome duplication, subfunction partitioning, and lineage divergence: Sox9 in stickleback and zebrafish. Dev Dyn. 2003;228:480.
95. Amores A, et al. Zebrafish hox clusters and vertebrate genome evolution. Science. 1998;282:1711.
96. Candy J, Collet C. Two tyrosine hydroxylase genes in teleosts. Biochim Biophys Acta. 2005;1727:35.
97. Semenova SA, Chen YC, Zhao X, Rauvala H, Panula P. The tyrosine hydroxylase 2 (TH2) system in zebrafish brain and stress activation of hypothalamic cells. Histochem Cell Biol. 2014;142:619.
98. Airaksinen MS, Panula P. The histaminergic system in the guinea pig central nervous system: an immunocytochemical mapping study using an antiserum against histamine. J Comp Neurol. 1988;273:163.
99. Panula P, Airaksinen MS, Pirvola U, Kotilainen E. A histamine-containing neuronal system in human brain. Neuroscience. 1990;34:127.
100. Panula P, et al. Histamine-immunoreactive nerve fibers in the mammalian spinal cord. Brain Res. 1989;484:234.
101. Panula P, Pirvola U, Auvinen S, Airaksinen MS. Histamine-immunoreactive nerve fibers in the rat brain. Neuroscience. 1989;28:585.
102. Lein ES, et al. Genome-wide atlas of gene expression in the adult mouse brain. Nature. 2007;445:168.
103. Jin CY, Panula P. The laminar histamine receptor system in human prefrontal cortex suggests multiple levels of histaminergic regulation. Neuroscience. 2005;132:137.
104. He M, Deng C, Huang XF. The role of hypothalamic H1 receptor antagonism in antipsychotic-induced weight gain. CNS Drugs. 2013;27:423.
105. Lin L, et al. Measurement of hypocretin/orexin content in the mouse brain using an enzyme immunoassay: the effect of circadian time, age and genetic background. Peptides. 2002;23:2203.
106. Dai H, et al. Selective cognitive dysfunction in mice lacking histamine H1 and H2 receptors. Neurosci Res. 2007;57:306.
107. Zlomuzica A, Ruocco LA, Sadile AG, Huston JP, Dere E. Histamine H1 receptor knockout mice exhibit impaired spatial memory in the eight-arm radial maze. Br J Pharmacol. 2009;157:86.
108. Jin C, Lintunen M, Panula P. Histamine H(1) and H(3) receptors in the rat thalamus and their modulation after systemic kainic acid administration. Exp Neurol. 2005;194:43.

109. Filby AL, Paull GC, Hickmore TF, Tyler CR. Unravelling the neurophysiological basis of aggression in a fish model. BMC Genomics. 2010;11:498.
110. Haas H, Panula P. The role of histamine and the tuberomamillary nucleus in the nervous system. Nat Rev Neurosci. 2003;4:121.
111. Meskanen K, et al. A randomized clinical trial of histamine 2 receptor antagonism in treatment-resistant schizophrenia. J Clin Psychopharmacol. 2013;33:472.
112. Arrang JM, Garbarg M, Schwartz JC. Auto-inhibition of brain histamine release mediated by a novel class (H3) of histamine receptor. Nature. 1983;302:832.
113. Nuutinen S, et al. Evidence for the role of histamine H3 receptor in alcohol consumption and alcohol reward in mice. Neuropsychopharmacology. 2011;36:2030–40.
114. Vanhanen J, Kinnunen M, Nuutinen S, Panula P. Histamine H3 receptor antagonist JNJ-39220675 modulates locomotor responses but not place conditioning by dopaminergic drugs. Psychopharmacology (Berl). 2015;232(1143).
115. Ferrada C, et al. Marked changes in signal transduction upon heteromerization of dopamine D1 and histamine H3 receptors. Br J Pharmacol. 2009;157:64.
116. Ferrada C, et al. Interactions between histamine H-3 and dopamine D-2 receptors and the implications for striatal function. Neuropharmacology. 2008;55:190.
117. Nuutinen S, Panula P. Histamine in neurotransmission and brain diseases. Adv Exp Med Biol. 2011;709:95.
118. Dauvilliers Y, et al. Pitolisant versus placebo or modafinil in patients with narcolepsy: a double-blind, randomised trial. Lancet Neurol. 2013;12:1068.
119. Drutel G, et al. Identification of rat H3 receptor isoforms with different brain expression and signaling properties. Mol Pharmacol. 2001;59:1.
120. Bakker RA, et al. Discovery of naturally occurring splice variants of the rat histamine H3 receptor that act as dominant-negative isoforms. Mol Pharmacol. 2006;69:1194.

# Chapter 5
# Histamine $H_3R$ Antagonists: From Scaffold Hopping to Clinical Candidates

**B. Sadek, D. Łażewska, S. Hagenow, K. Kieć-Kononowicz, and H. Stark**

**Abstract** The histamine $H_3$ receptor ($H_3R$), belonging to the family of G-protein-coupled receptors, is predominantly expressed in the central nervous system (CNS). Since its discovery by Arrang et al., it was related to several central nervous system diseases by playing a key role as actuator of neurotransmitter release for, e.g., dopamine, acetylcholine, noradrenaline, or serotonin. Therefore, a huge number of $H_3R$ antagonists have been investigated on their potential therapeutic applicability in obesity, depression, mood disorders, neuropathic pain, and sleep– wake disorders (including narcolepsy) as well as cognitive and CNS-linked sensorimotor deficit disorders such as Parkinson's disease, attention deficit hyperactivity disorder, Alzheimer's disease, schizophrenia, alcohol addiction, energy homeostasis, epilepsy, obstructive sleep apnea, diabetic neuropathic pain, Tourette's syndrome, and catalepsy. So far, many structurally diverse $H_3R$ antagonists have been synthesized and pharmacologically evaluated. Despite a high diversity of compounds, these structures share a similar construction pattern. The pharmacophore contains a tertiary basic amine (postulated to interact with the conserved aspartate 114 in helix 3), a linker (commonly a linear propyloxy chain or structurally constrained), a central core, and "the eastern" arbitrary region (with high diversity such as second basic, acidic, lipophilic, or polar moieties of different sizes).

**Keywords** Histamine • Antagonists • Inverse agonists • Pitolisant • Narcolepsy • Cognitive impairments • Drug development • GPCR • Clinical trials • Clinical candidates

---

B. Sadek, Ph.D.
Department of Pharmacology and Therapeutics, College of Medicine and Health Sciences, Al Ain, United Arab Emirates University, Al Ain, Abu Dhabi 17666, United Arab Emirates
e-mail: bassem.sadek@uaeu.ac.ae

D. Łażewska, Ph.D. • K. Kieć-Kononowicz, Ph.D.
Department of Technology and Biotechnology of Drugs, Jagiellonian University Medical College, Medyczna 9, Kraków 30-688, Poland
e-mail: dlazewska@cm-uj.krakow.pl; mfkonono@cyf-kr.edu.pl

S. Hagenow, M.Sc. • H. Stark, Ph.D. (✉)
Institute of Pharmaceutical and Medicinal Chemistry, Heinrich Heine University, Universitaetsstr. 1, Düsseldorf 40225, North Rhine-Westphalia, Germany
e-mail: Stefanie.hagenow@uni-duesseldorf.de; stark@hhu.de



P. Blandina, M.B. Passani (eds.), *Histamine Receptors*, The Receptors 28,
DOI 10.1007/978-3-319-40308-3_5

## 5.1 Introduction

The histamine $H_3$ receptor ($H_3R$), belonging to the family of G-protein-coupled receptors (GPCRs), is predominantly expressed in the central nervous system [1–3]. Since its discovery by Arrang et al., [4] it was related to several central nervous system (CNS) diseases by playing a key role as actuator of neurotransmitter release for, e.g., dopamine, acetylcholine, noradrenaline, or serotonin [1–3]. Therefore, a huge number of $H_3R$ antagonists have been investigated on their potential therapeutic applicability in obesity, depression, mood disorders, neuropathic pain, and sleep–wake disorders (including narcolepsy) as well as cognitive and CNS-linked sensorimotor deficit disorders such as Parkinson's disease (PD), attention deficit hyperactivity disorder (ADHD), Alzheimer's disease (AD), schizophrenia, alcohol addiction, energy homeostasis, epilepsy, obstructive sleep apnea, diabetic neuropathic pain, Tourette's syndrome, and catalepsy [1, 5–9]. So far, many structurally diverse $H_3R$ antagonists have been synthesized and pharmacologically evaluated. Despite a high diversity of compounds, these structures share a similar construction pattern. The pharmacophore (Fig. 5.1) contains a tertiary basic amine (postulated to interact with the conserved aspartate 114 in helix 3), a linker (commonly a linear propyloxy chain or structurally constrained), a central core, and "the eastern" arbitrary region (with high diversity such as second basic, acidic, lipophilic, or polar moieties of different sizes) [2, 10].

New "hits" for modifications were mostly found by high-throughput screening (HTS). In some cases amine moiety was modified but in others taken unchanged. Compounds can be roughly divided into (un)substituted azacycloalkanes (e.g., piperidines, pyrrolidines, piperazines) and "constrained" azacycloalkanes (fused azacycloalkanes and spiroazacycloalkanes) where linker is incorporated into the amine moiety. In some compounds, especially those with two basic centers, determination of the dominant basic amine is difficult without support of precise molecular and docking studies. The division presented here, in such cases, is a subjective experience of the authors. Changes were also enlarged to the central core, and instead of the phenyl ring, new scaffolds were introduced (e.g., aryl, heteroaryl, azacycloalkylamine, fused rings, and spiro derivatives). Greatest effort was given to modifications in the eastern (arbitrary) region of pharmacophore pattern, to overcome or mitigate obstacles such as phospholipidosis, hERG

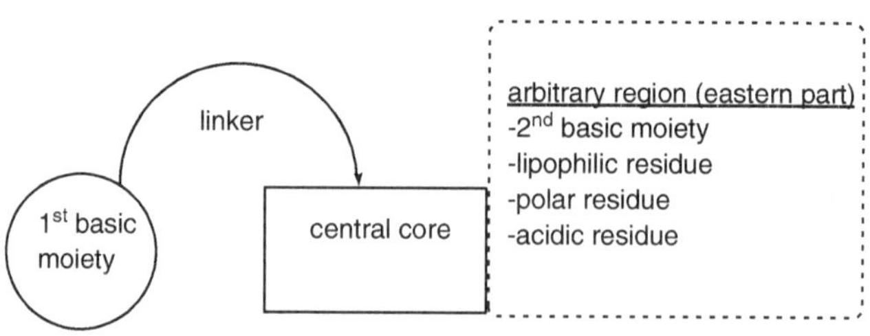


**Fig. 5.1** Pharmacophore model of histamine $H_3R$ antagonists/inverse agonists

potassium channel modulation, mutagenicity, weak oral availability, or undesired metabolism.

Among various structural $H_3R$ antagonists, several have figured markedly in preclinical studies, with evidenced clear capability to release a variety of central neurotransmitters, and have efficacy in preclinical animal models. As a result, this has reinvigorated ongoing research on $H_3R$ antagonists with enhanced potency, selectivity, and better drug-likeness to accelerate clinical evaluation. In the following chapter, the progress in recent scaffold-hopping strategies for $H_3R$ antagonist during the last 5 years (data from the literature 2010–2015; PubMed) will be addressed (receptor affinities are given in Table 5.1), while their preclinical potential, as given, will be highlighted. As of great interest in $H_3R$ antagonist research, further attention will be paid on compounds currently in clinical trial/stages.

**Table 5.1** Receptor affinities for selected $H_3R$ antagonists/inverse agonists

| | $K_i$ [nM] | | | |
|---|---|---|---|---|
| Compound | $hH_3R$ | $rH_3R$ | $hH_4R$ | Reference |
| **1** Thioperamide | 72 | 3.6 | 48 | [16] |
| **2** Clobenpropit | 0.4 | 0.2 | 42 | [16] |
| **3** Ciproxifan | 63 | 0.5 | 1,862 | [16] |
| **4** GT-2331 (Cipralisant) | 4.4 | 0.3 | 83 | [16] |
| **5** SCH-79687 | | 1.9 | | [33] |
| **6** UCL 1972 | | 39 | | [35] |
| **7** BF2.649 (Pitolisant) | 0.3–5.3 | 17 | >100,000 | [7, 133] |
| **8** NNC38-1049 | 1.2 | 5.1 | >10,000 | [38] |
| **9** A-317920 | 93 | 0.7 | >10,000 | [16] |
| **10** | 1.3 | | | [39] |
| **11** PF-03654746 | 2.3 | 37 | | [39] |
| **12** PF-03654764 | 1.4 | 19 | | [39] |
| **13** ABT-239 | 0.5 | 1.4 | | [32, 42] |
| **14** A-688057 | 0.5 | 3.1 | | [42] |
| **15** A-960656 | 21 | 76 | | [53] |
| **16** A-331440 | 3.2 | 6.3 | >10,000 | [57] |
| **17** CEP-26401 | 2.0 | 7.2 | | [59] |
| **18** | 3.1 | 9.9 | | [67] |
| **19** APD916 | 4.2 | 0.7 | >10,000 | [69] |
| **20** | 2.0 | 7.0 | >10,000 | [70] |
| **21** Ergoline | 8.3 | | | [72] |
| **22** | 19 | | | [72] |
| **23** | 3.8 | | | [73] |
| **24** | $IC_{50}$ 6.5 | | | [75] |
| **25** | $IC_{50}$ 0.9 | | >10,000 | [75] |
| **26** | 0.1 | 0.9 | | [76] |
| **27** SAR110068 | 1.0 | | | [77] |

(continued)

**Table 5.1** (continued)

| Compound | $K_i$ [nM] | | | Reference |
|---|---|---|---|---|
| | $hH_3R$ | $rH_3R$ | $hH_4R$ | |
| **28** ST-1283 | 0.2 | | >1,000,000 | [81] |
| **29** DL77 | 8.3 | | 48,978 | [87] |
| **30** | 10 | | | [89] |
| **31** | 11 | | | [90] |
| **32** JNJ-5207852 | 0.6 | 1.3 | | [91] |
| **33** JNJ-10181457 | 1.2 | 1.3 | | [52] |
| **34** BP1.4160 | 2.6 | | | [79] |
| **35** BP1.3432 | 0.4 | | | [97] |
| **36** ST-1036 | 2.0 | | >10,000 | [98] |
| **37** OUP186 | $IC_{50}$ 6.3 | | | [99] |
| **38** | 49 | | | [101] |
| **39** | 4.0 | | | [101] |
| **40** | 3.0 | | | [101] |
| **41** ST-1025 | 11 | | >30,000 | [104] |
| **42** | 1.3 | 6.6 | | [105] |
| **43** | 2.0 | | | [106] |
| **44** | $IC_{50}$ 11 | | | [107] |
| **45** | $IC_{50}$ 2.0 | | | [107] |
| **46** | 2.7 | 4.4 | | [108] |
| **47** | 3.8 | 7.3 | >10,000 | [109] |
| **48** | 17 | 26 | | [66] |
| **49** CEP-32215 | 2.0 | 3.6 | | [114] |
| **50** | 5.0 | 7.0 | | [115] |
| **51** | 5.0 | 12 | | [110] |
| **52** | gp $pA_2=8.47$ | | | [111] |
| **53** | gp 27 | | | [116] |
| **54** | gp 0.5 | | | [116] |
| **55** | gp 0.95 | | | [117] |
| **56** | 1.0 | | | [118] |
| **57** Bavisant | 5.4 | | >1,000 | [119] |
| **58** | 2.5 | | | [119] |
| **59** | 2.1 | | | [121] |
| **60** | $IC_{50}$ 1.6 | $IC_{50}$ 17 | >1,000 | [122] |
| **61** | | | >10,000 | [123] |
| **62** GSK-189254 | 0.2 | 0.7 | >3,000 | [120] |
| **63** GSK-207040 | 0.2 | 1.0 | >3,000 | [120] |
| **64** GSK-239512 | 0.2 | 0.2 | >3,000 | [120] |
| **65** GSK-334429 | 0.8 | 0.8 | | [124] |
| **66** JNJ-39220675 | 1.4 | 23 | | [127] |
| **67** | 4.0 | 9.0 | >10,000 | [131] |
| **68** | 7.0 | 17 | >10,000 | [132] |

*gp* guinea pig

## 5.2 Imidazole-Based $H_3R$ Antagonists

Since the discovery of the $H_3R$ in 1983 [4], the development of $H_3R$ antagonists has mainly been based on modifications of the endogenous neurotransmitter histamine [8, 11, 12]. The first $H_3R$ antagonists were imidazole based, and before the identification of the histamine $H_4$ receptor ($H_4R$) in 2001 [13], members of this class of compound were widely used as "$H_3R$ standards" for preclinical animal models. The ready availability of imidazole-containing ligands from commercial sources prompted their evaluation in numerous laboratories. As the first attempts to replace this ring by other aromatic heterocycles were unsuccessful (e.g., [14, 15]), the imidazole was thought to be obligatory for potency. Among many described imidazole compounds, the most prominent were **thioperamide** (**1**), **clobenpropit** (**2**), **ciproxifan** (**3**), and **GT-2331** (**cipralisant, perceptin**) (**4**) (Fig. 5.2) [16] (for review, see, e.g., [17]). **Thioperamide** is a highly potent $H_3R$ antagonist (human $H_3R$ ($hH_3R$) $pK_i=7.14$; rat $H_3R$ ($rH_3R$) $pK_i=8.44$; $rH_4R$ $pK_i=7.32$; $ED_{50}=1$ mg/kg p.o.) (Table 5.1) [16, 17] and has been found broadly effective in vivo in a large number of preclinical behavioral models, including elevated plus maze learning, Morris water maze, Barnes maze [18, 19], and a variety of epilepsy models [20, 21] (Table 5.2). Also, **thioperamide** demonstrated that $H_3Rs$ negatively regulate food intake in rodents, suggesting $H_3Rs$ may have therapeutic potential targeting the treatment of obesity and diabetes mellitus [22–24]. Interestingly, the latter finding was in agreement with previous studies in which mice with genes disrupted for the histamine $H_1$ receptor ($H_1R$) or histidine decarboxylase (HDC) are prone to becoming obese on a high-fat diet or at advanced age [22, 23]. Also, several antipsychotic drugs with high-antagonist affinities for $H_1Rs$ are known to cause weight gain in rodents and humans [25]. Another imidazole-based $H_3R$ antagonist is **ciproxifan** which has been found to demonstrate high $H_3R$ in vitro potency, high selectivity ($hH_3R$ $pK_i=7.20$; $rH_3R$ $pK_i=9.29$; $hH_4R$ $pK_i=5.73$; $ED_{50}=0.14$ mg/kg p.o.) (Table 5.1) [16, 17, 26], and oral bioavailability in a number of preclinical animal models and, thus, was used extensively as an in vivo reference $H_3R$ antagonist, for example, in

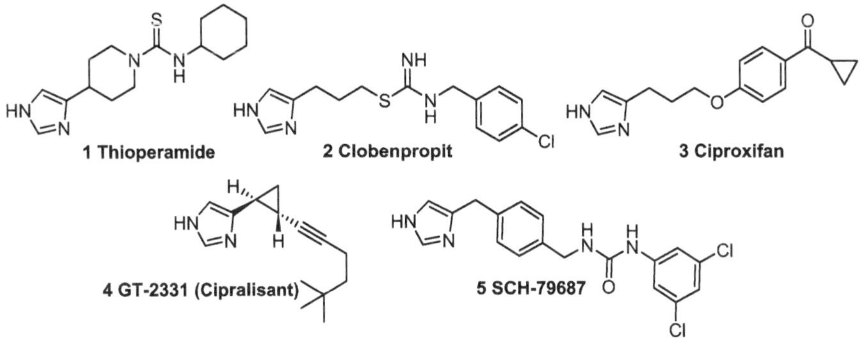


**Fig. 5.2** Structures of the most prominent imidazole histamine $H_3R$ ligands

**Table 5.2** Summary of pharmacokinetic and pharmacodynamic properties for selected $H_3R$ antagonists

| | Pharmacokinetics | | | Pharmacodynamics | | | | | |
|---|---|---|---|---|---|---|---|---|---|
| | | | | $K_i$ [nM] | | $ED_{50}$ (mg/kg) | $hH_3R$ | | |
| Compound | Oral F (%) | $C_{(brain)}/C_{(plasma)}$ ratio | $t_{1/2}$ (h) | $hH_3R$ | $rH_3R$ | | $K_b$ [nM] | $EC_{50}$ [nM] | Reference |
| Thioperamide | | $0.7^{*}$ | $2^{*}$ | 72 | 3.6 | 1 | 794 | 32 | [16, 17] |
| Ciproxifan | $62^{*}$ | $4.7^{*}$ | | 63 | 0.5 | 0.14 | 257 | 69 | [16, 17] |
| CEP-32215 | $51^{*}$<br>$78^{\#}$ | $3.3^{a*}$ | $1.8^{a*}$<br>$2.8^{b*}$<br>$9.0^{a\#}$<br>$8.7^{b\#}$ | 2.0 | 3.6 | | 0.3 | 0.6 | [114] |
| ABT-239 | $53^{*}$ | $>20^{*}$ | $5.3^{a*}$ | 0.5 | 1.4 | 0.37–0.54 | 8.8 | 1.9 | [32, 42] |
| A-331440 | $35^{*}$ | | $4^{b*}$ | 3.2 | 6.3 | | | | [57] |
| A-688057 | $26^{*}$ | 3.4 | $2.9^{a*}$ | 0.5 | 3.1 | | | 1.5 | [42] |
| A-960656 | $84^{*}$<br>$100^{\#}$ | $1.3^{*}$ | $1.9^{b*}$<br>$8.3^{b\#}$ | 21 | 76 | | 22 | | [53] |
| GSK-207040 | $88^{*}$ | $4.0^{*}$ | $2.6^{a*}$<br>$2.9^{b*}$ | 0.2 | 1–0 | 0.21 | 0.5 | 0.6 | [120, 124] |
| GSK-334429 | $91^{*}$ | $3.5^{*}$ | $2.0^{a*}$<br>$2.1^{b*}$ | 0.8 | 0.8 | 0.35 | 1.4 | 2.6 | [124] |
| NNC-38-1049 | | | $0.33^{*}$ | 1.2 | 5.1 | | 2.3 | | [38, 48] |
| JNJ-5207852 | $>85^{*}$ | | $>14^{b*}$<br>$>13^{c*}$ | 0.6 | 1.3 | 0.12 | 0.1 | | [91] |
| JNJ-10181457 | | $15^{*}$ | | 1.2 | 1.3 | | | | [52, 96] |
| JNJ-39220675 | $50^{*}$<br>$28^{**}$ | 3.7* | $1.2^{*}$<br>$1.1^{**}$ | 1.4 | 23 | | 0.4 | | [93, 127] |
| BF-2.649 (Pitolisant) | $84^{**}$ | 23.5 | $11^{b***}$<br>$\sim 2^{a**}$ | 0.3–1.0 | 17 | 1.6 | 0.3 | 1.5 | [7, 133] |
| PF-03654746 | $26^{*}$ | $2.1^{*}$ | $9–18^{***}$ | 2.3 | 37 | | | | [39, 47] |
| CEP-26401 (Irdabisant) | $83^{**}$<br>$83^{\#}$ | $2.6^{*}$ | $2.6^{a**}$<br>$2.9^{b**}$ | 2.0 | 7.2 | 0.06 | 0.4 | 1.1 | [59] |
| 18 (CEP-26401 analog) | $34^{*}$<br>$136^{\#}$ | $2^{*}$ | $1.1^{a*}$<br>$7.7^{a\#}$ | 3.1 | 9.9 | | 0.2 | 0.7 | [67] |
| ABT-288 | $37^{*}$<br>$66^{*}$ | $1.5^{*}$ | $1.3^{*}$ | 1.9 | 8.2 | | 8.7 | 3.8 | [49] |
| GSK-189254 | $83^{*}$ | $0.9^{*}$ | $1.6^{a*}$<br>$2.5^{b*}$ | 0.2 | 0.7 | 0.17 | 0.9 | 6.3 | [125] |
| GSK-239512 | $51^{*}$<br>$55^{\#}$ | $1.8^{a*}$ | $1.2^{a*}$<br>$1.7^{a\#}$ | 0.2 | 0.2 | | | | [120] |

(continued)

**Table 5.2** (continued)

| | Pharmacokinetics | | | Pharmacodynamics | | | | | |
|---|---|---|---|---|---|---|---|---|---|
| | | | | $K_i$ [nM] | | $ED_{50}$ | $hH_3R$ | | |
| Compound | Oral F (%) | $C_{(brain)}/C_{(plasma)}$ ratio | $t_{1/2}$ (h) | $hH_3R$ | $rH_3R$ | (mg/kg) | $K_b$ [nM] | $EC_{50}$ [nM] | Reference |
| AZD5213 | | | ~$5^{b***}$ | 0.5 | | | | | [146] |
| Bavisant | $112^{*}$ | | $0.9^{a*}$<br>14–$22^{b***}$ | 5.4 | | | | | [119] |

*Values for rat
**Values for mice
***Values for human
#Valued for monkey
a i.v.
b p.o.
c i.p.

attentional models, such as electroencephalogram (EEG)-assessed waking and impulsivity [19, 27, 28]. Interestingly, **ciproxifan** was found to improve attention and cognitive performance in naive animals and in several models of cognitive impairment and also to increase prepulse inhibition as a model of schizophrenia. **GT2331 (cipralisant) (4)** was the first $H_3R$ ligand ($hH_3R$ $pK_i=8.36$; $rH_3R$ $pK_i=9.60$; $hH_4R$ $pK_i=7.08$; $ED_{50}=0.08$ mg/kg i.p.) (Table 5.1) [16, 17, 26, 29] which entered into clinical studies and reached phase II for ADHD before being halted in 2002 [30]. Later it was shown that these compounds behaved as (partial) agonist in many assays [31], whereas antagonist behavior would be expected to demonstrate efficacy in ADHD. However, for the earliest imidazole-based compounds, it later became obvious that such compounds have shortcomings as clinical candidates due to several possible pharmacokinetic (PK) drawbacks like CYP450 inhibition, low brain penetration, incidence of off-target activity, or lack of subtype selectivity especially over $H_4R$ [6]. Nonetheless, these compounds are readily available from commercial sources and have a number of favorable properties, such as potent activity at rodent $H_3Rs$, and a history of in vivo use. Later, other imidazole-based compounds were described with activity in animal models, including **SCH-79876 (5)** ($rH_3R$ $-\log K_i=8.72$) (Fig. 5.2) [32, 33]. Regardless of the current focus on non-imidazole leads, these structures are very useful pharmacological tools for the sake of comparison, because considerable pharmacological data have been obtained, especially with **thioperamide (1)** and **ciproxifan (3)** as antagonist reference ligands. However, the exchange of the imidazole ring with other heterocycles was a milestone in the search for new histamine $H_3R$ ligands.

The problems of the early imidazole-based $H_3R$ antagonists drove immense efforts toward non-imidazoles that have become the chief focus of the $H_3R$ antagonist design work in the last decade, as these structures address the aforementioned class-related drawbacks noted with the early generation of imidazole-based agents, especially the off-target activity at $H_4Rs$ or other receptors [9, 30, 34]. In 1998, the first potent histamine $H_3R$ antagonists with the imidazole moiety replaced by, e.g.,

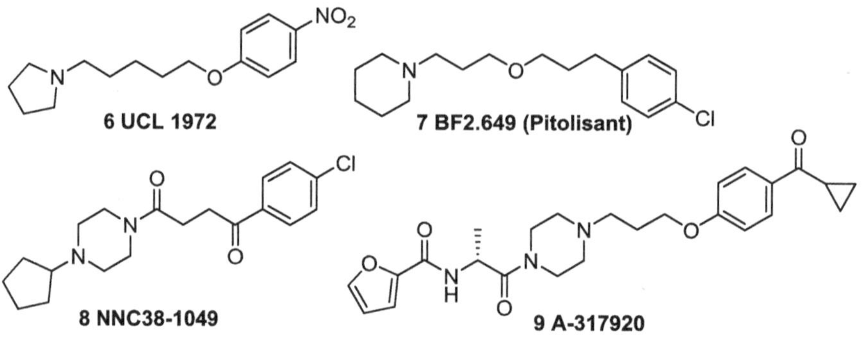


**Fig. 5.3** Structures of selected first-described potent non-imidazole histamine H3 receptor antagonists/inverse agonists

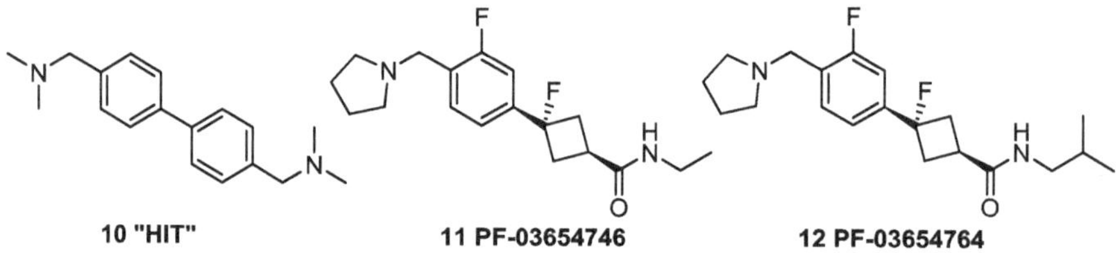


**Fig. 5.4** Structures of hit compounds and derived pyrrolidine derivatives PF-03654746 and PF-03654764

pyrrolidine or piperidine structures were synthesized [35], while later attempts would show that the imidazole moiety can be successfully replaced by a piperidine ring in several already-known imidazole-based $H_3R$ antagonists [36]. Early-described compounds with pyrrolidines, piperidines, and piperazines (with amide moiety) as imidazole replacement structures were **UCL 1972** (**6**) (r$H_3R$ $-\log K_i = 7.41$; $ED_{50} = 1.1$ mg/kg p.o.) [35], **BF2.649** (**pitolisant**) (**7**) (h$H_3R$ $-\log K_i = 9.52–9.00$; r$H_3R$ $-\log K_i = 7.77$; $ED_{50} = 1.6$ mg/kg p.o.) [7, 36], **NNC38-1049** (**8**) (h$H_3R$ $-\log K_i = 8.92$; r$H_3R$ $-\log K_i = 8.29$) (discussed later) [37, 38], or **A-317920** (**9**) (h$H_3R$ $pK_i = 7.03$; r$H_3R$ $pK_i = 9.15$) [16] (Fig. 5.3, Table 5.1). However, structure modifications with tertiary amines have been extended to more "complex" structures over the decades, although moieties of these first azacycloalkanes are still present in $H_3R$ antagonists/inverse agonists.

## 5.3 (Methyl)Pyrrolidine Derivatives

Several $H_3R$ antagonists were obtained having a pyrrolidine or methylated pyrrolidine moiety as the basic center. The HTS hit **10** with low-hERG modulation ($IC_{50} > 5600$ nM) led to two clinical candidates **PF-03654746** (**11**) (h$H_3R$ $pK_i = 8.64$; r$H_3R$ $pK_i = 7.43$) and **PF-03654764** (**12**) (h$H_3R$ $pK_i = 8.84$; r$H_3R$ $pK_i = 7.73$) (Fig. 5.4) [39], achieved by a strategy which combines medicinal chemistry knowledge

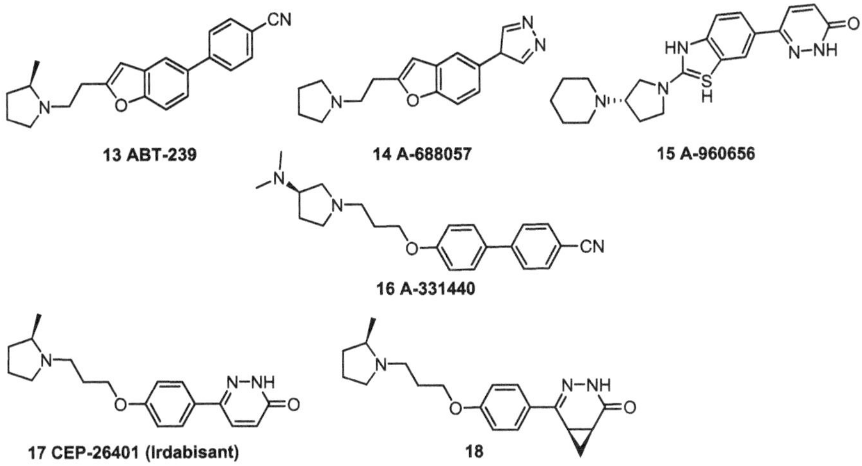


**Fig. 5.5** Selected (methyl)pyrrolidine derivates

and traditional in vivo assays with in silico and in vitro safety studies. These compounds were tested in patients with allergic rhinitis (discussed later) [40, 41].

The methylpyrrolidine derivative **ABT-239** (**13**) ($hH_3R$ $pK_i$=9.35; $rH_3R$ $pK_i$=8.87) (Fig. 5.5) [32, 42] with oral bioavailability of 37 % and brain-to-plasma ratio of more than 20-fold has been widely used as a standard $H_3R$ antagonist with high in vivo potency in numerous preclinical models of attention, ethanol-associated learning deficit, cognition, schizophrenia, osteoarthritic pain, and AD [5, 43–47]. In preclinical cognition studies, acquisition of a five-trial, inhibitory avoidance test in rat pups was improved with **ABT-239** (0.1–1.0 mg/kg, p.o.) with efficacy maintained for 3–6 h following repeated dosing [5, 48, 49]. Moreover, **ABT-239** improved social memory in adult (0.01–0.3 mg/kg) and aged (0.3–1.0 mg/kg) rats [50]. **ABT-239** (1.0–3.0 mg/kg) improved gating deficits in DBA/2 mice (Dilute Brown Non-Agouti/MHC Haplotype H2[d]) schizophrenia models using prepulse inhibition of startle and attenuated methamphetamine-induced hyperactivity in mice (1.0 mg/kg) [51]. Furthermore, it has been shown that **ABT-239** enhanced acetylcholine release (0.1–3.0 mg/kg, p.o.) in adult rat frontal cortex and hippocampus and enhanced dopamine release in the frontal cortex (3.0 mg/kg), but not the striatum in freely moving rat microdialysis studies and attenuated ketamine-induced deficits on spontaneous alternation in cross-maze [5]. Interestingly, **ABT-239** was found to reduce kainic acid-induced seizures and excitotoxicity [8]. A previous study demonstrated that **ABT-239** improved ethanol-induced deficits on hippocampal long-term potentiation, indicating that $H_3R$ antagonists can affect changes in synaptic plasticity related to cognitive processes [46]. However, **ABT-239** was found to potently bind to the cardiac hERG channel in vitro [47] and to induce phospholipidosis, a condition considered to be a potential toxicity issue and often observed in the case of cationic amphiphilic compounds bearing more than one basic center [52]. **A-688057** (**14**) ($hH_3R$ $pK_i$=9.31; $rH_3R$ $pK_i$=8.50) (Fig. 5.5), structurally strongly related to ABT-239, showed high brain penetration (brain/plasma ratio of 3.4) and has been found to improve attention and cognition when administered at 0.001–0.3 mg/kg, s.c. in animal models [42]. PK testing uncovered only moderate to poor oral bioavailability in rat (26 %), dog (30 %), and monkey (8 %) and only moderate blood half-lives after i.v. administration ($t_{1/2}$ in rat of 2.9 h, 1.7 h in dog, 1.8 h in monkey), suggesting poor human PKs. These data overall indicated that **A-688057** has an excellent profile for use as a pharmacological tool compound (Table 5.2) [42]. Notably, structurally related compound **A-960656** (**15**) ($hH_3R$ $-\log K_i = 7.68$ nM; $rH_3R$ $-\log K_i = 7.12$) (Fig. 5.5) [53] was found to be highly selective (>330-fold) for histamine $H_3Rs$ versus off-target sites screened in a commercial panel of >80 receptors, enzymes, and ion channels (including the hERG potassium channel, $K_i > 9000$ nM) and has no detectable interaction with other HR subtypes. Also, **A-960656** with its low molecular weight (295 g $mol^{-1}$), high solubility, moderate lipophilicity, and good CNS penetration (brain/plasma ratio of 1.3) proved to be of low potential for phospholipidosis, genotoxicity, and CYP450 inhibition in in vitro toxicological tests. **A-960656** was found to be effective in preclinical animal models of osteoarthritis and neuropathic pain [53]. It was evaluated in a spinal nerve ligation model of neuropathic pain, and its onset of efficacy was rapid and significant at 60 min after oral doses of 1 and 3 mg/kg. Furthermore, the efficacy of **A-960656** was maintained on sub-chronic dosing for 11 days, demonstrating even slight improvement in efficacy on sub-chronic dosing, so that doses of 0.3–3 mg/kg were statistically effective (Pb0.01). Thus **A-960656** is able to fully block mechanical allodynia, with doses of 1 and 3 mg/kg giving efficacy comparable to the clinically active control drug gabapentin dosed at 100 mg/kg [53]. Interestingly, **A-960656** was found to suppress neuropathic hypersensitivity in a preclinical rat model, suggesting that blocking the autoinhibitory $H_3R$ on histaminergic nerve terminals in the locus coeruleus facilitates release of histamine and thereby increases descending noradrenergic pain inhibition [54]. **A-331440** (**16**) ($hH_3R$ $-\log K_i = 7.64$; $rH_3R$ $-\log K_i = 7.66$) (Fig. 5.5) shows improved oral bioavailability (35 %) and longer $t_{1/2}$ (4 h) when compared to **A-960656** (**15**). When administered at 5 mg/kg in mice (stabilized on a high-fat diet; 45 kcal % lard), **A-331440** significantly decreased weight comparably to dexfenfluramine (10 mg/kg, p.o.) [27, 55, 56]. **A-331440** (15 mg/kg) reduced weight to a level comparable to mice, which were placed on the low-fat diet [55]. The two higher doses reduced body fat, and the highest dose also normalized an insulin tolerance test. These data showed that **A-331440** has potential as an anti-obesity agent [57]. However, **A-331440** was found to potentiate the sedative effects of ethanol, a paradoxical effect that might be due to the subtle regulatory actions related to the $H_3$ heteroreceptor function [58].

Rational-based synthesis preferring compounds with $hH_3R$ binding affinities $K_i < 15$ nM; subtype selectivity over $hH_1R$, $hH_2R$, and $hH_4R$; corresponding physicochemical parameters (e.g., water solubility); in vitro microsomal stability; low CYP450/hERG inhibition; and good PK properties (e.g., oral bioavailability, half-life time) led to the development of **CEP-26401** (**17**, **Irdabisant**) (Fig. 5.5). **CEP-26401** ($hH_3R$ $-\log K_i = 8.70$; $hH_3R$ $-\log K_i = 8.14$; $hH_4R$ inhibition <11 % at 10 μM; hERG $IC_{50} = 14$ μM) showed high subtype selectivity (>1000-fold selectivity) and excellent drug properties and safety [59, 60]. Consequently and based on its cognition-enhancing and wake-promoting activities, **CEP-26401** has entered into clinical trials for cognitive impairments (described later).

Further structural modifications concerning the pyridazin-3-one core in the "eastern region" of **CEP-26401** were introduced, e.g., replacement by a dihydropyridazinone core [61]; a pyridinone core [62]; a 5-pyridazin-3-one core [63, 64]; 4,5-fused derivatives, e.g., compound **18** [65–67]; or introduction of substituents into the pyridazinone moiety [68] (Fig. 5.5). Most of these developed compounds showed high $hH_3R$ affinities and subtype selectivity over $hH_1R$, $hH_2R$, and $hH_4R$ with acceptable PK properties and weak hERG and CYP450 inhibition. Interestingly, **18** also displayed potent $H_3R$ functional antagonism ($hH_3R$ $-\log K_i = 8.51$; $rH_3R$ $-\log K_i = 8.00$; hERG inhibition 23 % at 10 μM) [67] and robust wake-promoting activity in preclinical in vivo models in rodents. It revealed acceptable PKs and brain-to-plasma ratio of 3.1 2 h after administration of 10-mg/kg dose (i.p.) and $AUC_{0-t}$ of 5762, 4205, and 9574 (ng h/mL) in rat, dog, and monkey, respectively (Table 5.2). Therefore, the compound appears to be promising for further drug development [67].

Notably, **APD916 (19)** ($hH_3R$ $-\log K_i = 8.38$; $rH_3R$ $-\log K_i = 9.15$; hERG $IC_{50} = 11$ μM) (Fig. 5.6) [69], a derivative of (*R*)-2-methylpyrrolidinylethyl with a sulfone moiety, was evaluated for cross-species metabolism and PK properties as well as for its in vivo activity and safety profile (polysomnography test) [69–71]. Observed results indicated that **APD916** significantly increased total wake time during the first 2 h after administration at a dose of 0.6 mg/kg p.o.

Based on the knowledge of modafinil, an atypical wake-promoting agent, having a sulfinylacetamide moiety ($-SO_2CH_2CONH_2$) but lacking significant $H_3Rs$ affinity, researchers were encouraged to incorporate this moiety into known $H_3R$ structural motif. Accordingly, (*R*)-2-methylpyrrolidine derivatives with sulfinyl and

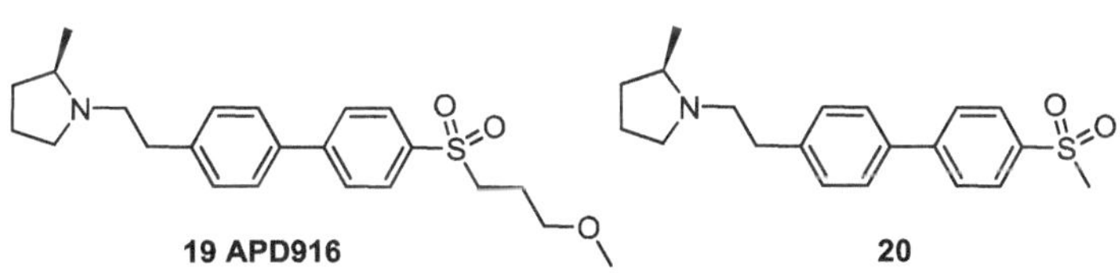


**Fig. 5.6** Methylpyrrolidine derivatives containing sulfone moieties

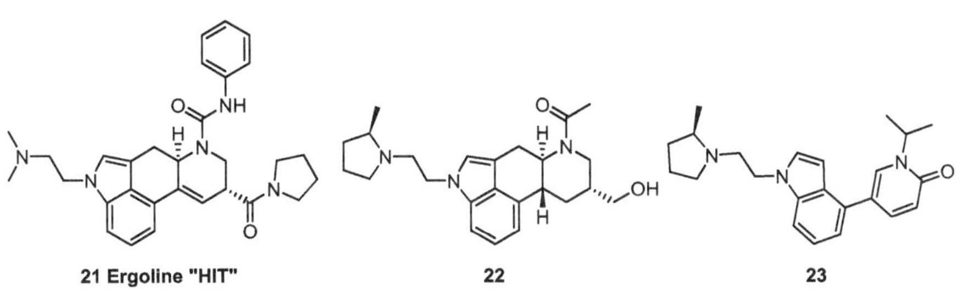


**Fig. 5.7** Ergoline derivatives showing $H_3R$ antagonistic properties

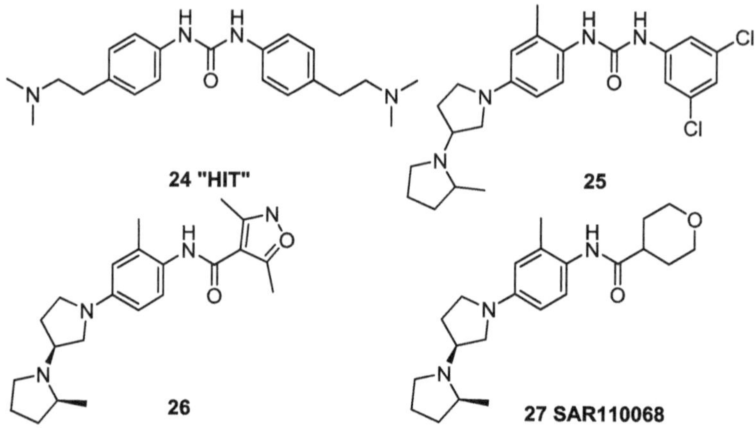


**Fig. 5.8** Selected bipyrrolidinyl derivatives

sulfone moieties attached via methylene group to a *para*-biphenyl central core were developed and displayed high $hH_3R$ affinity ($1 \leq K_i \leq 150$ nM), while among them only compound **20** ($hH_3R$ $-\log K_i = 8.70$) (Fig. 5.6) comprising a sulfone moiety showed promising results [70].

Interestingly, a series of $H_3R$ antagonists were obtained following modifications of a chemokine CXCR3 receptor antagonist **ergoline** (**21**) ($hH_3R$ $-\log K_i = 8.08$; hERG inhibition 17 % at 10 μM) (Fig. 5.7) [72, 73]. The ergoline moiety is well known as a pharmacophore in multiple targeting in which this class of alkaloids have been classified as dirty drugs due to multiple wanted and unwanted CNS receptor affinities [74]. Hereafter, the first structural changes in ergoline core led to compound **22** ($hH_3R$ $-\log K_i = 7.72$; hERG inhibition 16 % at 10 μM) with the desired PK profile with rapid penetration into the brain and enhanced wakefulness observed within a short duration of action [72]. Further structural simplifications of the ergoline core resulted in the indole derivatives **22** and **23** with high $H_3R$ affinity [73]. For compound **23** ($hH_3R$ $-\log K_i = 8.42$; hERG inhibition 26 % at 10 μM) (Fig. 5.7) significant increase in wakefulness (after injection lasting up to 2.5 h) was observed, and after this time, the sleep–wake pattern returned to a normal physiological level.

Following further structural modifications, a series of bipyrrolidinyl (2-methyl[1,3]bipyrrolidinyl-1-yl) derivatives with urea [75] and amide moiety [75–78] were derived from optimization of a HTS lead structure **24** ($hH_3R$ FLIPR $IC_{50} = 6.5$ nM) (Fig. 5.8). Consequently, efforts to remove the dibasic property of **24** along with modifications of the side chain (including rigidity) led to urea derivatives, e.g., **25** ($hH_3R$ FLIPR $IC_{50} = 0.9$ nM). As a bipyrrolidinyl moiety could be a mixture of four stereoisomers, pharmacological studies showed that stereoisomer (2S, 3′S) had the highest in vitro $hH_3R$ potency and strongest functional antagonism compared to the other stereoisomers combined with excellent PK properties. Therefore, (2S,3′S) bipyrrolidinyl derivatives were synthesized in the following studies with phenyl or naphthyl central core, e.g., **26** [76] or **SAR110068** (**27**) [77] (Fig. 5.8). Among several derivatives of this class, compound **26** displayed high affinity for $hH_3R$ ($hH_3R$ $-\log K_i=10.0$; hERG $IC_{50}=37$ μM), good PK properties, and safety in cardiovascular and neuropsychological/behavioral tests [76]. Interestingly, **26** was, also, active in forced swim test suggesting its utility as an antidepressive agent (reduction of immobility time was significant at 3 and 10 mg/kg po). Moreover, compound **27** (**SAR110068**) ($hH_3R$ $-\log K_i=9.00$; hERG $IC_{50}=19$ μM) exhibited a significant awakening effect in EEG studies, increasing wakefulness (for 4 h) and decreasing slow-wave and REM sleep [76]. The assessment of tested doses showed no cardiovascular or neuropsychological/behavioral side effects for compound **27.**

## 5.4 Piperidine Derivatives

Great efforts were spent on histamine $H_3R$ antagonists/inverse agonists with the piperidine moiety as basic center since the clinical candidate pitolisant (discussed later) contains a piperidine. Accordingly, piperidine was tethered to a propyloxyphenyl core in several compounds, while various modifications were introduced into the "eastern part" of the molecule [79–82]. Numerous compounds were prepared in some cases as bivalent or multi-target-directed ligands [81, 83]. Many other potent *N*-piperidinylpropyloxyphenyl derivatives comprise a second basic group. Interestingly, an introduction of a second amine moiety was found to increase the binding affinities for $hH_3Rs$ but can offer the risk of accumulation in the CNS and, therefore, the potential to induce phospholipidosis [84]. Consequently, the latter observations encouraged further structural modifications to lower the second amine basicity or to insert other moieties while keeping the same level of potency. Accordingly, several $H_3R$ antagonists were successfully developed including **ST-1283** (**28**) ($hH_3R$ $pK_i=9.62$; $hH_4R$ $pK_i=2.85$; $hH_1R$ $pK_i=4.98$) [81, 85, 86] and very recently **DL77** (**29**) ($hH_3R$ $pK_i=8.08$; $hH_4R$ $pK_i=4.31$; $hH_1R$ $pK_i=6.18$) [87] (Fig. 5.9), which were reported to reduce voluntary alcohol intake and ethanol-induced conditioned place preference in mice. Moreover, compound **ST-1283** demonstrated both anxiolytic-like and antidepressant-like effects at a dose of 7.5 mg/kg in mice [85]. **DL77** (5, 10, and 15 mg/kg, i.p.) has very recently been found to significantly and dose-dependently reduce MES-induced seizure duration and to improve retrieval, demonstrating that **DL77** provides anticonvulsant and procognitive properties which may have implications for the treatment of degenerative disorders associated with impaired memory function or to tackle cognitive problems associated with the chronic use of antiepileptic drugs [88].

As additional structural modification, compounds containing an acidic moiety connected to the piperidinylpropyloxyphenyl core were synthesized [89], showing that $H_3R$ tolerated acid properties to some extent such as **30** ($hH_3R$ $-\log K_i=8.00$) (Fig. 5.9), with interesting structure-activity relationships (SAR) observations that clearly demonstrated that less acidic derivatives are capable of showing better affinity for $H_3Rs$ [89]. Further modifications with kojic acid derivatives revealed that the bioisosteric replacement of the phenyl central core by a kojic acid moiety was not

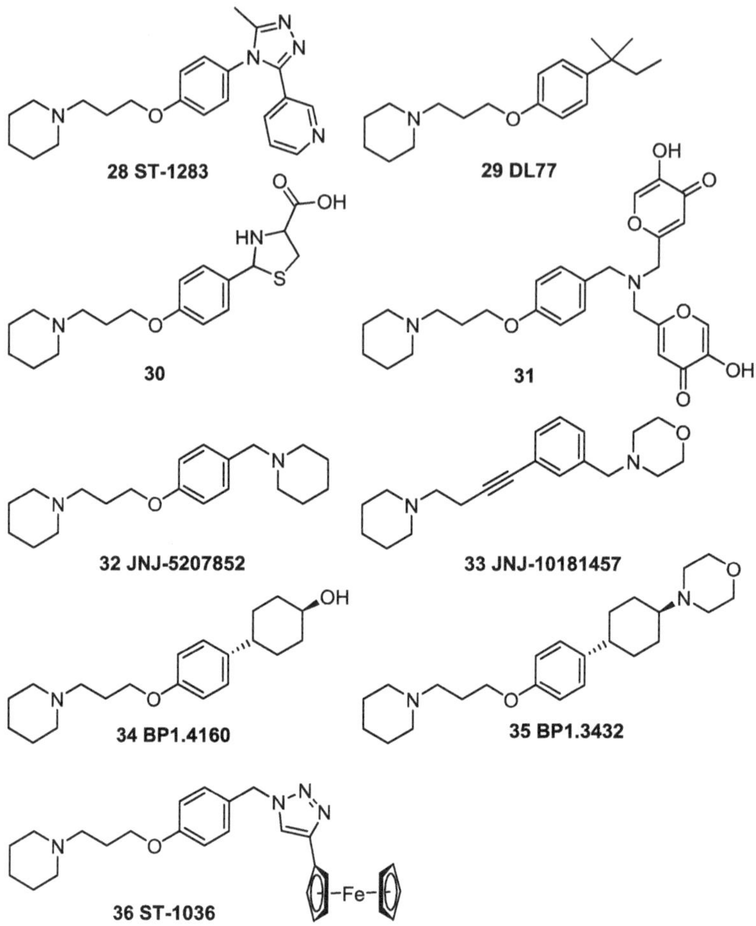


**Fig. 5.9** Propyl-piperidine-containing $H_3R$ antagonists

well tolerated ($hH_3R$ $-\log K_i < 5$ for most compounds), while it was well accepted in the eastern part of general structural pattern for $H_3R$ antagonists, e.g., **31** ($hH_3R$ $-\log K_i = 7.98$) (Fig. 5.9) [90].

Furthermore, highly potent $H_3R$ ligands **JNJ-5207852** (**32**) ($hH_3R$ $pK_i = 9.24$; $rH_3R$ $pK_i = 8.90$) [91] and **JNJ-10181457** (**33**) ($hH_3R$ $pK_i = 8.93$; $rH_3R$ $pK_i = 8.15$) [52] were described (Fig. 5.9) and reported to promote wakefulness in rat, mouse, and dog models [48, 52, 91], a property of $H_3R$ antagonists as a class [92]. **JNJ-5207852** (1–10 mg/kg, s.c.) increased time spent awake and decreases REM sleep and slow-wave sleep in mice and rats but failed to have an effect on wakefulness or sleep in $H_3R$ knockout mice [91]. Also, **JNJ-5207852** was found to prevent memory deficit induced by pentylenetetrazole kindling in weanling mice [93]. However, PK observations in rats and dogs revealed an undesirably long half-life for **JNJ-5207852** ($t_{1/2} > 13$ h) with a long brain residency time after single bolus administration

(10 mg/kg p.o.) which allowed a full receptor occupancy that was still observed after 24 h of administration [52, 91, 93, 94]. In addition, **JNJ-5207852** was found to induce phospholipidosis in rats [52]. Therefore, further structural modifications provided an understanding of the effect on PK properties and led to the development of several additional series of $H_3R$ antagonists including **JNJ-10181457**, a compound with improved PK properties and of rapid brain penetration and a maximum $H_3R$ occupancy ($H_3RO$) of 85 % achieved after 1 h following oral administration (10 mg/kg). Interestingly, a wake-promoting effect of **JNJ-10181457** was also demonstrated in mice and rat without association with an increase in locomotor activity up to 30 mg/kg i.p. in rat [52]. A study revealed that orexin/ataxin-3 narcoleptic mice were more sensitive to **JNJ-10181457** than their wild-type littermate as larger wake-promoting effects during the light period were observed in this animal model of narcolepsy [95]. In addition, **JNJ-10181457** (1.25–10 mg/kg, p.o.) significantly reduced cataplexy (number of cataplectic attacks and time spent in cataplexy) in familial narcoleptic Dobermans by using the standard food-elicited cataplexy test. Notably, the potency of the anticataplectic effects of **JNJ-10181457** was similar to that of desipramine (a tricyclic antidepressant currently used for the treatment of human cataplexy). While the definitive mechanism of action for the anticataplectic effects of $H_3R$ antagonists like **JNJ-10181457** remains to be determined, it is relatively well established that enhancement of central noradrenaline levels reduces cataplexy [95]. Microdialysis studies in freely moving rats demonstrated that **JNJ-10181457** increased extracellular noradrenaline level [52] and also increased extracellular acetylcholine levels in the frontal cortex (at 10 mg/kg, s.c.), explaining the significant memory improvement on the level of acquisition model conducted in spontaneously hypertensive rat pups subjected to a passive avoidance task [52]. In another preclinical study, **JNJ-10181457** (10 mg/kg, i.p.) significantly reversed the scopolamine-decreased percentage correct responding in the delayed nonmatching to position task, indicating that selective blockade of $H_3Rs$ might have therapeutic utility for the treatment of working memory deficits and learning disorders, especially those in which acetylcholine neurotransmission is compromised [96].

Also, further structural diversifications of **JNJ-10181457** led to the introduction of a hydroxyl group in the eastern part of the molecule with the potential goal to favor metabolic reactions and clearance of the developed compound **34** (**BP1.4160**) (Fig. 5.9). As a result, the cyclohexyl derivative **34** showed nanomolar binding affinity ($hH_3R$ $-\log K_i = 8.59$) and favorable PK profile with reduced half-life ($t_{1/2} = 2.6$ h). Moreover, compound **BP1.4160** demonstrated high in vivo efficacy ($ED_{50} = 0.3$ mg p.o.), satisfactory oral absorption, as well as brain penetration and exhibited no significant interaction with the hERG channel ($K_i = 5.8$ μM) and cytochrome P450 3A4, 2D6, and 2C9 ($IC_{50} > 30$ μM) [79]. Based on a QSAR (quantitative structure activity relationships) model for low-hERG-affinity scaffolds, cyclohexylamine derivatives were identified, where the most interesting structure **35** (**BP1.3432**) (Fig. 5.9) displayed high in vitro affinity ($hH_3R$ $-\log K_i = 9.39$) and in vivo efficacy ($ED_{50} = 1.4$ mg/kg p.o.) again without significant interaction with the hERG channel (34 % inhibition at 10 μM) and activity on CYP 3A4, 2D6, and 2C9 ($IC_{50} > 40$ μM) [97].

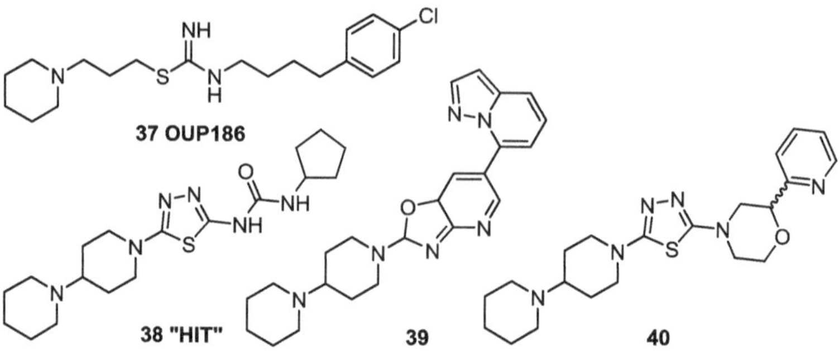


**Fig. 5.10** Isothiourea piperidinylpropyl derivative and fused bicyclic compounds

For the first metal-containing $hH_3R$ ligands, ferrocene sandwich complexes were developed through coupling via different linkers to the *N*-piperidinylpropyloxyphenyl moiety. One of the most potent compounds of this series was **36 (ST-1036)** (Fig. 5.9) with $hH_3R$ $pK_i$ value of 8.7 and no affinity for $hH_1R$ and $hH_4R$ ($pK_i < 5$) [98].

New series of isothiourea piperidinylpropyl derivatives, structurally strongly related to clobenpropit (**2**), showed antagonistic activities on $hH_3Rs$, while some of them exhibited high affinity for $hH_3R$, e.g., **37 (OUP186)** ($pIC_{50} = 8.2$; $pA_2 = 9.6$) (Fig. 5.10), and neither agonistic nor antagonistic activities toward $hH_4Rs$, even at a concentration of 1 μM. Unfortunately, these compounds were found to be inactive in vivo in the rat microdialysis test model [99].

Encouraged by earlier preclinical findings with imidazole-based $H_3R$ antagonist **thioperamide** showing involvement of central histaminergic system in regulation of food intake and obesity, extensive drug discovery efforts led to development of centrally acting $H_3R$ antagonists belonging to the fused bicycle class with the most promising $H_3R$ antagonist showing improved PK profile with good in vitro and confirmed oral in vivo potency in rat and mice obesity models [100]. Compounds obtained as a modifications of hit **38** ($hH_3R$ $-\log K_i = 7.31$) displayed high in vitro nanomolar binding affinities (e.g., **39** and **40**) (Fig. 5.10), but some derivatives showed weak PK properties and/or high hERG inhibition such as **39** ($hH_3R$ $-\log K_i = 8.40$; hERG 62 % inhibition at 10 μM) [101]. Interestingly, compound **39** (at a dose of 30 mg/kg) demonstrated both antidiabetic (63 mg/L, glucose level reduction) and anti-obesity effect (3.85 % food inhibition, 0.64 % body weight loss) in streptozotocin (STZ) diet-induced obesity (DIO) type 2 diabetic mice (STZ-DIO). The problem of hERG inhibition could be resolved with the development of compound **40** ($hH_3R$ $-\log K_i = 8.52$; 26 % hERG inhibition at 10 μM; 74 % ex vivo at 30 mpk p.o.) [101]. Compound **40** demonstrated antihyperglycemic effect in 2 days when compared with vehicle-treated mice, whereas it dose-dependently blocked the increase of glycated hemoglobin (HbA1c) following a 12-day treatment in STZ-DIO mice model [101–103].

Notably, several publications showed that thiazole or other heterocycles as a central core, also, are well tolerated by $hH_3R$ with the conclusion that the exact type of heterocycle shows moderate influence on the binding ability of the respective ligand

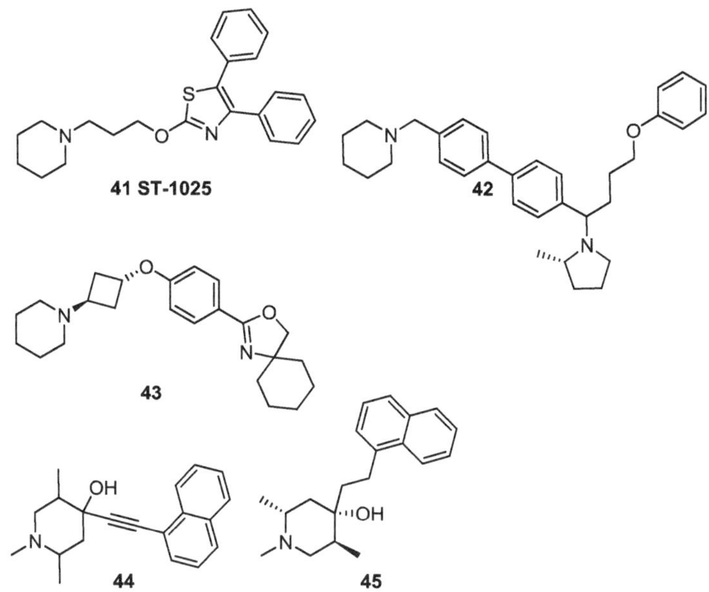


**Fig. 5.11** Structures of selected piperidine and piperidinol $H_3R$ antagonists

to $H_3Rs$. The largest influence on the affinity could be observed for substituents in the eastern part of the molecule. Generally, it has been found that large lipophilic or basic substituents increased receptor binding, e.g., **41 (ST-1025)** ($hH_3R$ $-\log K_i = 7.95$) (Fig. 5.11) [104]. Noticeably, dibasic piperidine derivatives with a biphenyl scaffold were described by Bordi et al. as a continuation of previous work. SAR analysis in this series revealed that two basic groups were important for $hH_3R$ binding as it has been observed for the most potent compound **42** ($hH_3R$ $pK_i = 8.90$; $rH_3R$ $pK_i = 8.18$) (Fig. 5.11) [105].

With a large number of histamine $H_3R$ antagonists/inverse agonists, a 3-propyloxy linker connected to an azacycloalkylamine moiety (e.g., piperidine, pyrrolidine, piperazine, or 1,4-diazepine) with a central core group is found. Consequently, rigidifying this n-alkyl linker by means of a cyclobutyl constraint in the series of oxazoline and oxazole derivatives was described. Accordingly and considering the two stereoisomers (*cis* and *trans*), *trans*-3-piperidinocyclobutanol derivative **43** ($hH_3R$ $pK_i = 8.70$) (Fig. 5.11) demonstrated an affinity-enhancing effect for $hH_3Rs$ when compared to unconstrained derivative ($hH_3R$ $pK_i = 7.10$; structure not shown) or the *cis* analog ($hH_3R$ $pK_i = 6.60$; structure not shown) [106].

In addition, interesting piperidinol derivatives were described by scientists from Athersys [107]. Following HTS campaign, the hit diastereoisomer **44** (Fig. 5.11) was found to be a potent $hH_3R$ antagonist ($hH_3R$ $IC_{50} = 11$ nM), however, with moderate affinity at mouse $H_3R$ ($mH_3R$) ($IC_{50} = 280$ nM). Further modifications on this track led to compounds like **45** (Fig. 5.11) with high $hH_3R$ potency ($hH_3R$ $IC_{50} = 2.0$ nM; hERG 16% inhibition at 10 μM), but still moderate $mH_3R$ activity ($mH_3R$

$IC_{50}$ = 85 nM). The discrepancy in potencies observed for these compounds excluded them from further studies as many of the in vivo assays were carried out in mice [107].

Concerning structural modifications addressing more rigidified compounds, some $hH_3R$ antagonists were developed and a constrained *N*-substituted piperidinyloxy moiety instead of the *N*-piperidinyl(pyrrolidinyl)propyloxy side chain achieve was incorporated [108–112]. Compounds with such basic amine core were described previously by other researchers (for review, see [11, 113]). Also, a reduced number of rotatable bonds have been found to improve PK parameters, especially in regard to oral bioavailability [11, 113]. Based on these observations, compounds possessing *N*-cyclobutyl-piperidinoxyl moiety were described, as a continuation of the work conducted with 4,5-dihydropyridazin-3(2*H*)-one moiety, in which the conformational restricted 4-phenoxypiperidine fragment, nonaromatic 4,5-dihydropyridazin-3-one, and cyclopropylfused pyridazin-3-one sub-series were investigated [66, 108, 109, 112] providing compounds **46** [108], **47** [109], and **48** [66] with good subtype selectivity and low-hERG inhibition (hERG $IC_{50}$ > 10 μM). The latter structural change barely influenced $hH_3R$ affinity (compare **47** versus **18**). The increase in bioavailability in rats was pronounced, whereas that in monkeys and dogs were already at a high level facilitating the transfer of different animal models (bioavailability in rat, F = 92 % at 5 mg/kg p.o. for **47** versus F = 34 % at 10 mg/kg p.o. for **18**; bioavailability in monkey, F = 114 % at 3 mg/kg p.o. for **47** versus F = 136 % at 3 mg/kg p.o. for **18;** bioavailability in dog, F = 94 % at 3 mg/kg p.o. for **47** versus F = 115 % at 3 mg/kg p.o. for **18** [67, 109]). Moreover, compound **47** (3 mg/kg i.p.) demonstrated a very potent wake-promoting effect in the rat EEG/EMG sleep–wake model and robust wake at higher doses (up to 30 mg/kg i.p.) without adverse events. In addition, compound **47** (0.03–0.3 mg/kg p.o.) enhanced short-term memory in the rat social recognition memory model [67, 109].

A further rigidified compound is **CEP-32215** (**49**) ($hH_3R$ $-\log K_i$ = 8.70; $rH_3R$ $-\log K_i$ = 8.44; hERG $IC_{50}$ = 16 μM) [114], a spiro-piperidine structure, which shows consistent oral bioavailability among rat, dog, and monkey with brain-to-

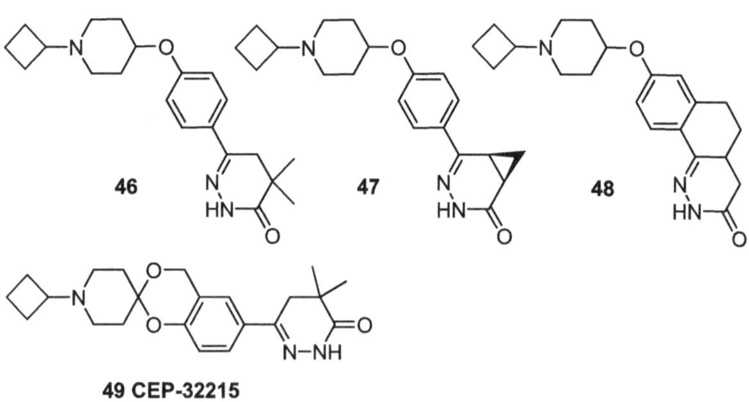


**Fig. 5.12** Selected N-substituted-piperidinoxyl derivatives and structurally related CEP-32215

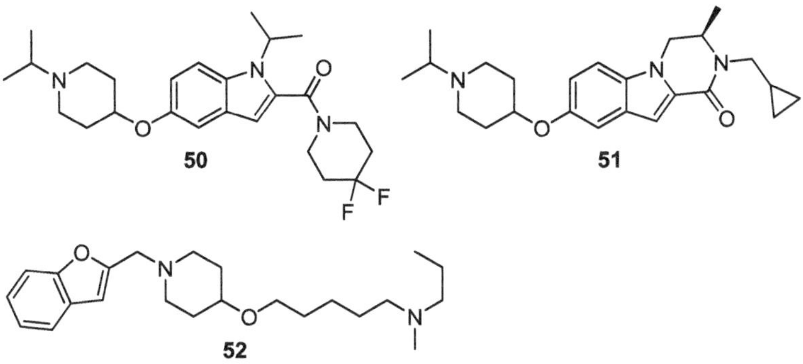


**Fig. 5.13** Piperidinyloxy compounds 50, 51, and 52

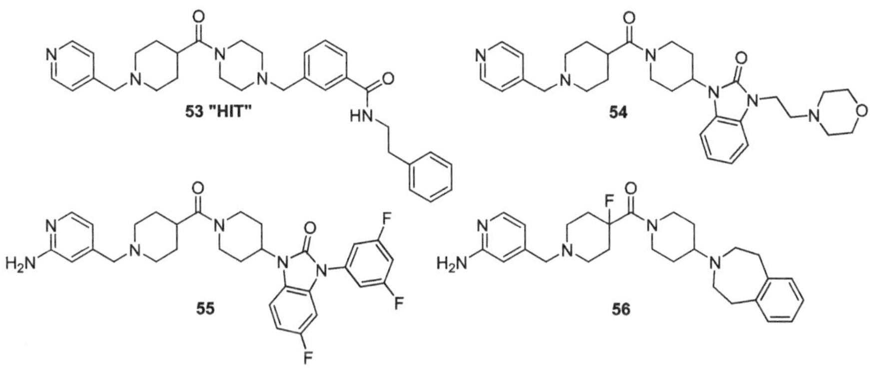


**Fig. 5.14** Structures of N-substituted piperidinylcarbonyl derivatives

plasma ratio of 3.3 at 2 h after administration of 10-mg/kg dose (i.p.). In a preclinical study, it significantly increased wake duration in the rat EEG model at 3–30 mg/kg p.o. without increasing motor activity or sleep rebound, also, following doses administrated up to 100 mg/kg p.o. of **CEP-32215**, indicating an acceptable therapeutic index for this compound (Table 5.2). Moreover and as demonstrated by the inhibition of (*R*)-alpha-methylhistamine-induced drinking in the rat dipsogenia model ($ED_{50}$ − 0.92 mg/kg), **CEP-32215** displays functional antagonism in the brain. Therefore, **CEP-32215** as a wake-promoting histamine $H_3R$ antagonist was proposed for potential utility in the treatment of a variety of sleep disorders [30, 114] (Fig. 5.12).

Additional conformationally restricted indole derivatives incorporating *N*-isopropyl-piperidin-4-yloxy moiety were described [110], while modifications were introduced to previously described indole compounds, e.g., **50** (Fig. 5.13) ($hH_3R$ $-\log K_i = 8.30$; $hH_3R$ $-\log K_i = 8.16$) [115]. Consequently, new compounds with tricyclic amide core (e.g., **51**) (Fig. 5.13) were designed and displayed high $hH_3R$ affinity as well as selectivity over other HR subtypes. Tested in a functional GTPγS assay, such compounds were characterized as potent full inverse agonists without significant inhibition of CYP450. When compared to compound **50,** the conformationally restricted structures did not afford improvements in potency or pharmacokinetic properties.

Later described dibasic 1-substituted methyl-piperidinyloxy derivatives with pentyl carbon chain linker and dialkylamine moiety were tested in vitro on isolated guinea pig jejunum preparations, and some of them, e.g., **52** (Fig. 5.13), exhibited high potency for histamine $H_3R$ (guinea pig $H_3R$ $pA_2=8.47$) without antagonistic activity at $H_1Rs$ (guinea pig $H_1R$ $pA_2<4$) [111].

A series of *N*-substituted piperidinylcarbonyl derivatives were obtained after structural modifications of hit **53** (guinea pig $H_3R$ $-\log K_i=7.57$) leading to benzimidazolyl derivatives, e.g., **54** (guinea pig $H_3R$ $-\log K_i=9.30$) (Fig. 5.14) [116]. Further optimization was undertaken with the aim to improve rat oral PK and to decrease CYP450 inhibition led to a new series of benzimidazolone [117]. Also, an introduction of the 2-amino group in the pyridine moiety improved CYP450 enzyme profile, e.g., **55** (guinea pig $H_3R$ $-\log K_i=9.02$; CYP3A4 and CYP2D6 $IC_{50}>30$ μM; hERG <10 % inhibition at 10 μM) (Fig. 5.14). Moreover, incorporation of substituents with increased lipophilic character significantly optimized PK profile, e.g., rat oral AUC for **55** was 12.9 μM h (at 10 mg/kg), and good plasma concentrations (1.85 μM after 6 h) [117]. Following further research efforts, replacement of the benzimidazolone moiety with other bicyclic or tricyclic rings led to the development of compound **56** ($hH_3R$ $-\log K_i=9.00$) (Fig. 5.14) with excellent $hH_3R$-binding affinity and an unusually high brain/plasma ratio of 97 in rat [118]. Based on the safety profile (clean hERG and CYP450 profile, no behavioral changes in the Irwin test), good rat PK parameters (rat 10 mg/kg: AUC 97 h ng/mL, $C_{max}$ 25 ng/mL, and $C_{6h}$ brain 1067 ng/g) and excellent receptor occupancy (98 %), compound **56** was selected for in vivo efficacy studies in a mouse model of diet-induced obesity (DIO). Following oral administration, compound **56** significantly reduced body weight relatively to control mice (inhibition: 89 %, 52 %, and 38 % at 10, 3, and 1 mg/kg, respectively) [118].

## 5.5 Piperazine/Homopiperazine and Benzoazepine Derivatives

Several $H_3R$ ligands described in the past contain a piperazine/homopiperazine (diazepane) or benzoazepine moiety as a tertiary amine moiety. In most cases the basicity of the second nitrogen was decreased by a neighboring carbonyl motif (piperazinyl-/homopiperazinyl-methanones). Optimization of physicochemical properties of this class and extensive preclinical studies led to the identification of several clinical candidates such as **bavisant** (**57**; **JNJ-31001074**) ($hH_3R$ $-\log K_i=8.27$; hERG $IC_{50}>10$ μM) (Fig. 5.15) [119], **GSK-239512** ($hH_3R$

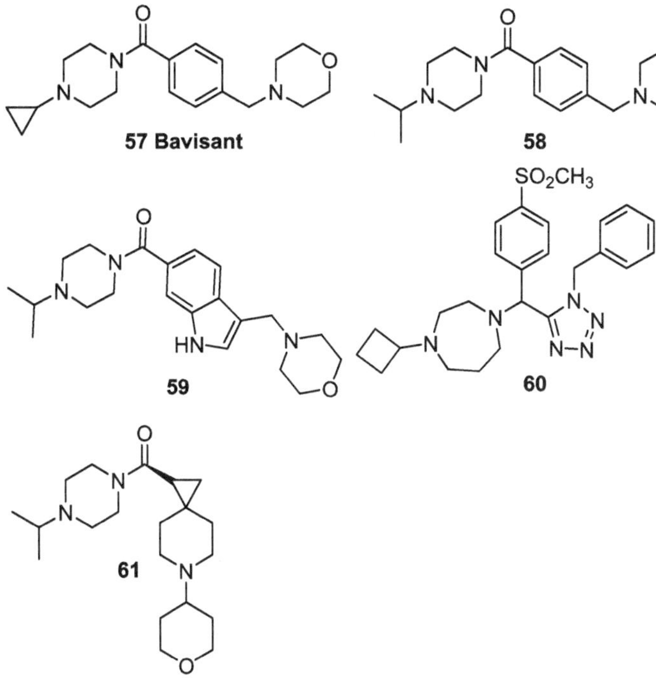


**Fig. 5.15** (Homo)piperazine and piperazinyl-methanone derivatives

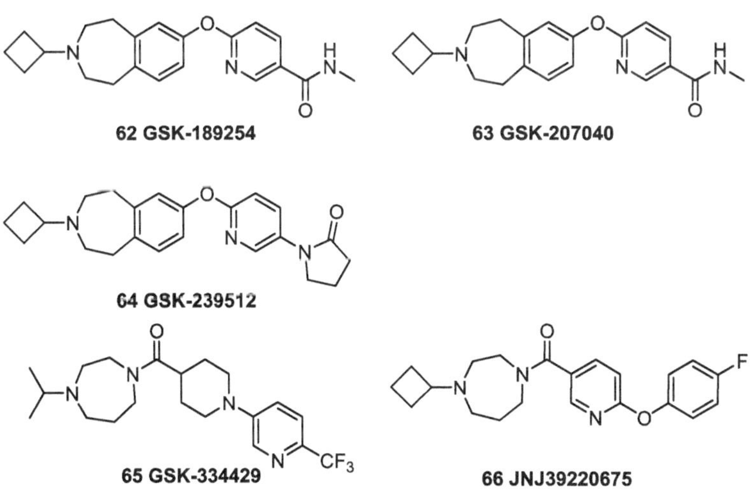


**Fig. 5.16** Potent benzazepine and homopiperazine derivatives

–log$K_i$=9.90; hERG inhibition <10 % at 10 μM) (**65**), or **GSK-189254** ($hH_3R$ –log$K_i$=8.27; hERG inhibition 36 % at 10 μM) (**63**) (Fig. 5.16) [120] (discussed later). Compound **58** (Fig. 5.15), structurally closely related to bavisant, showed comparable $hH_3R$ affinity and hERG modulation ($hH_3R$ –log$K_i$=8.60; hERG $IC_{50}$>10 μM), but higher plasma half-life times compared to **bavisant** [119].

In another series of piperazinyl-methanone derivatives, the replacement of the central phenyl core with a 6,5-bicyclic aromatic ring, namely, indole and benzothiophene, was investigated [121]. Most of the developed compounds exhibited good to excellent $hH_3R$ affinities ($hH_3R$ $K_i$<30 nM). Moreover, the indole-based analogs showed promising rat PK properties, e.g., 100 % bioavailability as observed for compound **59** ($hH_3R$ –log$K_i$=8.68) (Fig. 5.15) [121].

A series of methylene-linked benzyltetrazoles revealed that the substituent on the methylene bridge had a significant effect on the in vitro affinity and selectivity profile. Most promising compounds exhibited high $hH_3R$ affinities, moderate microsomal stability, and minimal CYP450 inhibition such as **60** ($hH_3R$ –log$K_i$=8.80) (Fig. 5.15) [122].

Notably, a preclinical study was conducted with already-mentioned **NNC38-1049** ($hH_3R$ –log$K_i$=8.92; $rH_3R$ –log$K_i$=8.29) (**8**) (Fig. 5.3) to elucidate whether antagonistic targeting of the $H_3Rs$ increases hypothalamic histamine levels, in parallel with decreases in food intake and body weight [38]. Following single dosing of normal rats with **NNC38-1049**, hypothalamic histamine levels, and acute effects on food intake and energy were assessed for 15 days during daily dosing. The results showed that **NNC38-1049** (20 mg/kg i.p.) significantly increased extracellular histamine concentrations and did not alter pica behavior and conditioned taste aversion [38]. However, reductions in food intake were seen very soon after administration and occurred in a dose-dependent fashion, demonstrating that twice daily administration of 20 mg/kg of **NNC38-1049** in old and dietary obese rats resulted in sustained reduction of food intake throughout a 2-week period without any changes in energy expenditure. These results strongly support the idea that an increase in the hypothalamic concentration of histamine produces a specific reduction of food intake and that this effect can be translated into a decrease in body weight [38, 48]. However, PK profile of **NNC38-1049** needs additional elucidation since with $t_{1/2}$ of 0.33 h in rats, it showed the shortest half-life among the so far preclinically investigated $H_3R$ antagonists.

Further structural modifications led to the development of piperazine and diazepine amide derivatives in which the amide functionality is tethered to spirofused rings (e.g., spirocyclopropyl, spirocyclobutyl, or five- and six-membered ring) such as **61** ($hH_3R$ –log$K_i$=9.10; $hH_3R$ $IC_{50}$=8.3 nM; hERG $IC_{50}$>33 μM) (Fig. 5.15) with high brain exposures and high oral bioavailability ($F$=99 %) [123]. Compound **61** tested at three different doses (1, 3, and 10 mg/kg s.c) demonstrated a statistically significant response compared with the control group in a mouse model of learning and recognition memory [123].

Further drug development efforts led to a benzazepine core, which can mimic a piperidinylpropyloxy group. Among numerous derivatives developed so far, a cyclobutyl substituent at the *N1* position of benzoazepine moiety was found to be

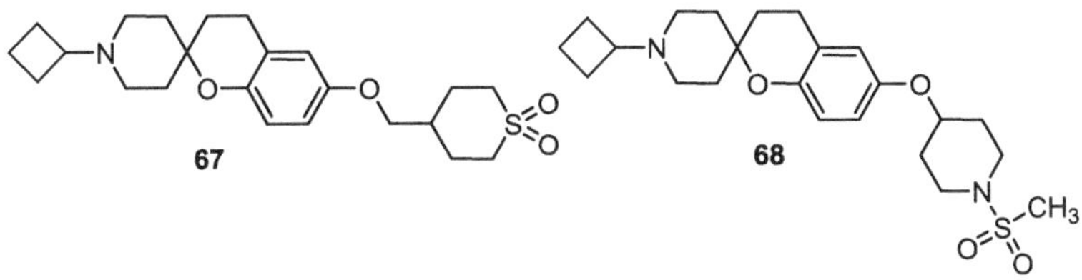

**Fig. 5.17** Azacycloalkanes with spiro-piperidine core

well tolerated by $hH_3Rs$ (**62–64**) (Fig. 5.16). The homopiperazinyl-methanone **GSK-334429** (**65**) ($hH_3R$ p*K*i = 9.08; $rH_3R$ p$K_i$ = 9.12) (Fig. 5.16) [124] and the benzazepine derivative **GSK-207040** (**63**) ($hH_3R$ –p*K*i = 9.72; $rH_3R$ p$K_i$ = 8.95) (Table 5.1) [120] showed excellent oral bioavailability of 91 % ($t_{1/2}$ 2.0 h) and 88 % ($t_{1/2}$ 2.6 h) (Table 5.2), respectively. Both potent $H_3R$ antagonists were confirmed for their effectiveness in preclinical models of scopolamine-induced memory impairment, neuropathic pain, and capsaicin-induced tactile allodynia [125, 126]. In this regard, **GSK-207040** and **GSK-334429** have been found to block RAMH-induced dipsogenia in rats ($ED_{50}$ = 0.02 and 0.11 mg/kg p.o. for **GSK-207040** and **GSK-334429**, respectively). In more pathophysiologically relevant pharmacodynamic models, **GSK-207040** (0.1, 0.3, 1 and 3 mg/kg p.o.) and **GSK-334429** (0.3, 1 and 3 mg/kg p.o.) significantly reversed scopolamine-induced amnesia in the passive avoidance paradigm in rats [125]. In addition, **GSK-207040** (0.1, 0.3, and 1 mg/kg, p.o.) and **GSK-334429** (3 and 10 mg/kg p.o.) significantly reversed capsaicin-induced reductions in paw withdrawal threshold, signifying for the first time that blockade of H3Rs may be able to reduce tactile allodynia [125]. A separate study revealed efficacy in an osteoarthritis preclinical pain model for both $H_3R$ antagonists **GSK-189254** (discussed later) and **GSK-334429** [45].

Likewise, low doses of the homopiperazinyl-methanone **JNJ-39220675** (**66**) ($hH_3R$ –log*K*i = 8.85; $rH_3R$ –log*K*i = 7.64) (Fig. 5.16) [127] were found to inhibit alcohol reward in experimental mice and rat models (as well as the above-described **JNJ-10181457** (**33**)) with no effect observed in histidine decarboxylase knockout (HDC KO) mice [128], while psychostimulant-induced reward remained unaffected [129]. These results suggest that histamine is required for the $H_3R$-mediated inhibition of alcohol-conditioned place preference and support the hypothesis that the brain histaminergic system has an inhibitory role in alcohol reward. These observations shed light on a potential and novel way of treating alcohol dependence [96, 127, 128]. Interestingly, **JNJ-39220675** was also found to have decongestant efficacy in an early clinical trial in subjects with allergic rhinitis [130].

Another interesting class of compounds with constrained azacycloalkanes by spiro-piperidine core (spiro-3,4-dihydrobenzopyran-2′,4′-piperidine) and structurally related to earlier described **CEP-32215** (**49**) was described by Becknell et al. [131] and Dandu et al. [132]. A series of alkyl and cyclic ethers or sulfone (**67**) ($hH_3R$ –log$K_i$ = 8.40; $rH_3R$ –log$K_i$ = 8.05; $hH_4R$ –log$K_i$ < 5.00; hERG $IC_{50}$ > 33 μM) [131] and sulfonamide (**68**) ($hH_3R$ –log$K_i$ = 8.16; $rH_3R$ –log$K_i$ = 7.77; $hH_4R$ inhibition <20 % at 10 μM) [132] derivatives (Fig. 5.17) revealed high $hH_3R$ affinity.

Both compounds exhibited subtype selectivity, metabolic stability, and favorable PK properties but only weak bioavailability in rat, e.g., $F=4\,\%$ for **69** [131, 132].

In conclusion, there have been immense research efforts that resulted in rapid progress over the last decade leading to several promising $H_3R$ antagonists with excellent in vivo efficacy in diverse preclinical models, and preclinical data observed so far for clinical candidates have been summarized in Table 5.2.

## 5.6 Clinical Candidates: Drawbacks and Progress in Drug development

In the past decade, numerous $H_3R$ antagonists have progressed to different phases of clinical trial stage, including **BF2.649 (pitolisant)**, **PF-03654746**, **PF-03654764**, **CEP-26401**, **ABT-288**, **ABT-652**, **GSK-189254**, **GSK-239512**, **GSK-1004723**, **AZD5213**, **SAR-110894**, **MK-0249**, **MK-3134**, **SCH-497079**, **JNJ-31001074**, **JNJ-17216498**, and **HPP-404** (Table 5.3, Figs. 5.18 and 5.19). Numerous compounds have completed safety and tolerability studies (phase 1), and some have been progressed to phase 2 studies on efficacy and beyond.

Of particular interest is **BF-2.649**, also known as **pitolisant** or **tiprolisant** [7, 36, 133, 134], a potent $H_3R$ antagonist ($hH_3R$ $-\log K_i=9.52$–$9.00$; $rH_3R$ $-\log K_i=7.77$) (Tables 5.1 and 5.2) with an $ED_{50}$ value of 1.6 mg/kg and a maximal effect elicited at 10 mg/kg (p.o.) and 3–5 mg/kg (i.p.) in an ex vivo binding assay (Table 5.2) [47, 48, 94, 133]. It shows good oral bioavailability (84 %) with AUC of 7,916 and $t_{1/2}$ of 2 h in mice at a dose of 10 mg/kg p.o. and good brain penetration ($C_{max,\ brain}/C_{max,plasma}=23.5$) [94, 135] (Table 5.2). Preclinically, **BF2.649** has shown to enhance dopamine and acetylcholine levels in microdialysates of the prefrontal cortex in rats [133]. Moreover, it markedly enhanced wakefulness at the expense of sleep states in cats and also enhanced fast cortical rhythms of the electroencephalogram, known to be associated with improved vigilance. A memory improvement effect was shown for **BF2.649** regarding either scopolamine-induced or natural memory impairment in the two-trial object recognition test in mice [133]. These preclinical observations suggested that **BF2.649** is a valuable drug candidate to be developed in wakefulness or memory deficits and other cognitive disorders [9, 94]. Notably, **BF2.649** completed several clinical phase 3 trials for a number of indications and has been found to promote wakefulness in patients with excessive daytime sleepiness (EDS) associated with narcolepsy (phase 3), PD (phase 3), obstructive sleep apnea syndrome (phase 3), and epilepsy (phase 3) (http://clinicaltrials.gov). It, also, completed a phase 2 clinical study addressing cognitive impairment associated with schizophrenia (CIAS) and showed in doses up to 20 mg/day procognitive effects in these patients and was effective in a proof-of-concept human photosensitivity model, suggesting that $H_3R$ antagonists are of therapeutic value in epilepsy [136]. **BF2.649** displayed efficacy in a 22-patient narcolepsy phase 3 trial in which a 40-mg daily dose reduced sleepiness versus placebo, with efficiency in the Epworth Sleepiness

**Table 5.3** Summary of $H_3R$ antagonists which are/have been in clinical trials

| Compound | Company | Number of trials | Clinical indication/condition | Phase | Status[a] | Clinical trials identifier |
|---|---|---|---|---|---|---|
| ABT-288 | Abbott | 3 | Schizophrenia | 1 | C | NCT00888693 |
| | | | AD | 2 | C | NCT01018875 |
| | | | CIAS | 2 | C | NCT01077700 |
| ABT-652 | Abbott | 4 | Diabetic neuropathic pain | 2 | T | NCT01579279 |
| | | | Osteoarthritis pain | 2 | C | NCT01207115 |
| | | | Sleep disorder | 1 | C | NCT01124851 |
| | | | Osteoarthritis pain (in combination with NSAIDs) | 2 | C | NCT01444365 |
| AZD5213 | AstraZeneca | 7 | Brain distribution | 1 | C | NCT01194986 |
| | | | Tolerability | 1 | C | NCT01121302 |
| | | | Healthy Japanese (safety and tolerability) | 1 | C | NCT01335451 |
| | | | Healthy (safety, tolerability, pharmacokinetic) | 1 | C | NCT01171105 |
| | | | Mild AD, mild cognitive impairments | 2 | C | NCT01548287 |
| | | | Tourette's syndrome | 2 | C | NCT01904773 |
| | | | Diabetic neuropathy (in combination with pregabalin) | 2 | C | NCT01928381 |

(continued)

**Table 5.3** (continued)

| Compound | Company | Number of trials | Clinical indication/condition | Phase | Status[a] | Clinical trials identifier |
|---|---|---|---|---|---|---|
| BF2.649 | Bioprojet | 14 | Narcolepsy, catalepsy, EDS (in combination with modafinil) | 3 | C | NCT01067235 |
| | | | Narcolepsy, catalepsy, EDS | 3 | C | NCT01800045 |
| | | | Treatment of EDS in narcolepsy | 3 | C | NCT01638403 |
| | | | Treatment of EDS in narcolepsy | 3 | C | NCT01067222 |
| | | | Narcolepsy (in combination with sodium oxybate) | 3 | C | NCT01789398 |
| | | | Narcolepsy (long term) | 3 | A | NCT01399606 |
| | | | EDS, PD | 2 | C | NCT00642928 |
| | | | EDS, PD | 3 | C | NCT01036139 |
| | | | EDS, PD | 3 | C | NCT01066442 |
| | | | Obstructive sleep apnea, EDS | 2 | C | NCT01620554 |
| | | | Obstructive sleep apnea, EDS (refusing CPAP) | 3 | C | NCT01072968 |
| | | | Obstructive sleep apnea, EDS (treated with CPAP) | 3 | C | NCT01071876 |
| | | | Schizophrenia | 2 | C | NCT00690274 |
| | | | Renal impairment | 1 | C | NCT01619033 |
| CEP-26401 | Teva Pharmaceutical Industries | 1 | Cognitive impairment | 1 | C | NCT01903824 |

| Compound | Company | Number of trials | Clinical indication/condition | Phase | Status[a] | Clinical trials identifier |
|---|---|---|---|---|---|---|
| GSK-189254 | GlaxoSmithKline | 3 | Hyperalgesia | 1 | C | NCT00387413 |
| | | | Narcolepsy | 2 | T | NCT00366080 |
| | | | Mild cognitive impairment, dementia | 1 | C | NCT00474513 |
| GSK-239512 | GlaxoSmithKline | 6 | AD | 1 | C | NCT00675090 |
| | | | Multiple sclerosis | 2 | C | NCT01772199 |
| | | | Schizophrenia | 2 | C | NCT01009060 |
| | | | Multiple sclerosis | 1 | C | NCT01802931 |
| | | | Mild cognitive impairment, dementia | 1 | C | NCT00474513 |
| | | | AD | 2 | C | NCT01009255 |
| GSK-835726 | GlaxoSmithKline | 3 | Allergic rhinitis | 2 | C | NCT00851344 |
| | | | Allergic rhinitis (proof of concept) | 2 | C | NCT00972504 |
| | | | Allergic rhinitis | 1 | C | NCT00605852 |
| GSK-1004723 | GlaxoSmithKline | 3 | Allergic rhinitis | 2 | C | NCT00824356 |
| | | | Allergic rhinitis(proof of concept) | 2 | C | NCT00972504 |
| | | | Allergic rhinitis | 1 | C | NCT00694993 |
| JNJ-17216498 | Johnson & Johnson | 1 | Narcolepsy | 2 | C | NCT00424931 |

(continued)

**Table 5.3** (continued)

| Compound | Company | Number of trials | Clinical indication/condition | Phase | Status[a] | Clinical trials identifier |
|---|---|---|---|---|---|---|
| JNJ-31001074 | Johnson & Johnson | 7 | ADHD (adults) | 2 | C | NCT00566449 |
| | | | Healthy (pharmacokinetics) | 1 | C | NCT00915434 |
| | | | Healthy (pharmacokinetics, drug interactions) | 1 | C | NCT00915746 |
| | | | ADHD (adults) | 2 | C | NCT00880217 |
| | | | ADHD (children 6–11) | 1 | C | NCT00890240 |
| | | | ADHD (children 12–17) | 1 | C | NCT00890292 |
| | | | Healthy (drug interaction) | 1 | C | NCT01159821 |
| JNJ-39220675 | Johnson & Johnson | 1 | Seasonal allergic rhinitis | 2 | C | NCT00804687 |
| MK-0249 | Merck | 5 | ADHD | 2 | C | NCT00475735 |
| | | | AD, dementia | 1 | T | NCT00874939 |
| | | | AD | 2 | C | NCT00420420 |
| | | | Sleep apnea, obstructive hypopnea syndrome, EDS | 2 | T | NCT00620659 |
| | | | Paranoid schizophrenia | 2 | C | NCT00506077 |
| MK-3134 | Merck | 3 | Dementia (scopolamine induced) | 1 | C | NCT01181310 |
| | | | Dementia (EEG) | 1 | C | NCT01110616 |
| | | | Dementia (fMRI and CBF) | 1 | C | NCT00887601 |

| Compound | Company | Number of trials | Clinical indication/condition | Phase | Status[a] | Clinical trials identifier |
|---|---|---|---|---|---|---|
| PF-03654746 | Pfizer | 7 | ADHD | 2 | C | NCT00531752 |
| | | | Healthy (PET) | 1 | C | NCT00730990 |
| | | | Mild–moderate AD | 1 | T | NCT01028911 |
| | | | Tourette's syndrome | 2 | T | NCT01475383 |
| | | | CIAS | 1 | C | NCT01346163 |
| | | | EDS, narcolepsy | 2 | C | NCT01006122 |
| | | | Allergic rhinitis | 2 | C | NCT00562120 |
| PF-03654764 | Pfizer | 2 | Allergic rhinitis | 2 | C | NCT01033396 |
| | | | Healthy (in combination with fexofenadine) | 1 | T | NCT01298505 |
| | | | Healthy | 1 | C | NCT00989391 |
| SAR-110894 | Sanofi-Aventis | 1 | Dementia (in combination with donepezil) | 2 | C | NCT01266525 |
| SCH-497079 | Schering-Plough | 2 | Obesity (in obese patients) | 2 | C | NCT00642993 |
| | | | Obesity (in patients with diabetes mellitus type 2) | 1 | C | NCT00673465 |
| HPP-404 | High Point Pharmaceuticals | 1 | Obesity | 2 | T | NCT01540864 |

[a]Status: *A* active not recruiting, *C* completed, *T* terminated, *AD* Alzheimer's disease, *ADHD* attention deficit hyperactivity disorder, *EDS* excessive day sleepiness, *PD* Parkinson's disease, *CIAS* cognitive impairment associated with Schizophrenia, *CPAP* continuous positive airway pressure, *EEG* electroencephalogram, *fMRI* functional magnetic resonance imaging, *CBF* cerebral blood flow, *PET* positron emission tomography

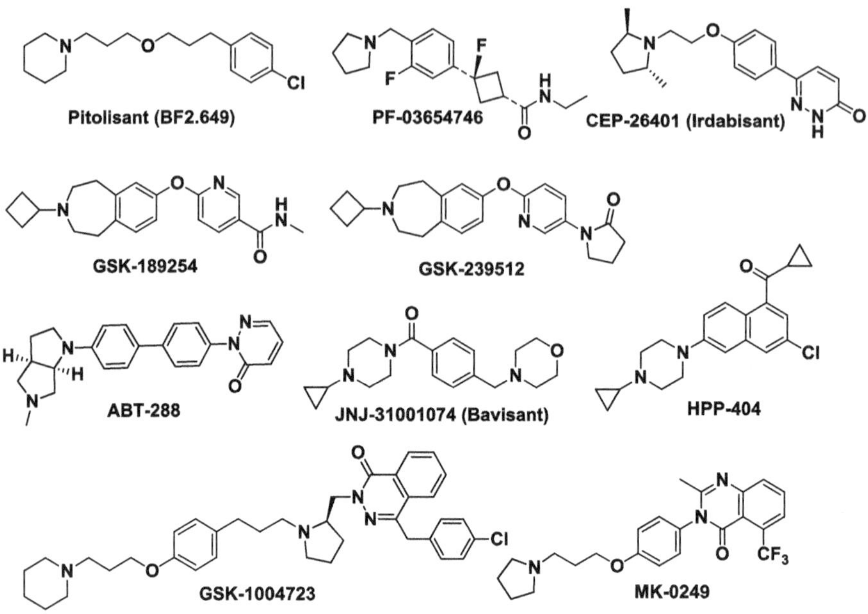


**Fig. 5.18** $H_3R$ antagonist/inverse agonist clinical candidates

Scale comparable to modafinil (ClinicalTrials.gov trial registration number: NCT01067235) [134]. Notably, most patients with PD have nighttime insomnia and excessive daytime sleepiness (EDS), and **BF2.649** was described to be effective in patients given 5–40 mg every day as measured in the Epworth Sleepiness Scale, while 20 mg was the optimal dose, targeted in a phase 3 trial (ClinicalTrials.gov trial registration number: NCT01036139). Furthermore, **BF2.649** completed two phase 3 studies showing efficacy in a dose range of 5–20 mg/day on EDS in patients with obstructive sleep apnea without (ClinicalTrials.gov trial registration number, NCT01072968) or with treatment with continuous positive airway pressure (CPAP) (ClinicalTrials.gov trial registration number, NCT01071876). Recently, a marketed application (Wakix$^R$) for **BF2.649** has been approved by the European Medical Agency (EMA) in 2015.

**PF-03654746** is a potent $H_3R$ antagonist ($hH_3R$ $-\log K_i = 8.49$; $rH_3R$ $-\log K_i = 7.43$) (Table 5.1, Fig. 5.18) and effective in novel object recognition model. It demonstrates a long $t_{1/2}$ (9–18 h) in humans with insomnia observed as the main adverse effect at the 3-mg dose reaching a serum concentration of about 15 ng/mL (Table 5.2) [9, 47, 94]. **PF-03654746** has completed a phase 1 study addressing CIAS (ClinicalTrials.gov trial registration number, NCT01346163) and a phase 2 study evaluating its efficacy and safety (1 and 2 mg/day) in adults with ADHD (ClinicalTrials.gov trial registration number, NCT00531752) without disclosure of the outcomes yet for both studies (Table 5.3, Fig. 5.19). Recently, **PF-03654746** completed a phase 2 study on EDS associated with narcolepsy with proved efficacy in two-drug groups versus placebo (ClinicalTrials.gov trial registration number:

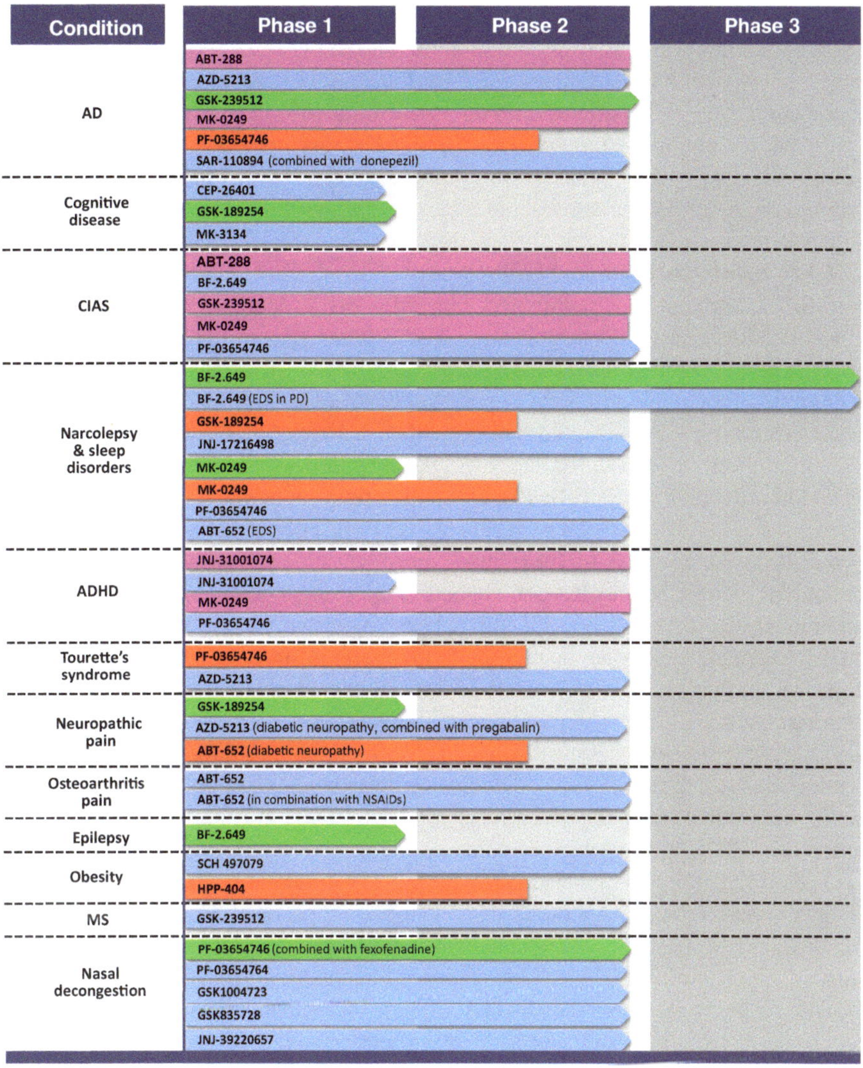


**Fig. 5.19** Histamine $H_3R$ antagonists in clinical development. *AD* Alzheimer's disease, *PD* Parkinson's disease, *ADHD* attention deficit hyperactivity disorder, *EDS* excessive daytime sleepiness, *CIAS* cognitive impairment associated with schizophrenia, *MS* multiple sclerosis. *Green* completed with positive outcomes, *blue* completed with no disclosed results, *pink* completed with no efficacy, *red* terminated

NCT01006122). However, phase 1 clinical trial with **PF-03654746** in volunteers with mild–moderate AD (ClinicalTrials.gov trial registration number, NCT01028911) and a phase 2 study in Tourette's syndrome volunteers (ClinicalTrials.gov trial registration number, NCT01475383) were terminated without disclosing the reasons for both terminations (www.clinicaltrials.gov). Interestingly, a preclinical study with a single dose of **PF-03654746** caused a reduction in allergen-induced nasal symptoms when used in combination with the $H_1R$ antagonist fexofenadine, an observation that sheds light on the importance of $H_3R$ antagonism as a potential therapeutic strategy in patients with allergic rhinitis [40, 137]. Therefore, a phase 2 study addressing the effect of **PF-03654764** on symptoms of allergic rhinitis (ClinicalTrials.gov trial registration number: NCT00562120) has been completed with no disclosure of the clinical outcomes yet (www.clinicaltrials.gov) (Table 5.3, Fig. 5.19).

**CEP-26401 (irdabisant**, $hH_3R$ $-\log K_i=8.70$; $rH_3R$ $-\log K_i=8.11$) (Table 5.1, Fig. 5.18), a potent $H_3R$ antagonist with drug-like properties, showed abrogated RAMH-induced drinking response ($ED_{50}=0.06$ mg/kg, p.o.) in a preclinical dipsogenia model in rats. **CEP-26401** displayed improved performance in the rat social recognition model of short-term memory at doses of 0.01–0.1 mg/kg p.o. and was wake promoting at 3–30 mg/kg p.o. [59, 60, 67]. Moreover, **CEP-26401** increased prepulse inhibition (PPI) at 10 and 30 mg/kg i.p in DBA/2NCrl mice, also observed in coadministration of **CEP-26401** with the antipsychotic risperidone at subeffective doses (3 and 0.1 mg/kg i.p., respectively) [60]. These results demonstrated the potent behavioral effects of **CEP-26401** in rodent models and suggested that this novel $H_3R$ antagonist may have therapeutic utility in the treatment of cognitive and attentional disorders [59, 60, 67]. It has been suggested that **CEP-26401** may also have therapeutic potential in treating schizophrenia or as adjunctive therapy to approved antipsychotics [60]. **CEP-26401** completed phase 1 of a clinical study addressing its PKs and pharmacodynamics in healthy subjects for cognition and sleep–wake indications (ClinicalTrials.gov trial registration number: NCT01903824) without disclosing the clinical outcomes of these studies (Table 5.3, Fig. 5.19). A recent preclinical study showed efficacy of **CEP-26401** in experimental models of narcolepsy, ADHD, and schizophrenia.

The aforementioned disadvantage of **ABT-239** to inhibit hERG has been solved in later series including **ABT-288** ($hH_3R$ $-\log K_i=8.72$, $rH_3R$ $-\log K_i=8.10$) (Table 5.2) which has shown memory-enhancing effects in numerous preclinical experiments [49]. Accordingly, **ABT-288** improved acquisition of a five-trial inhibitory avoidance test in rat pups (0.001–0.03 mg/kg), social recognition memory in adult rats (0.03–0.1 mg/kg), and spatial learning and reference memory in a rat water maze test (0.1–1.0 mg/kg) [49]. Moreover, **ABT-288** (0.01–10 mg/kg) attenuated methamphetamine-induced hyperactivity in an experimental mice model without observation of any potential abuse [138]. Despite the procognitive efficacy of **ABT-288** observed in rodents and, also, its excellent drug-like properties as well as its good PK profile (Table 5.2), two clinical phase 2 studies assessing **ABT-288** in patients with CIAS (ClinicalTrials.gov trial registration number, NCT01077700) as well as patients with mild–moderate AD (ClinicalTrials.gov trial registration number, NCT01018875) have been completed without disclosing the clinical outcomes so far (www.clinicaltrials.gov) (Table 5.3, Fig. 5.19). A phase 1 multiple-dose study addressing its safety, tolerability, PKs, and pharmacodynamics in stable subjects with schizophrenia has been completed without results being disclosed (ClinicalTrials.gov trial registration number: NCT00888693). Recently, **ABT-652**

($hH_3R$–log$K_i$=8.72, $rH_3R$–log$K_i$=8.10) (structure not disclosed), another $H_3R$ antagonist, has been reported, which has completed several clinical phase 2 studies addressing its efficacy when compared to or combined with naproxen in adults with osteoarthritis pain of the knee (www.clinicaltrials.gov). However, a phase 2 study comparing the efficacy and safety of **ABT-652** to placebo in subjects with diabetic neuropathic pain has been terminated without disclosure of the reasons (www.clinicaltrials.gov). In addition, a phase 1 clinical safety, tolerability, and pharmacokinetic/pharmacodynamic study of **ABT-652** in subjects with EDS has been reported (www.clinicaltrials.gov).

Several other heterocyclic-substituted benzofurans, namely, **A-331440**, **A-688057**, and **A-960656**, were developed and characterized for their in vitro and in vivo properties in preclinical animal models and are now in the pipeline for further future clinical investigations (Table 5.2, Fig. 5.5) [9].

**GSK-189254** ($hH_3R$ –log$K_i$=9.00; $rH_3R$ –log$K_i$=9.15) (Table 5.1) with an excellent selectivity profile over a large panel of targets [9, 10, 125] and good PK properties (Table 5.2) has been described to be predominantly effective in a number of preclinical animal experiments, including attentional, memory, pain, and narcolepsy models [125, 139, 140]. Accordingly, acute administration of **GSK-189254** (3–10 mg/kg, p.o.) in $Ox^{+/+}$ and $Ox^{-/-}$ mice increased wakefulness and decreased slow-wave and paradoxical sleep comparable to modafinil (64 mg/kg, p.o.), while it reduced narcoleptic episodes in $Ox^{-/-}$ mice [139]. These interesting observations provided further evidence to support the potential use of $H_3R$ antagonists in the treatment of narcolepsy and EDS. Despite its termination in a phase 2 clinical study examining its effectiveness in narcolepsy (ClinicalTrials.gov trial registration number, NCT00366080), **GSK-189254** has completed phase 1 study addressing neuropathic pain in healthy volunteers (ClinicalTrials.gov trial registration number, NCT00387413) (Table 5.3, Fig. 5.19). A study in human volunteers of both genders, **GSK-189254** (up to 100 μg/day) showed safety and tolerability comparable to that of the noradrenaline–serotonin reuptake inhibitor duloxetine (up to 60 mg/day) when used in an electrical hyperalgesia model (ClinicalTrials.gov trial registration number: NCT00387413). The low brain–blood ratio of 1.3 indicated a peripheral neuronal activity for **GSK-189254**, which is consistent with spinal expression and efficacy of $hH_3Rs$ in neuropathic pain [125]. Notably, **GSK-189254** has also been synthesized in an $^{11}$C-labeled form for use as a positron emission tomography (PET) tracer to quantify $H_3R$ occupancy of test compounds in intact animals and in the clinic [126]. Another successfully developed compound is **GSK-239512** which is a potent, selective, and highly brain-penetrant histamine $H_3R$ antagonist ($hH_3R$ –log$K_i$=9.70; $rH_3R$ log$K_i$=9.82) (Table 5.1) [126]. **GSK-239512** was found to be effective in several preclinical cognition models in rodents [120, 141–143] and has completed a phase 1 clinical trial with modest efficacy in mild–moderate AD patients (ClinicalTrials.gov trial registration number: NCT00675090). **GSK-239512** completed a phase 2 in patients with multiple sclerosis with no disclosure of the results yet (ClinicalTrials.gov trial registration number: NCT01772199) (Table 5.3, Fig. 5.19). Moreover, an open-label drug–drug interaction phase 1 study showed a 1.3-fold increase in **GSK-239512** exposure with ketoconazole, suggesting that in vivo metabolism of **GSK-239512** by CYP3A is unlikely to be the primary route of **GSK-239512** elimination (ClinicalTrials.gov trial registration number: NCT01802931). Furthermore, a phase 2 study evaluating the efficacy and safety of **GSK-239512** in patients with schizophrenia has been completed (ClinicalTrials.gov trial registration number: NCT01009060). Notably, a phase 1 imaging study using the technique of PET to investigate the distribution of **GSK-239512** in the brain has been completed without disclosure of results (ClinicalTrials.gov trial registration number: NCT00474513). Observations from a preclinical model of nasal congestion further supported the hypothesis that $H_3Rs$, in addition to $H_1Rs$, also participate in the histamine-induced nasal congestion and suggest a necessity to simultaneously block both $H_1Rs$ and $H_3Rs$ in such conditions [144, 145]. Based on these preclinical observations and recognizing the need for new treatment options in allergic rhinitis, **GSK-1004723** ($hH_3R/rH_3R$ $K_i$ not available) has been developed and proved to be of dual-acting $H_1R/H_3R$ antagonist properties with long duration of action and a profile suitable for once-a-day dosing to be delivered as topically application via the intranasal route (Fig. 5.19) [145]. Importantly, **GSK-1004723** has completed phase 2 of various clinical trials with the outcomes indicating its efficacy in human volunteers with allergic rhinitis (ClinicalTrials.gov trial registration number: NCT00824356) (Table 5.3, Fig. 5.19).

Another highly selective $H_3R$ antagonist is **AZD-5213** ($hH_3R$ $-\log K_i = 9.30$; structure undisclosed) [146] which was developed to achieve a PK profile permitting circadian fluctuations of $H_3R$ occupancy ($H_3RO$), since the therapeutic benefit of $H_3R$ antagonists may be disadvantaged by disruption of sleep that has been observed in humans with prolonged high $H_3RO$ extending into nighttime. The efficacy of **AZD-5213** has been demonstrated in rodent behavioral models of cognition, and it was safe and well tolerated following repeated doses (1–14 mg/day) with a short half-life (~5 h) in human subjects. In an in vivo PET study in seven young male volunteers using the radioligand (**[$^{11}$C]AZ12807110**), the $H_3RO$ was found to be ranging from 16 to 90 %. Dose as well as concentration dependency following a single dose of **AZD-5213** (0.05–30 mg) indicated that high daytime and low nighttime $H_3RO$ could be achieved following once daily oral dosing of **AZD-5213** [146]. These observed fluctuations of $H_3RO$ following circadian rhythm of the histamine system may be expected to reduce the risk of sleep disruption while maintaining daytime efficacy. Therefore, **AZD-5213** has been proposed as an optimal compound to evaluate the clinical benefit of selective $H_3R$ antagonism in cognitive disorders [146]. Interestingly, **AZD-5213** completed phase 2 stage of two clinical studies addressing its effectiveness in adolescent Tourette's syndrome (ClinicalTrials.gov trial registration number, NCT01904773) and in combination with pregabalin in diabetic neuropathic pain (ClinicalTrials.gov trial registration number, NCT01928381) with no disclosure of the results for both clinical trials (www.clinicaltrials.gov) (Table 5.3, Fig. 5.19). A clinical phase 1 stage assessing safety and tolerability of **AZD-5213** in healthy Japanese volunteers as well as its effect on sleep for patients with mild AD has successfully completed without mentioning the results yet (ClinicalTrials.gov trial registration number: NCT01548287).

The highly potent and selective compound **SAR-110894** ($hH_3R$ $-\log K_i = 10.0$; $rH_3R$ $-\log K_i = 9.40$; no information of the structure or patent can be disclosed) is a novel $H_3R$ antagonist which has been shown to improve performances in several variants of the object recognition task in mice behavioral models (0.3–3 mg/kg) or rats (0.3–1 mg/kg) addressing definite aspects of CIAS and ADHD [147]. **SAR-110894** (1 mg/kg) reversed a deficit in working memory in the Y-maze test [147]. Based on these findings, **SAR-110894** has been suggested to be of therapeutic interest for the treatment of the cognitive symptoms of AD, schizophrenia, and certain aspects of ADHD. Recently, **SAR-110894** completed a phase 2 of clinical study for mild–moderate AD where it was tested as an additive (0.5–5 mg/day) to the standard drug donepezil with no disclosure of the results (ClinicalTrials.gov trial registration number: NCT01266525) (Table 5.3, Fig. 5.19).

**MK-0249** ($hH_3R$ $-\log K_i = 8.20$) showed alertness-promoting effects following a single-dose administration in human experimental models [148] and improved alertness and objective psychomotor performance in a study of patients with obstructive sleep apnea receiving continuous positive airway pressure therapy with and diagnosed with EDS [149, 150] (Table 5.3, Fig. 5.19). However, a phase 1 clinical study has been terminated because **MK-0249** in two different doses (7.5 and 25 mg/day over 4 weeks) was not effective in improving cognitive function in mild to moderate AD patients who were on concomitant symptomatic AD treatment with donepezil (ClinicalTrials.gov trial registration number: NCT00874939) (Table 5.3, Fig. 5.19). Also, **MK-0249** (3–12 mg/day) has been terminated in phase 2 of a clinical study addressing refractory EDS in obstructive sleep apnea by using the maintenance of wakefulness test (MWT) that measures the ability of a patient to remain awake (ClinicalTrials.gov trial registration number: NCT00620659). The alertness-promoting effect with the use of **MK-0249** suggested that $H_3Rs$ may be effective in disorders involving excessive somnolence. However, the small numbers of participants (24 healthy young men) and technical problems with measurement itself were leading to unreliable or uninterpretable results given as reason for termination of phase 2 study (ClinicalTrials.gov trial registration number: NCT00620659). **MK-0249** has completed three different phase 2 studies in patients with CIAS (ClinicalTrials.gov trial registration number, NCT00506077), ADHD (ClinicalTrials.gov trial registration number, NCT00475735), and AD (ClinicalTrials.gov trial registration number, NCT00420420) showing no effect in doses of 10 mg/day in CIAS, 5–10 mg/day in ADHD, and 5 mg/day in AD [149, 151, 152] (Table 5.3, Fig. 5.19). Notably, **MK-0249** still remains as an interesting investigational potent $H_3R$ antagonist with a $t_{1/2}$ of approximately 14 h and a $T_{max}$ of approximately 4 h [153].

Another $H_3R$ antagonist is **MK-3134** ($hH_3R/rH_3R$ $K_i$ not available; structure not yet disclosed) which has completed phase 1 study on safety in a dose of 25 mg/day after preclinical screening on scopolamine-induced memory impairment models (ClinicalTrials.gov trial registration number: NCT01181310). The observed results show that peak effects were generally observed around 2 h after scopolamine administration, and **MK-3134** or donepezil improved performance on the Groton Maze Learning Task (GMLT) at the 2-h time point when compared with scopolamine alone [154]. Moreover, it appeared that the combination of **MK-3134** and donepezil as acetylcholine esterase inhibitor diminished the scopolamine effect to a greater extent than either drug alone. Therefore, the results provided evidence for cognitive improvement through $H_3R$ antagonism and an additive effect on simultaneous influence of the cholinergic and histaminergic neurotransmitter systems. Interestingly, **MK-3134** at three different doses (1, 5, 25 mg/day) successfully completed phase 1 study for bold functional magnetic resonance imaging (fMRI) and cerebral blood flow (CBF) measurements as biomarkers for cognition-enhancing drugs without disclosing the outcomes (ClinicalTrials.gov trial registration number: NCT00887601). A clinical phase 1 study addressing the ability of **MK-3134** to decrease electroencephalogram theta power, averaged across a topographical region of interest, has been completed, also, without disclosing the outcomes (ClinicalTrials.gov trial registration number: NCT01110616) (Table 5.3, Fig. 5.19).

Despite the encouraging preclinical results supporting the role of $H_3R$ antagonists as potential anti-obesity drugs, though, clinical trials with $H_3R$ antagonists for the treatment of obesity or feeding disorders were disappointing. However, a multicenter, randomized, placebo-controlled phase 2 clinical trial that assessed the efficacy of the $H_3R$ antagonist **SCH-497079** ($hH_3R/rH_3R$ $K_i$ not available) on weight loss in obese and overweight subjects was recently completed, but the results were not disclosed (ClinicalTrials.gov trial registration number: NCT00642993). Also, a randomized, placebo-controlled, three-way crossover study evaluating the effect of **SCH-497079** (100 mg/day, 28 days) on metabolic parameters and determining the influence of race/ethnic origin on therapeutic response in type 2 diabetes mellitus has successfully been completed without disclosure of the results (ClinicalTrials.gov trial registration number: NCT00673465).

The preclinical effectiveness of **JNJ-31001074** (**bavisant**; $hH_3R$ $-\log K_i = 8.27$) [119, 155] has been shown in different experimental models addressing wakefulness and attention, cognitive performance, and alcohol consumption (Table 5.1, Fig. 5.1) [156]. Initially, **JNJ-31001074** was withdrawn prior to enrollment of two phase 1 studies addressing its effects on urge to drink in alcohol-dependent adults (ClinicalTrials.gov registration number, NCT01362699) and its PK drug interaction between multiple doses of **JNJ-31001074** and a single dose of an oral contraceptive (ethinyl estradiol/levonorgestrel) in female healthy volunteers (ClinicalTrials.gov registration number, NCT00915668). Later **JNJ-31001074** has completed numerous several phase 1 studies evaluating PKs and the effect of food on different doses (1, 3, and 10 mg/day) in healthy volunteers (ClinicalTrials.gov registration number, NCT00915434), drug interactions (10 mg/day) and when combined with a dose of 400 mg/day of ketoconazole (ClinicalTrials.gov registration number, NCT00915746), and PKs of drug interaction with paroxetine in healthy volunteers (ClinicalTrials.gov registration number, NCT01159821). Also, a phase 1 study addressing the safety and PK tolerability after a single-dose administration (0.015–0.15 mg/day) of **JNJ-31001074** in pediatric patients (12–17 years) with ADHD has been performed (ClinicalTrials.gov registration number: NCT00890292) without disclosure of the results. Moreover, **JNJ-31001074** (1, 3, 10 mg/day) completed a randomized multicenter, double-blind, placebo, and active-controlled parallel phase 2 study on adult with ADHD syndrome where it did not differ from placebo, whereas atomoxetine and methylphenidate were effective (Table 5.3, Fig. 5.19) [156]. The results of these studies suggested an acceptable safety margin, with wakefulness as the primary target affected and no maximum tolerated dose being identified. However, dose-dependent insomnia was the most prominent adverse event, and no noteworthy cardiovascular effects were observed. Moreover, the results show that two lower dosages (1 and 3 mg/day) of **JN-31001074** demonstrated a good tolerability profile, but the higher dosage (10 mg/day) was less well tolerated, as evidenced by the incidence of total adverse events (61.8, 82.4, 89.0 %).

The $H_3R$ antagonist **JNJ-17216498** ($hH_3R/rH_3R$ $K_i$ not available) has completed phase 2 stage in the clinical area for its safety and effectiveness as a single dose (10, 50, 100 mg/day) in comparison with 200 mg of modafinil in patients with narcolepsy (ClinicalTrials.gov registration number: NCT00424931) (Table 5.3, Fig. 5.19).

Notably, a 26-week randomized, double-blind, placebo-controlled phase 2 study evaluating the safety and efficacy of various doses of $H_3R$ antagonist **HPP-404** (35 and 50 mg/day) on weight loss in overweight or obese subjects has been terminated without disclosure of the outcomes (ClinicalTrials.gov registration number: NCT01540864).

In conclusion, since the discovery of the $H_3R$ in the 1980s and later the cloning of $hH_3R$ more than 16 years ago, the search for innovative $H_3R$ antagonists as potential clinical candidates for the treatment of unmet medical needs within the CNS has immensely intensified. Consequently, the field of $H_3R$ antagonists still awaits with great interest more outcomes of continuing clinical trials to determine whether blockade of these receptors can deliver efficacy in various disorder states, such as narcolepsy, cataplexy, obstructive sleep apnea, diabetic neuropathic pain, Tourette's syndrome, ADHD, AD, EDS, PD, and CIAS. However, species-dependent differences in results (with largest between $hH_3R$ and $rH_3R$) have slowed down the translational interpretation of preclinical studies. Moreover, the existence of over 20 $hH_3R$ isoforms, with some having a different pharmacological profile, may hamper deeper understanding of the action of $H_3Rs$ and ligands thereof. The published initial clinical results for certain compounds showed not only the usefulness of $H_3R$ antagonists but also their undesirable effects. Therefore, more efforts should be made to overcome known (from clinical studies) adverse effects, for example, insomnia, to further facilitate the entry of $H_3R$ antagonists onto the pharmaceutical market. Also, it is worthwhile to denote that the lack of results or information about further clinical investigations for several tested $H_3R$ antagonists may indicate that they did not perform well in trials, as some projects were discontinued or terminated. Likewise, it is necessary to develop a good way of translational interpretation of preclinical results and achieve clinically active compounds with fewer drawbacks. So far, only **BF2.649**, known as **pitolisant** (or previously **tiprolisant**), completed phase 3 studies for narcolepsy, cataplexy, EDS, PD, and obstructive sleep apnea with the results confirming efficacy. Very recently (November 2015), the EMA has approved a marketed application (Wakix®) for **BF2.649** for the orphan indication of narcolepsy. In general, it appears to be most possible that $H_3R$ antagonists will be introduced into therapy as adjunctive drugs. Moreover and based on the knowledge of present clinical failures, it is expected that several research efforts will be able to obtain new $H_3R$ antagonists with improved PK profiles and reduced side effects. Taken together, the potential for symptoms as well as disease-altering effects through modulation of $H_3Rs$, therefore, still represents an exciting field for further research. The remarkable drug discovery efforts accomplished so far by $H_3R$ researchers worldwide are therefore appreciatively recognized, and the next few years we might therefore anticipate novel histamine $H_3R$ drugs advancing to the market.

**Acknowledgments** Support was kindly provided to BS by grants from the United Arab Emirates University, to KK by the Polish National Science Center funds, grants no. 2011/02/A/NZ4/00031 and 2011/01/N/NZ4/01126, and to HS by the European Union - European Cooperation of Science and Technology (EU COST) Actions CM1103, CM1207, and CA15135 as well by DFG INST 208/664-1 FUGG.

## References

1. Leurs R, Bakker RA, Timmerman H, de Esch IJP. The histamine H3 receptor: from gene cloning to H3 receptor drugs. Nat Rev Drug Discov. 2005;4:107–20.
2. Gemkow MJ, Davenport AJ, Harich S, Ellenbroek BA, Cesura A, Hallett D. The histamine H3 receptor as a therapeutic drug target for CNS disorders. Drug Discov Today. 2009;14:509–15.
3. Gao Z, Hurst WJ, Hall D, Hartung R, Reynolds W, Kang J, et al. Design and synthesis of a novel series of histamine H receptor antagonists through a scaffold hopping strategy. Bioorg Med Chem. 2014;23:429–38.
4. Arrang JM, Garbarg M, Schwartz JC. Auto-inhibition of brain histamine-release mediated by a novel class (H3) of histamine receptor. Nature. 1983;302:832–7.
5. Fox GB, Esbenshade TA, Pan JB, Radek RJ, Krueger KM, Yao BB, et al. Pharmacological properties of ABT-239 [4-(2-{2-[(2R)-2-Methylpyrrolidinyl]ethyl}-benzofuran-5-yl)benzonitrile]: II. Neurophysiological characterization and broad preclinical efficacy in cognition and schizophrenia of a potent and selective histamine H3 receptor antagonist. J Pharmacol Exp Ther. 2005;313:176–90.
6. Passani MB, Blandina P. Histamine receptors in the CNS as targets for therapeutic intervention. Trends Pharmacol Sci. 2011;32:242–9.
7. Schwartz JC. The histamine H3 receptor: from discovery to clinical trials with pitolisant. Br J Pharmacol. 2011;163:713–21.
8. Bhowmik M, Khanam R, Vohora D. Histamine H3 receptor antagonists in relation to epilepsy and neurodegeneration: a systemic consideration of recent progress and perspectives. Br J Pharmacol. 2012;167:1398–414.
9. Panula P, Chazot PL, Cowart M, Gutzmer R, Leurs R, Liu WL, et al. International union of basic and clinical pharmacology. XCVIII. Histamine receptors. Pharmacol Rev. 2015;67:601–55.
10. Wingen K, Stark H. Scaffold variations in amine warhead of histamine H3 receptor antagonists. Drug Discov Today Technol. 2013;10:e483–9.
11. Łażewska D, Kieć-Kononowicz K. Recent advances in histamine H3 receptor antagonists/inverse agonists. Expert Opin Ther Pat. 2010;20:1147–69.
12. Walter M, Stark H. Histamine receptor subtypes: a century of rational drug design. Front Biosci (Schol Ed). 2012;4:461–88.

13. Liu C, Ma X, Jiang X, Wilson S. Cloning and pharmacological characterization of a fourth histamine receptor (H4) expressed in bone marrow. Mol Pharmacol. 2001;59:420–6.
14. Kieć-Kononowicz K, Ligneau X, Schwartz JC, Schunack W. Potential histamine H3-receptor pyrazoles as potential histamine H3-receptor antagonists. Arch Pharm (Weinheim). 1995;328:469–72.
15. Kieć-Kononowicz K, Ligneau X, Stark H, Schwartz J, Schunack W. Azines and diazines as potential histamine H3-receptor antagonists. Arch Pharm (Weinheim). 1995;328:445–50.
16. Fox G, Pan J, Radek R, Lewis A. Two novel and selective nonimidazole H3 receptor antagonists A-304121 and A-317920: II. In vivo behavioral and neurophysiological characterization. J Pharmacol Exp Ther. 2003;305:897–908.
17. Stark H, Kathmann M, Schlicker E, Schunack W, Schlegel B, Sippl W. Medicinal chemical and pharmacological aspects of imidazole-containing histamine H3 receptor antagonists. Mini Rev Med Chem. 2004;4:965–77.
18. Miyazaki S, Imaizumi M, Onodera K. Effects of thioperamide on the cholinergic system and the step-through passive avoidance test in mice. Methods Find Exp Clin Pharmacol. 1995;17:653–8.
19. Komater VA, Buckley MJ, Browman KE, Pan JB, Hancock AA, Decker MW, et al. Effects of histamine H3 receptor antagonists in two models of spatial learning. Behav Brain Res. 2005;159:295–300.
20. Arrang JM, Garbarg M, Lancelot JC, Lecomte JM, Pollard H, Robba M, et al. Highly potent and selective ligands for histamine H3-receptors. Nature. 1987;327:117–23.
21. Hill S, Ganellin C, Timmerman H. International Union of Pharmacology. XIII. Classification of histamine receptors. Pharmacol Rev. 1997;49:253–78.
22. Masaki T, Yoshimatsu H, Chiba S, Watanabe T, Sakata T. Targeted disruption of histamine H1-receptor attenuates regulatory effects of leptin on feeding, adiposity, and UCP family in mice. Diabetes. 2001;50:385–91.
23. Mollet A, Lutz TA, Meier S, Riediger T, Rushing PA, Scharrer E. Histamine H1 receptors mediate the anorectic action of the pancreatic hormone amylin. Am J Physiol Regul Integr Comp Physiol. 2001;281:R1442–8.
24. Yoshimoto R, Miyamoto Y, Shimamura K, Ishihara A, Takahashi K, Kotani H, et al. Therapeutic potential of histamine H3 receptor agonist for the treatment of obesity and diabetes mellitus. Proc Natl Acad Sci U S A. 2006;103:13866–71.
25. Wetterling T. Bodyweight gain with atypical antipsychotics. A comparative review. Drug Saf. 2001;24:59–73.
26. Ligneau X, Lin J, Vanni-Mercier G, Jouvet M, Muir JL, Ganellin CR, et al. Neurochemical and behavioral effects of ciproxifan, a potent histamine H3-receptor antagonist. J Pharmacol Exp Ther. 1998;287:658–66.
27. Hancock AA. The challenge of drug discovery of a GPCR target: analysis of preclinical pharmacology of histamine H3 antagonists/inverse agonists. Biochem Pharmacol. 2006;71:1103–13.
28. Day M, Pan JB, Buckley MJ, Cronin E, Hollingsworth PR, Hirst WD, et al. Differential effects of ciproxifan and nicotine on impulsivity and attention measures in the 5-choice serial reaction time test. Biochem Pharmacol. 2007;73:1123–34.
29. Liu H, Kerdesky FA, Black LA, Fitzgerald M, Henry R, Esbenshade TA, et al. An efficient multigram synthesis of the potent histamine H3 antagonist GT-2331 and the reassessment of the absolute configuration. J Org Chem. 2004;69:192–4.
30. Berlin M, Boyce CW, de Lera Ruiz M. Histamine H3 receptor as a drug discovery target. J Med Chem. 2011;54:26–53.
31. Ito S, Yoshimoto R, Miyamoto Y, Mitobe Y, Nakamura T, Ishihara A, et al. Detailed pharmacological characterization of GT-2331 for the rat histamine H3 receptor. Eur J Pharmacol. 2006;529:40–6.
32. Cowart M, Altenbach R, Black L, Faghih R, Zhao C, Hancock AA. Medicinal chemistry and biological properties of non-imidazole histamine H3 antagonists. Mini Rev Med Chem. 2004;4:979–92.

33. McLeod RL, Rizzo CA, West Jr RE, Aslanian R, McCormick K, Bryant M, et al. Pharmacological characterization of the novel histamine H3-receptor antagonist N-(3,5-dichlorophenyl)-N'-[[4-(1H-imidazol-4-ylmethyl)phenyl]-methyl]-urea (SCH 79687). J Pharmacol Exp Ther. 2003;305:1037–44.
34. Miller TR, Baranowski JL, Estvander BR, Witte DG, Carr TL, Manelli AM, et al. A robust and high-capacity [(35)S]GTPgammaS binding assay for determining antagonist and inverse agonist pharmacological parameters of histamine H(3) receptor ligands. Assay Drug Dev Technol. 2008;6:339–49.
35. Ganellin CR, Leurquin F, Piripitsi A, Arrang JM, Garbarg M, Ligneau X, et al. Synthesis of potent non-imidazole histamine H3 receptor antagonists. Arch Pharm (Weinheim). 1998;331:395–404.
36. Meier G, Apelt J, Reichert U, Grassmann S, Ligneau X, Elz S, et al. Influence of imidazole replacement in different structural classes of histamine H(3)-receptor antagonists. Eur J Pharm Sci. 2001;13:249–59.
37. Zaragoza F, Stephensen H, Knudsen SM, Pridal L, Wulff BS, Rimvall K. 1-Alkyl-4-acylpiperazines as a new class of imidazole-free histamine H3 receptor antagonists. J Med Chem. 2004;47:2833–8.
38. Malmlof K, Zaragoza F, Golozoubova V, Refsgaard HH, Cremers T, Raun K, et al. Influence of a selective histamine H3 receptor antagonist on hypothalamic neural activity, food intake and body weight. Int J Obes (Lond). 2005;29:1402–12.
39. Wager TT, Pettersen BA, Schmidt AW, Spracklin DK, Mente S, Butler TW, et al. Discovery of Two Clinical Histamine H(3) Receptor Antagonists: *trans*-N-Ethyl-3-fluoro-3-[3-fluoro-4-(pyrrolidinylmethyl)phenyl]cyclobutanecarboxamide (PF-03654746) and trans-3-Fluoro-3-[3-fluoro-4-(pyrrolidin-1-ylmethyl)phenyl]-N-(2-methylpropyl)cyclobutanecarboxamide (PF-03654764). J Med Chem. 2011;54:7602–20.
40. Stokes JR, Romero FA, Allan RJ, Phillips PG, Hackman F, Misfeldt J, et al. The effects of an H3 receptor antagonist (PF-03654746) with fexofenadine on reducing allergic rhinitis symptoms. J Allergy Clin Immunol. 2012;129:409–12.
41. North ML, Walker TJ, Steacy LM, Hobsbawn BG, Allan RJ, Hackman F, et al. Add-on histamine receptor-3 antagonist for allergic rhinitis: a double blind randomized crossover trial using the environmental exposure unit. Allergy Asthma Clin Immunol. 2014;10:33.
42. Cowart M, Gfesser GA, Browman KE, Faghih R, Miller TR, Milicic I, et al. Novel heterocyclic-substituted benzofuran histamine H3 receptor antagonists: in vitro properties, drug-likeness, and behavioral activity. Biochem Pharmacol. 2007;73:1243–55.
43. Cowart M, Faghih R, Gfesser G, Curtis M, Sun M, Zhao C, et al. Achievement of behavioral efficacy and improved potency in new heterocyclic analogs of benzofuran H3 antagonists. Inflamm Res. 2005;54:S25–6.
44. Bitner RS, Markosyan S, Nikkel AL, Brioni JD. In-vivo histamine H3 receptor antagonism activates cellular signaling suggestive of symptomatic and disease modifying efficacy in Alzheimer's disease. Neuropharmacology. 2010;60:460–6.
45. Hsieh GC, Chandran P, Salyers AK, Pai M, Zhu CZ, Wensink EJ, et al. H4 receptor antagonism exhibits anti-nociceptive effects in inflammatory and neuropathic pain models in rats. Pharmacol Biochem Behav. 2010;95:41–50.
46. Varaschin RK, Akers KG, Rosenberg MJ, Hamilton DA, Savage DD. Effects of the cognition-enhancing agent ABT-239 on fetal ethanol-induced deficits in dentate gyrus synaptic plasticity. J Pharmacol Exp Ther. 2010;334:191–8.
47. Brioni J, Esbenshade T. Discovery of histamine H3 antagonists for the treatment of cognitive disorders and Alzheimer's disease. J Pharmacol Exp Ther. 2011;336:38–46.
48. Swanson DM, Shah CR, Lord B, Morton K, Dvorak LK, Mazur C, et al. Heterocyclic replacement of the central phenyl core of diamine-based histamine H3 receptor antagonists. Eur J Med Chem. 2009;44:4413–25.
49. Esbenshade TA, Browman KE, Miller TR, Krueger KM, Komater-Roderwald V, Zhang M, et al. Pharmacological properties and procognitive effects of ABT-288, a potent and selective histamine H3 receptor antagonist. J Pharmacol Exp Ther. 2012;343:233–45.

50. Othman AA, Haig G, Florian H, Locke C, Zhang J, Dutta S. Safety, tolerability and pharmacokinetics of the histamine H3 receptor antagonist, ABT-288, in healthy young adults and elderly volunteers. Br J Clin Pharmacol. 2013;75:1299–311.
51. Brown JW, Whitehead CA, Basso AM, Rueter LE, Zhang M. Preclinical evaluation of non-imidazole histamine H3 receptor antagonists in comparison to atypical antipsychotics for the treatment of cognitive deficits associated with schizophrenia. Int J Neuropsychopharmacol. 2013;16:889–904.
52. Bonaventure P, Letavic M, Dugovic C, Wilson S, Aluisio L, Pudiak C, et al. Histamine H3 receptor antagonists: from target identification to drug leads. Biochem Pharmacol. 2007;73:1084–96.
53. Cowart M, Hsieh G, Black LA, Zhan C, Gomez EJ, Pai M, et al. Pharmacological characterization of A-960656, a histamine H(3) receptor antagonist with efficacy in animal models of osteoarthritis and neuropathic pain. Eur J Pharmacol. 2012;684:87–94.
54. Wei H, Jin CY, Viisanen H, You HJ, Pertovaara A. Histamine in the locus coeruleus promotes descending noradrenergic inhibition of neuropathic hypersensitivity. Pharmacol Res. 2014;90:58–66.
55. Browman KE, Komater VA, Curzon P, Rueter LE, Hancock AA, Decker MW, et al. Enhancement of prepulse inhibition of startle in mice by the H3 receptor antagonists thioperamide and ciproxifan. Behav Brain Res. 2004;153:69–76.
56. Coruzzi G, Adami M, Pozzoli C, de Esch IJ, Smits R, Leurs R. Selective histamine H(3) and H(4) receptor agonists exert opposite effects against the gastric lesions induced by HCl in the rat stomach. Eur J Pharmacol. 2011;669:121–7.
57. Hancock AA, Bennani YL, Bush EN, Esbenshade TA, Faghih R, Fox GB, et al. Antiobesity effects of A-331440, a novel non-imidazole histamine H3 receptor antagonist. Eur J Pharmacol. 2004;487:183–97.
58. Didone V, Quoilin C, Nyssen L, Closon C, Tirelli E, Quertemont E. Effects of L-histidine and histamine H3 receptor modulators on ethanol-induced sedation in mice. Behav Brain Res. 2013;238:113–8.
59. Hudkins RL, Raddatz R, Tao M, Mathiasen JR, Aimone LD, Becknell NC, et al. Discovery and Characterization of 6-{4-[3-(R)-2-Methylpyrrolidin-1-yl)propoxy]phenyl}-2H-pyridazin-3-one (CEP-26401, Irdabisant): a Potent, Selective Histamine H(3) Receptor Inverse Agonist. J Med Chem. 2011;54:4781–92.
60. Raddatz R, Hudkins RL, Mathiasen JR, Gruner JA, Flood DG, Aimone LD, et al. CEP-26401 (irdabisant), a potent and selective histamine H3 receptor antagonist/inverse agonist with cognition-enhancing and wake-promoting activities. J Pharmacol Exp Ther. 2012;340:124–33.
61. Hudkins RL, Aimone LD, Dandu RR, Dunn D, Gruner JA, Huang Z, et al. 4,5-Dihydropyridazin-3-one derivatives as histamine H(3) receptor inverse agonists. Bioorg Med Chem Lett. 2011;22:4–10.
62. Becknell NC, Lyons JA, Aimone LD, Gruner JA, Mathiasen JR, Raddatz R, et al. Synthesis and evaluation of pyridone-phenoxypropyl-R-2-methylpyrrolidine analogues as histamine H3 receptor antagonists. Bioorg Med Chem Lett. 2011;21:7076–80.
63. Tao M, Aimone LD, Huang Z, Mathiasen J, Raddatz R, Lyons J, et al. Optimization of 5-pyridazin-3-one phenoxypropylamines as potent, selective histamine H3 receptor antagonists with potent cognition enhancing activity. J Med Chem. 2012;55:414–23.
64. Becknell NC, Lyons JA, Aimone LD, Huang Z, Gruner JA, Raddatz R, et al. Synthesis and evaluation of 4- and 5-pyridazin-3-one phenoxypropylamine analogues as histamine-3 receptor antagonists. Bioorg Med Chem. 2012;20:3880–6.
65. Tao M, Raddatz R, Aimone LD, Hudkins RL. Synthesis and structure-activity relationships of 4,5-fused pyridazinones as histamine H3 receptor antagonists. Bioorg Med Chem Lett. 2011;21:6126–30.
66. Josef KA, Aimone LD, Lyons J, Raddatz R, Hudkins RL. Synthesis of constrained benzocinnolinone analogues of CEP-26401 (irdabisant) as potent, selective histamine H3 receptor inverse agonists. Bioorg Med Chem Lett. 2012;22:4198–202.

67. Hudkins RL, Becknell NC, Lyons JA, Aimone LD, Olsen M, Haltiwanger RC, et al. 3,4-Diaza-bicyclo[4.1.0]hept-4-en-2-one phenoxypropylamine analogs of irdabisant (CEP-26401) as potent histamine-3 receptor inverse agonists with robust wake-promoting activity. Eur J Med Chem. 2015;95:349–56.
68. Hudkins RL, Aimone LD, Bailey TR, Bendesky RJ, Dandu RR, Dunn D, et al. Identification of pyridazin-3-one derivatives as potent, selective histamine H3 receptor inverse agonists with robust wake activity. Bioorg Med Chem Lett. 2011;21:5493–7.
69. Semple G, Santora VJ, Smith JM, Covel JA, Hayashi R, Gallardo C, et al. Identification of biaryl sulfone derivatives as antagonists of the histamine H3 receptor: discovery of (R)-1-(2-(4′-(3-methoxypropylsulfonyl)biphenyl-4-yl)ethyl)-2-methylpyrrolidine (APD916). Bioorg Med Chem Lett. 2012;22:71–5.
70. Dunn D, Raddatz R, Ator MA, Bacon ER, Chatterjee S. From an atypical wake-promoting agent to potent histamine-3 receptor inverse agonists. Chem Biol Drug Des. 2013;81:433–5.
71. Covel JA, Santora VJ, Smith JM, Hayashi R, Gallardo C, Weinhouse MI, et al. Design and evaluation of novel biphenyl sulfonamide derivatives with potent histamine H(3) receptor inverse agonist activity. J Med Chem. 2009;52:5603–11.
72. Auberson YP, Troxler T, Zhang X, Yang CR, Fendt M, Feuerbach D, et al. Ergoline-derived inverse agonists of the human H3 receptor for the treatment of narcolepsy. ChemMedChem. 2014;9:1683–96.
73. Auberson YP, Troxler T, Zhang X, Yang CR, Feuerbach D, Liu Y-C, et al. From ergolines to indoles: improved inhibitors of the human h3 receptor for the treatment of narcolepsy. ChemMedChem. 2015;10:266–75.
74. Kvernmo T, Härtter S, Burger E. A review of the receptor-binding and pharmacokinetic properties of dopamine agonists. Clin Ther. 2006;28:1065–78.
75. Gao Z, Hurst WJ, Guillot E, Czechtizky W, Lukasczyk U, Nagorny R, et al. Discovery of aryl ureas and aryl amides as potent and selective histamine H3 receptor antagonists for the treatment of obesity (Part I). Bioorg Med Chem Lett. 2013;23:3416–20.
76. Gao Z, Hurst WJ, Czechtizky W, Hall D, Moindrot N, Nagorny R, et al. Identification and profiling of 3,5-dimethyl-isoxazole-4-carboxylic acid [2-methyl-4-((2S,3′S)-2-methyl-[1,3′]bipyrrolidinyl-1′-yl)phenyl] amide as histamine H(3) receptor antagonist for the treatment of depression. Bioorg Med Chem Lett. 2013;23:6269–73.
77. Gao Z, Hurst WJ, Czechtizky W, Francon D, Griebel G, Nagorny R, et al. Discovery of a potent, selective, and orally bioavailable histamine H3 receptor antagonist SAR110068 for the treatment of sleep–wake disorders. Bioorg Med Chem Lett. 2013;23:6141–5.
78. Gao Z, Hurst WJ, Guillot E, Nagorny R, Pruniaux M-P, Hendrix JA, et al. Synthesis, characterization, and biological assessment of the four stereoisomers of the H(3) receptor antagonist 5-fluoro-2-methyl-N-[2-methyl-4-(2-methyl[1,3′]bipyrrolidinyl-1′-yl)phenyl]benzamide. Bioorg Med Chem Lett. 2013;23:4044–7.
79. Labeeuw O, Levoin N, Poupardin-Olivier O, Calmels T, Ligneau X, Berrebi-Bertrand I, et al. Novel and highly potent histamine H3 receptor ligands. Part 3: an alcohol function to improve the pharmacokinetic profile. Bioorg Med Chem Lett. 2013;23:2548–54.
80. Levoin N, Labeeuw O, Calmels T, Poupardin-Olivier O, Berrebi-Bertrand I, Lecomte J-M, et al. Novel and highly potent histamine H3 receptor ligands. Part 1: withdrawing of hERG activity. Bioorg Med Chem Lett. 2011;21:5378–83.
81. Sadek B, Schwed JS, Subramanian D, Weizel L, Walter M, Adem A, et al. Non-imidazole histamine H3 receptor ligands incorporating antiepileptic moieties. Eur J Med Chem. 2014;77:269–79.
82. Lipani L, Odadzic D, Weizel L, Schwed J-S, Sadek B, Stark H. Studies on molecular properties prediction and histamine H3 receptor affinities of novel ligands with uracil-based motifs. Eur J Med Chem. 2014;86:578–88.
83. Huang W, Tang L, Shi Y, Huang S, Xu L, Sheng R, et al. Searching for the Multi-Target-Directed Ligands against Alzheimer's disease: discovery of quinoxaline-based hybrid compounds with AChE, H3R and BACE 1 inhibitory activities. Bioorg Med Chem. 2011;19:7158–67.

84. Ratcliffe AJ. Medicinal chemistry strategies to minimize phospholipidosis. Curr Med Chem. 2009;16:2816–23.
85. Bahi A, Schwed JS, Walter M, Stark H, Sadek B. Anxiolytic and antidepressant-like activities of the novel and potent non-imidazole histamine H3 receptor antagonist ST-1283. Drug Des Devel Ther. 2014;8:627–37.
86. Bahi A, Sadek B, Schwed SJ, Walter M, Stark H. Influence of the novel histamine H3 receptor antagonist ST1283 on voluntary alcohol consumption and ethanol-induced place preference in mice. Psychopharmacology (Berl). 2013;228:85–95.
87. Bahi A, Sadek B, Nurulain SM, Łażewska D, Kieć-Kononowicz K. The novel non-imidazole histamine H3 receptor antagonist DL77 reduces voluntary alcohol intake and ethanol-induced conditioned place preference in mice. Physiol Behav. 2015;151:189–97.
88. Sadek B, Saad A, Subramanian D, Shafiullah M, Łażewska D, Kieć-Kononowiczc K. Anticonvulsant and procognitive properties of the non-imidazole histamine H3 receptor antagonist DL77 in male adult rats. Neuropharmacology. 2015.
89. Sander K, Von Coburg Y, Camelin J, Ligneau X, Rau O, Schubert-zsilavecz M, et al. Acidic elements in histamine H3 receptor antagonists. Bioorg Med Chem Lett. 2010;20:1581–4.
90. Sander K, Kottke T, Weizel L, Stark H. Kojic acid derivatives as histamine H(3) receptor ligands. Chem Pharm Bull (Tokyo). 2010;58:1353–61.
91. Barbier AJ, Berridge C, Dugovic C, Laposky AD, Wilson SJ, Boggs J, et al. Acute wake-promoting actions of JNJ-5207852, a novel, diamine-based H3 antagonist. Br J Pharmacol. 2004;143:649–61.
92. Barbier AJ, Bradbury MJ. Histaminergic control of sleep-wake cycles: recent therapeutic advances for sleep and wake disorders. CNS Neurol Disord Drug Targets. 2007;6:31–43.
93. Letavic MA, Aluisio L, Atack JR, Bonaventure P, Carruthers NI, Dugovic C, et al. Pre-clinical characterization of aryloxypyridine amides as histamine H3 receptor antagonists: identification of candidates for clinical development. Bioorg Med Chem Lett. 2010;20:4210–4.
94. Nikolic K, Filipic S, Agbaba D, Stark H. Procognitive properties of drugs with single and multitargeting H3 receptor antagonist activities. CNS Neurosci Ther. 2014;20:613–23.
95. Fujiki N. Associated professional sleep societies—SLEEP 2006 20th Anniversary Meeting. 17–22 June 2006, Salt Lake City, UT, USA. IDrugs. 2006;9:610–3.
96. Galici R, Boggs JD, Aluisio L, Fraser IC, Bonaventure P, Lord B, et al. JNJ-10181457, a selective non-imidazole histamine H(3) receptor antagonist, normalizes acetylcholine neurotransmission and has efficacy in translational rat models of cognition. Neuropharmacology. 2009;56:1131–7.
97. Labeeuw O, Levoin N, Poupardin-Olivier O, Calmels T, Ligneau X, Berrebi-Bertrand I, et al. Novel and highly potent histamine H3 receptor ligands. Part 2: exploring the cyclohexylamine-based series. Bioorg Med Chem Lett. 2011;21:5384–8.
98. Sander K, Kottke T, Hoffend C, Walter M, Weizel L, Camelin J-C, et al. First metal-containing histamine H3 receptor ligands. Org Lett. 2010;12:2578–81.
99. Harusawa S, Sawada K, Magata T, Yoneyama H, Araki L, Usami Y, et al. Synthesis and evaluation of N-alkyl-S-[3-(piperidin 1-yl)propyl]isothioureas: high affinity and human/rat species-selective histamine H(3) receptor antagonists. Bioorg Med Chem Lett. 2013;23: 6415–20.
100. Xiao D, Palani A, Sofolarides M, Huang Y, Aslanian R, Vaccaro H, et al. Discovery of a series of potent arylthiadiazole H3 antagonists. Bioorg Med Chem Lett. 2011;21:861–4.
101. Rao A, Shao N, Aslanian R. Discovery of a potent thiadiazole class of histamine H3 receptor antagonist for the treatment of diabetes. ACS Med Chem Lett. 2012;3:198–202.
102. Shao N, Aslanian R, West RE, Williams SM, Wu R-L, Hwa J, et al. Synthesis and structure-activity relationship (SAR) study of 4-azabenzoxazole analogues as H3 antagonists. Bioorg Med Chem Lett. 2012;22:2075–8.
103. Xiao J, Free RB, Barnaeva E, Conroy JL, Doyle T, Miller B, et al. Discovery, optimization, and characterization of novel D2 dopamine receptor selective antagonists. J Med Chem. 2014;57:3450–63.

104. Walter M, Von Coburg Y, Isensee K, Sander K, Ligneau X, Camelin J, et al. Azole derivatives as histamine H3 receptor antagonists, Part I: Thiazol-2-yl ethers. Bioorg Med Chem Lett. 2010;20:5879–82.
105. Bordi F, Rivara S, Dallaturca E, Carmi C, Pala D, Lodola A, et al. Dibasic biphenyl H3 receptor antagonists: steric tolerance for a lipophilic side chain. Eur J Med Chem. 2012;48:214–30.
106. Wijtmans M, Denonne F, Celanire S, Gillard M, Hulscher S, Delaunoy C, et al. Histamine H 3 receptor ligands with a 3-cyclobutoxy motif: a novel and versatile constraint of the classical 3-propoxy linker. Med Chem Commun. 2010;1:39–44.
107. Anderson JT, Campbell M, Wang J, Brunden KR, Harrington JJ, Stricker-Krongrad A, et al. Investigation of 4-piperidinols as novel H3 antagonists. Bioorg Med Chem Lett. 2010;20:6246–9.
108. Hudkins RL, Zulli AL, Dandu R, Tao M, Josef KA, Aimone LD, et al. 4-Phenoxypiperidine pyridazin-3-one histamine H3 receptor inverse agonists demonstrating potent and robust wake promoting activity. Bioorg Med Chem Lett. 2012;22:1504–9.
109. Hudkins RL, Josef KA, Becknell NC, Aimone LD, Lyons JA, Mathiasen JR, et al. Discovery of (1R,6S)-5-[4-(1-cyclobutyl-piperidin-4-yloxy)-phenyl]-3,4-diaza-bicyclo[4.1.0]hept-4-en-2-one (R, S-4a): histamine H(3) receptor inverse agonist demonstrating potent cognitive enhancing and wake promoting activity. Bioorg Med Chem Lett. 2014;24:1303–6.
110. Richter HGF, Freichel C, Huwyler J, Nakagawa T, Nettekoven M, Plancher J-M, et al. Discovery of potent and selective histamine H3 receptor inverse agonists based on the 3,4-dihydro-2H-pyrazino[1,2-a]indol-1-one scaffold. Bioorg Med Chem Lett. 2010;20: 5713–7.
111. Masłowska-Lipowicz I, Walczyński K. Structure-Activity Relationships of New 1-substituted methyl-4-[5-(N -methyl- N -propylamino)pentyloxy]piperidines and Selected 1-[(N -substituted- N- methyl)-3-propyloxy]-5-(N -methy-l- N -propyl)-pentanediamin. Chem Biol Drug Des. 2014;83:106–18.
112. Tao M, Aimone LD, Gruner JA, Mathiasen JR, Huang Z, Lyons J, et al. Synthesis and structure–activity relationship of 5-pyridazin-3-one phenoxypiperidines as potent, selective histamine H3 receptor inverse agonists. Bioorg Med Chem Lett. 2012;22:1073–7.
113. Łażewska D, Kieć-Kononowicz K. New developments around histamine H3 receptor antagonists/inverse agonists: a patent review (2010-present). Expert Opin Ther Pat. 2014;24:89–111.
114. Hudkins RL, Gruner JA, Raddatz R, Mathiasen JR, Aimone LD, Marino MJ, et al. 3-(1′-Cyclobutylspiro[4H-1,3-benzodioxine-2,4′-piperidine]-6-yl)-5,5-dimethyl-1,4-dihydropyridazin-6-one (CEP-32215), a new wake-promoting histamine H3 antagonist/inverse agonist. Neuropharmacology 2015;in press. doi:10.1016/j.neuropharm.2015.09.025.
115. Pierson PD, Fettes A, Freichel C, Gatti-McArthur S, Hertel C, Huwyler J, et al. 5-hydroxyindole-2-carboxylic acid amides: novel histamine-3 receptor inverse agonists for the treatment of obesity. J Med Chem. 2009;52:3855–68.
116. Ting PC, Lee JF, Albanese MM, Wu J, Aslanian R, Favreau L, et al. The synthesis and structure-activity relationship of 4-benzimidazolyl-piperidinylcarbonyl-piperidine analogs as histamine H3 antagonists. Bioorg Med Chem Lett. 2010;20:5004–8.
117. Zeng Q, Rosenblum SB, Yang Z, Jiang Y, McCormick KD, Aslanian RG, et al. Synthesis and SAR studies of benzimidazolone derivatives as histamine H3-receptor antagonists. Bioorg Med Chem Lett. 2013;23:6001–3.
118. de Lera RM, Zheng J, Berlin MY, McCormick KD, Aslanian RG, West R, et al. Bicyclic and tricyclic heterocycle derivatives as histamine H3 receptor antagonists for the treatment of obesity. Bioorg Med Chem Lett. 2013;23:6004–9.
119. Letavic MA, Aluisio L, Apodaca R, Bajpai M, Barbier AJ, Bonneville A, et al. Novel benzamide-based histamine H3 receptor antagonists: the identification of two candidates for clinical development. ACS Med Chem Lett. 2015;6:450–4.
120. Wilson DM, Apps J, Bailey N, Bamford MJ, Beresford IJ, Brackenborough K, et al. Identification of clinical candidates from the benzazepine class of histamine H3 receptor antagonists. Bioorg Med Chem Lett. 2013;23:6890–6.

121. Santillan A, McClure KJ, Allison BD, Lord B, Boggs JD, Morton KL, et al. Indole- and benzothiophene-based histamine H3 antagonists. Bioorg Med Chem Lett. 2010;20: 6226–30.
122. Davenport AJ, Stimson CC, Corsi M, Vaidya D, Glenn E, Jones TD, et al. Discovery of substituted benzyl tetrazoles as histamine H3 receptor antagonists. Bioorg Med Chem Lett. 2010;20:5165–9.
123. Brown DG, Bernstein PR, Griffin A, Wesolowski S, Labrecque D, Tremblay MC, et al. Discovery of spirofused piperazine and diazepane amides as selective histamine-3 antagonists with in vivo efficacy in a mouse model of cognition. J Med Chem. 2014;57:733–58.
124. Medhurst AD, Briggs MA, Bruton G, Calver AR, Chessell I, Crook B, et al. Structurally novel histamine H3 receptor antagonists GSK207040 and GSK334429 improve scopolamine-induced memory impairment and capsaicin-induced secondary allodynia in rats. Biochem Pharmacol. 2007;73:1182–94.
125. Medhurst AD, Atkins AR, Beresford IJ, Brackenborough K, Briggs MA, Calver AR, et al. GSK189254, a novel H3 receptor antagonist that binds to histamine H3 receptors in Alzheimer's disease brain and improves cognitive performance in preclinical. J Pharmacol Exp Ther. 2007;321:1032–45.
126. Ashworth S, Rabiner EA, Gunn RN, Plisson C, Wilson AA, Comley RA, et al. Evaluation of 11C-GSK189254 as a novel radioligand for the H3 receptor in humans using PET. J Nucl Med. 2010;51:1021–9.
127. Galici R, Rezvani AH, Aluisio L, Lord B, Levin ED, Fraser I, et al. JNJ-39220675, a novel selective histamine H3 receptor antagonist, reduces the abuse-related effects of alcohol in rats. Psychopharmacology (Berl). 2011;214:829–41.
128. Vanhanen J, Nuutinen S, Lintunen M, Maki T, Ramo J, Karlstedt K, et al. Histamine is required for H(3) receptor-mediated alcohol reward inhibition, but not for alcohol consumption or stimulation. Br J Pharmacol. 2013;170:177–87.
129. Vanhanen J, Kinnunen M, Nuutinen S, Panula P. Histamine H3 receptor antagonist JNJ-39220675 modulates locomotor responses but not place conditioning by dopaminergic drugs. Psychopharmacology (Berl). 2015;232:1143–53.
130. Barchuk WT, Salapatek AM, Ge T, D'Angelo P, Liu X. A proof-of-concept study of the effect of a novel H3-receptor antagonist in allergen-induced nasal congestion. J Allergy Clin Immunol. 2013;132:836–8.
131. Becknell NC, Lyons JA, Aimone LD, Gruner JA, Mathiasen JR, Raddatz R, et al. Synthesis and evaluation of 4-alkoxy-[10-cyclobutyl-spiro(3,4-dihydrobenzopyran-2,40-piperidine)] analogues as histamine-3 receptor antagonists. Bioorg Med Chem Lett. 2012;22:186–9.
132. Dandu RR, Lyons JA, Raddatz R, Huang Z, Aimone LD, Hudkins RL. Synthesis and evaluation of a new series of 1′-cyclobutyl-6-(4-piperidyloxy)spiro[benzopyran-2,4′ piperidine] derivatives as high affinity and selective histamine-3 receptor (H3R) antagonists. Bioorg Med Chem Lett. 2012;22:2151–3.
133. Ligneau X, Perrin D, Landais L, Camelin JC, Calmels TP, Berrebi-Bertrand I, et al. BF2.649 [1-{3-[3-(4-Chlorophenyl)propoxy]propyl}piperidine, hydrochloride], a nonimidazole inverse agonist/antagonist at the human histamine H3 receptor: preclinical pharmacology. J Pharmacol Exp Ther. 2007;320:365–75.
134. Lin JS, Dauvilliers Y, Arnulf I, Bastuji H, Anaclet C, Parmentier R, et al. An inverse agonist of the histamine H(3) receptor improves wakefulness in narcolepsy: studies in orexin–/– mice and patients. Neurobiol Dis. 2008;30:74–83.
135. Vohora D. Histamine-selective H3 receptor ligands and cognitive functions: an overview. IDrugs. 2004;7:667–73.
136. Kasteleijn-Nolst Trenite D, Parain D, Genton P, Masnou P, Schwartz JC, Hirsch E. Efficacy of the histamine 3 receptor (H3R) antagonist pitolisant (formerly known as tiprolisant; BF2.649) in epilepsy: dose-dependent effects in the human photosensitivity model. Epilepsy Behav. 2013;28:66–70.
137. Stokes JR, Casale TB. Allergic rhinitis and asthma: celebrating 100 years of immunotherapy. Curr Opin Immunol. 2011;23:808–13.

138. Hudzik TJ, Basso A, Boyce-Rustay JM, Bracken W, Browman KE, Drescher K, et al. Assessment of the abuse liability of ABT-288, a novel histamine H(3) receptor antagonist. Psychopharmacology (Berl). 2013;228:187–97.
139. Guo RX, Anaclet C, Roberts JC, Parmentier R, Zhang M, Guidon G, et al. Differential effects of acute and repeat dosing with the H3 antagonist GSK189254 on the sleep-wake cycle and narcoleptic episodes in Ox–/– mice. Br J Pharmacol. 2009;157:104–17.
140. McGaraughty S, Chu KL, Cowart MD, Brioni JD. Antagonism of supraspinal histamine H3 receptors modulates spinal neuronal activity in neuropathic rats. J Pharmacol Exp Ther. 2012;343:13–20.
141. Holtz NA, Anker JJ, Regier PS, Claxton A, Carroll ME. Cocaine self-administration punished by i.v. histamine in rat models of high and low drug abuse vulnerability: effects of saccharin preference, impulsivity, and sex. Physiol Behav. 2013;122:32–8.
142. Wilson DM, Apps J, Bailey N, Bamford MJ, Beresford IJ, Briggs MA, et al. The discovery of the benzazepine class of histamine H3 receptor antagonists. Bioorg Med Chem Lett. 2013;23:6897–901.
143. RA G, Harrington C, Mahler A, Beresford I, Maruff P, Lowy M, et al. A randomized, double-blind, placebo-controlled, 16-week study of the H3 receptor antagonist, GSK239512 as a monotherapy in subjects with mild-to-moderate Alzheimer's disease. Curr Alzheimer Res. 2014;11:47–58.
144. Taylor-Clark T, Sodha R, Warner B, Foreman J. Histamine receptors that influence blockage of the normal human nasal airway. Br J Pharmacol. 2005;144:867–74.
145. Slack RJ, Russell LJ, Hall DA, Luttmann MA, Ford AJ, Saunders KA, et al. Pharmacological characterization of GSK1004723, a novel, long-acting antagonist at histamine H(1) and H(3) receptors. Br J Pharmacol. 2011;164:1627–41.
146. Jucaite A, Takano A, Bostrom E, Jostell KG, Stenkrona P, Halldin C, et al. AZD5213: a novel histamine H3 receptor antagonist permitting high daytime and low nocturnal H3 receptor occupancy, a PET study in human subjects. Int J Neuropsychopharmacol. 2013;16:1231–9.
147. Griebel G, Pichat P, Pruniaux MP, Beeske S, Lopez-Grancha M, Genet E, et al. SAR110894, a potent histamine H(3)-receptor antagonist, displays procognitive effects in rodents. Pharmacol Biochem Behav. 2012;102:203–14.
148. Iannone R, Palcza J, Renger JJ, Calder N, Cerchio K, Gottesdiener K, et al. Acute alertness-promoting effects of a novel histamine subtype-3 receptor inverse agonist in healthy sleep-deprived male volunteers. Clin Pharmacol Ther. 2010;88:831–9.
149. Herring WJ, Wilens TE, Adler LA, Baranak C, Liu K, Snavely DB, et al. Randomized controlled study of the histamine H3 inverse agonist MK-0249 in adult attention-deficit/hyperactivity disorder. J Clin Psychiatry. 2012;73:e891–8.
150. Herring WJ, Liu K, Hutzelmann J, Snavely D, Snyder E, Ceesay P, et al. Alertness and psychomotor performance effects of the histamine-3 inverse agonist MK-0249 in obstructive sleep apnea patients on continuous positive airway pressure therapy with excessive daytime sleepiness: a randomized adaptive crossover study. Sleep Med. 2013;14:955–63.
151. Egan M, Yaari R, Liu L, Ryan M, Peng Y, Lines C, et al. Pilot randomized controlled study of a histamine receptor inverse agonist in the symptomatic treatment of AD. Curr Alzheimer Res. 2012;9:481–90.
152. Egan M, Zhao X, Gottwald R, Harper-Mozley L, Zhang Y, Snavely D, et al. Randomized crossover study of the histamine H3 inverse agonist MK-0249 for the treatment of cognitive impairment in patients with schizophrenia. Schizophr Res. 2013;146:224–30.
153. Nagase T, Mizutani T, Ishikawa S, Sekino E, Sasaki T, Fujimura T, et al. Synthesis, structure-activity relationships, and biological profiles of a quinazolinone class of histamine H3 receptor inverse agonists. J Med Chem. 2008;51:4780–9.
154. Cho W, Maruff P, Connell J, Gargano C, Calder N, Doran S, et al. Additive effects of a cholinesterase inhibitor and a histamine inverse agonist on scopolamine deficits in humans. Psychopharmacology (Berl). 2011;218:513–24.

155. Barrett K, Carruthers N, Ennis M, Pearce FL, Bell D, Chazot PL, et al. The European Histamine Research Society 43rd Annual Meeting, May 7–10, 2014 Lyon, France. Inflamm Res. 2014;63:1–48.
156. Weisler RH, Pandina GJ, Daly EJ, Cooper K, Gassmann-Mayer C, Investigators ATTS. Randomized clinical study of a histamine H3 receptor antagonist for the treatment of adults with attention-deficit hyperactivity disorder. CNS Drugs. 2012;26:421–34.

# Chapter 6
# Clinical Significance of Histamine $H_1$ Receptor Gene Expression and Drug Action of Antihistamines

**Hiroyuki Fukui, Hiroyuki Mizuguchi, Yoshiaki Kitamura, and Noriaki Takeda**

**Abstract** Histamine $H_1$ receptor (H1R) has a special upregulation mechanism by the stimulation of H1R, in addition to the downregulation mechanism often observed in many G protein-coupled receptors (GPCRs). The upregulation was mediated by the activation of protein kinase C-delta (PKCδ) and H1R gene expression, resulting in increase in H1R signaling.

Increase in H1R mRNA in nasal mucosa was induced after the provocation of nasal hypersensitivity in rat models and suppressed by the pretreatment of antihistamines. Improvement of nasal symptoms and suppression of H1R mRNA level in nasal mucosa were also observed by the pretreatment with antihistamines in patients with pollinosis. A correlation between symptoms and mRNAs of H1R, histidine decarboxylase (HDC), and interleukin-5 (IL-5) was elucidated. H1R gene was upregulated through PKCδ-MEK-poly(ADP-ribose) polymerase 1 (PARP-1) signaling. Binding sites for transcriptional control elements including AP-1, ETS-1, and Ku86/Ku70 complex were located in the H1R gene promoter region. Similar to the antihistamines, improvement of nasal symptoms and suppression of H1R mRNA were observed in nasal hypersensitivity in rat models by the pretreatment with *Kujin* extract, an antiallergic *Kampo* medicine. H1R-mediated activation of PKCδ was also suppressed by *Kujin* extract. The accumulated pieces of evidences have suggested that H1R, HDC, and IL-5 form a group of allergic-sensitive genes and the antihistamines exert its action by the suppression of H1R-mediated activation of H1R gene expression.

---

H. Fukui, M.D., Ph.D. (✉)
Department of Molecular Studies for Incurable Diseases, Institute of Biomedical Sciences, Tokushima University Graduate School, 3-18-15 Kuramotocho, Tokushima 770-8503, Japan
e-mail: hfukui@tokushima-u.ac.jp

H. Mizuguchi
Department of Molecular Pharmacology, Institute of Biomedical Sciences, Tokushima University Graduate School, Tokushima 770-8505, Japan

Y. Kitamura • N. Takeda
Department of Otorhinolaryngology, Institute of Biomedical Sciences, Tokushima University Graduate School, Tokushima 770-8503, Japan

 
P. Blandina, M.B. Passani (eds.), *Histamine Receptors*, The Receptors 28,
DOI 10.1007/978-3-319-40308-3_6

**Keywords** Allergic disease-sensitive gene • Histamine $H_1$ receptor • *Kampo* medicine • Pollinosis • Protein kinase C

## Abbreviations

| | |
|---|---|
| GPCR | G protein-coupled receptor |
| H1R | Histamine $H_1$ receptor |
| HDC | Histidine decarboxylase |
| IL-5 | Interleukin-5 |
| PARP-1 | Poly(ADP-ribose) polymerase 1 |
| PKC | Protein kinase C |
| PMA | Phorbol 12-myristate 13-acetate |
| TDI | Toluene 2,4-diisocyanate |

## 6.1 Introduction

Histamine is a key mediator of immune hypersensitivity (type I allergy). Major symptoms of allergy representative diseases, such as allergic rhinitis, allergic conjunctivitis, and urticaria, are induced by type I allergic reaction mainly though the histamine $H_1$ receptor (H1R). Thus, the antihistamines that work mainly through the target molecule of H1R are the first therapeutic choice for allergic diseases. However, the multiple mechanisms through which antihistamines exert their clinical effects are not completely understood.

The structure of H1R was revealed to be a G protein-coupled receptor (GPCR) with seven transmembrane-spanning domains by the receptor cloning study [1] and by the crystal structural study [2]. Many GPCRs have been reported to be desensitized by the stimulation of the receptor. However, H1R was upregulated by the stimulation of H1R [3]. Because the number of receptor molecules is prominently small, compared to that of G protein molecules and effector enzyme molecules, the upregulation mechanism of H1R was thought to play a key role in H1R-mediated signal transduction. Indeed, it was reported that the strength of H1R signaling depends on the H1R expression level [4]. Furthermore, upregulation of H1R has been observed in patients with allergic rhinitis [5, 6]. Thus, the upregulation mechanism was thought to play an important role in allergic diseases.

In the present review, the mechanism of H1R upregulation through the activation of H1R gene expression is described. Correlative upregulation of H1R gene expression with nasal symptoms in human patients with pollinosis is also described.

## 6.2 Downregulation and Upregulation of H1R

### 6.2.1 H1R Downregulation

Like many other GPCRs, H1R desensitization was abundantly documented. The final step of desensitization is the receptor downregulation. Wild-type and mutant human H1R genes subcloned into a mammalian expression vector, pdKCR-dhfr, were transfected, and recombinant wild-type and mutant H1Rs were expressed in Chinese hamster ovary (CHO) cells [7]. Because the vector pdKCR-dhfr is free from H1R signaling, the H1R gene is not activated. Time-dependent downregulation of recombinant wild-type H1Rs was induced by H1R stimulation in CHO cells (Fig. 6.1). Wild-type H1R was phosphorylated by protein kinases such as protein kinase A, protein kinase C, protein kinase G, and calcium calmodulin-dependent protein kinase II [8]. Many serine and threonine residues were located in the intracellular domain of H1R. Then synthetic peptides corresponding to the partial sequences of intracellular domains and similar synthetic peptides lacking serine and/or threonine residues were examined to evaluate their phosphorylation by protein kinases. Finally, three threonine residues ($Thr^{140}$,

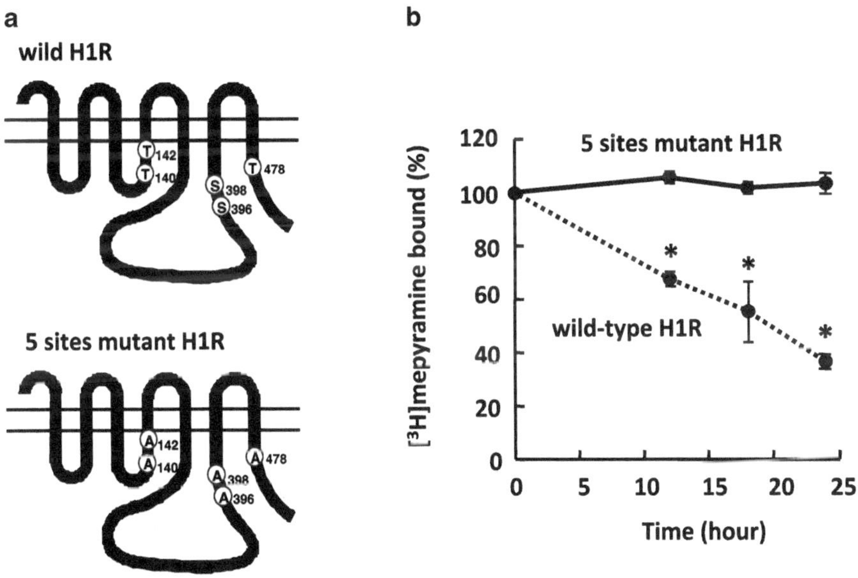


**Fig. 6.1** Downregulation of histamine $H_1$ receptors (H1R) by the stimulation of H1R. Schematic drawing of wild-type H1R (upper in (**a**)) and mutant H1R (five sites of mutant H1R) lacking five putative phosphorylation sites (lower in (**a**)). Putative phosphorylation sites are three threonine residues ($T^{140}$, $T^{142}$, and $T^{478}$) and two serine residues ($S^{396}$ and $S^{398}$). A wild-type and mutant H1R gene subcloned into a mammalian expression vector, pdKCR-dhfr, was transfected, and recombinant H1Rs were expressed in Chinese hamster ovary (CHO) cells. The vector, pdKCR-dhfr, is not affected by H1R stimulation. Expression levels of wild-type H1R and five sites mutant H1R are shown by *dotted line* and *solid line*, respectively (**b**). $*p<0.001$ vs. five sites of mutant H1R ($n=6$) [7]

Thr[142], and Thr[478]) and two serine residues (Ser[396] and Ser[398]) were strongly suggested as phosphorylation sites. The mutant H1R lacking five putative phosphorylation sites showed no downregulation at all [7]. The results indicate that protein kinase activation is the final step. Downregulation of H1R was demonstrated to be dependent on ubiquitin/proteasome system following clathrin-mediated internalization [9].

## 6.2.2 H1R Upregulation

H1Rs are naturally expressed in HeLa cells, a human uterine cervical cancer cell line [10]. The H1R gene promoter is thought to work naturally in HeLa cells [3]. Stimulation of HeLa cells by histamine induced H1R upregulation in a time-dependent manner (Fig. 6.2a). Preceding upregulation elevated H1R mRNA expression level, and H1R gene promoter activity was also induced (Fig. 6.2b, c). The results strongly suggest that H1R upregulation is mediated through the activation of H1R gene expression, although other stimulators than histamine induced the elevation of H1R gene expression. Stimulation with diesel exhaust particulates, platelet-activating factor, $\beta_2$-adrenergic receptor agonist, $M_3$ muscarinic receptor agonist, and interleukin-4 was reported to induce the elevation of H1R gene expression [11–15].

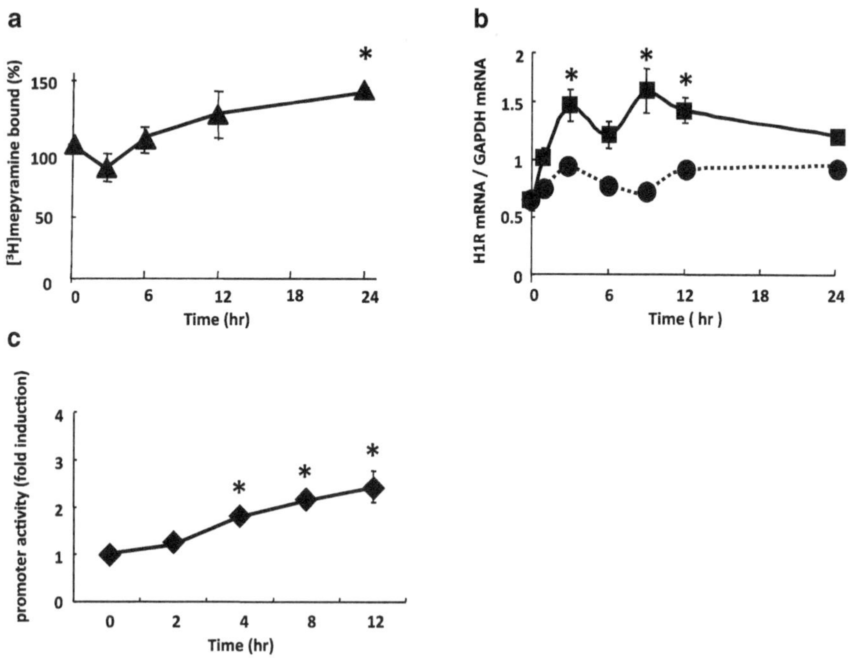


**Fig. 6.2** H1R stimulation-induced H1R upregulation through H1R gene expression in HeLa cells. (**a**) H1R-mediated H1R upregulation. *$p<0.05$ vs. control ($n=3$). (**b**) H1R-mediated elevation of H1R mRNA (*solid line*). Control (*dotted line*). *$p<0.05$ vs. control ($n=3$). (**c**) H1R-mediated activation of H1R gene promoter. *$p<0.05$ vs. control ($n=3$) [3]

### 6.2.3 Significance of H1R Upregulation

The number of GPCR molecules is prominently small compared to that of G proteins, adenylyl cyclase, and phospholipase C-β. Then the regulatory mechanism of the receptor level is thought to play an important role in signal transduction. Upregulation of H1R has been observed in HeLa cells not only by H1R stimulation with histamine but also protein kinase C stimulation with phorbol 12-myristate 13-acetate (PMA) (Fig. 6.3a) [16]. H1R-mediated accumulation of inositol phosphates was larger in H1R-upregulated HeLa cells than that of control cells (Fig. 6.3b). It was also demonstrated that calcium mobilization was correlated to the density of H1Rs [4]. These findings suggest that H1R signaling depends on the H1R expression level, and upregulation of H1R is very important for the increase in H1R signaling.

In general, downregulation of H1R and many other GPCRs are commonly accepted as the final step of receptor desensitization. Downregulation of H1R was clearly demonstrated in H1R-overexpressing CHO cells [7]. H1R expressed in HeLa cells has five putative phosphorylation sites involved in the receptor downregulation. Protein kinases that phosphorylate these residues are also expressed in HeLa cells. Thus, the mechanism of histamine-induced H1R downregulation is thought to be possessed even in HeLa cells. Conclusively, the net expression level of H1Rs is thought to be determined by the balance, i.e., summation of down- and upregulations. H1R upregulation in HeLa cells is an uncommon observation (Fig. 6.4). However, from the results of the H1R studies in HeLa cells, H1R upregulation mechanism seems dominant compared to the mechanism of H1R downregulation [3]. This regulatory mechanism of gene expression characterizes

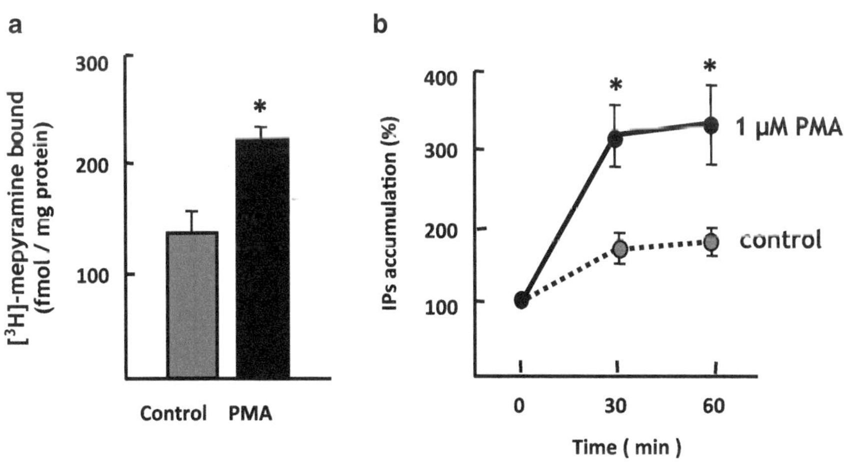


**Fig. 6.3** H1R-induced inositol phosphate (IP) accumulation in phorbol 12-myristate 13-acetate (PMA)-treated HeLa cells. (**a**) H1R upregulation by 1 μM PMA treatment for 24 h. $^*P<0.01$ vs. control ($n=3$). (**b**) H1R-mediated IPs accumulation are shown by *solid line* and *dotted line* in PMA-treated cells and control cells, respectively. $^*P<0.05$ vs. control ($n=3$) [16]

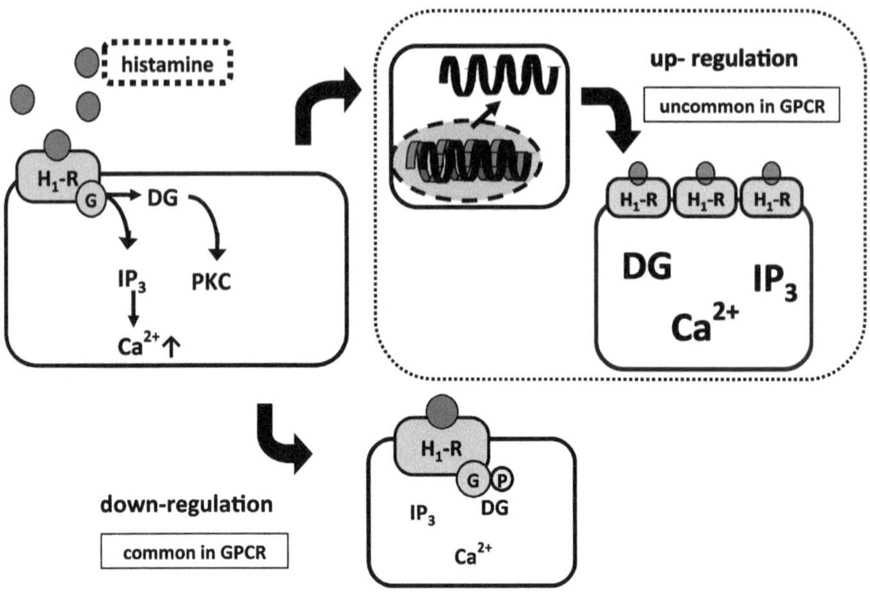


**Fig. 6.4** Up- and downregulation of H1Rs. Stimulated H1Rs were downregulated through the common receptor desensitization mechanism. In addition, stimulation of H1Rs upregulated the H1Rs through the H1R gene expression as an atypical mechanism

to H1R. Then H1R gene is suggested to be closely related to pathological function of allergic diseases. H1R gene expression may also participate in the pathological function of atherosclerosis [17, 18]. Downregulation of H1R reported in the frontal cortex of brains from patients with chronic schizophrenia may be due to low activation of H1R gene expression in the central nervous system [19].

## 6.3 Pathological Significance of H1R Gene Expression

### *6.3.1 Elevation of H1R Gene Expression and H1R Upregulation in Nasal Hypersensitivity Model Rats*

Brown Norway rats were sensitized with toluene 2,4-diisocyanate (TDI) (Fig. 6.5a) and used for pathological studies as nasal hypersensitivity in rat models [20]. Two sets of sensitization to nasal vestibule were performed for 5 days by the application of TDI once a day with 2 days' interval (Fig. 6.5b). After 9 days' interval, TDI provocation was performed to induce allergic rhinitis-like symptoms (Fig. 6.5c). H1R mRNA level was elevated, and H1R upregulation appeared after the provocation in nasal mucosa with TDI (Fig. 6.5d, e).

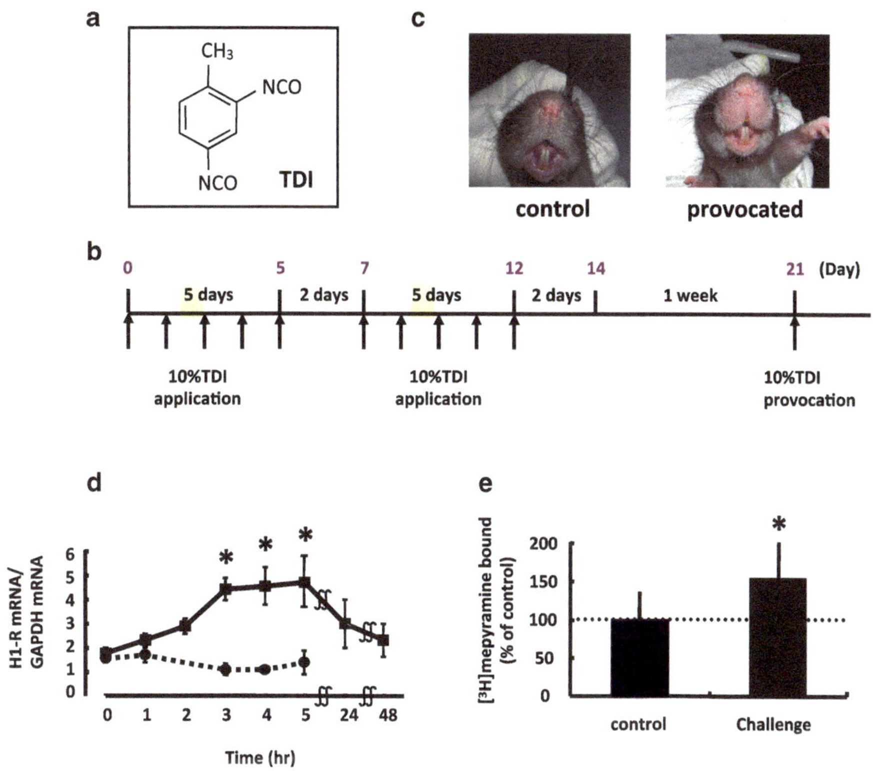


**Fig. 6.5** H1R mRNA elevation and H1R (determined by [$^3$H]mepyramine binding) upregulation in the nasal mucosa of toluene 2,4-diisocyanate (TDI)-sensitized nasal hypersensitivity model rats after provocation. (**a**) Chemical structure of TDI. (**b**) TDI-sensitized brown Norway rat, nonsensitized rat (*left*), provoked with TDI (*right*). (**c**) Protocol of TDI sensitization: brown Norway rats were sensitized with 10 % TDI solution in ethyl acetate by the application of 5 μL solution to nasal vestibule once a day for two sets of continual 5 days with 2 interval days. Nasal provocation was induced by the application of 5 μL TDI solution to nasal vestibule after 9 days of sensitization. (**d**) H1R mRNA elevation in control (*dotted line*) and TDI provocation (*solid line*). $*p<0.01$ vs. control ($n=4$). (**e**) H1R upregulation 24 h after provocation. $*p<0.05$ vs. control ($n=4$) ([20] with modification)

### *6.3.2 Suppression of H1R mRNA Elevation in Nasal Hypersensitivity Model Rats by Long-Term Antihistamine Pretreatment*

Short-term and long-term pretreatments with epinastine, an antihistamine, were performed to TDI-sensitized nasal hypersensitivity model rats (Fig. 6.6a). Short-term pretreatment suppressed the number of sneezing weakly (Fig. 6.6b). However, long-term pretreatment suppressed sneezing more strongly. Correlatively, elevation

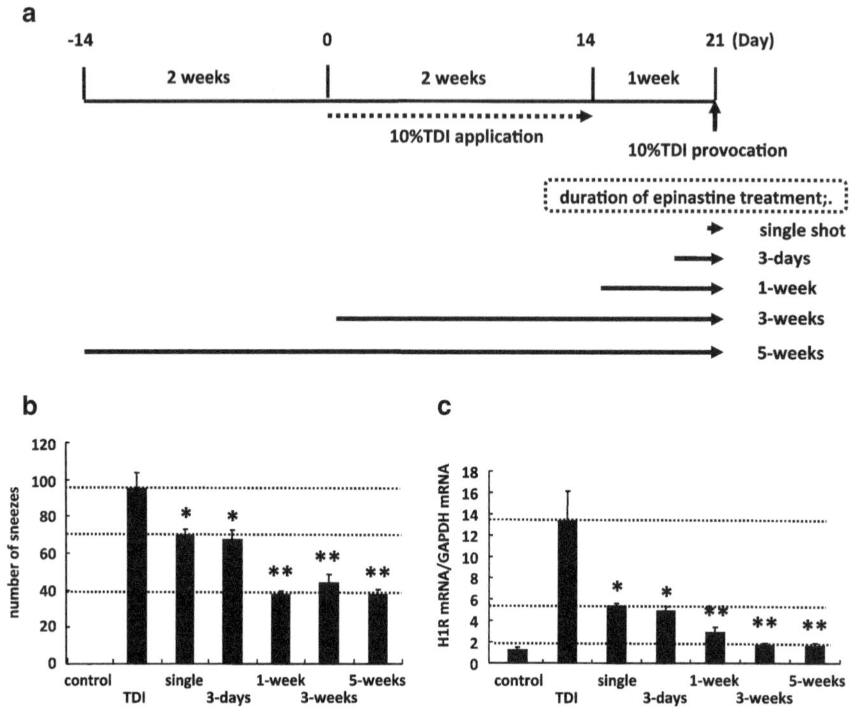


**Fig. 6.6** Effects of long-term pretreatment with epinastine on sneezing and H1R mRNA level in nasal mucosa of TDI-sensitized nasal hypersensitivity model rats. (**a**) Protocol of TDI sensitization and epinastine treatment (30 mg/kg p.o.). TDI treatment was performed once a day for 2 weeks indicated by a *dotted line*. Pretreatments with epinastine by a single shot once a day were performed by a single shot, 3 days, 1 week, 3 weeks, and 5 weeks prior to the provocation. (**b**) Effects of epinastine treatment for different periods on sneezing. *$p<0.05$ vs. TDI and **$p<0.05$ vs. single treatment ($n=3$). (**c**) Effects of epinastine treatment for different periods on H1R mRNA level. *$p<0.05$ vs. TDI and **$p<0.05$ vs. single treatment ($n=4$) [21]

of H1R mRNA level induced by TDI provocation was weakly suppressed by short-term pretreatment of epinastine (Fig. 6.6c). However, long-term pretreatment suppressed it almost equal to the control level. Other antihistamines such as *d*-chlorpheniramine and olopatadine also showed similar results.

A good correlation between sneezing and H1R mRNA level was observed in nasal hypersensitivity model rats [21]. H1R gene was suggested to be an allergic disease-sensitive gene. In addition, long-term pretreatment of antihistamines for more than 1 week showed better improvement of sneezing and suppression of H1R mRNA level. Prophylactic treatment of antihistamines for pollinosis, recommended by the Japanese Society of Allergology [22], also reported good results. The results using nasal hypersensitivity model rats suggested an important clinical significance of prophylactic treatment with antihistamines.

### 6.3.3 *Prophylactic Antihistamine Treatment of Pollinosis Patients*

Improvement of symptoms by prophylactic treatment with antihistamines for the therapy of pollinosis in Japan was reported [23]. However, the multiple mechanisms through which antihistamines exert their clinical effects are not completely understood. On the other hand, clinical study of prophylactic treatment with antihistamines showed improvement of sneezing and score of water rhinorrhea (Fig. 6.7a, b). In association with the improvement of symptoms, H1R mRNA level was suppressed by prophylactic treatment with antihistamines (Fig. 6.7c). A good correlation between nasal symptom score and H1R mRNA level in nasal mucosa was observed (Fig. 6.7d). These results clearly showed that H1R mRNA elevation was

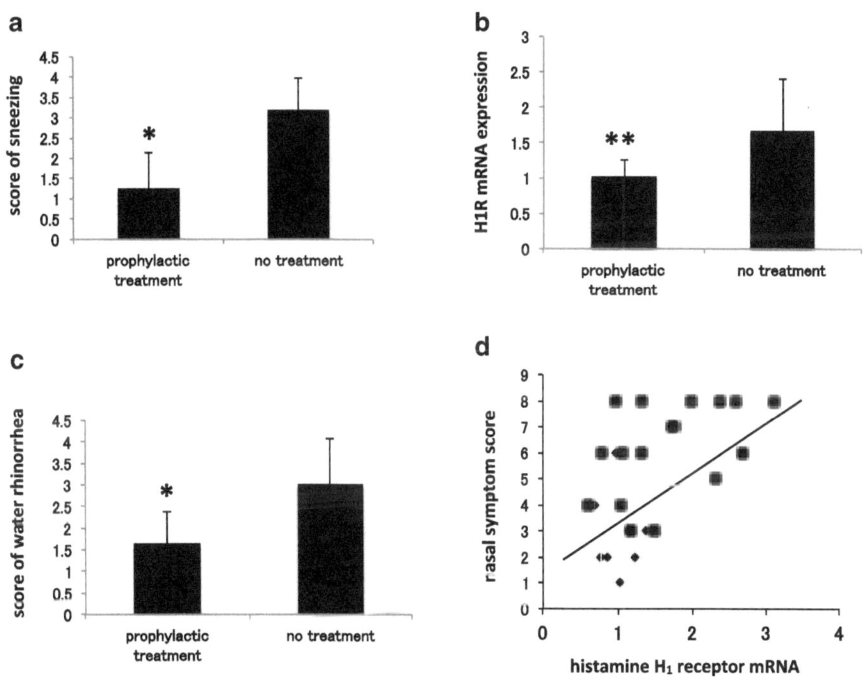


**Fig. 6.7** Effects of prophylactic treatment of antihistamines on symptoms and nasal mucosal histamine $H_1$ receptor (H1R) mRNA level in pollinosis patients. Prophylactic treatment ($n=8$) and non-treatment ($n=17$). Improvement of sneezing (**a**) and nasal symptom score (**b**). Nasal symptoms in all patients were evaluated according to Practical Guideline for the Management of Allergic Rhinitis in Japan [23]. Sneezing, watery rhinorrhea, and nasal obstruction were separately scored on a 0- to 5-point scale. *$p<0.01$ vs. nontreatment. Suppression of H1R mRNA elevation (**c**). **$p<0.05$ vs. nontreatment. Correlation between nasal symptom score and H1R mRNA level (**d**). (*Filled square*) nontreated patients. (*Filled diamond*) patients with prophylactic treatment with ebastine 10 mg/day p.o. or fexofenadine 120 mg/day p.o. Correlation coefficient: $r=0.51$, $p<0.01$ [24]

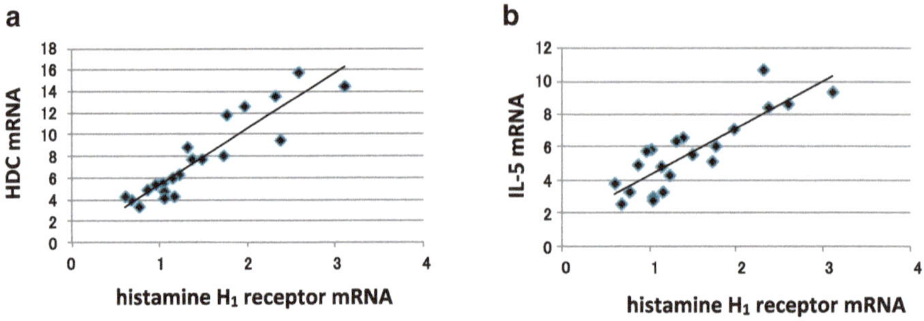


**Fig. 6.8** Correlation between H1R mRNA (**a**) and histidine decarboxylase (HDC) mRNA or interleukin-5 (IL-5) mRNA (**b**) in nasal mucosa of pollinosis patients. Correlation coefficient: $r=0.923$, $p<0.01$ (**a**), $r=0.788$, $p<0.01$ (**b**) [25]

suppressed correlatively with symptoms and H1R gene is strongly suggested to be an allergic disease-sensitive gene [24].

In addition to H1R mRNA, levels of histidine decarboxylase (HDC) mRNA and interleukin-5 (IL-5) mRNA were also elevated in the nasal mucosa of patients with pollinosis [25]. Degree of HDC mRNA elevation was well correlated with that of H1R mRNA (Fig. 6.8). Similarly, a good correlation between levels of IL-5 mRNA and H1R mRNA was also observed. Antihistamines seem to regulate not only H1R mRNA level but also levels of HDC mRNA and IL-5 mRNA. Collective results suggest that H1R gene, HDC gene, and IL-5 genes form a same category of allergic disease-sensitive gene and levels of H1R mRNA, HDC mRNA, and IL-5 mRNA may be regulated by the same mechanism. Upregulation of H1R mRNA was also reported in the nasal mucosa of patients with allergic rhinitis [5, 6, 26, 27].

## 6.4 Improvement of Symptoms by *Kujin* Extract with the Suppression of H1R mRNA Elevation

*Kujin* is the root of *Sophora flavescens*, with *taxonomy of ID* 49840 and *inherited blast name of* eudicots. Sho-hu-san, a *Kampo* prescription containing *Kujin*, is used for the therapy of urticaria and eczema. The number of sneezing was significantly decreased by the treatment with *Kujin* extract of TDI-sensitized rats (Fig. 6.9) [28]. Score of rhinorrhea showed a tendency of improvement (Table 6.1). Elevation of H1R mRNA and upregulation of H1R in nasal mucosa were almost completely suppressed. The profile of improvement of symptoms and suppression of H1R mRNA by *Kujin* extract was comparable to that of antihistamines.

An active substance that suppresses H1R mRNA elevation was successfully isolated and identified from *Kujin* extract (manuscript in preparation). The active

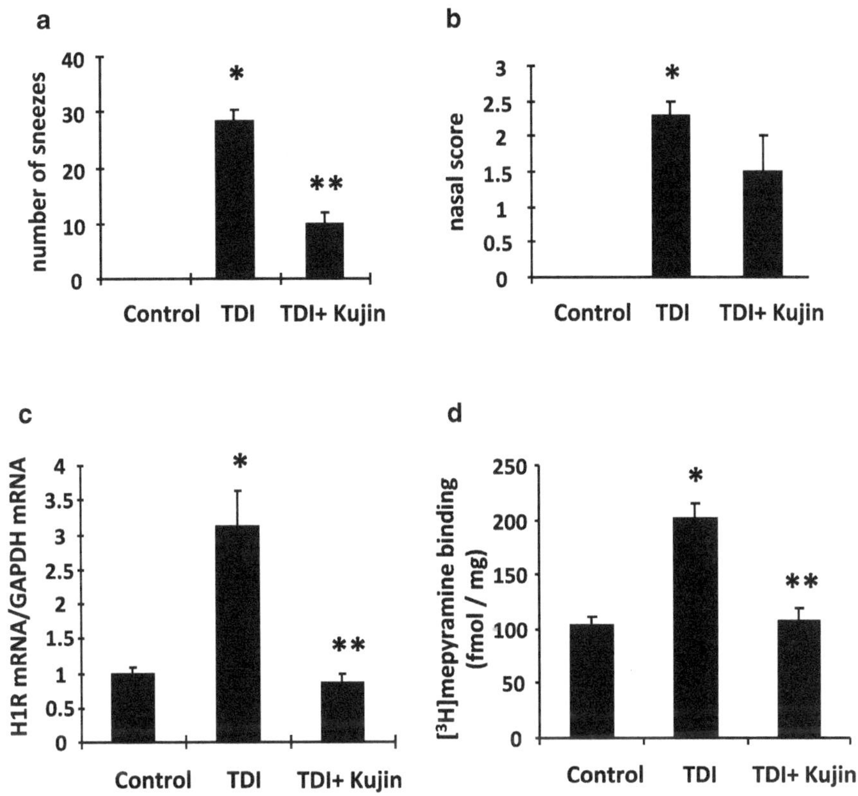


**Fig. 6.9** Suppressive effects of *Kujin* extract on symptoms and H1R expression level in the nasal mucosa of TDI-sensitized nasal hypersensitivity model rats. Sneezing (**a**) and nasal scores (**b**) were determined 10 min after provocation. Determinations of H1R mRNA level by real-time PCR and H1R expression level by [$^3$H]mepyramine binding were performed 4 h and 24 h, respectively, after provocation. *$p<0.05$ vs. control and **$p<0.05$ vs. TDI ($n=4$) [28]

**Table 6.1** Criteria for grading the severity of TDI-induced nasal responses in rats

| Response | Score | | | |
|---|---|---|---|---|
| Nasal response | 0 | 1 | 2 | 3 |
| Number of sneezes | (–) | 1–4 | 5–11 | 12< |
| Watery rhinorrhea | (–) | At the nostril | Between 1 and 3 | Drops of discharge from the nose |

Criteria were described by Kitamura et al. [20]

compound suppressed the phosphorylation of PKC induced by either histamine or PMA. These results strongly suggest that *Kujin* and antihistamines improve the symptoms of pollinosis by the suppression of the same signaling pathway, i.e., PKC activation-induced elevation of H1R gene expression.

## 6.5 Mechanism of H1R-Mediated H1R Upregulation Through the Activation of Gene Expression

Elevation of H1R mRNA induced by histamine was suppressed by Ro31-8220 (a universal inhibitor of PKC) and rottlerin (a PKCδ-selective inhibitor), but not by Go-6976 (an inhibitor of PKCα, β, γ) in HeLa cells (Fig. 6.10a) [29]. Similar results were obtained with the H1R promoter assay (Fig. 6.10b). Phosphorylation of two amino acid residues, $Tyr^{311}$ and $Thr^{505}$, which were reported to be involved in PKCδ activation, was induced by the stimulation with histamine. Elevation of H1R mRNA was induced by $H_2O_2$, an activator of PKCδ. Histamine-induced elevation of H1R mRNA was suppressed by PKCδ-specific siRNA and enhanced by PKCδ overexpression [29]. These results strongly suggest that PKCδ plays a key role in H1R-mediated H1R gene expression. PKCδ was translocated from cytosol to Golgi by the stimulation with histamine or PMA (Fig. 6.11). In the presence of rottlerin, the PKCδ translocation induced by histamine and PMA was completely suppressed. Suppression by U0126, a MEK inhibitor or DPQ, a PARP-1 inhibitor, of PMA-induced H1R mRNA elevation and H1R gene promoter activation indicated the presence of PKCδ-ERK-PARP-1 signaling (Fig. 6.12).

H1R gene expression was regulated through two regions in the H1R gene promoter [30]. Three binding sites for two transcription factors, two AP-1, and an ETS-1 were located in the upstream region. A unique mechanism through a transcription factor complex, Ku86 and Ku70, was discovered through the downstream region. Suppression of H1R gene promoter activity due to the binding of the complex of Ku86 and Ku70 to the downstream region was liberated by the activation of PKCδ.

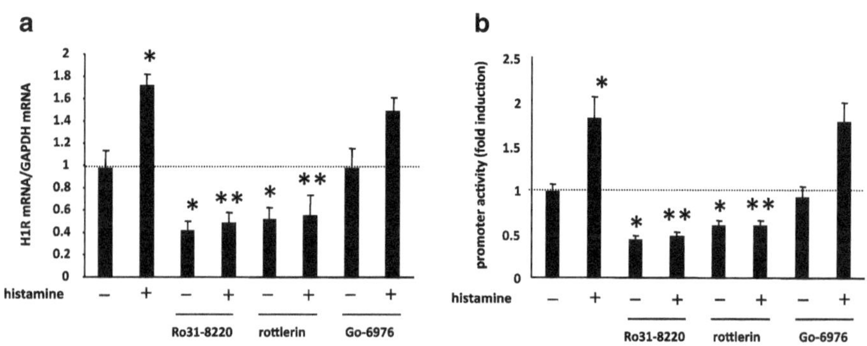


**Fig. 6.10** Effects of protein kinase inhibitors on H1R-mediated H1R mRNA elevation (**a**) and H1R gene promoter activation in HeLa cells. *$p<0.05$ vs. control ($n=4$) and **$p<0.05$ vs. histamine ($n=4$) ([29] with modification)

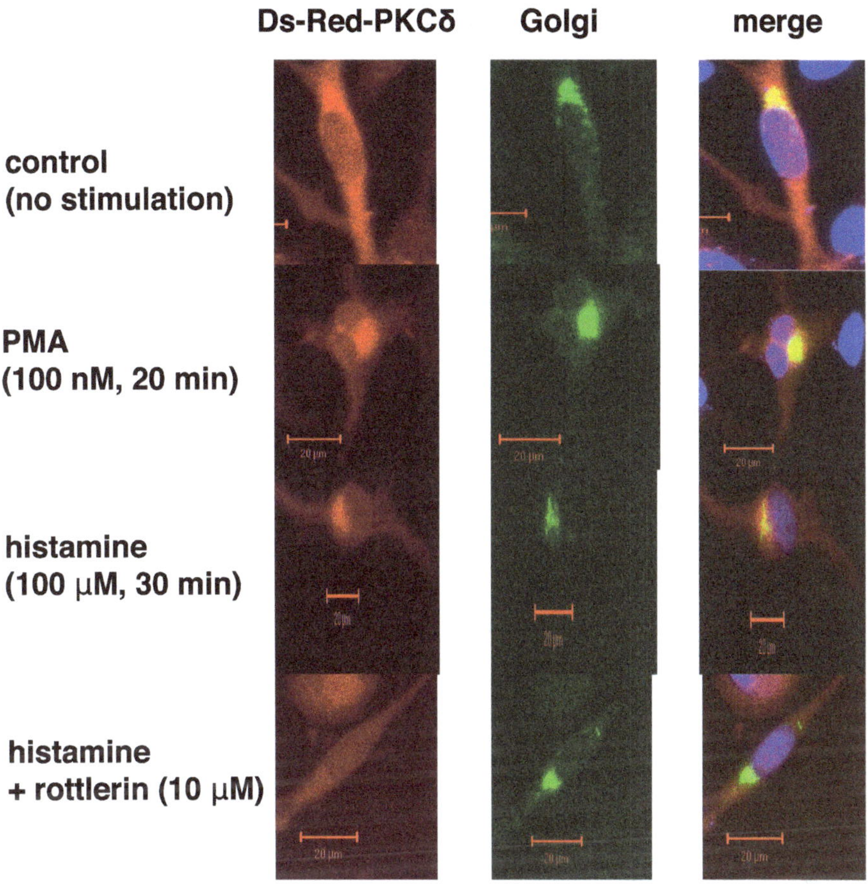


**Fig. 6.11** Translocation of PKCδ by histamine or PMA stimulation. The expressing plasmid encoding Ds-Red-PKCδ and pAcGFP-Golgi (for labeling of the Golgi) were co-transfected into HeLa cells attached onto 35-mm glass-bottomed dishes. The cells were serum starved for 24 h and then stimulated with 100 μM histamine or 100 nM PMA for the indicated times in the same medium. The cells were treated with the PKCδ-specific inhibitor rottlerin (10 μM) for 1 h before histamine stimulation. The subcellular localization of Ds-Red-PKC was determined using a confocal laser microscope. *Bars* = 20 μm ([29], with modification)

## 6.6 Conclusion

Stimulation of H1R elevated H1R gene expression and induced H1R upregulation by PKCδ activation with its translocation to Golgi and activation of MEK-ERK-PARP-1 signaling. Antihistamines suppressed the elevation of H1R mRNA in the mucosa of both nasal hypersensitivity model rats and patients with pollinosis. Correlation among nasal symptoms and mRNA levels of H1R, HDC, and IL-5 was

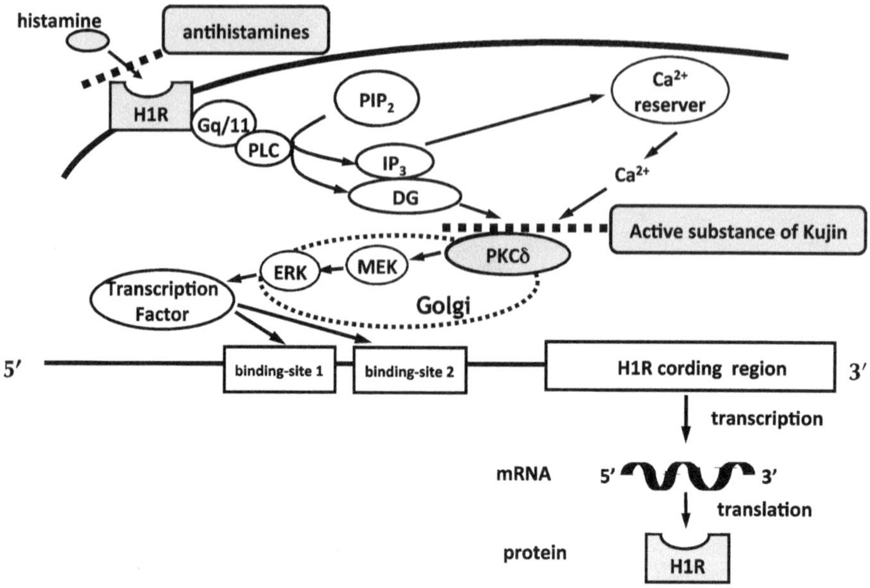


**Fig. 6.12** Mechanism of H1R-mediated H1R gene expression through PKCδ activation

observed in patients with pollinosis. The extract of *Kujin*, an antiallergic *Kampo* medicine, showed improvement of symptoms and suppression of H1R mRNA elevation in nasal hypersensitivity model rats. Cumulative results strongly suggest that PKCδ-mediated signaling induces an elevation of H1R gene expression that is suppressed by antihistamines and *Kujin* indicating their important role in pollinosis, allergic rhinitis, and other allergic diseases.

Acknowledgment

**Conflict of Interest** The authors declare that there are no conflicts of interest.

## References

1. Yamashita M, Fukui H, Sugama K, Horio Y, Ito S, Mizuguchi H, et al. Expression cloning of a cDNA encoding the bovine histamine $H_1$ receptor. Proc Natl Acad Sci U S A. 1991;88:11515–9.
2. Shimamura T, Shiroishi M, Weyand S, Tsujimoto H, Winter G, Katritch V, Abagyan R, Cherezov V, Liu W, Han GW, Kobayashi T, Stevens RC, Iwata S. Structure of the human histamine H1 receptor complex with doxepin. Nature. 2011;475:65–70.
3. Das AK, Yoshimura S, Mishima R, Fujimoto K, Dev S, Wakayama Y, et al. Stimulation of histamine $H_1$ receptor up-regulates histamine $H_1$ receptor itself through activation of receptor gene transcription. J Pharmacol Sci. 2007;103:374–82.
4. Ohuchi Y, Yanai K, Sakurai E, Fukui H, Yanagisawa T, Watanabe T. Histamine-induced calcium mobilization in single cultured cells expressing histamine $H_1$ receptors: A relationship between its sensitivity and the density of $H_1$ receptors. Int J Mol Med. 1998;1:355–60.

5. Dinh QT, Cryer A, Dinh S, Peiser C, Wu S, Springer J, Hamelmann E, Klapp BF, Heppt W, Fischer A. Transcriptional up-regulation of histamine receptor-1 in epithelial, mucus and inflammatory cells in perennial allergic rhinitis. Clin Exp Allergy. 2005;35:1443–8.
6. Iriyoshi N, Takeuchi K, Yuta A, Ukai K, Sakakura Y. Increased expression of histamine H1 receptor mRNA in allergic rhinitis. Clin Exp Allergy. 1996;26:379–85.
7. Horio S, Ogawa M, Kawakami N, Fujimoto K, Fukui H. Identification of amino acid residues responsible for agonist-induced down-regulation of histamine $H_1$ receptors. J Pharmacol Sci. 2004;94:410–9.
8. Horio S, Kato T, Ogawa M, Fujimoto K, Fukui H. Two threonine residues and two serine residues in the second and third loops are both involved in histamine $H_1$ receptor down-regulation. FEBS Lett. 2004;573:226–30.
9. Hishinuma S, Komazaki H, Fukui H, Shoji M. Ubiquitin/proteasome-dependent down-regulation following clathrin-mediated internalization of histamine H1-receptors in Chinese hamster ovary cells. J Neurochem. 2010;113:990–1001.
10. Tilly BC, Tertoolen LG, Lambrechts AC, Remorie R, de Laat SW, Moolenaar WH. Histamine-H1-receptor-mediated phosphoinositide hydrolysis, $Ca^{2+}$ signalling and membrane-potential oscillations in human HeLa carcinoma cells. Biochem J. 1990;266:235–43.
11. Horio S, Fujimoto K, Mizuguchi H, Fukui H. Interleukin-4 up-regulates histamine H1 receptors by activation of H1 receptor gene transcription. Naunyn Schmiedebergs Arch Pharmacol. 2010;381:305–13.
12. Mak JC, Roffel AF, Katsunuma T, Elzinga CR, Zaagsma J, Barnes PJ. Up-regulation of airway smooth muscle histamine H(1) receptor mRNA, protein, and function by beta(2)-adrenoceptor activation. Mol Pharmacol. 2000;57:857–64.
13. Miyoshi K, Kawakami N, Das AK, Fujimoto K, Horio S, Fukui H. Heterologous up-regulation of the histamine H1 receptor by M3 muscarinic receptor-mediated activation of H1-receptor gene transcription. J Pharm Pharmacol. 2007;59:843–8.
14. Nakasaki T, Masuyama K, Fukui H, Ogino S, Eura M, Samejima Y, Ishikawa T, Yumoto E. Effects of PAF on histamine H1 receptor mRNA expression in rat trigeminal ganglia. Prostaglandins Other Lipid Mediat. 1999;58:29–41.
15. Terada N, Hamano N, Maesako KI, Hiruma K, Hohki G, Suzuki K, Ishikawa K, Konno A. Diesel exhaust particulates upregulate histamine receptor mRNA and increase histamine-induced IL-8 and GM-CSF production in nasal epithelial cells and endothelial cells. Clin Exp Allergy. 1999;29:52–9.
16. Fukui H, Horio Y. Regulation of the histamine $H_1$ receptor signaling by protein kinase C. In: Uchida MK, editor. Receptor desensitization and $Ca^{2+}$-signaling. Tokyo: Japan Science Society Press; 1996. p. 107–19.
17. Hao F, Tan M, Xu X, Cui MZ. Histamine induces Egr-1 expression in human aortic endothelial cells via the H1 receptor mediated protein kinase Cdelta-dependent ERK activation pathway. J Biol Chem. 2008;283:26928–36.
18. Takagishi T, Sasaguri Y, Nakano R, Arima N, Tanimoto A, Fukui H, Morimatsu M. Expression of the histamine H1 receptor gene in relation to atherosclerosis. Am J Pathol. 1995;146:981–8.
19. Nakai T, Kitamura N, Hashimoto T, Kajimoto Y, Nishino N, Mita T, Tanaka C. Decreased histamine H1 receptors in the frontal cortex of brains from patients with chronic schizophrenia. Biol Psychiatry. 1991;30:349–56.
20. Kitamura Y, Miyoshi A, Murata Y, Kalubi B, Fukui H, Takeda N. Effect of glucocorticoid on upregulation of histamine $H_1$ receptor mRNA in nasal mucosa of rats sensitized by exposure to toluene diisocyanate. Acta Otolaryngol. 2004;124:1053–8.
21. Mizuguchi H, Hatano M, Matsushita C, Umehara H, Kuroda W, Kitamura Y, et al. Repeated pre-treatment with antihistamines suppresses transcriptional up-regulations of histamine $H_1$ receptor and interleukin-4 genes in toluene-2,4-diisocyanate–sensitized rats. J Pharmacol Sci. 2008;108:480–6.

22. The Japanese Society of Allergology. Practical guidelines for the management of allergic rhinitis in Japan. 7th ed. Tokyo: Life Science; 2013.
23. Baba K, Konno A, Takenaka H, editors. Practical guideline for the management of allergic rhinitis in Japan. Tokyo: Life Science; 2009.
24. Mizuguchi H, Kitamura Y, Kondo Y, Kuroda W, Yoshida H, Miyamoto Y, et al. Pre-seasonal prophylactic treatment with antihistamines suppresses nasal symptoms and expression of histamine $H_1$ receptor mRNA in the nasal mucosa of patients with pollinosis. Methods Find Exp Clin Pharmacol. 2010;32:745–8.
25. Kitamura Y, Mizuguchi H, Ogishi H, Kuroda W, Hattori M, Fukui H, et al. Pre-seasonal prophylactic treatment with antihistamines suppresses IL-5, but not IL-33 mRNA expression in the nasal mucosa of patients with seasonal allergic rhinitis caused by Japanese cedar pollen. Acta Otolaryngol. 2012;132:434–8.
26. Fujikura T, Shimosawa T, Yakuo I. Regulatory effect of histamine H1 receptor antagonist on the expression of messenger RNA encoding CC chemokines in the human nasal mucosa. J Allergy Clin Immunol. 2001;107:123–8.
27. Shirasaki H, Kanaizumi E, Seki N, Himi T. Localization and upregulation of the nasal histamine H1 receptor in perennial allergic rhinitis. Mediators Inflamm. 2012;2012:951316. doi:10.1155/2012/951316.
28. Dev S, Mizuguchi H, Das AK, Maeyama K, Horinaga S, Kato S, et al. *Kujin* suppresses histamine signaling at transcriptional level in toluene 2,4-diisocyanate-sensitized rats. J Pharmacol Sci. 2009;109:606–17.
29. Mizuguchi H, Terao T, Kitai M, Ikeda M, Yoshimura Y, Das AK, et al. Involvement of PKC{delta}/ERK/Poly(ADP-ribose) polymerase-1 (PARP-1) signaling pathway in histamine-induced up-regulation of histamine $H_1$ receptor gene expression in HeLa cells. J Biol Chem. 2011;286:30542–51.
30. Mizuguchi H, Miyagi K, Terao T, Sakamoto N, Yamawaki Y, Adachi T, et al. PMA-induced dissociation of Ku86 from the promoter causes transcriptional up-regulation of histamine $H_1$ receptor. Sci Rep. 2012;2:916. doi:10.1038/srep00916.

# Chapter 7
# Histamine and Its Receptors as a Module of the Biogenic Amine Diseasome

**Rocío Rodríguez-López, María Morales, and Francisca Sánchez-Jiménez**

**Abstract** Biogenic amines play important roles in most important physiological processes, from cell proliferation and differentiation to nutrition, immune response, and neurobiology and reproduction. These effects are spread through a wide variety of cell-specific receptors, cell-specific signaling, and metabolic pathways. However, the biochemical events underlying these effects conform very complex networks of interactions that are far from being completely understood in most cases. In addition, two or more biogenic amines can coexist in the same physiological scenarios keeping cross talk events with influence in their respective physiological functions. In this respect, histamine seems to be the most pleiotropic biogenic amine keeping biochemical and functional interactions with both growth-related polyamines and neurotransmitters in different cell models and tissues. As diseases are the consequence of a biochemical imbalance in one or more tissues, the physiological importance of these compounds and their multiple relationships must have a reflection in the human *diseasome*, the scope of which is not yet known. This fact impedes development of new solutions for diagnosis, prognosis, and treatment of the multiple diseases involving the action of biogenic amines. This work is a further effort of our group to integrate genetic, functional, and clinical information about biogenic amine-related diseases assisted by text mining and network theory-based tools with the aim of helping to advance in personalized biomedical strategies.

R. Rodríguez-López
Departamento de Biología Molecular y Bioquímica, Facultad de Ciencias, Universidad de Málaga, Campus Andalucía-Tech, Campus de Teatinos, Malaga 29071, Spain

Instituto de Biomedicina IBIMA, Campus de Teatinos, Malaga 29071, Spain
e-mail: rorodriguez@uma.es; kika@uma.es

M. Morales
Departamento de Biología Molecular y Bioquímica, Facultad de Ciencias, Universidad de Málaga, Campus Andalucía-Tech, Campus de Teatinos, Malaga 29071, Spain
e-mail: mariamoralesmar@hotmail.com

F. Sánchez-Jiménez, Ph.D. (✉)
Departamento de Biología Molecular y Bioquímica, Facultad de Ciencias, Universidad de Málaga, Campus Andalucía-Tech, Campus de Teatinos, Malaga 29071, Spain

Unidad 741 CIBER de Enfermedades Raras, Campus de Teatinos, Malaga 29071, Spain

Instituto de Biomedicina IBIMA, Campus de Teatinos, Malaga 29071, Spain



P. Blandina, M.B. Passani (eds.), *Histamine Receptors*, The Receptors 28,
DOI 10.1007/978-3-319-40308-3_7

**Keywords** Histamine • G-protein-coupled receptors • Dopamine • Serotonin • Polyamines • Amine oxidases • Cancer • Inflammation • Neurotransmission • Rare diseases • Systems medicine

## Abbreviations

| | |
|---|---|
| 5′-HT | 5′-Hydroxitriptamine or serotonin |
| Ac | Acetyl moiety |
| ADHD | Attention-deficit hyperactivity disorder |
| DA | Dopamine |
| DFMO | Difluoromethylornithine |
| GABA | Gamma-aminobutyric acid |
| GO | Gene Ontology |
| GWAS | Genome-wide association studies |
| $H_1R$ | Histamine receptor type 1 (protein) |
| $H_2R$ | Histamine receptor type 2 (protein) |
| $H_3R$ | Histamine receptor type 3 (protein) |
| $H_4R$ | Histamine receptor type 4 (protein) |
| Hia | Histamine |
| HPO | Human phenotype ontology |
| ODC | Ornithine decarboxylase (protein) |
| OMIM | Online Mendelian inheritance in man |
| PA | Polyamines |
| PLP | Pyridoxal 5′-phosphate |
| Put | Putrescine |
| ROS | Reactive oxygen species |
| Spd | Spermidine |
| Spm | Spermine |
| | Genes and their encoded proteins are abbreviated by their official symbol as recommended in the NCBI gene database and listed in Table 6.1. |

## 7.1 Introduction

Metabolism of amino acid derivatives has almost been neglected for years in general biochemistry books as it was considered part of the secondary metabolism. However, the progressive knowledge integration of different biomedical areas and the more systemic view of the biological processes are progressively revealing the great importance of these derivatives in human physiopathology. For instance, decarboxylation products of amino acids, commonly known as biogenic amines, are

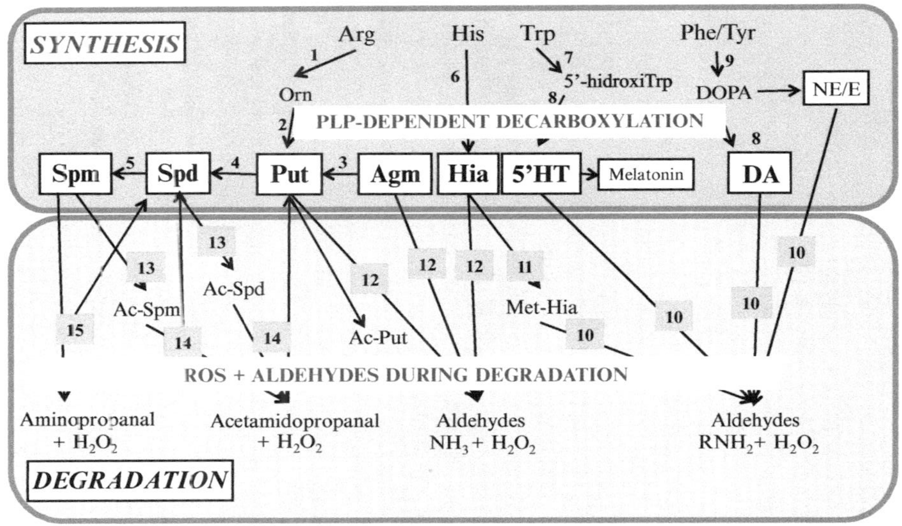


**Fig. 7.1** Scheme of synthesis and degradation pathways for biogenic amines derived from cationic and aromatic L-amino acids. *White* cases crossing the pathways label common steps among them, and *bold* lettering points out the biogenic amines considered in the frame of this chapter. The enzymes considered in this work are located by numbers in the figure, with the correspondence to the official symbols of the respective genes (see Table 7.1) as follows: 1, ARG; 2, ODC; 3, AGMAT; 4, SMS; 5, SRM; 6, HDC; 7, TPH; 8, DDC; 9, TH; 10, MAO; 11, HNMT; 12; DAO; 13, SAT; 14, PAOX; 15, SAOX

essential biomolecules for all the most important physiological functions of a human being. A synthetic scheme of the synthesis and degradation pathways of biogenic amines is depicted in Fig. 7.1, showing some of the common features among them. Biogenic amine synthesis includes a decarboxylation step that, in some cases, is the only reaction required for their synthesis; i.e., putrescine (Put), histamine (Hia), and gamma-aminobutyric acid (GABA). In mammals, L-amino acid decarboxylases are pyridoxal 5′-phosphate (PLP)-dependent enzymes [1]. Degradation of biogenic amines includes the action of different amine oxidases, which can lead to the overproduction of ROS and toxic aldehydes [2]. Moreover, Hia and Put share the possibility of being degraded by diamine oxidase, the AOC1 encoding protein (Table 7.1) [3]. Further information on the enzymes involved in these pathways can be obtained from previous references [4–8].

Among biogenic amines, only polyamines that are putrescine, spermidine, and spermine (Put, Spd, and Spm, respectively) are synthetized in nearly all human cell types, at least in those with the capability to proliferate. The so-called higher polyamines (Spd and Spm) are essential for nucleic acid synthesis and conformation, Spd being the precursor of hypusine, an essential moiety for activity of the translation factor eIF-5A [37]. In fact, the enzyme responsible for Put synthesis, ornithine decarboxylase (ODC), is being used as an anticancer drug target. The ornithine analog difluoromethylornithine (DFMO) or eflornithine is a suicide ODC inhibitor that is being used in clinical works against several types of cancer (i.e., as chemopreventive in colon cancer and in therapies against neuroblastoma) [38, 39] and has antiparasitic properties [40]. In addition, it has been proposed as potentially useful for many other neoplasias. Evidence is also claiming for a role of polyamines as neuroactive compounds [41] as well as required for the correct maturation of some immune cell types (i.e., mast cells) [42]. No structure of any specific mammalian polyamine transport system has been characterized, in spite of valuable efforts by several research groups [43, 44]. Nevertheless, it is accepted that the simultaneous inhibition of PA synthesis and transport could provide a good strategy against pathologies involving undesirable cell proliferation [29]. Although, several key questions on polyamine biochemistry, molecular biology, and physiology are still open in spite of the demonstrated relevance of the compounds in cell life/death equilibrium [45].

Histamine (Hia) is the product of the histamine decarboxylase reaction. Hia is the ligand of at least four specific membrane receptor types that are members of the G-protein-coupled receptor (GPCR) family and named as $H_1R$–$H_4R$ (encoded by HRH1–4 genes, Table 7.1) [17, 46]. Histamine, along to polyamines, is a ligand of the N-methyl aspartate receptors (encoded by GRIN genes, Table 7.1), but uses a different binding site in the target [47]. Histamine can be considered the most pleiotropic biogenic amine. On the one hand, it is clear that histamine is able to modulate cell proliferation [48, 49], sometimes showing antagonic synthesis with respect to polyamines [50, 51]. In vitro, it is demonstrated that Hia is able to bind DNA changing its structure [52], and it has been detected in human breast cancer cell nuclei [53]. On the other hand, histamine is a neurotransmitter that is functionally connected to other biogenic amine neurotransmitters [54–57]. In fact, it has been proven that $H_3R$ is a therapeutic target for several emergent neurological diseases, and there are advanced clinical trials checking the pharmacological usefulness of

**Table 7.1** Biogenic amine-related genes (considered in this work) identified by the encoded protein/polypeptide names, the gene official symbols, and the respective Entrez Gene ID

| Protein name | Official gene symbol | Gene ID | Reference |
|---|---|---|---|
| *Polyamines* | | | |
| Arginase 1 | ARG1 | 383 | [9] |
| Arginase 2 | ARG2 | 384 | [10] |
| Ornithine decarboxylase | ODC1 | 4953 | [11] |
| Ornithine decarboxylase antizyme 1 | OAZ1 | 4946 | [12] |
| Ornithine decarboxylase antizyme 2 | OAZ2 | 4947 | [12] |
| Ornithine decarboxylase antizyme 3 | OAZ3 | 51686 | [12] |
| Antizyme inhibitor 1 | AZIN1 | 51582 | [12] |
| Antizyme inhibitor 2 | AZIN2 | 113451 | [13] |
| Spermidine synthase | SRM | 6723 | [4] |
| Spermine synthase | SMS | 6611 | [14] |
| Spermidine/spermine acetyl transferase | SAT1 | 6303 | [7] |
| Polyamine oxidase | PAOX | 196743 | [7] |
| Spermine oxidase | SMOX | 54498 | [7] |
| *Histamine* | | | |
| Histidine decarboxylase | HDC | 3067 | [15] |
| Histamine N-methyl transferase | HNMT | 3176 | [16] |
| Histamine receptor 1 | HRH1 | 3269 | [17] |
| Histamine receptor 2 | HRH2 | 3274 | [17] |
| Histamine receptor 3 | HRH3 | 11255 | [17] |
| Histamine receptor 4 | HRH4 | 59340 | [17] |
| *Dopamine/serotonin* | | | |
| Tyrosine hydroxylase | TH | 7054 | [18] |
| Tryptophan hydroxylase 1 | TPH1 | 7166 | [19] |
| Tryptophan hydroxylase 2 | TPH2 | 121278 | [19] |
| Aromatic L-amino acid decarboxylase | DDC | 1644 | [20] |
| Dopamine receptor 1 | DRD1 | 1812 | [21] |
| Dopamine receptor 2 | DRD2 | 1813 | [21] |
| Dopamine receptor 3 | DRD3 | 1814 | [21] |
| Dopamine receptor 4 | DRD4 | 1815 | [21] |
| Dopamine receptor 5 | DRD5 | 1816 | [21] |
| 5-Hydroxytryptamine receptor 1 | HTR1A | 3350 | [22] |
| | HTR1B | 3351 | |
| 5-Hydroxytryptamine receptor 2 | HTR2A | 3356 | [23, 24] |
| | HTR2B | 3357 | |
| | HTR2C | 3358 | |

(continued)

**Table 7.1** (continued)

| Protein name | Official gene symbol | Gene ID | Reference |
|---|---|---|---|
| 5-Hydroxytryptamine receptor 3 | HTR3A | 3359 | [23] |
| | HTR3B | 9177 | |
| | HTR3C | 170572 | |
| | HTR3D | 200909 | |
| | HTR3E | 285242 | |
| 5-Hydroxytryptamine receptor 4 | HTR4 | 3360 | [23] |
| 5-Hydroxytryptamine receptor 5 | HTR5A | 336 | [23] |
| 5-Hydroxytryptamine receptor 6 | HTR6 | 3362 | [23] |
| 5-Hydroxytryptamine receptor 7 | HTR7 | 3363 | [23] |
| *Shared elements between biogenic amines (*)* | | | |
| Diamine oxidase | AOC1 | 26 | [3, 25] |
| Retina amine oxidase | AOC2 | 314 | [26] |
| Semicarbazide sensitive amine oxidase | AOC3 | 8639 | [27] |
| Monoamine oxidase A | MAOA | 4128 | [28] |
| Monoamine oxidase B | MAOB | 4129 | [28] |
| Organic cation transporter 2 | SLC22A2 | 6582 | [29] |
| Organic cation transporter 3 | SLC22A3 | 6581 | [29] |
| Solute carrier family 3 | SLC3A2 | 6520 | [29] |
| Solute carrier family 6 | SLC6A3 | 6531 | [29] |
| | SLC6A4 | 6532 | |
| Solute carrier family 8 | SLC12A8A | 84561 | [29] |
| Vesicular amine transporter 1 | SLC18A1 | 6570 | [30] |
| Vesicular amine transporter 2 | SLC18A2 | 6571 | [31] |
| Transglutaminase 1 | TMG1 | 7051 | [32] |
| Transglutaminase 2 | TMG2 | 7052 | [33, 34] |
| N-Methyl aspartate receptor 1 | GRIN1 | 2902 | [35] |
| N-Methyl aspartate receptor 2 | GRIN2A | 2903 | [36] |
| | GRIN2B | 14811 | |

* Elements involved in metabolism of more than one biogenic amine subset

$H_3R$ antagonist and inverse agonists against them [58]. Hia is also a well-known immune mediator with a major role in allergies among other immune pathologies [6, 59–61], which in turn could take part in the inflammation-carcinogenesis interplay [62]. In addition, Hia plays an important role in gastric physiology and is responsible for gastric acid secretion [63, 64]. A role in Leydig cell functions has also been suggested for Hia [65]. Thus, the most important and complex human physiological functions are modulated by this biogenic amine (neurology, immunology, nutrition, reproduction, proliferation, and differentiation).

The products of aromatic L-amino acid decarboxylase or DOPA decarboxylase (DDC), mainly serotonin (5′-HT) and dopamine (DA), are also neurotransmitters and neuroendocrine compounds also transmitting their signals through a series of members of the GPCR family [21]. It is known that disturbances in their synthesis, transport, degradation, or reception are in the bases of many emergent neurological disorders (i.e., schizophrenia, Parkinson's, anxiety and depression, attention-deficit hyperactive disorder, bipolar disorder, etc.), circulatory and immunological problems (i.e., hypertension, allergies, psoriasis), as well as rare diseases (i.e., aromatic L-amino acid decarboxylase deficiency, Lesch-Nyhan syndrome, Prader-Willi syndrome, among many others) [20, 30, 66–68].

There are other biogenic amines derived from L-aromatic amino acids playing very key roles in our neurophysiology, i.e., melatonin, epinephrine, and norepinephrine, which play very important roles as modulators of our circadian cycle and coordination of physical activity, alert/relax shift, etc. [68, 69]. In addition, glutamate decarboxylase produces gamma-aminobutyric acid (GABA), the most important neurotransmitter that reduces neuronal excitability and controls muscle tone. Alterations of GABA-related molecular elements are related to many human diseases (i.e., fragile X syndrome, Rett syndrome, Down syndrome, schizophrenia, Tourette's syndrome, neurofibromatosis, tremor, epilepsy, etc.) [70, 71]. Many pharmaceutical investments are currently devoted to drug development against both GABA- and catecholamine-related diseases. This fact has contributed to a higher degree of information about gene-disease-drug relationships with respect to PA and Hia, as we will see later.

In this chapter, we will focus our attention on the biogenic amines derived from cationic amino acids (Put, Spd, Spm, and Hia) and the DDC product (5′-HT and DA) and their related macromolecules (Table 7.1); our expertise is mainly with biological problems related to these amines and their related elements [72, 73]. Nevertheless, our opinion is that the biogenic amine physiopathology needs a full integration of the information concerning the entire family of biogenic amine-related elements. Thus, this chapter should be considered as the starting point for a more ambitious integrative project on the molecular and biomedical information of all biogenic amines.

When the physiological responses associated with the different amines are observed, Hia appears as a structural and functional connector among them. Hia is the product of a cationic amino acid and is able to modulate cell growth as well as share neurological functions with biogenic amines derived from L-aromatic amino acids. Our group has obtained multiple evidences about the cross talk between PA and Hia summarized in several previous reviews [73, 74]. This experience led us to the following perspective: The biomedical universe of biogenic amines derived from cationic and aromatic amino acids consists of multiple subnetworks of interactions among biomolecular elements (genes, proteins, and metabolites), each one involving hundreds of molecular elements synthetized in a cell-type-specific manner. In addition, these subnetworks also keep cross interactions among them through different events in different tissues/organs. This competition for the same ligands or targets has metabolic and physiological consequences that are not well characterized so far. Nevertheless, all of these interactions must be coordinated to keep a healthy state of a human organism. Thus, further characterization of these complex and intertwined biogenic amine-related physiological scenarios is essential to fully understand a long list of pathological symptoms and diseases and requires an effort to integrate all the biochemical, molecular, and phenotypic data around the elements related to biogenic amine metabolism and signaling [72, 73]. This is to say that we need to advance toward a more holistic view of the problem. These efforts should help for future and more efficient intervention strategies. Taking this into account,

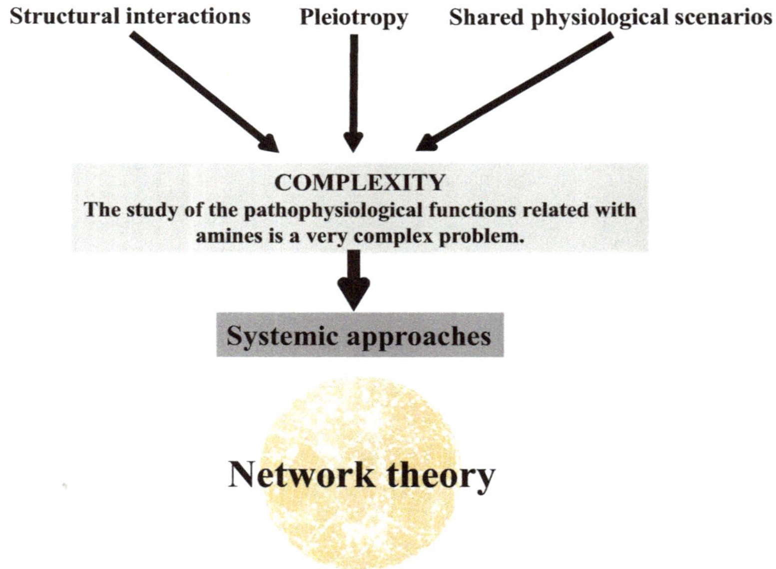


**Fig. 7.2** A scheme of the hypothesis and strategy of the present work

we need to develop biocomputational support (databases, analytical tools) to organize, prioritize, and curate the molecular and clinical information. Figure 7.2 is a scheme of the hypothesis and strategy of the present work.

Thus, herein we locate and present a first set of integrative information on relationships of biogenic amine-related elements in the context of human pathologies. This information comes from computer-assisted searches and statistical calculations, in addition to our own experience. Nevertheless, we encourage the organization of a multinational open platform to be progressively enriched and curated by the "aminer community." This platform could include structural and functional cross talk events among all biogenic amine-related elements, with the aim of understanding the network topology better, as well as genetic and pharmacological data. It would consequently help the characterization and intervention of still obscure biomedical problems as important as behavior abnormalities, psychosomatic problems, brain-gut axis abnormalities, roles of immune cells in neurodegenerative diseases, additive/synergistic effects among genetic variants, and so on.

## 7.2 Histamine: A Systemic Controller Synthetized by Just a Few Selected Cell Types

Histamine is able to scatter intercellular communication signals to a wide variety of cell types of a human body by using different tissue-specific receptor targets (see other chapters of this book), but synthetized and stored by a very reduced set of cells, known as histamine-producing cells: histaminergic neurons, enterochromaffin-like cells, and mast cells [75–78]. Other immune cells and some tumor types can synthetize but not store histamine into specific endosomes [79].

It is also known that histamine metabolism-related elements share functions and associated pathologies with elements of other biogenic amine subnetworks (Fig. 7.3). With respect to PA and cell growth, we have observed that non-producing cells (i.e., HEK-293, derived from human embryo kidney), transfected to overproduce Hia, reduce their ODC activity, PA levels, as well as cell viability and cell cycle progression [80], with concomitant induction of caspases 3/7 and alpha-synuclein [81]. In fact, it is hard to get stable transfected cells overexpressing human histidine decarboxylase (HDC), which is coherent with the lack of experimental models in literature overexpressing HDC. Maybe it is related to the fact that HDC activity is sorted and secluded in lumen of endoplasmic reticulum before maturation/activation in histamine-producing cells [82]. It is worth mentioning that malignant forms of human mastocytosis express high levels of HDC; curiously this type of neoplasia does not exhibit a high rate of cell proliferation [83, 84].

It has also been observed that elevation in histamine levels reduces the levels of PA (and/or ODC activity/expression) in different mouse cultured mast cells and during mast cell differentiation in vivo [42, 50, 51, 85, 86]. In turn, human myeloid leukemia cell differentiation to macrophage is negatively regulated by Spm [87]. At a physiological level, they also establish a cross talk in other different scenarios, for instance, progression of several human cancer types and gastrointestinal and neurological functions

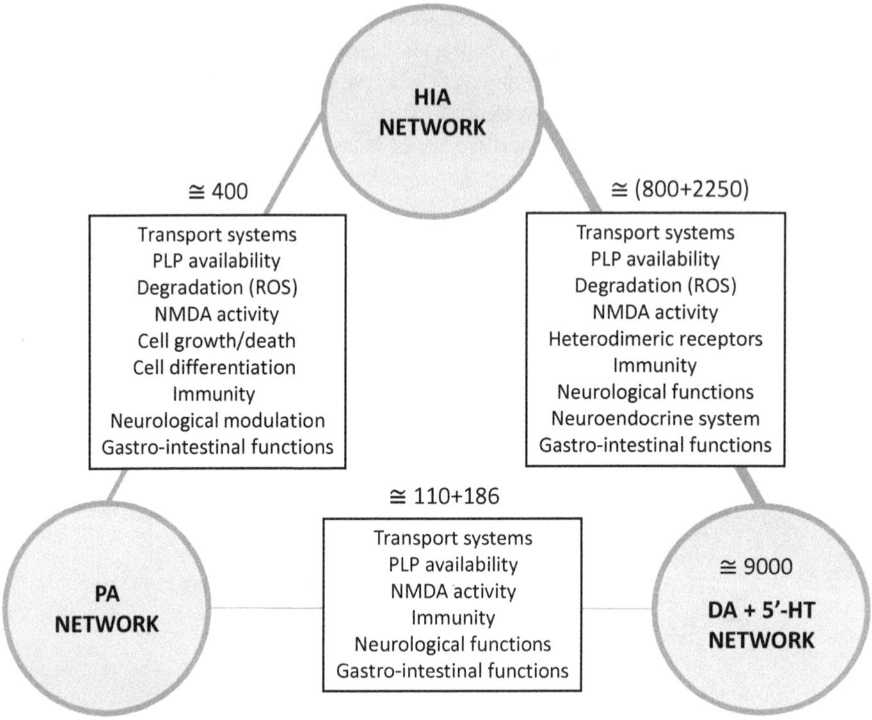


**Fig. 7.3** Biochemical and physiological processes shared among biogenic amine subnetworks. Numbers on the edges or inside the DA/5′-HT subnetwork indicate the number of PubMed database publications retrieved under the order "human plus the name of two of them"

[61, 74, 88]. At metabolic levels, Hia and PA coincide in several metabolic points: receptors (i.e., NMDA) [47] and transport systems (i.e., several organic cation transporters and vesicular membrane amine transporters) [29], amino oxidases [3, 89], transglutaminase activity [90], and detox elements (i.e., Cytochrome P450) [91].

These facts should be taken into account in order to characterize the role of Hia-PA interplay in different cancer models, as well as in genesis and/or evolution of Parkinson's and other neurological and neurodegenerative diseases [92]. Several international groups are producing very interesting results on Hia implication in these pathologies [93–97].

Hia and 5′-HT are two immune mediators stored by mast cell granules [66]. In fact, an equilibrium is kept inside the granule between proteoglycans (anionic biomolecules) and Hia or 5′-HT (organic cations) [98]. This suggests that both amines could exclude each other as intravesicular components of mast cells. Both amines also play important roles in appetite and digestion. Serotonin is considered to be a member of the gut-brain axis linked to orexigenic signals [99]. Hia is also related to the orexin/hypocretin system [100]. Hia is the key inducer of gastric acid secretion during digestion and is included in the ghrelin-gastrin-hia-HCl axis [101, 102]. In turn, ghrelin increases the concentration of dopamine in the substantia nigra [103].

Respecting both 5′-HT and DA, it is well known that Hia shares roles with these amines in similar scenarios inside and outside the brain [61]. Both amines are related to many neurological and neuroendocrine disorders, for instance, schizophrenia; Parkinson's; Alzheimer's; affective disorders; hyperactive, addictive, and aggressive behaviors; ADHD; and appetite disorders [78, 104–108]. Physical interaction has been reported between D1 or D2 receptors and $H_3R$ in striatal postsynaptic membranes [109, 110], which have also been proposed as being important for these neurologic disorders [56]. In addition, as mentioned above, recent results indicate that an excess of newly nascent Hia in cytosol induces the synthesis of α-synuclein in an HDC transfected model (human embryo kidney cells-297) [81]. Increased levels of intracranial PA have also been detected in Alzheimer's patients [111].

Other molecular and/or functional tripartite cross interactions have been suggested among PA, Hia, and/or DA/5′-HT. Human HDC and DDC share more than 50 % of the protein sequence; in fact, they share some ligands (i.e., PLP, histamine, and EGCG) [112, 113]. In spite of ODC and HDC/DDC apoenzymes not being homologous proteins, they share PLP as the cofactor, PLP acting as a chaperone of their respective apoenzyme native conformations [114]. Therefore under conditions of vitamin B6 deficiency, PLP availability could affect many steps of amino acid metabolism including synthesis of all biogenic amines mentioned in this chapter. Degradation of all biogenic amines produces ROS through activity of different polyamine, diamine, and monoamine oxidases (Fig. 7.1) with deleterious effects in different tissues, as mentioned above [89, 115].

Gastrointestinal microbiota (including pathogen organisms) produces biogenic amines, which can be an important source of these compounds for human beings. PA, Hia, DA, and 5′-HT play important roles in gastric and intestinal functions. PA is important for gastrointestinal epithelial proliferation [116]; Hia is needed for gastric acid secretion but deleterious in the case of inflammatory bowel diseases [117]. A competence between Hia and Put incorporation into rat enterocytes involving transglutaminase activity has been reported [118]. As well as DDC products, Hia plays a role in the brain-gut axis, as mentioned before [119]. Food and microbiota seem to be the sources of agmatine for human beings. This biogenic amine is the product of arginine decarboxylase activity (apparently absent in human cells) and a precursor of putrescine [120]. Beneficial effects have been assigned to agmatine in human health, and its therapeutic use has been proposed for a wide spectrum of pathologies, i.e., diabetes mellitus, neurodegenerative diseases, opioid addiction, mood disorders, cognitive disorders, and cancer [121–123].

The effects of dietary biogenic amines have been the subjects of two European COST Actions: COST 917 [124] and 922 [125]. There were several clear conclusions from these communication forums, one of them being that further efforts are required to clarify the absorption rates and transport systems determining the real concentrations and consequently the influence of microbiota-derived biogenic amines versus endogenous synthesis in human physiopathology. In addition, to evaluate the neurological effects of dietary amines, we should get more data on blood-brain barrier permeability to amines and amine derivatives [126].

## 7.3 Need and Strategy for a First Biocomputational Information Integration Effort in the Biogenic Amine-Related Human Pathology Field

Figure 7.3 is a scheme of the biochemical and physiological processes shared by Hia subnetwork with the other two biogenic amine subnetworks considered in this chapter, as synthetized in the previous section. Each one of the mentioned processes involves hundreds or even thousands of molecular, metabolic, genetic, and cellular elements, that is to say, too much information to be managed just for a human brain. Something similar occurs for references. For instance, in PubMed database (http://www.ncbi.nlm.nih.gov/pubmed/) more than 40,000 references are retrieved under the order "human polyamines" or "human histamine" and around 60,000 asking for "human dopamine" or "human serotonin." Even when further restricted by including the word "diseases" in the order, we can get from 6000 to 20,000 references depending on the amine. The number of papers containing "human and the name of two of these amines" and retrieved by PubMed database is also shown in Fig. 7.3. It seems to be very low considering the similarity among the biochemical and functional list of processes, suggesting that there must be an important quantity of information on physiopathological relationships among biogenic amines still undisclosed. It is therefore clear that the full characterization of the biogenic amine "universe", in general, needs the help of integrative bioinformatics assisted resources. The effort requires not only the development of the biocomputational tools but also a cooperative frame among international experts; we have been claiming this for years [72, 73].

As a paradox, in spite of the huge quantity of phenomenological and biochemical information on the roles of biogenic amines in human disease models and samples, the most complete databases on human diseases (OMIM, Orphanet, Decipher) provide incomplete information on relationships between diseases and biogenic amine-related elements as they usually consider genetic information only. This fact is specially marked in the cases of Hia and PA.

Among polyamine-related elements, inactivating mutations of the spermidine synthase gene codification causes Snyder-Robinson syndrome [14]. Ornithine decarboxylase genetic variations are related to APC-dependent colon cancer risk. In pediatric neuroblastoma, PA metabolism also plays a key role in the metabolic remodeling, which is essential for tumor survival and proliferation [127]. In fact, there are ongoing clinical trials on both types of neoplasias based on these findings [128, 129]. Nevertheless, there is an impressive quantity of information on polyamine and other types of cancer for which the current databases are almost blind.

In the case of histamine, OMIM (http://omim.org/) relates expression of truncated forms of human HDC to Gilles de la Tourette's syndrome [130]. The lack of histamine N-methyl transferase (HNMT), an enzyme that participates in histamine degradation, is related to susceptibility to asthma [131]. The role of $H_1R$ (HRH1) in susceptibility to encephalomyelitis/multiple sclerosis is still controversial (Table 7.2) [132, 133]. Again these facts indicate important gaps of physiopathological information in the current repositories on multigenic and complex diseases as those involving cationic biogenic amines. This delay in systematic integration of biogenic

**Table 7.2** Diseases associated with histamine-related elements at NCBI gene and/or OMIM databases

| Gene symbols | Related disease names | Disease OMIM |
|---|---|---|
| HDC | Tourette's syndrome | 137580 |
| HNMT | Susceptibility to asthma | 600807 |
| HRH1 | Susceptibility to multiple sclerosis | 126200 |
| AOC1 | Cystic fibrosis | 219700[a] |
| MAOB | Parkinson's disease | 168600[a] |

[a]Not fully validated by the databases

**Table 7.3** Functional Hia element-disease relationships revealed previously by text mining tools [61]

| Disease name | Disease OMIM |
|---|---|
| *Neurological diseases* | |
| Hereditary essential tremor | 190300 |
| Myoclonic dystonia | 159900 |
| Narcolepsy and cataplexy | 161400 |
| *Neuroinflammatory diseases* | |
| Hereditary sensory and autonomic neuropathies IV and V | 256800 and 608654 |
| Multiple sclerosis | 126200 |
| *Immune/inflammatory diseases* | |
| Crohn diseases/ulcerative colitis | 266600 |
| Familial cold autoinflammatory syndrome | 120100 |
| Idiopathic aplastic anemia | 609135 |
| Infantile neurologic cutaneous articular syndrome | 607115 |
| Muckle-Wells syndrome | 191900 |
| Psoriatic arthritis | 607507 |
| Systemic juvenile psoriatic arthritis | 604302 |
| *Rare diseases* | |
| Acute myeloid leukemia | 252270 |
| Brugada syndrome | 601144 |
| Congenital adrenal hyperplasia | 145295 |
| Familiar long QT syndrome | 152427 |
| Mastocytosis | 154800 |
| Vitamin D-dependent rickets type 2A | 277440 |
| Von Willebrand disease | 193400 |
| Zollinger-Ellison syndrome | 131100 |

amine information is blocking the advance of biomedical knowledge in the field and consequently the development of new intervention strategies.

On the one hand, in a first attempt to reduce the "dark matter" of the histamine network, our group located around 20 diseases for which clear evidence exists in the involvement of histamine-related elements; this work was assisted by text mining tools [61] (Table 7.3). It is proof of concept that further biocomputational integrative efforts will give rise to emergent information on biogenic amine physiopathology.

On the other hand, in the last few years, our group has developed the tool PhenUMA [134]. This tool takes the advantages provided by biomedical ontologies, Gene Ontology (GO, http://geneontology.org) and Human Phenotype Ontology (HPO, http://human-phenotype-ontology.github.io/) [135]. These structures are standardized vocabularies organized in a hierarchical structure. Each one of the elements of these ontologies (called terms) is ordered from the most general (terms placed close to the root) to the more specific ones (terms placed close to the leaves). The use of standardized vocabularies allows the definition of functional profiles (GO) or phenotypic profiles (HPO) for both genes and diseases. A profile is built selecting the terms of the ontology that provided the best description of the functional processes or the phenotypic manifestation for a gene or a disease. Several approaches can be used over these profiles to establish similarities between them, and these approaches are called semantic similarity measures [134]. The objective of these measures is to score the similarity between two genes or diseases using terms of the ontology related to them.

PhenUMA uses HPO to establish phenotypic relationships among genes and diseases and integrates these relationships with functional and physical information. Additionally, PhenUMA allows the query of a set of genes, diseases, or phenotypes to retrieve networks that integrate different kinds of relationships with respect to the input data. Thus, in addition to GO and HPO, the tool works with data from international open source ontologies and databases, for instance, OMIM (http://omim.org/), Orphanet (ww.orpha.net), and STRING (http://string-db.org/). Currently PhenUMA is also open to the web. Figure 7.4 provides a scheme of PhenUMA database. Thus, it integrates known relationships among genes from several interactomes and public resources and, in addition, similarities among genes and diseases combining biomedical ontologies and semantic similarity measures [136, 137].

By using PhenUMA, the abovementioned information gap with respect to involvement of both PA- and Hia-related elements in human pathologies is inherited by our tool, making impossible to establish any relationships among these genes. As an example, Fig. 7.5 (panel A) shows the results obtained from the tool when asking for all DA receptor-related genes, compared to those obtained when asking for specific Hia-related elements listed in Table 7.1 (Fig. 7.5, panel B).

On the bases of this previous experience, we decided to combine both text mining and biomedical ontologies. Briefly the working plan was as follows. Firstly, text mining resources were used to retrieve diseases associated with amine-related genes from literature, and, secondly, new gene-gene relationships were predicted using the phenotypic profile associated with these diseases. Figure 7.6 is a scheme of the whole procedure used in the present work, including biocomputational workflows, steps of manual curation, and tests. The molecular elements (genes) considered as the seed of the search are shown in Table 7.1.

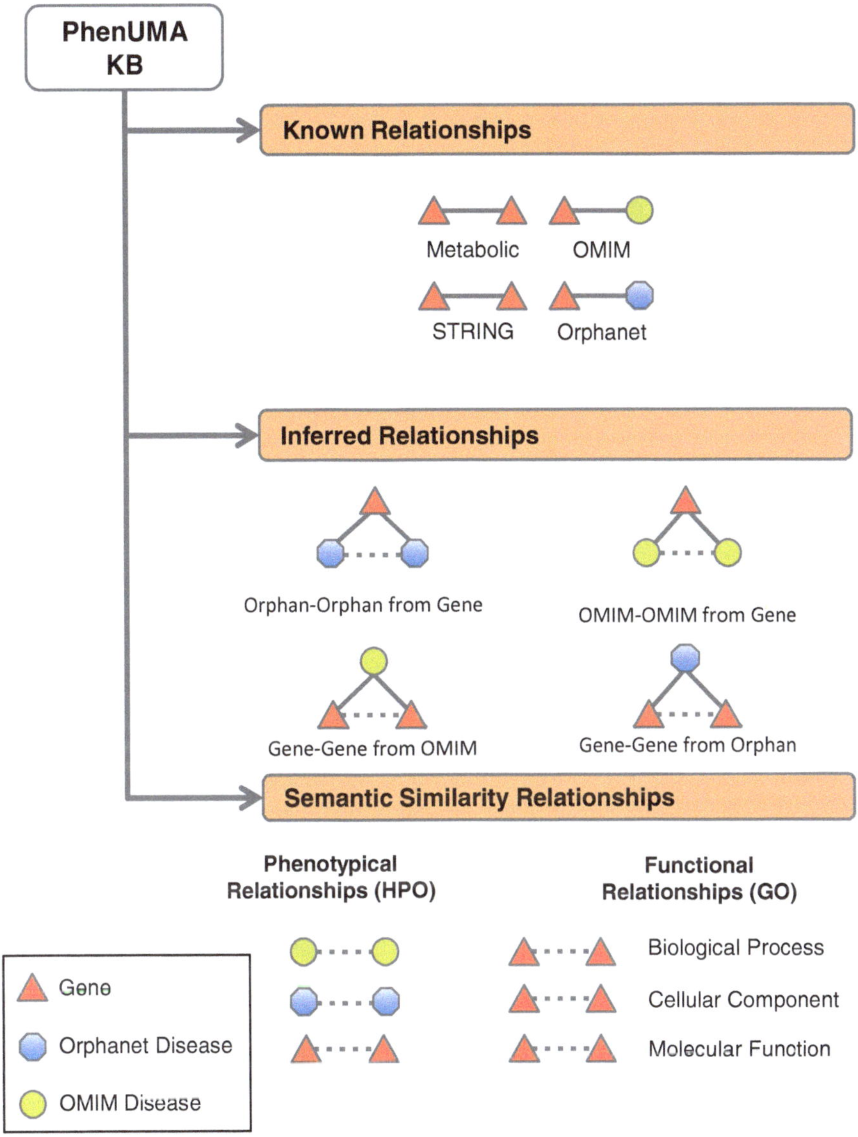


**Fig. 7.4** PhenUMA working scheme. The tool integrates: known relationships between pairs of genes (STRING and metabolic relationships) and gene-disease relationships (OMIM and Orphanet), inferred relationships (*dashed lines*) between genes and/or diseases, and semantic similarity relationships among genes or diseases by using Gene Ontology (functional relationships) and Human Phenotype Ontology (phenotypic relationships)

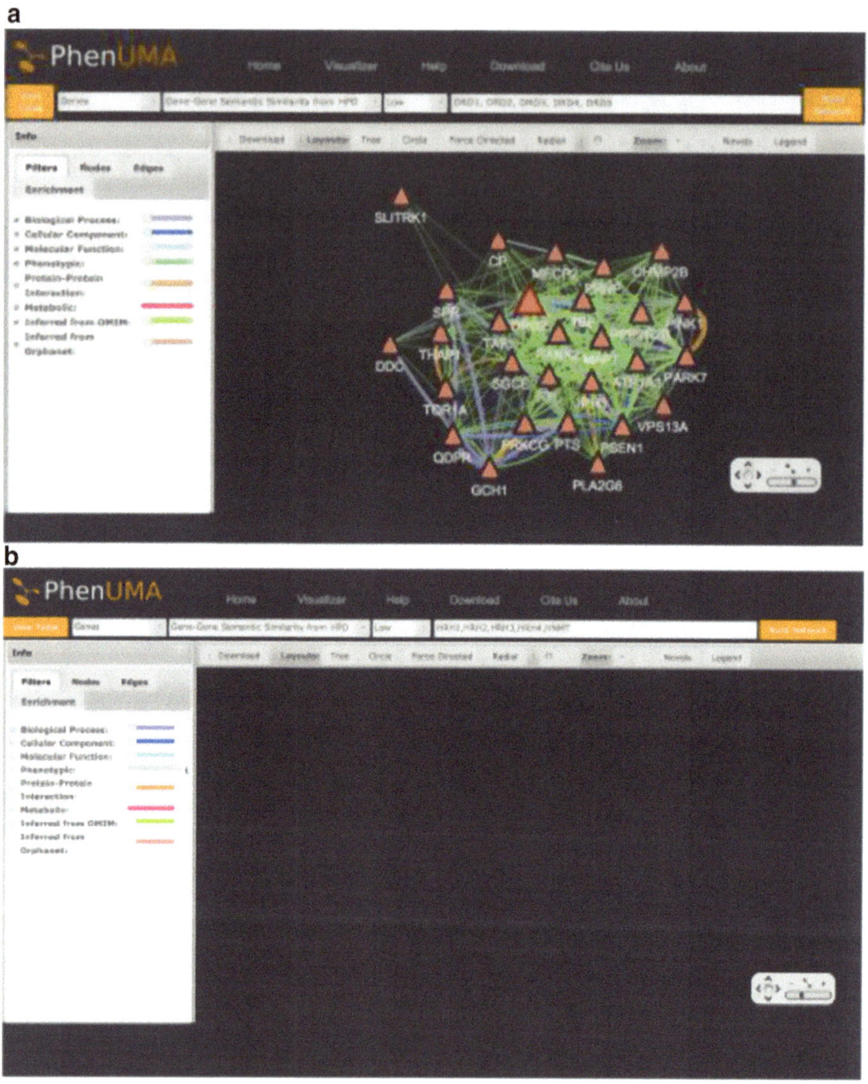


**Fig. 7.5** Representation of the network retrieved from PhenUMA initially asking for gene-gene pathologic relationships of the five dopamine receptors DRD1-5 (*panel A*) or for gene-gene pathological relationships of the four histamine receptors (HRH1-4 and HNMT) (*panel B*)

## *7.3.1 Text Mining Procedures*

The first step of the workflow (Fig. 7.6) was the usage of tools that use text mining to retrieve gene-disease associations. Those genes considered as the seed of the search are shown in Table 7.1. Two tools were used in this stage: DisGeNET [138] and DISEASES [139].

DisGeNET contains 429,111 curated relationships among 17,181 genes and 14,619 diseases with a continuous updating system. The information came from

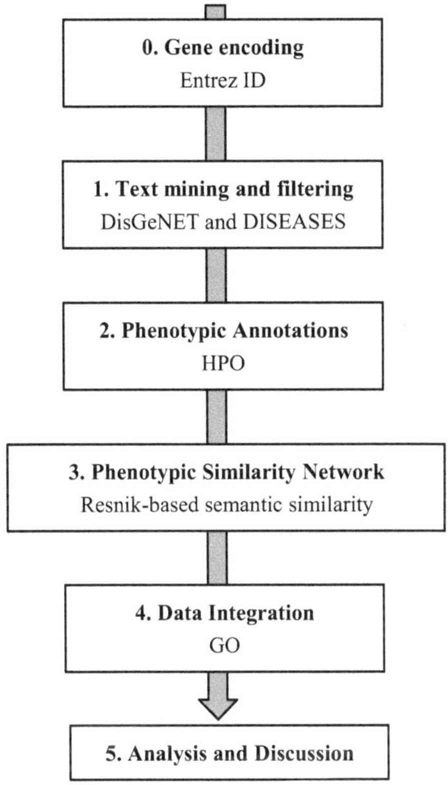


**Fig. 7.6** Workflow used to detect relationships established by amine-related genes. The biocomputational resources used in each step are mentioned

other resources such as Comparative Toxicogenomics Database (CTD), UniProt, Rat Genome Database (RGD), and Mouse Genome Database (MGD) and from previous text mining initiatives like revised articles of genome-wide association studies (GWAS) and Genetic Association Database (GAD), Literature-derived Human Gene-Disease Network (LHGDN), and BeFree [140].

DISEASES provides an approach for retrieving gene-disease relationships from abstracts coming from Genetics Home Reference (GHR), UniProtKB, results of genome-wide association studies (GWAS) and DistiLD (http://distild.jensenlab.org), and mutation data from the Catalog of Somatic Mutations in Cancer (COSMIC).

Score values provided by both tools were normalized. Then a filter was applied; only those relationships within the highest 10 % of the normalized score ranking were considered from here on.

## 7.3.2 *Phenotypic Annotation*

At this stage of the workflow, the next objective is to define the phenotypic profiles associated with the genes related to the diseases gathered in the previous section. To do that, we used two types of relationships: gene-disease relationships (obtained from the previous step) and disease-phenotype relationships. The latter were downloaded from HPO website, which provided the phenotypic profile associated with each OMIM disease. So, using the gene-disease associations, we are able to assign a set of phenotypes (HPO terms) for each amine-related gene. Some of these genes are associated with more than one disease; in these cases, the phenotypic profile is determined by the union of the profiles of the diseases associated with the gene.

### 7.3.3 Semantic Similarity Relationships

The next step was the calculation of the semantic similarity among amine-related genes and the rest of the genes annotated to the ontology. This measure allowed assigning a score to the overlapping between the phenotypic profiles of two genes; i.e., the similarity between the symptoms of two genes or diseases. In this case, the measure used [137] is based on the information content (IC) concept. IC is defined by *−log(probability(t)),* where *t* is a term of the ontology, and gives us an idea of the specificity of each phenotype, as explained previously [134]. The similarity between two genes is determined through the comparison of all the elements included in the phenotypic profile of both genes. All the genes with pathological information are compared, and the most significant values (in our case, those over the 98th percentile) are taken into account.

### 7.3.4 Data Integration

All these phenotypic similarity relationships among genes obtained in step 3 (Fig. 7.6) are integrated with functional information coming from the resources mentioned above. The objective of this part of the workflow is to highlight the relationships among genes that are involved in both the same functional processes and the same phenotypic characteristics. For this purpose, the functional semantic similarity among these genes was calculated using the Gene Ontology (GO) and its three sub-ontologies (biological process, cellular component, and molecular function).

### 7.3.5 Data Processing

The retrieved information was manually analyzed. The resulting data and networks are briefly described and discussed in the next sections.

## 7.4 Retrieved Information: The Starting Point for AMINETWORKING 1.0

Figure 7.7 is a representation of the network obtained following the procedure described in point 7.3.1 (text mining, normalization, and filtering). Elements of the different gene subgroups of Table 7.1 are differentially colored (see Fig. 7.7 caption). As the resulting edges between gene-disease pairs are too many to be properly observed as a network, they are also listed in Table 7.4. Nevertheless, the network topology is also informative as discussed below.

Results include many of the previously known relationships mentioned in Tables 7.2 and 7.3; they are represented as red edges. It is an internal validation of our strategy, as it is indeed able to automatically locate validated information from bibliography. Many other new relationships are inferred from our strategy when compared with the information present in the most common databanks of gene-

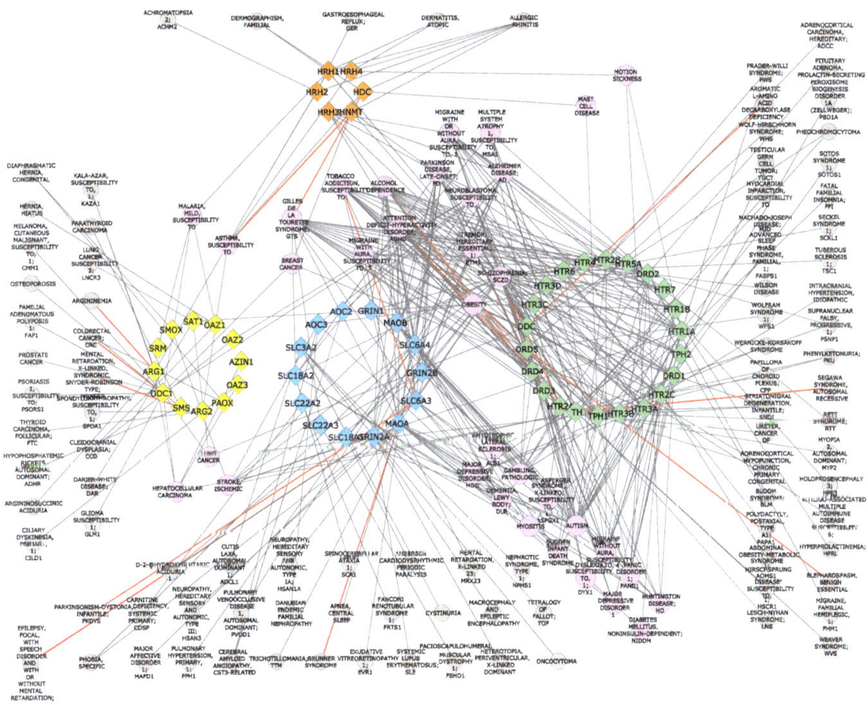


**Fig. 7.7** Network obtained from our workflow (Fig. 7.6) by asking for relationships between any of the amine-related genes of Table I and human diseases. Elements of the different gene subgroups of Table 7.1 are differentially colored: *yellow*, PA-related elements; *orange*, Hia-related elements; *green*, DA/5′-HT; *blue*, shared-elements. Diseases related to more than one group of amine-related genes listed in Table 7.1 are colored in magenta. *Red edges* represent gene-disease relationships previously located and included in Tables 7.2 or 7.3

**Table 7.4** Biogenic amine-related genes associated with human diseases deduced from the text mining procedure described in Sect. 7.3.1

| Disease names | OMIM | Entrez gene | Gene symbols |
|---|---|---|---|
| Abdominal obesity-metabolic syndrome 1 | 605552 | 3358 | HTR2C |
| Achromatopsia 2 | 216900 | 3274 | HRH2 |
| Adrenocortical carcinoma, hereditary | 202300 | 3357 | HTR2B |
| Adrenocortical hypofunction, chronic primary congenital | 103230 | 1644; 7166 | DDC; TPH1 |
| Advanced sleep phase syndrome, familial, 1 | 604348 | 121278 | TPH2 |
| Alcohol dependence | 103780 | 1812; 1813; 1814; 1815; 2902; 2903; 14811; 3350; 3351; 3356; 3358; 3359; 9177; 3363; 4128; 4129; 6531; 6532; 7166; 121278 | DRD1; DRD2; DRD3; DRD4; GRIN1; GRIN2A; GRIN2B; HTR1A; HTR1B; HTR2A; HTR2C; HTR3A; HTR3B; HTR7; MAOA; MAOB; SLC6A3; SLC6A4; TPH1; TPH2 |
| Allergic rhinitis | 607154 | 3067; 3176; 3269; 3274; 59340 | HDC; HNMT; HRH1; HRH2; HRH4 |
| Alzheimer's disease | 104300 | 314; 8639; 2903; 14811; 11255; 3350; 4128; 4129; 6532; 7054 | AOC2; AOC3; GRIN2A; GRIN2B; HRH3; HTR1A; MAOA; MAOB; SLC6A4; TH |
| Amyotrophic lateral sclerosis 1 | 105400 | 4129; 7054 | MAOB; TH |
| Andersen cardiodysrhythmic periodic paralysis | 170390 | 6570 | SLC18A1 |
| Apnea, central sleep | 107640 | 4128 | MAOA |
| Argininemia | 207800 | 383 | ARG1 |
| Argininosuccinic aciduria | 207900 | 383 | ARG1 |
| Aromatic L-amino acid decarboxylase deficiency | 608643 | 1644 | DDC |
| Asperger's syndrome, x-linked, susceptibility to, 1 | 300494 | 3356; 6532 | HTR2A; SLC6A4 |
| Asthma, susceptibility to | 600807 | 383; 384; 3067; 3176; 3269; 3274; 59340 | ARG1; ARG2; HDC; HNMT; HRH1; HRH2; HRH4 |
| Attention-deficit hyperactivity disorder | 143465 | 1644; 1812; 1813; 1814; 1815; 1816; 2903; 11255; 3351; 3356; 3358; 3360; 3363; 4128; 4129; 6531; 6532; 7054; 7166; 121278 | DDC; DRD1; DRD2; DRD3; DRD4; DRD5; GRIN2A; HRH3; HTR1B; HTR2A; HTR2C; HTR4; HTR7; MAOA; MAOB; SLC6A3; SLC6A4; TH; TPH1; TPH2 |

(continued)

**Table 7.4** (continued)

| Disease names | OMIM | Entrez gene | Gene symbols |
|---|---|---|---|
| Autism | 209850 | 1812; 1813; 1814; 1815; 1816; 2903; 14811; 3350; 3351; 3356; 3357; 3358; 3359; 170572; 3361; 3363; 4128; 4129; 6570; 6531; 6532; 7166; 121278 | DRD1; DRD2; DRD3; DRD4; DRD5; GRIN2A; GRIN2B; HTR1A; HTR1B; HTR2A; HTR2B; HTR2C; HTR3A; HTR3C; HTR5A; HTR7; MAOA; MAOB; SLC18A1; SLC6A3; SLC6A4; TPH1; TPH2 |
| Blepharospasm, benign essential | 606798 | 1816 | DRD5 |
| Bloom syndrome | 210900 | 3361 | HTR5A |
| Breast cancer | 114480 | 3176; 3359; 4953 | HNMT; HTR3A; ODC1 |
| Brunner syndrome | 300615 | 4128 | MAOA |
| Carnitine deficiency, systemic primary | 212140 | 6582 | SLC22A2 |
| Cerebral amyloid angiopathy, cst3 related | 105150 | 314; 8639 | AOC2; AOC3 |
| Ciliary dyskinesia, primary, 1 | 244400 | 51686 | OAZ3 |
| Cleidocranial dysplasia | 119600 | 4946; 4947 | OAZ1; OAZ2 |
| Colorectal cancer | 114500 | 4953; 196743 | ODC1; PAOX |
| Cutis laxa, autosomal dominant 1 | 123700 | 6520 | SLC3A2 |
| Cystinuria | 220100 | 6520 | SLC3A2 |
| D-2-Hydroxyglutaric aciduria 1 | 600721 | 2903; 14811 | GRIN2A; GRIN2B |
| Danubian endemic familial nephropathy | 124100 | 8639 | AOC3 |
| Darier-White disease | 124200 | 6303 | SAT1 |
| Dementia, Lewy body | 127750 | 4128; 4129; 6531; 6532; 7054 | MAOA; MAOB; SLC6A3; SLC6A4; TH |
| Dermatitis, atopic | 603165 | 3269; 59340 | HRH1; HRH4 |
| Dermographism, familial | 125635 | 3067; 3269 | HDC; HRH1 |
| Diabetes mellitus, noninsulin-dependent | 125853 | 8639; 3358 | AOC3; HTR2C |
| Diaphragmatic hernia, congenital | 142340 | 196743; 54498 | PAOX; SMOX |
| Dyslexia, susceptibility to, 1 | 127700 | 1814; 1815; 1816; 6531 | DRD3; DRD4; DRD5; SLC6A3 |
| Epilepsy, focal, with speech disorder and with or without mental retardation | 245570 | 2903 | GRIN2A |
| Exudative vitreoretinopathy 1 | 133780 | 4128 | MAOA |
| Facioscapulohumeral muscular dystrophy 1 | 158900 | 6581 | SLC22A3 |

(continued)

**Table 7.4** (continued)

| Disease names | OMIM | Entrez gene | Gene symbols |
|---|---|---|---|
| Familial adenomatous polyposis 1 | 175100 | 4953 | ODC1 |
| Fanconi renotubular syndrome 1 | 134600 | 6582 | SLC22A2 |
| Fatal familial insomnia | 600072 | 7166 | TPH1 |
| Gambling, pathologic | 606349 | 1812; 1813; 1814; 1815; 1816; 2902; 3351; 3356; 4128; 4129; 6531; 6532 | DRD1; DRD2; DRD3; DRD4; DRD5; GRIN1; HTR1B; HTR2A; MAOA; MAOB; SLC6A3; SLC6A4 |
| Gastroesophageal reflux | 109350 | 3274 | HRH2 |
| Gilles de la Tourette's syndrome | 137580 | 1812; 1813; 1814; 1815; 1816; 3067; 3350; 3351; 3356; 3357; 3358; 9177; 3361; 3363; 4128; 51686; 6581; 6531; 6532; 121278 | DRD1; DRD2; DRD3; DRD4; DRD5; HDC; HTR1A; HTR1B; HTR2A; HTR2B; HTR2C; HTR3B; HTR5A; HTR7; MAOA; OAZ3; SLC22A3; SLC6A3; SLC6A4; TPH2 |
| Glioma susceptibility 1 | 137800 | 4953 | ODC1 |
| Hepatocellular carcinoma | 114550 | 383; 51582; 51686; 4953; 7054 | ARG1; AZIN1; OAZ3; ODC1; TH |
| Hernia, hiatus | 142400 | 4946 | OAZ1 |
| Heterotopia, periventricular, x-linked dominant | 300049 | 2903 | GRIN2A |
| Hirschsprung disease, susceptibility to, 1 | 142623 | 7054 | TH |
| Holoprosencephaly 3 | 142945 | 3361 | HTR5A |
| Huntington disease | 143100 | 1812; 1813; 1816; 2903; 14811; 4129; 7054 | DRD1; DRD2; DRD5; GRIN2A; GRIN2B; MAOB; TH |
| Hyperprolactinemia | 615555 | 1813 | DRD2 |
| Hypophosphatemic rickets, autosomal dominant; ADHR | 193100 | 6611 | SMS |
| Intracranial hypertension, idiopathic | 243200 | 3358 | HTR2C |
| Kala-azar, susceptibility to, 1 | 608207 | 383; 4953 | ARG1; ODC1 |
| Lesch-Nyhan syndrome | 300322 | 1816 | DRD5 |
| Lung cancer | 211980 | 1644; 3359; 4953; 6303 | DDC; HTR3A; ODC1; SAT1 |
| Lung cancer susceptibility 3 | 612571 | 6303 | SAT1 |
| Machado-Joseph disease | 109150 | 3350 | HTR1A |
| Macrocephaly and epileptic encephalopathy | 606369 | 2903 | GRIN2A |
| Major affective disorder 1 | 125480 | 6531 | SLC6A3 |
| Major depressive disorder | 608516 | 3350; 3358; 3359; 4128 | HTR1A; HTR2C; HTR3A; MAOA |

(continued)

**Table 7.4** (continued)

| Disease names | OMIM | Entrez gene | Gene symbols |
|---|---|---|---|
| Major depressive disorder 1 | 608520 | 1814; 1815; 3358; 4128 | DRD3; DRD4; HTR2C; MAOA |
| Malaria, mild, susceptibility to | 609148 | 3269; 4946; 4953; 6723 | HRH1; OAZ1; ODC1; SRM |
| Mast cell disease | 154800 | 3067; 7166 | HDC; TPH1 |
| Melanoma, cutaneous malignant, susceptibility to, 1 | 155600 | 4953 | ODC1 |
| Mental retardation, x-linked 23 | 300046 | 4128 | MAOA |
| Mental retardation, x-linked, syndromic, Snyder-Robinson type | 309583 | 6611 | SMS |
| Migraine with aura, susceptibility to, 7 | 609179 | 1813; 1815; 3357; 3358; 4128; 6532 | DRD2; DRD4; HTR2B; HTR2C; MAOA; SLC6A4 |
| Migraine with or without aura, susceptibility to, 1 | 157300 | 1812; 1813; 1814; 1815; 1816; 11255; 3350; 3351; 3356; 3358; 3359; 3363; 4128; 6532; 7166 | DRD1; DRD2; DRD3; DRD4; DRD5; HRH3; HTR1A; HTR1B; HTR2A; HTR2C; HTR3A; HTR7; MAOA; SLC6A4; TPH1 |
| Migraine without aura, susceptibility to, 4 | 607501 | 1812; 1813; 1814; 1815; 1816; 3350; 3351; 3356; 3357; 3358; 3362; 4128; 6532 | DRD1; DRD2; DRD3; DRD4; DRD5; HTR1A; HTR1B; HTR2A; HTR2B; HTR2C; HTR6; MAOA; SLC6A4 |
| Migraine, familial hemiplegic, 1 | 141500 | 1813 | DRD2 |
| Motion sickness | 158280 | 3269; 3350; 3359 | HRH1; HTR1A; HTR3A |
| Multiple system atrophy 1, susceptibility to | 146500 | 1644; 1813; 7054; 7166 | DDC; DRD2; TH; TPH1 |
| Myocardial infarction, susceptibility to | 608446 | 7054 | TH |
| Myopia 2, autosomal dominant | 160700 | 7054 | TH |
| Myositis | 160750 | 3359; 6532 | HTR3A; SLC6A4 |
| Nephrotic syndrome, type 1 | 256300 | 6520 | SLC3A2 |
| Neuroblastoma, susceptibility to | 256700 | 1644; 3359; 4128; 4129; 4947; 4953; 7054 | DDC; HTR3A; MAOA; MAOB; OAZ2; ODC1; TH |
| Neuropathy, hereditary sensory and autonomic, type 1A | 162400 | 4128 | MAOA |
| Neuropathy, hereditary sensory and autonomic, type 3 | 223900 | 4128 | MAOA |

(continued)

Table 7.4 (continued)

| Disease names | OMIM | Entrez gene | Gene symbols |
|---|---|---|---|
| Obesity | 601665 | 8639; 1813; 11255; 3351; 3356; 3358; 6303 | AOC3; DRD2; HRH3; HTR1B; HTR2A; HTR2C; SAT1 |
| Oncocytoma | 553000 | 6520 | SLC3A2 |
| Osteoporosis | 166710 | 6611 | SMS |
| Panic disorder 1 | 167870 | 1812; 1813; 1815; 3350; 3351; 3356; 3358; 3359; 4128; 6532; 7166; 121278 | DRD1; DRD2; DRD4; HTR1A; HTR1B; HTR2A; HTR2C; HTR3A; MAOA; SLC6A4; TPH1; TPH2 |
| Papilloma of choroid plexus | 260500 | 3358 | HTR2C |
| Parathyroid carcinoma | 608266 | 6611 | SMS |
| Parkinson's disease, late onset | 168600 | 1644; 1812; 1813; 1814; 1815; 2903; 14811; 3176; 11255; 3350; 3359; 4128; 4129; 6531; 6532; 7054; 7166 | DDC; DRD1; DRD2; DRD3; DRD4; GRIN2A; GRIN2B; HNMT; HRH3; HTR1A; HTR3A; MAOA; MAOB; SLC6A3; SLC6A4; TH; TPH1 |
| Parkinsonism-dystonia, infantile | 613135 | 6531 | SLC6A3 |
| Peroxisome biogenesis disorder 1a (Zellweger) | 214100 | 121278 | TPH2 |
| Phenylketonuria | 261600 | 7054; 7166 | TH; TPH1 |
| Pheochromocytoma | 171300 | 1644; 7054 | DDC; TH |
| Phobia, specific | 608251 | 4128; 6532 | MAOA; SLC6A4 |
| Pituitary adenoma, prolactin secreting | 600634 | 1813 | DRD2 |
| Polydactyly, postaxial, type A1 | 174200 | 3362 | HTR6 |
| Prader-Willi syndrome | 176270 | 3357 | HTR2B |
| Prader-Willi syndrome | 176270 | 3358 | HTR2C |
| Prostate cancer | 176807 | 4953 | ODC1 |
| Psoriasis 1, susceptibility to | 177900 | 4953 | ODC1 |
| Pulmonary hypertension, primary, 1 | 178600 | 6532 | SLC6A4 |
| Pulmonary veno-occlusive disease 1, autosomal dominant | 265450 | 6532 | SLC6A4 |
| Rett syndrome | 312750 | 3363; 7054 | HTR7; TH |
| Schizophrenia | 181500 | 1644; 1812; 1813; 1814; 1815; 1816; 2902; 2903; 14811; 3269; 11255; 3350; 3351; 3356; 3358; 3359; 9177; 200909; 3360; 3361; 3362; 3363; 4128; 4129; 6570; 6531; 6532; 7054; 7166; 121278; 6520 | DDC; DRD1; DRD2; DRD3; DRD4; DRD5; GRIN1; GRIN2A; GRIN2B; HRH1; HRH3; HTR1A; HTR1B; HTR2A; HTR2C; HTR3A; HTR3B; HTR3D; HTR4; HTR5A; HTR6; HTR7; MAOA; MAOB; SLC18A1; SLC6A3; SLC6A4; TH; TPH1; TPH2; SLC3A2 |

(continued)

**Table 7.4** (continued)

| Disease names | OMIM | Entrez gene | Gene symbols |
|---|---|---|---|
| Seckel syndrome 1 | 210600 | 3361 | HTR5A |
| Segawa syndrome, autosomal recessive | 605407 | 7054 | TH |
| Sotos syndrome 1 | 117550 | 1812 | DRD1 |
| Spondyloarthropathy, susceptibility to, 1 | 106300 | 384 | ARG2 |
| Striatonigral degeneration, infantile | 271930 | 1644; 7054; 7166 | DDC; TH; TPH1 |
| Stroke, ischemic | 601367 | 8639; 2903; 14811; 3350; 4129; 6532; 54498; 7054 | AOC3; GRIN2A; GRIN2B; HTR1A; MAOB; SLC6A4; SMOX; TH |
| Sudden infant death syndrome | 272120 | 3350; 4128; 6532; 7054; 7166; | HTR1A; MAOA; SLC6A4; TH; TPH1; |
| Supranuclear palsy, progressive, 1 | 601104 | 1813; 7054 | DRD2; TH |
| Systemic lupus erythematosus | 152700 | 2903 | GRIN2A |
| Testicular germ cell tumor | 273300 | 1812 | DRD1 |
| Thyroid carcinoma, follicular | 188470 | 384 | ARG2 |
| Tobacco addiction, susceptibility to | 188890 | 1644; 1812; 1813; 1814; 1814; 1815; 1816; 3350; 3356; 9177; 3363; 4128; 6531; 6532; 7054; 7166 | DDC; DRD1; DRD2; DRD3; DRD4; DRD5; HTR1A; HTR2A; HTR3B; HTR7; MAOA; SLC6A3; SLC6A4; TH; TPH1 |
| Tremor, hereditary essential, 1 | 190300 | 1814; 3176; 6531 | DRD3; HNMT; |
| Trichotillomania | 613229 | 4128 | MAOA |
| Tuberous sclerosis 1 | 191100 | 3362 | HTR6 |
| Ureter, cancer of | 191600 | 7166 | TPH1 |
| Vitiligo-associated multiple autoimmune disease susceptibility 6 | 193200 | 1644 | DDC |
| Weaver syndrome | 277590 | 7054 | TH |
| Wernicke-Korsakoff syndrome | 277730 | 7166 | TPH1 |
| Wilson disease | 277900 | 1813 | DRD2 |
| Wolf-Hirschhorn syndrome | 194190 | 1816 | DRD5 |
| Wolfram syndrome 1 | 222300 | 1816 | DRD5 |

diseases relationships, as far as we know. Diseases represented as magenta circles are at least shared by two biogenic amine elements. After this text mining effort, it is clear that Hia and PA are not unplugged modules of the human *diseasome*.

Among these results there is an impressive quantity of inferred information that, of course, needs to be analyzed and curated by the "aminer community". We invite all our colleagues interested in biogenic amines to cooperate in the task. We are open to receive proof of concepts (references, personal communication, etc.), suggestions, and comments through the email account aminetworking@uma.es.

When focusing on the subsets of diseases related to the PA-related elements (Fig. 7.7 and Table 7.4), we can observe that, as expected, this module is mainly related to an important list of cancer types. The module of genes related to DDC-derived biogenic amines, as expected, is linked to a huge catalog of neurological and neuroendocrine disorders. Hia-related elements also establish multiple relationships with neurological disorders and the subnetwork of DDC-derived elements, but they seem to present a wider spectrum of affected organs/tissues. In the next paragraph, we will mainly focus our discussion on diseases involving elements from at least two different modules.

In fact, results remark the molecular complexity of several neurological diseases involving ten (or more) biogenic amine-related genes from different subsets. This is the case of (in alphabetic order) alcohol dependence, Alzheimer's diseases, attention-deficit hyperactivity disorder, autism, gambling (pathological), Gilles de la Tourette's syndrome, migraine (susceptibility with or without aura), panic disorder, Parkinson's disease, schizophrenia, and tobacco addiction (susceptibility) (Table 7.4). The concurrence of Hia-related genes and DDC product-related genes associated with different addictive behaviors is remarkable. It is a very interesting field that, for sure, involves both functional and even physical interactions between the gene products that are not fully characterized yet [141].

Taking into account the density of concurrent relationships between Hia-related genes and DDC-product-related genes under different neurological circumstances, it is clear that discussion of genetic/biochemical data of a single element in the context of a patient (or experimental model) of one of these diseases (or susceptibilities) should not ignore the probable contribution of the concurrent elements of its own module or the others. This strategy would help a personalized location of molecular contributors to a given patient and disease, as well as the efficient advance in the molecular characterization, prevention, and intervention of many other potential patients. This hypothesis encourages us to claim for the convenience of an integrative multinational project that incorporates information on patients of prevalent and emergent neurological abnormalities or susceptibilities to them.

Several interesting concurrence events on the same disease (or susceptibility to a given disease) are worth mentioning since it enriches our information of common physiopathological scenarios and could contribute to getting new insights for discussion of high-throughput results and personalized medicine initiatives. Briefly, we will mention them in the paragraphs below.

Our text mining search gave a few but very interesting diseases described as involving both PA and Hia-related elements (Table 7.4). Breast cancer has been related to elements of both modules [53, 142, 143] and different elements of their metabolic pathways have been proposed as markers and/or therapeutic targets. DA/5′-HT-related elements have also been related to breast cancer progression, and DRD1 ligands have been proposed for breast cancer chemotherapy [144]. It is clear that the degree of knowledge on PA roles in cancer growth and progression is much higher for PA than for the other biogenic amines, in spite of the suggested roles and mechanisms for different Hia-related elements and tumor types from over 30 years ago [145]. This topic is being helped by recent findings of correlation between polymorphisms of Hia-related genes (HDC, HNMT, and HRH3) and breast cancer [48, 143]. As several metabolic points of mutual PA-Hia interference have been detected (as mentioned in introduction), we think that the PA-Hia-DA/5′-HT metabolic interplay in the context of human breast cancer deserves more attention from the "aminer community". This interest could be extended to other cancer types where PA and other biogenic amine elements could share the same environment (for instance, mastocytosis, myeloid leukemias, melanomas, and brain, lung, and gastrointestinal cancer types). The pattern of Hia receptors expressed in different human cancer types will be determinant for the Hia effects induced on each type. They could even be antagonic depending on both the receptor expression pattern and also on the specific proteome of each tumor. Fortunately, multiple ongoing initiatives are trying to integrate and classify cell/tissue-specific "omic" information that could be used to clarify the pleiotropic effects of Hia not only in cancer but in other biomedical scenarios [146–148].

A common phenotype to all biogenic amine subgroups is "susceptibility to asthma". Up to 5 Hia-related elements were associated. They include elements taking part in synthesis, degradation, and signaling [149–153], as well as arginase [154], which is the first and essential enzyme for both NO and PA synthesis in mammalian tissues [155]. Interestingly, it is well known that amino acids, arginine, and histidine, are considered essential amino acids during the first years of human life, just when the immune system is being conformed. It is tempting to speculate that there is a putative link between children's diets (in terms of Arg/His content) and immune characteristics of human beings. In fact, it is proposed that PA is important for the ideal children's immune system development [156]. Unfortunately, it indeed is not a well-explored scenario, in spite of the recent increasing evidence of the essential role of PA in nonmalignant myeloid cell differentiation [42].

Literature describes the PA-related element named antizyme inhibitor 2 (AZIN2) as a regulator of the intracellular vesicle trafficking involved in secretory processes (immune cells and others) [86, 157]. AZIN2 is an inactive ODC paralog/pseudogene, capable of binding ODC antizymes (AOZ). AZIN2 is also expressed in the brain (specific neurons of the hippocampus and cerebellum) [157]; in fact it is one of the few human organs expressing AZIN2 [158]. It is reported that AZIN2 accumulates during Alzheimer's disease progression [159]. Several membrane amine transporters such as SLC6A3, SLC6A4 (known as serotonin transporters), and SLC22A3 (or OCT3), which are also able to accept PA as ligands [29, 160], appear as related to multiple neurological diseases; for instance, addictions, ADHD [161], autism [162], dyslexia, dementia associated to Lewy body [163], Gilles de la

Tourette's syndrome [164], major affective disorders [165], panic [166], Parkinson's disease (late onset) [167], phobias [168], schizophrenia [169], susceptibility to Asperger's syndrome [170], susceptibility to migraine [171], and tremor (Table 7.4) [164]. This concurrence of PA and neurotransmitters at the level of transport systems suggests the possibility of still undisclosed aspects of PA influence on neurological problems and neurodegenerative diseases, as well as in other physiological scenarios where they also coincide (for instance, intestine).

We can also find interesting intermodule relationships outside the brain. From Table 7.4 and even Fig. 7.7, it can be deduced that all gene subgroups have elements related to strokes. These are solute carrier 6A4, the NMDA-polypeptides GRIN2A and GRIN2B, and several amine oxidases, MAOB (monoamine oxidase), semicarbazide-sensitive amine oxidase (AOC3), and SMOX (a specific spermine oxidase); all of them are ROS sources covering a wide spectrum of amine substrates (Fig. 7.1). In fact, Igarashi and Kashigagi propose the use of polyamine metabolites as markers for stroke and renal failure [172]. Ictus is also a pathological event involving arginine/NO metabolism-, immune-, and stress response-related genes. Consequently, a cross talk among different biogenic amines can be hypothesized [173].

Concurrence of different biogenic amine degradation systems could also play importance in other pathological processes involving simultaneously inflammation and proliferation, for instance, inflammation-associated carcinogenesis, tumor growth progression, hepatic injury, and so on, as discussed elsewhere [73]. This is also the case of parasitic infections (i.e., malaria, Table 7.4). Our search found the phenotype "susceptibility to malaria" as related to several PA elements, spermidine synthase (SRM), antizyme 1 (OAZ1), and ODC as well as histamine receptor type 1 (HRH1). Parasite requires PA to proliferate, and the surrounding inflammatory cells constitute an important source of other biogenic amines and ROS, with consequences that are still not well evaluated [72, 174].

Regarding the different malignancies mentioned in Table 7.4 associated with several biogenic amines, some of them had been reviewed in previous works, for instance, mast cell and rare gastric malignancies [61]. Up to seven biogenic amine-related elements coming from all four gene subsets (Tables 7.1 and 7.4) were related to "susceptibility to neuroblastoma." As mentioned above, it is one of the diseases where PA synthesis inhibition is getting promising success in clinical work [39], which is in agreement with the recently demonstrated tight coordination between energy metabolism, protein synthesis, and PA synthesis in this pediatric cancer [127]. At present, we are interested in locating metabolic/genetic features contributing to the differential sensitivity to PA synthesis inhibitors observed among patients [39]. In the data analyses, integration of molecular and functional information of the other biogenic amine gene subsets is considered very convenient.

In addition to neuroblastoma, Table 7.4 includes many other low-prevalence and rare diseases. Approaches capable of saving biological samples and experiments are especially valuable for these diseases. Many of them had been located previously [61, 175], but the present work adds several to the list. As members of "Centro de Investigación Biomédica En Red en Enfermedades Raras," the Spanish institute for research in rare diseases, we hope this information can be beneficial to many biomedical research and clinical groups working on these pathologies that are currently worldwide considered a health priority.

As mentioned in the previous section, the computer-assisted procedure described in Sect. 7.3 had as its objective the location of gene-gene interactions involving both functional and phenotypic concurrence between biogenic amine-related and other human genes, as they must reflect biochemical, molecular, and cellular interactions that could be involved in human diseases. This step could provide useful insights for genomic result analyses and initiatives of personalized medicine.

Figure 7.8 shows the crude results of the phenotypic similarity network of all gene-gene relationships inferred from the procedures described in Sects. 7.3.3 and 7.3.4. Nodes are human genes sharing phenotypic similarities with any of the genes listed in Table 7.4 and scored over the 98th percentile. In bright blue, the biogenic amine-related genes are shown. The other human genes phenotypically related to biogenic amine genes are depicted in gray. Among the results there is an impressive quantity of inferred information. The full analysis and validation of these sets of results is out of the frame of the present work.

The phenotypic relationships were therefore enriched with functional information as described in Sect. 7.3.4. Thus, Fig. 7.9 represents gene-gene relationships among genes sharing both phenotypic and annotated functional characteristics. The quality

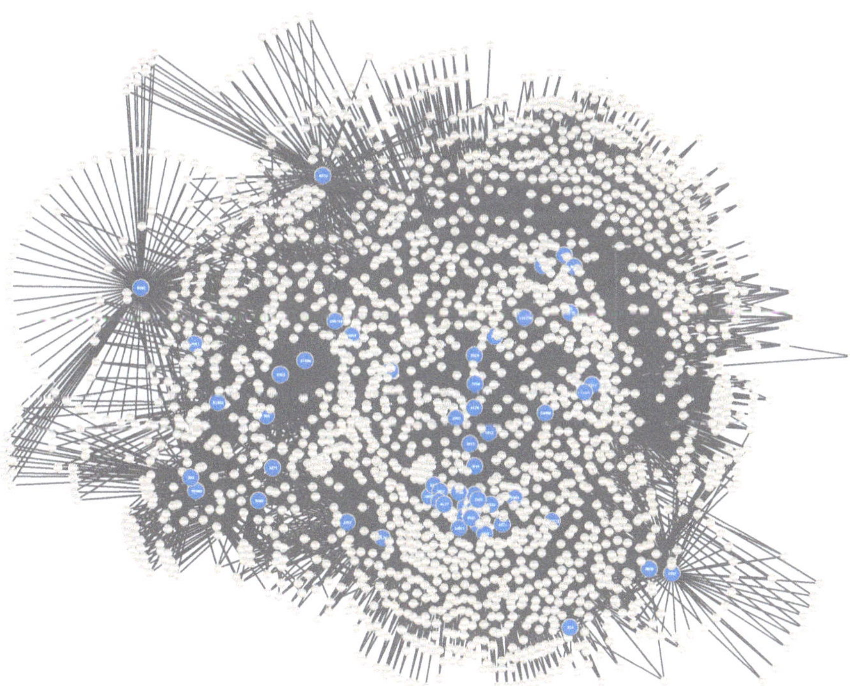

**Fig. 7.8** Network of phenotypic relationships between biogenic amine genes and any other human gene. Biogenic amine genes included in Table 7.1 are colored as *blue spheres*

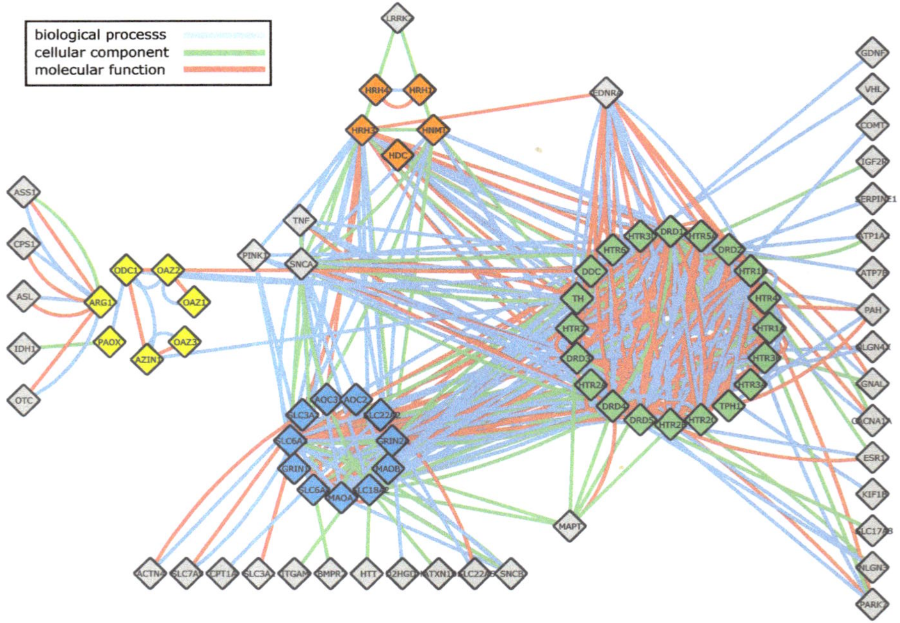


**Fig. 7.9** Network of biogenic amine-related genes related to other human genes by sharing both phenotypic and functional characteristics. Elements of the different gene subgroups (Table 7.1) are differentially colored: *yellow*, PA-related elements; *orange*, Hia-related elements; *green*, DA/5′-HT; *blue*, shared-elements. The functional characteristics (*edges*), named as in the Gene Ontology, are also distinguished with different colors

of the functional interactions (edges in the figure) is distinguished by colors, as well as the different amine modules (as specified in the lettering included in the figure and its footnote). Many shared (GO-defined) biological processes (blue edges) can be observed, as well as many (GO-defined) molecular functions (pink edges) that are shared specially among neurotransmitter-related elements. Interactions with other human genes (gray diamonds) are also detected.

The full set of information will be included in the first version of an integrative platform on biogenic amine-related biomedical information named AMINETWORKING 1.0, which is an ongoing project at present. These sets of information should be curated and validated by the "aminer community". We invite all our colleagues interested in biogenic amines to cooperate in the task. We are open to receive proof of concepts (references, personal communication, etc.), suggestions, and comments at present through the email account aminetworking@uma.es.

## 7.5 Concluding Remarks

Brain function and associated diseases, immunology and immune diseases, and cancer and rare diseases are all currently biomedical international priorities. Biogenic amines are involved in all of them, and sometimes the cross talk events among the different amine subnetworks are proposed to play an important role in the cross talk among the processes, for instance, in the case of the brain-gut axis and several immune, neurological, and neuroendocrine diseases and in the immune system-cancer/parasite interaction events during cancer/infection progression.

The present work underlies our claim to go toward the construction of a database specialized on the whole biogenic amine network. In the future, it should include information on genes, their variants and associated RNA species, tissue-specific expression data, information of the encoded protein structures and their kinetic data, protein-protein interaction data, intracellular location, as well as gene associations with human phenotypes/diseases, and gene (protein target)-drug associations. This work is just a first step that could really grow if it were supported by the collaboration of international groups with expertise in different biogenic amine modules and/or diseases. We could contribute with our previous experience in the development of biocomputational predictive models and tools for location and integration of metabolic and enzymatic data [176, 177], gene-phenotype/disease and protein/ligand association tools and analyses, [134, 178–180], and development of social curation tools [181]. As mentioned, the data sets behind the figures shown in the present work could be named as AMINETWORKING 1.0 and will be available to any researcher or clinician interested in the relationships mentioned/depicted in Table 7.4 and Figs. 7.7, 7.8, and 7.9. We have opened the email address aminetworking@uma.es to encourage the interchange of information and comments on this initiative with other research groups.

As a consequence of the exponentially growing *omics* initiatives, this kind of biocomputational support is becoming essential to get an efficient yield of the analytic investments. On the one hand, data on human genomic variants and transcript variability on the elements included in Table 7.1 is rapidly increasing, and the integration of this information could provide valuable support for personalized (predictive) medicine in a wide catalog of diseases for which more than a single biogenic amine module is involved. In fact, it could help to combine therapies or more accurate disease susceptibility predictions. On the other hand, the integration of drug-target information in a single repository could give light to new drug discovery initiatives and combined therapies, due to the similarity among structures, reaction mechanisms, interactions, and requirements of elements of different subnetworks, as experienced by our group on different occasions [113, 182]. In this sense, taking into account the intracellular location and the tissue-specific molecular and functional characteristic of the different targets is a very important issue to progress toward the right direction as also claimed in a previous work by our group [183] Unfortunately many of these translational possibilities are being currently delayed without the support of an integrative platform as the one we propose here.

**Acknowledgments** This work was supported by Grants SAF2011-26518 (MINECO, Spain) and PAIDI Grant P10-CVI6585 (Andalusian Government). We also thank the support (RRHH) from CIBERER, as well as from the University of Málaga and Andalucia Tech facilities. CIBERER is an initiative of Instituto de Salud Carlos III.

## References

1. Amadasi A, Bertoldi M, Contestabile R, Bettati S, Cellini B, di Salvo ML, et al. Pyridoxal 5′-phosphate enzymes as targets for therapeutic agents. Curr Med Chem. 2007;14:1291–324. doi:10.2174/092986707780597899.
2. Agostinelli E, Seiler N. Non-irradiation-derived reactive oxygen species (ROS) and cancer: therapeutic implications. Amino Acids. 2006;31:341–55. doi:10.1007/s00726-005-0271-8.
3. Kirschner KM, Braun JFW, Jacobi CL, Rudigier LJ, Persson AB, Scholz H. Amine oxidase copper-containing 1 (AOC1) is a downstream target gene of the Wilms tumor protein, WT1, during kidney development. J Biol Chem. 2014;289:24452–62. doi:10.1074/jbc.m114.564336.
4. Pegg AE. Mammalian polyamine metabolism and function. IUBMB Life. 2009;61:880–94. doi:10.1002/iub.230.
5. Casero RA, Pegg AE. Polyamine catabolism and disease. Biochem J. 2009;421:323–38. doi:10.1042/bj20090598.
6. Stegaev V, Nies AT, Porola P, Mieliauskaite D, Sánchez-Jiménez F, Urdiales JL, et al. Histamine transport and metabolism are deranged in salivary glands in Sjogren's syndrome. Rheumatology (Oxford). 2013;52:1599–608. doi:10.1093/rheumatology/ket188.
7. Battaglia V, DeStefano SC, Murray-Stewart T, Casero RA. Polyamine catabolism in carcinogenesis: potential targets for chemotherapy and chemoprevention. Amino Acids. 2014;46:511–9. doi:10.1007/s00726-013-1529-6.
8. Bertoldi M. Mammalian Dopa decarboxylase: structure, catalytic activity and inhibition. Arch Biochem Biophys. 2014;546:1–7. doi:10.1016/j.abb.2013.12.020.
9. Cederbaum SD, Yu H, Grody WW, Kern RM, Yoo P, Iyer RK. Arginases I and II: do their functions overlap? Mol Genet Metab. 2004;81 Suppl 1:S38–44. doi:10.1016/j.ymgme.2003.10.012.

10. Satriano J. Arginine pathways and the inflammatory response: interregulation of nitric oxide and polyamines: review article. Amino Acids. 2004;26:321–9. doi:10.1007/s00726-004-0078-4.
11. Perez-Leal O, Merali S. Regulation of polyamine metabolism by translational control. Amino Acids. 2012;42:611–7. doi:10.1007/s00726-011-1036-6.
12. Kahana C. Regulation of cellular polyamine levels and cellular proliferation by antizyme and antizyme inhibitor. Essays Biochem. 2009;46:47–61. doi:10.1042/bse0460004.
13. López-Contreras AJ, Ramos-Molina B, Cremades A, Peñafiel R. Antizyme inhibitor 2: molecular, cellular and physiological aspects. Amino Acids. 2010;38:603–11. doi:10.1007/s00726-009-0419-4.
14. Pegg AE. The function of spermine. IUBMB Life. 2014;66:8–18. doi:10.1002/iub.1237.
15. Komori H, Nitta Y, Ueno H, Higuchi Y. Structural study reveals that Ser-354 determines substrate specificity on human histidine decarboxylase. J Biol Chem. 2012;287:29175–83. doi:10.1074/jbc.m112.381897.
16. Heidari A, Tongsook C, Najafipour R, Musante L, Vasli N, Garshasbi M, et al. Mutations in the histamine N-methyltransferase gene, HNMT, are associated with nonsyndromic autosomal recessive intellectual disability. Hum Mol Genet. 2015;24:5697–710. doi:10.1093/hmg/ddv286.
17. Panula P, Chazot PL, Cowart M, Gutzmer R, Leurs R, Liu WLS, et al. International union of basic and clinical pharmacology. XCVIII. Histamine receptors. Pharmacol Rev. 2015;67:601–55. doi:10.1124/pr.114.010249.
18. Tekin I, Roskoski R, Carkaci-Salli N, Vrana KE. Complex molecular regulation of tyrosine hydroxylase. J Neural Transm. 2014;121:1451–81. doi:10.1007/s00702-014-1238-7.
19. Bennett PJ, McMahon WM, Watabe J, Achilles J, Bacon M, Coon H, et al. Tryptophan hydroxylase polymorphisms in suicide victims. Psychiatr Genet. 2000;10:13–7.
20. Montioli R, Dindo M, Giorgetti A, Piccoli S, Cellini B, Voltattorni CB. A comprehensive picture of the mutations associated with aromatic amino acid decarboxylase deficiency: from molecular mechanisms to therapy implications. Hum Mol Genet. 2014;23:5429–40. doi:10.1093/hmg/ddu266.
21. Beaulieu J-M, Espinoza S, Gainetdinov RR. Dopamine receptors—IUPHAR review 13. Br J Pharmacol. 2015;172:1–23. doi:10.1111/bph.12906.
22. Hartig PR, Hoyer D, Humphrey PP, Martin GR. Alignment of receptor nomenclature with the human genome: classification of 5-HT1B and 5-HT1D receptor subtypes. Trends Pharmacol Sci. 1996;17:103–5. doi:10.1016/0165-6147(96)30002-3.
23. Hoyer D, Clarke DE, Fozard JR, Hartig PR, Martin GR, Mylecharane EJ, et al. International Union of Pharmacology classification of receptors for 5-hydroxytryptamine (Serotonin). Pharmacol Rev. 1994;46:157–203.
24. Tohda M. Serotonin 2C receptor as a superhero: diversities and talents in the RNA universe for editing, variant, small RNA and other expected functional RNAs. J Pharmacol Sci. 2014;126:321–8. doi:10.1254/jphs.14r06cr.
25. García-Martín E, Martínez C, Serrador M, Alonso-Navarro H, Ayuso P, Navacerrada F, et al. Diamine oxidase rs10156191 and rs2052129 variants are associated with the risk for migraine. Headache. 2015;55:276–86. doi:10.1111/head.12493.
26. Kaitaniemi S, Elovaara H, Grön K, Kidron H, Liukkonen J, Salminen T, et al. The unique substrate specificity of human AOC2, a semicarbazide-sensitive amine oxidase. Cell Mol Life Sci. 2009;66:2743–57. doi:10.1007/s00018-009-0076-5.
27. Buffoni F. Semicarbazide-sensitive amine oxidases: some biochemical properties and general considerations. Prog Brain Res. 1995;106:323–31.
28. Bortolato M, Chen K, Shih JC. Monoamine oxidase inactivation: from pathophysiology to therapeutics. Adv Drug Deliv Rev. 2008;60:1527–33. doi:10.1016/j.addr.2008.06.002.
29. Abdulhussein AA, Wallace HM. Polyamines and membrane transporters. Amino Acids. 2014;46:655–60. doi:10.1007/s00726-013-1553-6.
30. German CL, Baladi MG, McFadden LM, Hanson GR, Fleckenstein AE. Regulation of the dopamine and vesicular monoamine transporters: pharmacological targets and implications for disease. Pharmacol Rev. 2015;67:1005–24. doi:10.1124/pr.114.010397.

31. Lazarov NE, Reindl S, Fischer F, Gratzl M. Histaminergic and dopaminergic traits in the human carotid body. Respir Physiol Neurobiol. 2009;165:131–6. doi:10.1016/j.resp.2008.10.016.
32. Johnson KB, Petersen-Jones H, Thompson JM, Hitomi K, Itoh M, Bakker ENTP, et al. Vena cava and aortic smooth muscle cells express transglutaminases 1 and 4 in addition to transglutaminase 2. Am J Physiol Heart Circ Physiol. 2012;302:H1355–66. doi:10.1152/ajpheart.00918.2011.
33. Ientile R, Currò M, Caccamo D. Transglutaminase 2 and neuroinflammation. Amino Acids. 2015;47:19–26. doi:10.1007/s00726-014-1864-2.
34. Qiao S-W, Piper J, Haraldsen G, Oynebråten I, Fleckenstein B, Molberg O, et al. Tissue transglutaminase-mediated formation and cleavage of histamine-gliadin complexes: biological effects and implications for celiac disease. J Immunol. 2005;174:1657–63.
35. Williams K. Extracellular Modulation of NMDA Receptors. In: Van Dongen AM, editor. Biology of the NMDA receptor. Boca Raton: CRC Press/Taylor & Francis; 2009. http://www.ncbi.nlm.nih.gov/books/NBK5272/.
36. Hirose T, Saiki R, Yoshizawa Y, Imamura M, Higashi K, Ishii I, et al. Spermidine and Ca(2+), but not Na(+), can permeate NMDA receptors consisting of GluN1 and GluN2A or GluN2B in the presence of Mg(2+). Biochem Biophys Res Commun. 2015;463:1190–5. doi:10.1016/j.bbrc.2015.06.081.
37. Landau G, Bercovich Z, Park MH, Kahana C. The role of polyamines in supporting growth of mammalian cells is mediated through their requirement for translation initiation and elongation. J Biol Chem. 2010;285:12474–81. doi:10.1074/jbc.m110.106419.
38. Raj KP, Zell JA, Rock CL, McLaren CE, Zoumas-Morse C, Gerner EW, et al. Role of dietary polyamines in a phase III clinical trial of difluoromethylornithine (DFMO) and sulindac for prevention of sporadic colorectal adenomas. Br J Cancer. 2013;108:512–8. doi:10.1038/bjc.2013.15.
39. Saulnier Sholler GL, Gerner EW, Bergendahl G, MacArthur RB, VanderWerff A, Ashikaga T, et al. A phase I trial of DFMO targeting polyamine addiction in patients with relapsed/refractory neuroblastoma. PLoS One. 2015;10, e0127246. doi:10.1371/journal.pone.0127246.
40. Birkholtz L-M, Williams M, Niemand J, Louw AI, Persson L, Heby O. Polyamine homoeostasis as a drug target in pathogenic protozoa: peculiarities and possibilities. Biochem J. 2011;438:229–44. doi:10.1042/bj20110362.
41. Fiori LM, Turecki G. Implication of the polyamine system in mental disorders. J Psychiatry Neurosci. 2008;33:102–10.
42. García-Faroldi G, Rodríguez CE, Urdiales JL, Pérez-Pomares JM, Dávila JC, Pejler G, et al. Polyamines are present in mast cell secretory granules and are important for granule homeostasis. PLoS One. 2010;5:e15071. doi:10.1371/journal.pone.0015071.
43. Masuko T, Kusama-Eguchi K, Sakata K, Kusama T, Chaki S, Okuyama S, et al. Polyamine transport, accumulation, and release in brain. J Neurochem. 2003;84:610–7. doi:10.1046/j.1471-4159.2003.01558.x.
44. Poulin R, Casero RA, Soulet D. Recent advances in the molecular biology of metazoan polyamine transport. Amino Acids. 2012;42:711–23. doi:10.1007/s00726-011-0987-y.
45. Miller-Fleming L, Olin-Sandoval V, Campbell K, Ralser M. Remaining mysteries of molecular biology: the role of polyamines in the cell. J Mol Biol. 2015;427:3389–406. doi:10.1016/j.jmb.2015.06.020.
46. Thurmond RL. The histamine H4 receptor: from orphan to the clinic. Front Pharmacol. 2015;6:65. doi:10.3389/fphar.2015.00065.
47. Burban A, Faucard R, Armand V, Bayard C, Vorobjev V, Arrang J-M. Histamine potentiates N-methyl-D-aspartate receptors by interacting with an allosteric site distinct from the polyamine binding site. J Pharmacol Exp Ther. 2010;332:912–21. doi:10.1124/jpet.109.158543.
48. Martinel Lamas DJ, Rivera ES, Medina VA. Histamine H4 receptor: insights into a potential therapeutic target in breast cancer. Front Biosci (Schol Ed). 2015;7:1–9. doi:10.2741/420.
49. Glatzer F, Gschwandtner M, Ehling S, Rossbach K, Janik K, Klos A, et al. Histamine induces proliferation in keratinocytes from patients with atopic dermatitis through the histamine 4 receptor. J Allergy Clin Immunol. 2013;132:1358–67. doi:10.1016/j.jaci.2013.06.023.

50. Fajardo I, Urdiales JL, Medina MA, Sanchez-Jimenez F. Effects of phorbol ester and dexamethasone treatment on histidine decarboxylase and ornithine decarboxylase in basophilic cells. Biochem Pharmacol. 2001;61:1101–6. doi:10.1016/s0006-2952(01)00567-6.
51. García-Faroldi G, Correa-Fiz F, Abrighach H, Berdasco M, Fraga MF, Esteller M, et al. Polyamines affect histamine synthesis during early stages of IL-3-induced bone marrow cell differentiation. J Cell Biochem. 2009;108:261–71. doi:10.1002/jcb.22246.
52. Ruiz-Chica AJ, Soriano A, Tuñón I, Sánchez-Jiménez FM, Silla E, Ramírez FJ. FT-Raman and QM/MM study of the interaction between histamine and DNA. Chem Phys. 2006;324:579–90. doi:10.1016/j.chemphys.2005.11.022.
53. Medina V, Croci M, Crescenti E, Mohamad N, Sanchez-Jiménez F, Massari N, et al. The role of histamine in human mammary carcinogenesis: H3 and H4 receptors as potential therapeutic targets for breast cancer treatment. Cancer Biol Ther. 2008;7:28–35. doi:10.4161/cbt.7.1.5123.
54. Panula P, Nuutinen S. The histaminergic network in the brain: basic organization and role in disease. Nat Rev Neurosci. 2013;14:472–87. doi:10.1038/nrn3526.
55. Panula P, Sundvik M, Karlstedt K. Developmental roles of brain histamine. Trends Neurosci. 2014;37:159–68. doi:10.1016/j.tins.2014.01.001.
56. Ellenbroek BA, Ghiabi B. The other side of the histamine H3 receptor. Trends Neurosci. 2014;37:191–9. doi:10.1016/j.tins.2014.02.007.
57. Flik G, Folgering JHA, Cremers TIHF, Westerink BHC, Dremencov E. Interaction between brain histamine and serotonin, norepinephrine, and dopamine systems: in vivo microdialysis and electrophysiology study. J Mol Neurosci. 2015;56:320–8. doi:10.1007/s12031-015-0536-3.
58. Passani MB, Blandina P. Histamine receptors in the CNS as targets for therapeutic intervention. Trends Pharmacol Sci. 2011;32:242–9. doi:10.1016/j.tips.2011.01.003.
59. Thurmond RL, Gelfand EW, Dunford PJ. The role of histamine H1 and H4 receptors in allergic inflammation: the search for new antihistamines. Nat Rev Drug Discov. 2008;7:41–53. doi:10.1038/nrd2465.
60. Zampeli E, Tiligada E. The role of histamine H4 receptor in immune and inflammatory disorders. Br J Pharmacol. 2009;157:24–33. doi:10.1111/j.1476-5381.2009.00151.x.
61. Pino-Ángeles A, Reyes-Palomares A, Melgarejo E, Sánchez-Jiménez F. Histamine: an undercover agent in multiple rare diseases? J Cell Mol Med. 2012;16:1947–60. doi:10.1111/j.1582-4934.2012.01566.x.
62. Yang XD, Ai W, Asfaha S, Bhagat G, Friedman RA, Jin G, et al. Histamine deficiency promotes inflammation-associated carcinogenesis through reduced myeloid maturation and accumulation of CD11b+ Ly6G+ immature myeloid cells. Nat Med. 2011;17:87–95. doi:10.1038/nm.2278.
63. Chen D, Aihara T, Zhao C-M, Håkanson R, Okabe S. Differentiation of the gastric mucosa. I. Role of histamine in control of function and integrity of oxyntic mucosa: understanding gastric physiology through disruption of targeted genes. Am J Physiol Gastrointest Liver Physiol. 2006;291:G539–44. doi:10.1152/ajpgi.00178.2006.
64. Nozaki K, Weis V, Wang TC, Falus A, Goldenring JR. Altered gastric chief cell lineage differentiation in histamine-deficient mice. Am J Physiol Gastrointest Liver Physiol. 2009;296:G1211–20. doi:10.1152/ajpgi.90643.2008.
65. Pagotto RM, Monzón C, Moreno MB, Pignataro OP, Mondillo C. Proliferative effect of histamine on MA-10 Leydig tumor cells mediated through HRH2 activation, transient elevation in cAMP production, and increased extracellular signal-regulated kinase phosphorylation levels. Biol Reprod. 2012;87:150. doi:10.1095/biolreprod.112.102905.
66. Arreola R, Becerril-Villanueva E, Cruz-Fuentes C, Velasco-Velázquez MA, Garcés-Alvarez ME, Hurtado-Alvarado G, et al. Immunomodulatory effects mediated by serotonin. J Immunol Res. 2015;2015:1–21. doi:10.1155/2015/354957.
67. Choi MR. Renal dopaminergic system: Pathophysiological implications and clinical perspectives. World J Nephrol. 2015;4:196–212. doi:10.5527/wjn.v4.i2.196.
68. Gratwicke J, Jahanshahi M, Foltynie T. Parkinson's disease dementia: a neural networks perspective. Brain. 2015;138:1454–76. doi:10.1093/brain/awv104.

69. Johnston JD, Skene DJ. 60 YEARS OF NEUROENDOCRINOLOGY: Regulation of mammalian neuroendocrine physiology and rhythms by melatonin. J Endocrinol. 2015;226:T187–98. doi:10.1530/joe-15-0119.
70. Deidda G, Bozarth IF, Cancedda L. Modulation of GABAergic transmission in development and neurodevelopmental disorders: investigating physiology and pathology to gain therapeutic perspectives. Front Cell Neurosci. 2014;8:119. doi:10.3389/fncel.2014.00119.
71. Yuan H, Low C-M, Moody OA, Jenkins A, Traynelis SF. Ionotropic GABA and glutamate receptor mutations and human neurologic diseases. Mol Pharmacol. 2015;88:203–17. doi:10.1124/mol.115.097998.
72. Sánchez-Jiménez F, Ruiz-Pérez MV, Urdiales JL, Medina MA. Pharmacological potential of biogenic amine-polyamine interactions beyond neurotransmission. Br J Pharmacol. 2013;170:4–16. doi:10.1111/bph.12109.
73. Sánchez-Jiménez F, Montañez R, Correa-Fiz F, Chaves P, Rodríguez-Caso C, Urdiales JL, et al. The usefulness of post-genomics tools for characterization of the amine cross-talk in mammalian cells. Biochem Soc Trans. 2007;35:381–5. doi:10.1042/bst0350381.
74. Medina MA, Correa-Fiz F, Rodríguez-Caso C, Sánchez-Jiménez F. A comprehensive view of polyamine and histamine metabolism to the light of new technologies. J Cell Mol Med. 2005;9:854–64. doi:10.1111/j.1582-4934.2005.tb00384.x.
75. Seifert R, Strasser A, Schneider EH, Neumann D, Dove S, Buschauer A. Molecular and cellular analysis of human histamine receptor subtypes. Trends Pharmacol Sci. 2013;34:33–58. doi:10.1016/j.tips.2012.11.001.
76. Micallef S, Stark H, Sasse A. Polymorphisms and genetic linkage of histamine receptors. Life Sci. 2013;93:487–94. doi:10.1016/j.lfs.2013.08.012.
77. Schneider EH, Neumann D, Seifert R. Modulation of behavior by the histaminergic system: Lessons from HDC-, H3R- and H4R-deficient mice. Neurosci Biobehav Rev. 2014;47:101–21. doi:10.1016/j.neubiorev.2014.07.020.
78. Schneider EH, Neumann D, Seifert R. Modulation of behavior by the histaminergic system: lessons from H(1)R-and H(2)R-deficient mice. Neurosci Biobehav Rev. 2014;42:252–66. doi:10.1016/j.neubiorev.2014.03.009.
79. Schneider E, Leite-de-moraes M, Dy M. Histamine, immune cells and autoimmunity. Adv Exp Med Biol. 2010;709:81–94. doi:10.1007/978-1-4419-8056-4_9.
80. Abrighach H, Fajardo I, Sánchez-Jiménez F, Urdiales JL. Exploring polyamine regulation by nascent histamine in a human-transfected cell model. Amino Acids. 2010;38:561–73. doi:10.1007/s00726-009-0417-6.
81. Caro-Astorga J, Fajardo I, Ruiz-Pérez MV, Sánchez-Jiménez F, Urdiales JL. Nascent histamine induces α-synuclein and caspase-3 on human cells. Biochem Biophys Res Commun. 2014;451:580–6. doi:10.1016/j.bbrc.2014.08.022.
82. Furuta K, Nakayama K, Sugimoto Y, Ichikawa A, Tanaka S. Activation of histidine decarboxylase through post-translational cleavage by caspase-9 in a mouse mastocytoma P-815. J Biol Chem. 2007;282:13438–46. doi:10.1074/jbc.m609943200.
83. Krauth M-T, Agis H, Aichberger KJ, Simonitsch-Klupp I, Müllauer L, Mayerhofer M, et al. Immunohistochemical detection of histidine decarboxylase in neoplastic mast cells in patients with systemic mastocytosis. Hum Pathol. 2006;37:439–47. doi:10.1016/j.humpath.2005.11.015.
84. Garcia-Montero AC, Jara-Acevedo M, Teodosio C, Sanchez ML, Nunez R, Prados A, et al. KIT mutation in mast cells and other bone marrow hematopoietic cell lineages in systemic mast cell disorders: a prospective study of the Spanish Network on Mastocytosis (REMA) in a series of 113 patients. Blood. 2006;108:2366–72. doi:10.1182/blood-2006-04-015545.
85. Fajardo I, Urdiales JL, Paz JC, Chavarría T, Sánchez-Jiménez F, Medina MA. Histamine prevents polyamine accumulation in mouse C57.1 mast cell cultures. Eur J Biochem. 2001;268:768–73. doi:10.1046/j.1432-1327.2001.01930.x.
86. Kanerva K, Lappalainen J, Mäkitie LT, Virolainen S, Kovanen PT, Andersson LC. Expression of antizyme inhibitor 2 in mast cells and role of polyamines as selective regulators of serotonin secretion. PLoS One. 2009;4, e6858. doi:10.1371/journal.pone.0006858.

87. Gavin IM, Glesne D, Zhao Y, Kubera C, Huberman E. Spermine acts as a negative regulator of macrophage differentiation in human myeloid leukemia cells. Cancer Res. 2004;64:7432–8. doi:10.1158/0008-5472.can-04-0051.
88. Ding XQ, Chen D, Rosengren E, Persson L, Hakanson R. Comparison between activation of ornithine decarboxylase and histidine decarboxylase in rat stomach. Am J Physiol. 1996;270:G476–86.
89. Seiler N. Catabolism of polyamines. Amino Acids. 2004;26:217–33. doi:10.1007/s00726-004-0070-z.
90. Ballas SK, Mohandas N, Clark MR, Embury SH, Smith ED, Marton LJ, et al. Reduced transglutaminase-catalyzed cross-linking of exogenous amines to membrane proteins in sickle erythrocytes. Biochim Biophys Acta. 1985;812:234–42. doi:10.1016/0005-2736(85)90543-7.
91. LaBella FS, Brandes LJ. Interaction of histamine and other bioamines with cytochromes P450: implications for cell growth modulation and chemopotentiation by drugs. Semin Cancer Biol. 2000;10:47–53. doi:10.1006/scbi.2000.0307.
92. Kallweit U, Aritake K, Bassetti CL, Blumenthal S, Hayaishi O, Linnebank M, et al. Elevated CSF histamine levels in multiple sclerosis patients. Fluids Barriers CNS. 2013;10:19. doi:10.1186/2045-8118-10-19.
93. Rinne JO, Anichtchik OV, Eriksson KS, Kaslin J, Tuomisto L, Kalimo H, et al. Increased brain histamine levels in Parkinson's disease but not in multiple system atrophy. J Neurochem. 2002;81:954–60. doi:10.1046/j.1471-4159.2002.00871.x.
94. Ballerini C, Aldinucci A, Luccarini I, Galante A, Manuelli C, Blandina P, et al. Antagonism of histamine H4 receptors exacerbates clinical and pathological signs of experimental autoimmune encephalomyelitis. Br J Pharmacol. 2013;170:67–77. doi:10.1111/bph.12263.
95. Büttner S, Broeskamp F, Sommer C, Markaki M, Habernig L, Alavian-Ghavanini A, et al. Spermidine protects against α-synuclein neurotoxicity. Cell Cycle. 2014;13:3903–8. doi:10.4161/15384101.2014.973309.
96. Benetti F, Furini CRG, de Carvalho MJ, Provensi G, Passani MB, Baldi E, et al. Histamine in the basolateral amygdala promotes inhibitory avoidance learning independently of hippocampus. Proc Natl Acad Sci. 2015;112:E2536–42. doi:10.1073/pnas.1506109112.
97. Shan L, Dauvilliers Y, Siegel JM. Interactions of the histamine and hypocretin systems in CNS disorders. Nat Rev Neurol. 2015;11:401–13. doi:10.1038/nrneurol.2015.99.
98. Ringvall M, Rönnberg E, Wernersson S, Duelli A, Henningsson F, Abrink M, et al. Serotonin and histamine storage in mast cell secretory granules is dependent on serglycin proteoglycan. J Allergy Clin Immunol. 2008;121:1020–6. doi:10.1016/j.jaci.2007.11.031.
99. Smitka K, Papezova H, Vondra K, Hill M, Hainer V, Nedvidkova J. The role of "mixed" orexigenic and anorexigenic signals and autoantibodies reacting with appetite-regulating neuropeptides and peptides of the adipose tissue-gut-brain axis: relevance to food intake and nutritional status in patients with anorexia nervosa. Int J Endocrinol. 2013;2013:483145. doi:10.1155/2013/483145.
100. Sundvik M, Panula P. Interactions of the orexin/hypocretin neurones and the histaminergic system. Acta Physiol (Oxf). 2015;213:321–33. doi:10.1111/apha.12432.
101. Ai W, Liu Y, Langlois M, Wang TC. Kruppel-like factor 4 (KLF4) represses histidine decarboxylase gene expression through an upstream Sp1 site and downstream gastrin responsive elements. J Biol Chem. 2004;279:8684–93. doi:10.1074/jbc.m308278200.
102. Sakurada T, Ro S, Onouchi T, Ohno S, Aoyama T, Chinen K, et al. Comparison of the actions of acylated and desacylated ghrelin on acid secretion in the rat stomach. J Gastroenterol. 2010;45:1111–20. doi:10.1007/s00535-010-0269-6.
103. Andrews ZB, Erion D, Beiler R, Liu Z-W, Abizaid A, Zigman J, et al. Ghrelin promotes and protects nigrostriatal dopamine function via a UCP2-dependent mitochondrial mechanism. J Neurosci. 2009;29:14057–65. doi:10.1523/jneurosci.3890-09.2009.
104. Lee M, Ryu YH, Cho WG, Kang YW, Lee SJ, Jeon TJ, et al. Relationship between dopamine deficit and the expression of depressive behavior resulted from alteration of serotonin system. Synapse. 2015;69:453–60. doi:10.1002/syn.21834.

105. Seyedabadi M, Fakhfouri G, Ramezani V, Mehr SE, Rahimian R. The role of serotonin in memory: interactions with neurotransmitters and downstream signaling. Exp Brain Res. 2014;232:723–38. doi:10.1007/s00221-013-3818-4.
106. Naoi M, Riederer P, Maruyama W. Modulation of monoamine oxidase (MAO) expression in neuropsychiatric disorders: genetic and environmental factors involved in type A MAO expression. J Neural Transm (Vienna). 2015;123(2):91–106. doi:10.1007/s00702-014-1362-4.
107. Nelson DL, Gehlert DR. Central nervous system biogenic amine targets for control of appetite and energy expenditure. Endocrine. 2006;29:49–60. doi:10.1385/endo:29:1:149.
108. Idova GV, Alperina EL, Cheido MA. Contribution of brain dopamine, serotonin and opioid receptors in the mechanisms of neuroimmunomodulation: evidence from pharmacological analysis. Int Immunopharmacol. 2012;12:618–25. doi:10.1016/j.intimp.2012.02.010.
109. Ferrada C, Ferré S, Casadó V, Cortés A, Justinova Z, Barnes C, et al. Interactions between histamine H3 and dopamine D2 receptors and the implications for striatal function. Neuropharmacology. 2008;55:190–7. doi:10.1016/j.neuropharm.2008.05.008.
110. Moreno E, Hoffmann H, Gonzalez-Sepúlveda M, Navarro G, Casadó V, Cortés A, et al. Dopamine D 1 -histamine H 3 receptor heteromers provide a selective link to MAPK signaling in GABAergic neurons of the direct striatal pathway. J Biol Chem. 2011;286:5846–54. doi:10.1074/jbc.m110.161489.
111. Inoue K, Tsutsui H, Akatsu H, Hashizume Y, Matsukawa N, Yamamoto T, et al. Metabolic profiling of Alzheimer's disease brains. Sci Rep. 2013;3:2364. doi:10.1038/srep02364.
112. Moya-Garcia AA, Medina MA, Sánchez-Jiménez F. Mammalian histidine decarboxylase: from structure to function. Bioessays. 2005;27:57–63. doi:10.1002/bies.20174.
113. Ruiz-Pérez MV, Pino-Ángeles A, Medina MA, Sánchez-Jiménez F, Moya-García AA. Structural perspective on the direct inhibition mechanism of EGCG on mammalian histidine decarboxylase and DOPA decarboxylase. J Chem Inf Model. 2012;52:113–9. doi:10.1021/ci200221z.
114. Cellini B, Montioli R, Oppici E, Astegno A, Voltattorni CB. The chaperone role of the pyridoxal 5′-phosphate and its implications for rare diseases involving B6-dependent enzymes. Clin Biochem. 2014;47:158–65. doi:10.1016/j.clinbiochem.2013.11.021.
115. Meiser J, Weindl D, Hiller K. Complexity of dopamine metabolism. Cell Commun Signal. 2013;11:34. doi:10.1186/1478-811x-11-34.
116. Timmons J, Chang ET, Wang J-Y, Rao JN. Polyamines and gut mucosal homeostasis. J Gastrointest Dig Syst. 2012;S7:001.
117. Kotlyar DS, Shum M, Hsieh J, Blonski W, Greenwald DA. Non-pulmonary allergic diseases and inflammatory bowel disease: a qualitative review. World J Gastroenterol. 2014;20:11023–32. doi:10.3748/wjg.v20.i32.11023.
118. Guihot G, Blachier F. Histidine and histamine metabolism in rat enterocytes. Mol Cell Biochem. 1997;175:143–8.
119. Bertrand PP, Bertrand RL. Serotonin release and uptake in the gastrointestinal tract. Auton Neurosci. 2010;153:47–57. doi:10.1016/j.autneu.2009.08.002.
120. Kanerva K, Mäkitie LT, Pelander A, Heiskala M, Andersson LC. Human ornithine decarboxylase paralogue (ODCp) is an antizyme inhibitor but not an arginine decarboxylase. Biochem J. 2008;409:187–92. doi:10.1042/bj20071004.
121. Piletz JE, Aricioglu F, Cheng J-T, Fairbanks CA, Gilad VH, Haenisch B, et al. Agmatine: clinical applications after 100 years in translation. Drug Discov Today. 2013;18:880–93. doi:10.1016/j.drudis.2013.05.017.
122. Arndt MA, Battaglia V, Parisi E, Lortie MJ, Isome M, Baskerville C, et al. The arginine metabolite agmatine protects mitochondrial function and confers resistance to cellular apoptosis. Am J Physiol Cell Physiol. 2009;296:C1411–9. doi:10.1152/ajpcell.00529.2008.
123. Moretti M, Matheus FC, de Oliveira PA, Neis VB, Ben J, Walz R, et al. Role of agmatine in neurodegenerative diseases and epilepsy. Front Biosci (Elite Ed). 2014;6:341–59. doi:10.2741/710.

124. Morgan DML, Bauer F, White A, editors. COST Action 917: biogenically active amines in food. Vol. VII. Luxembourg: Office for official publications of the European Communities, Luxembourg; 2005.
125. Wallace HY, Hughes A, editors. COST Action 922: health implications of dietary amines. Review of current status. Luxembourg: Official publications of the European Commission; 2004.
126. Fogel WA, Lewinski A, Jochem J. Histamine in food: is there anything to worry about? Biochem Soc Trans. 2007;35:349–52. doi:10.1042/bst0350349.
127. Ruiz-Pérez MV, Medina MÁ, Urdiales JL, Keinänen TA, Sánchez-Jiménez F. Polyamine metabolism is sensitive to glycolysis inhibition in human neuroblastoma cells. J Biol Chem. 2015;290:6106–19. doi:10.1074/jbc.m114.619197.
128. Rial NS, Meyskens FL, Gerner EW. Polyamines as mediators of APC-dependent intestinal carcinogenesis and cancer chemoprevention. Essays Biochem. 2009;46:111–24. doi:10.1042/bse0460008.
129. Lozier AM, Rich ME, Grawe AP, Peck AS, Zhao P, Chang AT, et al. Targeting ornithine decarboxylase reverses the LIN28/Let-7 axis and inhibits glycolytic metabolism in neuroblastoma. Oncotarget. 2015;6:196–206. doi:10.18632/oncotarget.2768.
130. Castellan Baldan L, Williams KA, Gallezot J-D, Pogorelov V, Rapanelli M, Crowley M, et al. Histidine decarboxylase deficiency causes tourette syndrome: parallel findings in humans and mice. Neuron. 2014;81:77–90. doi:10.1016/j.neuron.2013.10.052.
131. Yamauchi K. Regulation of gene expression of L-histidine decarboxylase and histamine N-methyl-transferase, and its relevance to the pathogenesis of bronchial asthma. Nihon Rinsho. 1996;54:377–88.
132. Saligrama N, Case LK, del Rio R, Noubade R, Teuscher C. Systemic lack of canonical histamine receptor signaling results in increased resistance to autoimmune encephalomyelitis. J Immunol. 2013;191:614–22. doi:10.4049/jimmunol.1203137.
133. Lu C, Diehl SA, Noubade R, Ledoux J, Nelson MT, Spach K, et al. Endothelial histamine H1 receptor signaling reduces blood-brain barrier permeability and susceptibility to autoimmune encephalomyelitis. Proc Natl Acad Sci U S A. 2010;107:18967–72. doi:10.1073/pnas.1008816107.
134. Rodríguez-López R, Reyes-Palomares A, Sánchez-Jiménez F, Medina M. PhenUMA: a tool for integrating the biomedical relationships among genes and diseases. BMC Bioinformatics. 2014;15:375. doi:10.1186/s12859-014-0375-1.
135. Robinson PN, Köhler S, Bauer S, Seelow D, Horn D, Mundlos S. The Human Phenotype Ontology: a tool for annotating and analyzing human hereditary disease. Am J Hum Genet. 2008;83:610–5. doi:10.1016/j.ajhg.2008.09.017.
136. Köhler S, Schulz MH, Krawitz P, Bauer S, Dölken S, Ott CE, et al. Clinical diagnostics in human genetics with semantic similarity searches in ontologies. Am J Hum Genet. 2009;85:457–64. doi:10.1016/j.ajhg.2009.09.003.
137. Resnik P. Semantic similarity in a taxonomy: an information-based measure and its application to problems of ambiguity in natural language. J Artif Intel Res. 1999;11:95–130. doi:10.1613/jair.514.
138. Pinero J, Queralt-Rosinach N, Bravo A, Deu-Pons J, Bauer-Mehren A, Baron M, et al. DisGeNET: a discovery platform for the dynamical exploration of human diseases and their genes. Database. 2015;2015:bav028. doi:10.1093/database/bav028.
139. Pletscher-Frankild S, Pallejà A, Tsafou K, Binder JX, Jensen LJ. DISEASES: text mining and data integration of disease-gene associations. Methods. 2014;74:83–9. doi:10.1016/j.ymeth.2014.11.020.
140. Bravo À, Cases M, Queralt-Rosinach N, Sanz F, Furlong LI. A knowledge-driven approach to extract disease-related biomarkers from the literature. Biomed Res Int. 2014;2014:253128. doi:10.1155/2014/253128.
141. Vanhanen J, Nuutinen S, Lintunen M, Mäki T, Rämö J, Karlstedt K, et al. Histamine is required for $H_3$ receptor-mediated alcohol reward inhibition, but not for alcohol consumption or stimulation. Br J Pharmacol. 2013;170:177–87. doi:10.1111/bph.12170.

142. Xu T, Du L, Zhou Y. Evaluation of GO-based functional similarity measures using S cerevisiae protein interaction and expression profile data. BMC Bioinformatics. 2008;9:472. doi:10.1186/1471-2105-9-472.
143. He G-H, Lin J-J, Cai W-K, Xu W-M, Yu Z-P, Yin S-J, et al. Associations of polymorphisms in histidine decarboxylase, histamine N-methyltransferase and histamine receptor H3 genes with breast cancer. PLoS One. 2014;9, e97728. doi:10.1371/journal.pone.0097728.
144. Borcherding DC, Tong W, Hugo ER, Barnard DF, Fox S, LaSance K, et al. Expression and therapeutic targeting of dopamine receptor-1 (D1R) in breast cancer. Oncogene. 2015. doi:10.1038/onc.2015.369.
145. Medina VA, Rivera ES. Histamine receptors and cancer pharmacology. Br J Pharmacol. 2010;161:755–67. doi:10.1111/j.1476-5381.2010.00961.x.
146. Mele M, Ferreira PG, Reverter F, DeLuca DS, Monlong J, Sammeth M, et al. The human transcriptome across tissues and individuals. Science. 2015;348:660–5. doi:10.1126/science.aaa0355.
147. Kim M-S, Pinto SM, Getnet D, Nirujogi RS, Manda SS, Chaerkady R, et al. A draft map of the human proteome. Nature. 2014;509:575–81. doi:10.1038/nature13302.
148. Wilhelm M, Schlegl J, Hahne H, Moghaddas Gholami A, Lieberenz M, Savitski MM, et al. Mass-spectrometry-based draft of the human proteome. Nature. 2014;509:582–7. doi:10.1038/nature13319.
149. Gervasini G, Agúndez JAG, García-Menaya J, Martínez C, Cordobés C, Ayuso P, et al. Variability of the L-Histidine decarboxylase gene in allergic rhinitis. Allergy. 2010;65:1576–84. doi:10.1111/j.1398-9995.2010.02425.x.
150. North ML, Grasemann H, Khanna N, Inman MD, Gauvreau GM, Scott JA. Increased ornithine-derived polyamines cause airway hyperresponsiveness in a mouse model of asthma. Am J Respir Cell Mol Biol. 2013;48:694–702. doi:10.1165/rcmb.2012-0323oc.
151. Szczepankiewicz A, Bręborowicz A, Sobkowiak P, Popiel A. Polymorphisms of two histamine-metabolizing enzymes genes and childhood allergic asthma: a case control study. Clin Mol Allergy. 2010;8:14. doi:10.1186/1476-7961-8-14.
152. Welter D, Gray WA, Kille P, Ontology TG. Investigation of semantic similarity as a tool for comparative genomics. In: Wagner R, Revell N, Pernul G, editors. Database and expert systems applications. Berlin: Springer; 2007. p. 772–9.
153. Weltman JK. Histamine as a regulator of allergic and asthmatic inflammation. Allergy Asthma Proc. 2003;24:227–9.
154. Zhang R, Kubo M, Murakami I, Setiawan H, Takemoto K, Inoue K, et al. l-Arginine administration attenuates airway inflammation by altering l-arginine metabolism in an NC/Nga mouse model of asthma. J Clin Biochem Nutr. 2015;56:201–7. doi:10.3164/jcbn.14-140.
155. Montañez R, Sánchez-Jiménez F, Aldana-Montes JF, Medina MA. Polyamines: metabolism to systems biology and beyond. Amino Acids. 2007;33:283–9. doi:10.1007/s00726-007-0521-4.
156. Dandrifosse G, Peulen O, El Khefif N, Deloyer P, Dandrifosse AC, Grandfils C. Are milk polyamines preventive agents against food allergy? Proc Nutr Soc. 2000;59:81–6. doi:10.1017/S0029665100000100.
157. López-Garcia C, Ramos-Molina B, Lambertos A, López-Contreras AJ, Cremades A, Peñafiel R. Antizyme inhibitor 2 hypomorphic mice. New patterns of expression in pancreas and adrenal glands suggest a role in secretory processes. PLoS One. 2013;8, e69188. doi:10.1371/journal.pone.0069188.
158. Ramos-Molina B, López-Contreras AJ, Cremades A, Peñafiel R. Differential expression of ornithine decarboxylase antizyme inhibitors and antizymes in rodent tissues and human cell lines. Amino Acids. 2012;42:539–47. doi:10.1007/s00726-011-1031-y.
159. Mäkitie LT, Kanerva K, Polvikoski T, Paetau A, Andersson LC. Brain neurons express ornithine decarboxylase-activating antizyme inhibitor 2 with accumulation in Alzheimer's disease. Brain Pathol. 2010;20:571–80. doi:10.1111/j.1750-3639.2009.00334.x.
160. Li DC, Nichols CG, Sala-Rabanal M. Role of a hydrophobic pocket in polyamine interactions with the polyspecific organic cation transporter OCT3. J Biol Chem. 2015;290:27633–43. doi:10.1074/jbc.M115.668913.

161. Tong JHS, Cummins TDR, Johnson BP, McKinley L-A, Pickering HE, Fanning P, et al. An association between a dopamine transporter gene (SLC6A3) haplotype and ADHD symptom measures in nonclinical adults. Am J Med Genet B Neuropsychiatr Genet. 2015;168B:89–96. doi:10.1002/ajmg.b.32283.
162. Sutcliffe JS, Delahanty RJ, Prasad HC, McCauley JL, Han Q, Jiang L, et al. Allelic heterogeneity at the serotonin transporter locus (SLC6A4) confers susceptibility to autism and rigid-compulsive behaviors. Am J Hum Genet. 2005;77:265–79. doi:10.1086/432648.
163. O'Brien JT, Colloby S, Fenwick J, Williams ED, Firbank M, Burn D, et al. Dopamine transporter loss visualized with FP-CIT SPECT in the differential diagnosis of dementia with Lewy bodies. Arch Neurol. 2004;61:919–25. doi:10.1001/archneur.61.6.919.
164. Gunther J, Tian Y, Stamova B, Lit L, Corbett B, Ander B, et al. Catecholamine-related gene expression in blood correlates with tic severity in tourette syndrome. Psychiatry Res. 2012;200:593–601. doi:10.1016/j.psychres.2012.04.034.
165. Horschitz S, Hummerich R, Lau T, Rietschel M, Schloss P. A dopamine transporter mutation associated with bipolar affective disorder causes inhibition of transporter cell surface expression. Mol Psychiatry. 2005;10:1104–9. doi:10.1038/sj.mp.4001730.
166. Strug LJ, Suresh R, Fyer AJ, Talati A, Adams PB, Li W, et al. Panic disorder is associated with the serotonin transporter gene (SLC6A4) but not the promoter region (5-HTTLPR). Mol Psychiatry. 2010;15:166–76. doi:10.1038/mp.2008.79.
167. Kelada SNP, Checkoway H, Kardia SLR, Carlson CS, Costa-Mallen P, Eaton DL, et al. 5′ and 3′ region variability in the dopamine transporter gene (SLC6A3), pesticide exposure and Parkinson's disease risk: a hypothesis-generating study. Hum Mol Genet. 2006;15:3055–62. doi:10.1093/hmg/ddl247.
168. Furmark T, Tillfors M, Garpenstrand H, Marteinsdottir I, Långström B, Oreland L, et al. Serotonin transporter polymorphism related to amygdala excitability and symptom severity in patients with social phobia. Neurosci Lett. 2004;362:189–92. doi:10.1016/j.neulet.2004.02.070.
169. Sáiz PA, García-Portilla MP, Arango C, Morales B, Arias B, Corcoran P, et al. Genetic polymorphisms in the dopamine-2 receptor (DRD2), dopamine-3 receptor (DRD3), and dopamine transporter (SLC6A3) genes in schizophrenia: data from an association study. Prog Neuropsychopharmacol Biol Psychiatry. 2010;34:26–31. doi:10.1016/j.pnpbp.2009.09.008.
170. Wendland JR, DeGuzman TB, McMahon F, Rudnick G, Detera-Wadleigh SD, Murphy DL. SERT Ileu425Val in autism, Asperger syndrome and obsessive-compulsive disorder. Psychiatr Genet. 2008;18:31–9. doi:10.1097/ypg.0b013e3282f08a06.
171. Liu H, Liu M, Wang Y, Wang X-M, Qiu Y, Long J-F, et al. Association of 5-HTT gene polymorphisms with migraine: a systematic review and meta-analysis. J Neurol Sci. 2011;305:57–66. doi:10.1016/j.jns.2011.03.016.
172. Igarashi K, Kashiwagi K. Use of polyamine metabolites as markers for stroke and renal failure. Methods Mol Biol. 2011;720:395–408.
173. Ortega SB, Noorbhai I, Poinsatte K, Kong X, Anderson A, Monson NL, et al. Stroke induces a rapid adaptive autoimmune response to novel neuronal antigens. Discov Med. 2015;19.381–92.
174. Mecheri S. Contribution of allergic inflammatory response to the pathogenesis of malaria disease. Biochim Biophys Acta. 1822;2012:49–56. doi:10.1016/j.bbadis.2011.02.005.
175. Correa-Fiz F, Reyes-Palomares A, Medina MA, Sanchez-Jiménez F. Roles of biogenic amines in emergent and rare diseases. In: Dandrifosse G, editor. Biological aspects of biogenic amines and polyamine conjugates. Kerala: Research Signpost; 2009. p. 339–419.
176. Reyes-Palomares A, Montañez R, Real-Chicharro A, Chniber O, Kerzazi A, Navas-Delgado I, et al. Systems biology metabolic modeling assistant: an ontology-based tool for the integration of metabolic data in kinetic modeling. Bioinformatics. 2009;25:834–5. doi:10.1093/bioinformatics/btp061.
177. Schlüter A, Real-Chicharro A, Gabaldón T, Sánchez-Jiménez F, Pujol A. PeroxisomeDB 2.0: an integrative view of the global peroxisomal metabolome. Nucleic Acids Res. 2010;38:D800–5. doi:10.1093/nar/gkp935.

178. Real-Chicharro A, Ruiz-Mostazo I, Navas-Delgado I, Kerzazi A, Chniber O, Sánchez-Jiménez F, et al. Protopia: a protein-protein interaction tool. BMC Bioinformatics. 2009;10 Suppl 1:S17. doi:10.1186/1471-2105-10-s12-s17.
179. Moya-García AA, Ranea JAG. Insights into polypharmacology from drug-domain associations. Bioinformatics. 2013;29:1934–7. doi:10.1093/bioinformatics/btt321.
180. Reyes-Palomares A, Rodríguez-López R, Ranea JAG, Sánchez Jiménez F, Medina MA. Global analysis of the human pathophenotypic similarity gene network merges disease module components. PLoS One. 2013;8, e56653. doi:10.1371/journal.pone.0056653.
181. Navas-Delgado I, Real-Chicharro A, Medina MÁ, Sánchez-Jiménez F, Aldana-Montes JF. Social pathway annotation: extensions of the systems biology metabolic modelling assistant. Brief Bioinform. 2011;12:576–87. doi:10.1093/bib/bbq061.
182. Castro-Oropeza R, Pino-Ángeles A, Khomutov MA, Urdiales JL, Moya-García AA, Vepsäläinen J, et al. Aminooxy analog of histamine is an efficient inhibitor of mammalian L-histidine decarboxylase: combined in silico and experimental evidence. Amino Acids. 2014;46:621–31. doi:10.1007/s00726-013-1589-7.
183. Sánchez-Jiménez F, Reyes-Palomares A, Moya-García AA, Ranea JAG, Medina MÁ. Biocomputational resources useful for drug discovery against compartmentalized targets. Curr Pharm Des. 2014;20:293–300. doi:10.2174/13816128113199990030.

# Chapter 8
# Histaminergic Regulation of Blood–Brain Barrier Activity

**R. Ferreira, C. Ballerini, Maria Beatrice Passani, and L. Bernardino**

**Abstract** The histaminergic signalling system is very complex and several new roles for this small amine are still being explored. Histamine is known for its peripheral immune system functions and for its regulation of the central nervous system (CNS). In the present book chapter, we will detail histaminergic signalling on central and peripheral pathophysiological processes and highlight the role of histamine on the blood–brain barrier (BBB) function, an important therapeutic target, albeit a common obstacle, for the treatment of cerebrovascular diseases. Additionally, the modulation of BBB properties by histamine may constitute a means to deliver drugs more efficiently to the diseased brain tissue. In particular, we will discuss the role of histamine on key elements of the healthy and disrupted BBB, namely, brain endothelial cells, astrocytes, microglia, pericytes and smooth muscle cells. A deeper insight on the mechanisms underlying brain pathology will be a determinant to develop the most efficient therapeutic platforms targeting the histaminergic system for BBB repair and ultimately to CNS-related diseases.

**Keywords** Histamine • Receptors • Blood–brain barrier • Neurodegenerative diseases • Endothelial cells • Glial cells

---

R. Ferreira, Ph.D. • L. Bernardino, Ph.D. (✉)
Health Sciences Research Centre, Faculty of Health Sciences, University of Beira Interior, Av. Infante D. Henrique, Covilhã 6200-506, Portugal
e-mail: raquelmargarida@gmail.com; libernardino@gmail.com

C. Ballerini, Ph.D.
Dipartimento di Neuroscienze, Psicologia, Area del Farmaco e Salute del Bambino-Universitá di Firenze, Viale Pieraccini 6, Florence 50139, Italy
e-mail: clara.ballerini@unifi.it

M.B. Passani, Ph.D.
Department of Health Science, University of Florence, Florence, Italy
e-mail: beatrice.passani@unifi.it

 
P. Blandina, M.B. Passani (eds.), *Histamine Receptors*, The Receptors 28,
DOI 10.1007/978-3-319-40308-3_8

## 8.1 An Overview of the Histaminergic System in the Brain and Periphery

The physiological role of β-imidazolethylamine was first described by Dale and Laidlaw in the early 1900s, namely, in muscle contraction and vasodilation [1]. This biogenic amine naturally occurring in tissues (*histos* in Greek) became addressed as histamine. Mainly known for its role in allergic inflammatory responses and in the regulation of gastric acid secretion, histamine was later found in the central nervous system (CNS) [2, 3]. The existence of histaminergic neurons was unequivocally shown in the brain in the early 1980s [4, 5]. In the CNS, histamine functions as a neurotransmitter that modulates vital physiological and neurological processes such as thermoregulation, circadian rhythm, mood, appetite, learning and memory [6]. For instance, rats with reduced levels of histamine or lacking histamine receptors have a significantly decreased cognitive and behavioural function [7]. Changes in the brain histaminergic system in post-mortem specimens of patients with neuropsychiatric disorders such as Parkinson's disease (PD), Alzheimer's disease (AD), Huntington's disease (HD), depression and narcolepsy have been detected [8]. In the CNS, histamine is released by histaminergic neurons that project fibres from the tuberomammillary nucleus (TMN), located in the hypothalamus [9], and by mast cells, which accounts for approximately half of non-neuronal histamine content. To the best of our knowledge, there are only two research articles claiming histamine production by an additional cellular type in the brain, namely, microglia [10, 11]. Nevertheless, these cells are responsive to histamine and express all of its receptors, playing a role in cell migration and inflammatory responses [12, 13]. Histamine is produced by decarboxylation of L-histidine by histidine decarboxylase (HDC) activity [14]. HDC expression is tightly controlled in peripheral tissues but little is known about the regulation of its expression in the brain [15]. Histamine is then released into the extracellular space and methylated by neuronal histamine N-methyltransferase (HMT) into tele-methylhistamine, hence inactivating histamine neurotransmission [16, 17]. Tele-methylhistamine finally undergoes oxidative deamination via monoamine oxidase-B to t-methyl-imidazole acetic acid. In peripheral tissues and invertebrates, histamine is alternatively degraded by diamine oxidase (DAO) originating imidazole acetic acid [18].

The majority of peripheral histamine pools are stored in cytosolic granules of mast cells but they can also be found in a wide range of other circulating immune cells, including basophils and eosinophils and to a lesser extent in neutrophils, macrophages, natural killer (NK) cells and T cells. In turn, histamine modulates their activity [19–21]. Interestingly, macrophages can remove extracellular histamine as a mechanism of elimination of a noxious agent [22].

## 8.2 Histamine Receptors' Signalling and Function

Histamine exerts its actions through four G protein-coupled receptor subtypes, namely, $H_1R$, $H_2R$, $H_3R$ and $H_4R$ [23–26] (Fig. 8.1). $H_1R$ operates via coupling to $G\alpha_{q/11}$ proteins which trigger phospholipase C activation, inositol phosphate production and

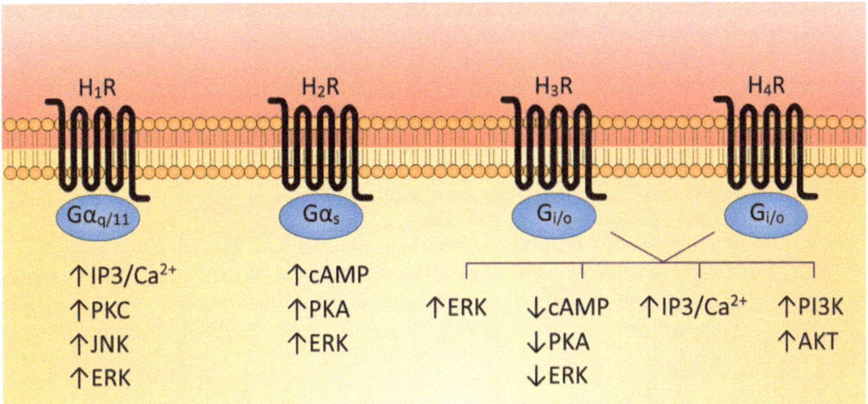


**Fig. 8.1** Histaminergic signalling cascade. Histamine acts via four G protein-coupled receptor subtypes, namely, histamine receptors 1, 2, 3 and 4 ($H_1R$, $H_2R$, $H_3R$ and $H_4R$, respectively). Depending on the receptor subtype, downstream signalling may activate different secondary messenger and effector molecules. For instance, activation of inositol trisphosphate (IP3) and intracellular calcium ($Ca^{2+}$) mobilization can trigger apoptosis, cell growth, migration and differentiation. These same cellular processes can be elicited by Akt (or protein kinase B) or by phosphatidylinositol-4,5-bisphosphate 3-kinase (PI3K). The activation of c-Jun N-terminal kinases (JNK) and extracellular signal-regulated kinases (ERK) is mainly associated to inflammatory processes and cell growth, respectively. Protein kinase C (PKC) activation commonly promotes vasoconstriction, while signal transduction molecule cAMP and protein kinase A (PKA), a class of cAMP-dependent enzymes, are involved in the regulation of basic metabolic processes. Artwork provided by Tiago Santos

calcium ($Ca^{2+}$) mobilization, a series of steps implicated in several cellular functions such as apoptosis, cell growth, migration and differentiation. $H_1R$ activation may also stimulate cyclic adenosine monophosphate (cAMP) production in the brain [27, 28]. Due to its role in immune reactions, $H_1R$ antagonists have become popular therapeutic agents and are commonly used drugs (e.g. carbinoxamine, diphenhydramine, brompheniramine). $H_1R$ is found in diverse peripheral tissues such as the bronchi, intestine and blood vessels, where its activation produces bronchoconstriction, intestinal motility and vasodilation. In the CNS, $H_1R$ activation is responsible for effects such as motion sickness, and its blockade of weight gain and sleepiness. In accordance, $H_1R$ knockout mice have disturbed diurnal feeding rhythms [29], impaired locomotor activity and exploratory behaviour [30]. In the brain, $H_1R$ is expressed by neurons and glial and endothelial cells.

$H_2R$ operates via $G\alpha_s$ subunit and has a pattern of expression similar to $H_1R$ emphasizing the concept of synergistic action between these two receptors. $H_2R$ is expressed by various cell types. In gastric parietal cells in the stomach, $H_2R$ stimulates the release of gastric acid [31]. $H_2R$ is also found on cells of the immune response [32] and has a dual impact in the immune system since it can induce cytokine production and accumulation or exert antioxidant properties by inhibiting ROS production in neutrophils or macrophages [33]. This latter protective effect has been observed in myeloid cells of leukaemia patients where $H_2R$ activation also induces NK and T-cell survival [34]. Also, it has been recently demonstrated that $H_2R$ activation influences the innate intestinal immune response to microbes [35, 36]. The development of $H_2R$ knockout mice allowed a better understanding on the involvement of this receptor in several pathologies, including experimental autoimmune encephalopathy (EAE), a model for multiple sclerosis (MS). $H_2R$ knockout mice are less susceptible to acute early phase EAE since this receptor appears to regulate antigen-presenting cell function [37]. $H_2R$ activation in the hippocampus improves memory for aversive events [38, 39], further attesting the role of this receptor in the CNS [6]. Early studies have also shown that the $H_2R$ is expressed by endothelial cells [27, 40, 41] and regulates blood–brain barrier (BBB) permeability [42, 43].

$H_3R$ acts via coupling to $G_{i/o}$ proteins and inhibits cAMP production while alternatively increasing $Ca^{2+}$ mobilization via other signalling mechanisms. This receptor is the most abundant in the CNS where it is constitutively expressed by TMN neurons. $H_3R$ is not normally expressed by haematopoietic cells; it is mostly confined to the CNS where it limits histamine synthesis and release [44] and regulates neurotransmitter release [45]. It modulates neurotransmission since it can act as an inhibitory autoreceptor or heteroreceptor controlling the release of several other neurotransmitters such as acetylcholine [46], glutamate [47] and gamma-aminobutyric acid [48]. These features generated great expectation for the therapeutic potentials of $H_3R$ ligands in the treatment of neurodegenerative diseases and sleep disorders [45, 49, 50]. An $H_3R$ antagonist (FUB181) blocks presynaptic autoreceptors and increases histamine release; this boost in histamine content can significantly improve cognitive function [50, 51]. Pitolisant, another selective non-imidazole histamine $H_3R$ inverse agonist that activates histaminergic neurons, has been approved by the European Medicines Agency (EMA) in patients with narcolepsy. $H_3R$ is also expressed by brain endothelial cells and possibly it may modulate BBB permeability [41]. In this regard, $H_3R$ deletion leads to more severe EAE, an effect associated with altered BBB permeability and an unexpected increased expression of chemokines/chemokine receptors that promote CNS entry of peripheral T cells that do not express $H_3R$ themselves [52]. Several molecules targeting $H_3R$ (immepip, clobenpropit and imidazole derivatives) have shown their ability to cross the BBB in vivo [53, 54].

$H_4R$, a homologue of $H_3R$, operates via coupling to $G_{i/o}$ proteins. Activation of $H_4R$ causes the rise of intracellular calcium, activation of both PI3K/AKT pathways and decrease of cAMP production and PKA activation (Fig. 8.1). It is highly expressed in cells involved in immune and inflammatory responses such as eosinophils, basophils, T cells, dendritic cells and mast cells. Activation of $H_4R$ plays a relevant role in chemotaxis by affecting cytoskeleton organization [55, 56]. In the CNS, $H_4R$ is mainly found in the spinal cord, hippocampus and cerebral cortex [57–59]. At the cellular level, some of us recently detected $H_4R$ expression in microglia [13]. We demonstrated that $H_4R$ activation stimulated microglia motility upon treatment with histamine or histamine-loaded microparticles. Importantly, histamine or an $H_4R$ agonist inhibited LPS-induced IL-1β release, suggesting an anti-inflammatory role for $H_4R$. Of note, histamine-loaded microparticles had a 200-fold increased efficiency compared to free histamine. Recent data have now emphasized the role of $H_4R$ in inflammatory processes. Treatment with an $H_4R$ antagonist, JNJ7777120, improved symptoms in animal models of rheumatoid arthritis [60] and prevented cytokine release and consequently inflammation. This effect can be achieved by inhibiting the signalling cascade induced by toll-like receptors on the surface of dendritic cells (TLR2, TLR4, TLR7, TLR9, depending on the dendritic cell subtype) leading to the inhibition of pro-inflammatory factors like tumour necrosis factor-α (TNF-α) [60–63]. $H_4R$ has also been linked to the modulation of ROS production, albeit using elevated concentrations of receptor antagonist [64]. Unexpectedly though, $H_4R$ antagonism is detrimental for EAE as it increased inflammation and demyelination in the spinal cord of EAE mice and increased IFN-γ expression in lymph nodes [65]. Accordingly, $H_4R$ knockout mice develop a more severe EAE compared to wild-type mice [66].

In addition to these metabotropic receptors, there are two histamine-gated chloride channels (HisCl α1 and α2), which are expressed in invertebrates [67, 68]. The existence of ionotropic histamine receptors has also been described in oxytocin neurons in the mouse supraoptic nucleus [69], but these findings remain to be further substantiated.

## 8.3 Histamine Actions on BBB Properties

Histamine has a pivotal role on vascular function by regulating endothelial cell activation, blood vessel diameter and blood flow [70]. These properties are very important for the maintenance of the BBB, whose disruption is a critical event in the pathogenesis of several brain disorders such as AD, MS and ischemic stroke [71–75]. In broad terms, the BBB consists of endothelial cells connected by close-fitting junctions that separate the flowing blood from the brain extracellular fluid, thus controlling the entrance of biomolecules into the brain and protecting it against pathological microorganisms and/or particles [76] (Fig. 8.2). Likewise, a major problem in treating CNS-related diseases is the difficulty in delivering drugs across the BBB or even to treat/repair the BBB without causing collateral damage to the brain parenchyma [77]. As such, there are several lines of research and several clinical trials targeting BBB disruption that occurs in neurodegenerative disorders and inflammation-related diseases in the brain. Common processes shared by these pathologies include increased BBB permeability caused by decreased tight junction protein expression, induction of proteases that degrade the extracellular matrix, formation of ROS and leucocyte extravasation. In the affected area, apoptotic pathways and induction of pro-inflammatory cytokines and chemokines further intensify leucocyte recruitment and extravasation and accumulation of free radicals [76]. For instance, in stroke, tissue reperfusion can further activate the endothelium, inducing excessive production of oxygen free radicals and cytokines, leucocyte recruitment and oedema formation (reperfusion injury) [78]. This continuous cycle of tissue damage accelerates disease progression. However it is still debatable whether BBB disruption is a cause or a consequence of CNS injury/disorder. A better understanding of the cascading events that occur during BBB disruption will allow the development of more effective therapeutic tools against these effects. Since histamine can regulate BBB endothelial permeability, it is of great interest to explore the application of histamine receptor

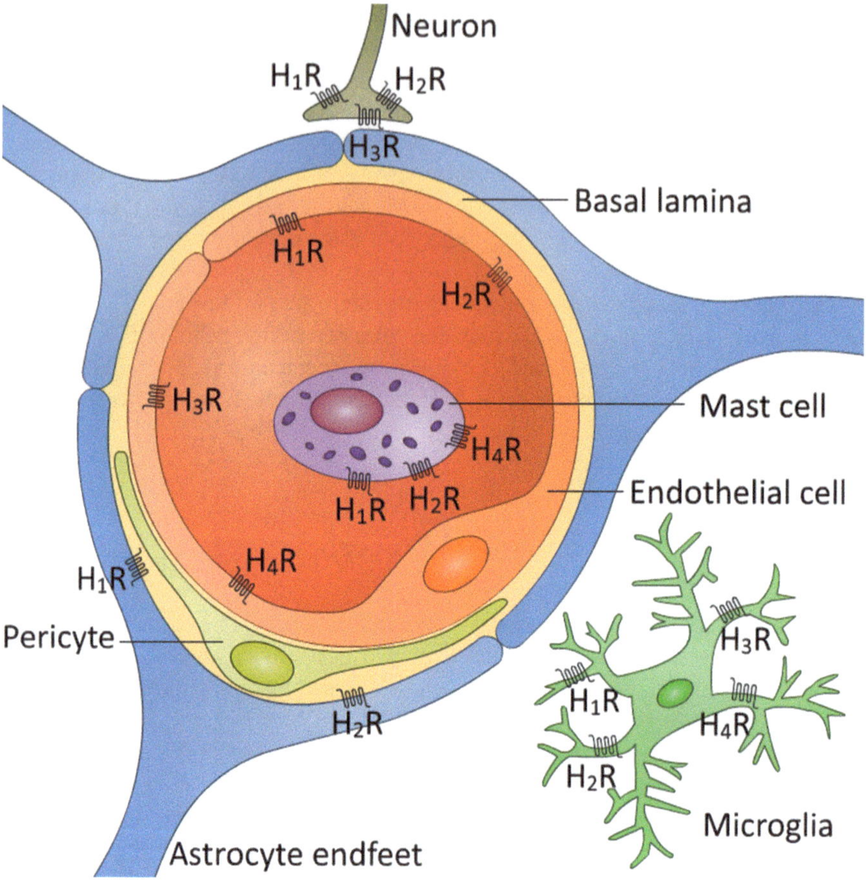


**Fig. 8.2** Histamine receptor distribution in the blood–brain barrier (BBB). The BBB comprises tightly connected endothelial cells that are covered by pericytes and a basal lamina, and a protecting glial barrier composed by astrocytes and microglia. This vascular and glial complex is in close contact with neurons. All of these cellular elements express histamine receptors and may release histamine. Peripheral histamine can also reach and regulate BBB activity/properties via circulating blood cells such as mast cells that in turn express histamine receptors. Artwork provided by Tiago Santos

agonists/inverse agonists/antagonists alone or in combination with drugs that cannot cross the BBB to treat CNS disorders or to recuperate BBB integrity [79, 80]. The therapeutic relevance of these molecules will depend on how well characterized is the pathophysiological state of endothelial cells. Histamine appears to have opposite roles depending on whether it acts on resting endothelial cells or on cells preactivated by other inflammatory stimuli. For instance, histamine has an inhibitory role in platelet–endothelium interaction when acting on cultured human vascular endothelial cells previously activated by TNF-α, whereas it favours platelet adhesion in nonstimulated endothelial cells [27, 81]. This observation may have clinical relevance for treating platelet dysfunction occurring in high-risk surgeries or during the application of thrombolytic drugs and anticoagulants such as aspirin. Furthermore, in in vitro experiments using human umbilical vein endothelial cells, histamine appears to up-regulate the expression of TLR2 and TLR4 and to amplify endothelial cell inflammatory responses to Gram-negative and Gram-positive bacteria [27, 81]. Interestingly, in human monocyte-derived dendritic cells (mDC), histamine did not affect TLR2 and TLR4 expression (Aldinucci et al., J Biol Chem, in press), further suggesting the heterogeneity of responses induced by histamine and the histaminergic receptor(s) involved.

### *8.3.1 Histamine Effects on Endothelial Cells*

In a physiological state, histamine binds to endothelial $H_1R$ and activates nitric oxide synthase (eNOS) expression triggering nitric oxide (NO) production and vasodilation [82, 83]. Blood vessel dilation prevents leucocyte and platelet adhesion; hence, it may have a protective effect [84]. However, this process occurs in a short period of time, and its effectiveness is not remarkable when facing a pathological context. In atherosclerosis, histamine has the opposite effect by increasing the expression of adhesion molecules in endothelial cells and therefore increases the interaction between endothelial cells and leucocytes. In this context, eNOS expression may also become dysfunctional and lead to the generation of superoxide rather than NO [82]. Histamine receptors can also modulate blood flow in ischaemic brain injury [85, 86]. In an animal model of brain ischaemia, treatment with an $H_2R$ antagonist (ranitidine) maintained normal permeability of BBB and prevented cerebral oedema [87]. Several other antagonists targeting $H_1R$ (terfenadine, chlorpheniramine, oxatomide or triprolidine), $H_2R$ (cimetidine) and $H_3R$ (thioperamide or clobenpropit) have since been tested to prevent cerebral oedema associated to stroke [86]. These receptors are implicated in the reduction of the infarct volume and neuronal damage by reducing neurotransmitter release (glutamine and dopamine) and oxidative stress and encouraging autophagy. $H_1R$ promotes glutamate transporter GLT-1 activity and glutamine synthetase expression, while $H_2R$ inhibits glial scar formation and inflammation. Both receptors stimulate neurogenesis. Additionally, simultaneous administration of an $H_1R$ antagonist (diphenhydramine) with L-histidine reduced the infarct size in rats subjected to middle cerebral artery occlusion (MCAO), an animal model of stroke [88]. In the ischaemic brain, histaminergic neurons and mast cells release histamine, and several studies claim that this promotes neuronal recovery and neurogenesis, by decreasing infarct volume and excitotoxicity via the suppression of glutamate and dopamine release. We have previously demonstrated that histamine acting via $H_1R$ is responsible for promoting neuronal differentiation in the subventricular zone (SVZ), one of the main neurogenic niches of the adult mammalian brain [89]. Histamine maintains neural stem cell proliferation via $H_2R$ activation [90]. In our work we also tested the application of histamine-loaded microparticles that offered a 2000-fold increased efficiency compared to free histamine. The key advantage in using these microparticles consists on them being docked on the cell surface and potentiating a localized effect with increased target specificity. Additionally, this effect on immature cells can have an enormous potential in the development of a multirestorative treatment for neurodegenerative diseases whose recovery also depends on the efficiency of the neurogenic process.

### 8.3.2 *Histamine Effects on Astrocytes and Microglial Cells*

In the BBB, the endothelium is protected by a supporting network of pericytes and a glial barrier formed by astrocytes and microglia [76]. Accordingly, these cells affect endothelial activity and physiology and can have a strong impact on BBB overall integrity. Human astrocytes are known to promote histamine clearance and inactivation as a means to maintain normal histaminergic neuronal activity [91]. These cells are in fact highly responsive to histamine, which is known to induce the secretion of nerve growth factor (NGF) by astrocytes [92, 93]. Astrocyte-derived secretion of trophic factors increases upon brain insult serving as a means to counteract an immune response by infiltrating cells [94]. Microglial cells are highly sensitive to their surrounding environment in order to quickly mount an immune response. Microglia are a heterogeneous population of cells and, as such, only a small subset responds to histamine [95]. Additionally, neonatal and adult primary cultures show a different pattern of response to an extensive list of neurotransmitters, including histamine, when primed by different stimuli [95]. Work performed with primary microglia cultures from neonatal rats has shown that 0.1 μM histamine induces the release of TNF-α and IL-6 via $H_1R$ and $H_4R$ activation [12], while others, using primary mouse microglia, have shown that 10 μM histamine, acting via $H_3R$, inhibits chemotaxis, phagocytosis and cytokine secretion [11]. These seemingly opposing reports use significantly different concentrations of histamine from what we used in a murine cell line (100 μM histamine) [13]. The response to histamine may also vary due to species variability (rat vs. mouse), cell origin (primary cultures vs. cell line), parameters under analysis (e.g. inflammatory mediator release, migration and phagocytosis) and ligand affinity since different receptors are activated depending on the concentration range of histamine, receptor antagonists, agonists or partial agonists that is tested. In amyotrophic lateral sclerosis (ALS), histamine stimulates microglia to produce NO and ROS and the release of pro-inflammatory factors TNF-α and IL-6 by binding to both $H_1R$ and $H_4R$ [96]. Therefore, in this sense, histamine-activated microglia contributes to the severity of ALS suggesting yet again that antihistamines may be used to treat neurodegenerative diseases and neuroinflammation. However, it is of paramount importance to know which cell type is being targeted. An $H_1R$ antagonist (clemastine) has been tested in ALS disease progression and was shown to reduce both macrophage and microglia activation and consequently boost an anti-inflammatory response [97]. In these patients there is also significant mast cell activation, detected in the cerebrospinal fluid. In MS patients, another study has found decreased serum levels of histamine-degrading DAO and elevated levels of histamine [98]. These data further support the role of histamine as a BBB permeabilizer that causes immune cell infiltration and inflammation in MS, although the histaminergic signalling pathways used remain unclear. However, in EAE, $H_1R$ overexpression in endothelial cells controlled leucocyte migration and extravasation to tissues and reduced susceptibility to the disease [40, 99]. In another study, authors found that $H_1R$ activation decreased cytokine production and endothelial adhesiveness leading to a less severe EAE [99]. However, a conflicting report has described that an $H_1R$ antagonist leads to a decrease in disease severity [100]. There is also controversy regarding $H_2R$ activation in EAE with different authors showing either an amelioration or deterioration of this pathology [101, 102]. A recent report has however hinted at the importance of a combined pharmacological targeting $H_1R/H_2R$ and $H_3R/H_4R$. The authors developed double-knockout mice and showed that the concomitant absence of $H_1R/H_2R$ leads to a less severe clinical disease course, while the opposite effect was observed in $H_3R/H_4R$ knockout mice [103]. Of note, interrupting $H_1R/H_2R$ signalling caused less disrupted BBB permeability. Using $H_2R$ knockout mice and a pharmacological approach with receptor antagonists and agonists, Luo and colleagues demonstrated that $H_2R$ activation elicits an inflammatory response that causes F-actin rearrangement, increased permeability and neutrophil infiltration [104]. These events were shown to amplify myocardial ischaemia/reperfusion injury and indicate a potential use of $H_2R$ antagonists in diseases with impaired BBB integrity. Altogether, in a pathological context, increased BBB permeability caused by increased histamine levels allows cells of the peripheral immune system to become activated and cross the BBB contributing to brain oedema and inflammation [105].

### *8.3.3 Histamine Effects on Pericytes and Smooth Muscle Cells*

Pericytes are contractile and phagocytic cells that partially cover and modulate endothelial cell activity [106]. Few reports exist on a modulatory role of histamine on pericyte activity, but it is known that $H_1R$ activation increases pericyte contraction, therefore reducing the area of contact with blood vessels [107]. Additionally, pericytes form a more cohesive coverage of the endothelium when responding to histamine-induced inflammation, probably to counteract an increase in permeability [108]. In larger brain blood vessels, pericytes are replaced by a layer of smooth muscle cells (SMC) [109]. Histamine stimulates SMC proliferation [110, 111] and can also trigger inducible nitric oxide synthase (iNOS) expression, and consequently NO, via $H_1R$ activation [112]. These events can increase BBB permeability.

## 8.4 Indirect Regulation of Histamine Functions

There are also molecules of potential vascular interest that modulate histamine activity. Relaxin, a small hormone mainly associated to blood flow regulation during pregnancy, was shown to inhibit histamine release from mast cells and basophils and to reduce oxidative stress in a swine model of ischaemia/reperfusion myocardial injury and dysfunction [113]. Relaxin could be used peripherally to halt BBB disruption, while other BBB-permeable drugs could be co-administered to act centrally on neuronal repair. Recently, relaxin was shown to cross the BBB and to reduce hypertension occurring in cerebral small vessel disease, a risk factor for stroke [114]. In this sense, an enzyme that could have an interesting application on the disrupted BBB is DAO, which can be extracted from plants (histaminase). For instance, when given intravenously, histaminase can significantly reduce leucocyte infiltration, ROS production and endothelial activation in an in vivo model of intestinal ischaemia [115]. Moreover, a similar protective effect was observed on an in vivo model of cardiac ischaemia/reperfusion [116]. The putative therapeutic application of histaminase extends to cardiac anaphylaxis and allergic asthma-like reaction [117]. Other agents of interest include molecules that indirectly affect histamine actions, such as endothelin. This vasoconstrictor increases blood pressure at the BBB by inhibiting cAMP accumulation induced by histamine in the brain [118].

## 8.5 Future Prospects

Histamine is a simple molecule with a complex role acting as an inflammatory mediator and regulator of a series of neurological processes. The BBB represents one of the most important structures for maintaining CNS homeostasis and integrity, but it is also an obstacle to the passage of therapeutic agents, therefore impairing the treatment of several brain diseases. Clearly there are two physically separated systems where histamine exerts its actions, but it can also represent a common and useful link between the two. In this sense, histaminergic modulation of the BBB may represent an efficient approach to effectively treat CNS disorders where this structure is compromised or a means to treat it to control further damage. In this regard, histamine- or histamine receptor ligand-loaded microparticles offer several advantages by promoting the gradual release of a lower concentration of the encapsulated molecule. The functionalization of these formulations will allow increased target specificity. For instance, it would be interesting to modulate BBB permeability in PD patients to facilitate drug delivery without affecting the substantia nigra and striatum with histamine that may induce dopaminergic cell death [119]. On the other hand, this approach would help to restore BBB permeability by restraining inflammatory events occurring in stroke and MS associated with BBB disruption and would dampen neuronal damage. However, tests in animal models have delivered conflicting results, and researchers have not settled on a definite effect for a given antihistamine. Understanding the cellular and molecular mechanisms underlying pathology will clearly bring us a better understanding on which histamine target(s) to choose.

In the last few years, there were some clinical trials focusing on the application of antihistamines. For instance, famotidine ($H_2R$ antagonist) was tested on levodopa-induced dyskinesia in PD patients (phase II), and the effect of GSK239512 ($H_3R$ antagonist) was evaluated on mild-to-moderate AD patients (phase II). Another clinical trial measured the basal activity of $H_3R$ in the brain of AD patients (phase 0). No study results were posted. One clinical trial is currently recruiting healthy volunteers to test the effect of hydroxyzine, a strong $H_1R$ receptor inverse agonist, on cognition. Overall, there is a great interest in developing therapy-relevant research focused on the histaminergic system for BBB repair so that we can envision the development of a more efficient and comprehensive treatment of CNS-related diseases.

**Acknowledgement** The authors acknowledge Tiago Santos for his help with the artwork. This work was funded, in part, by SFRH/BPD/94228/2013 (RF), from Fundação para a Ciência e a Tecnologia (FCT) and FISM project 2010/R27. This work was also supported by FEDER funds through the POCI - COMPETE 2020 - Operational Programme Competitiveness and Internationalization in Axis I - Strengthening research, technological development and innovation (Project No.007491) and National Funds by FCT (Project UID/Multi /00709).

## References

1. Dale HH, Laidlaw PP. The physiological action of beta-iminazolylethylamine. J Physiol. 1910;41:318–44.
2. Kwiatkowski H. Histamine in nervous tissue. J Physiol. 1943;102:32–41.
3. Schwartz JC, et al. [Development of histaminergic systems in the newborn rat brain]. J Physiol Paris. 1970;62 Suppl 3:447.
4. Watanabe T, et al. Evidence for the presence of a histaminergic neuron system in the rat brain: an immunohistochemical analysis. Neurosci Lett. 1983;39:249–54.
5. Panula P, Yang HY, Costa E. Histamine-containing neurons in the rat hypothalamus. Proc Natl Acad Sci U S A. 1984;81:2572–6.
6. Haas HL, Sergeeva OA, Selbach O. Histamine in the nervous system. Physiol Rev. 2008;88:1183–241. doi:10.1152/physrev.00043.2007.
7. Huston JP, Wagner U, Hasenohrl RU. The tuberomammillary nucleus projections in the control of learning, memory and reinforcement processes: evidence for an inhibitory role. Behav Brain Res. 1997;83:97–105.
8. Shan L, Bao AM, Swaab DF. The human histaminergic system in neuropsychiatric disorders. Trends Neurosci. 2015;38:167–77. doi:10.1016/j.tins.2014.12.008.
9. Haas H, Panula P. The role of histamine and the tuberomamillary nucleus in the nervous system. Nat Rev Neurosci. 2003;4:121–30. doi:10.1038/nrn1034.
10. Katoh Y, et al. Histamine production by cultured microglial cells of the mouse. Neurosci Lett. 2001;305:181–4.
11. Iida T, et al. Histamine H3 receptor in primary mouse microglia inhibits chemotaxis, phagocytosis, and cytokine secretion. Glia. 2015;63:1213–25.
12. Dong H, et al. Histamine induces upregulated expression of histamine receptors and increases release of inflammatory mediators from microglia. Mol Neurobiol. 2014;49:1487–500.
13. Ferreira R, et al. Histamine modulates microglia function. J Neuroinflammation. 2012;9:90. doi:10.1186/1742-2094-9-90.
14. White T. Formation and catabolism of histamine in brain tissue in vitro. J Physiol. 1959;149:34–42.
15. Karlstedt K, Nissinen M, Michelsen KA, Panula P. Multiple sites of L-histidine decarboxylase expression in mouse suggest novel developmental functions for histamine. Dev Dyn. 2001;221:81–91.
16. Bowsher RR, Verburg KM, Henry DP. Rat histamine N-methyltransferase. Quantification, tissue distribution, purification, and immunologic properties. J Biol Chem. 1983;258:12215–20.

17. Matuszewska B, Borchardt RT. Guinea pig brain histamine N-methyltransferase: purification and partial characterization. J Neurochem. 1983;41:113–8.
18. Mondovi B, et al. Effects of amine oxidases in allergic and histamine-mediated conditions. Recent Pat Inflamm Allergy Drug Discov. 2013;7:20–34.
19. Hellstrand K, Asea A, Dahlgren C, Hermodsson S. Histaminergic regulation of NK cells. Role of monocyte-derived reactive oxygen metabolites. J Immunol. 1994;153:4940–7.
20. Khan MM, Keaney KM, Melmon KL, Clayberger C, Krensky AM. Histamine regulates the generation of human cytolytic T lymphocytes. Cell Immunol. 1989;121:60–73.
21. Takahashi H, et al. Histamine inhibits high mobility group box 1-induced adhesion molecule expression on human monocytes. Eur J Pharmacol. 2013;718:305–13.
22. Tanaka S, Deai K, Inagaki M, Ichikawa A. Uptake of histamine by mouse peritoneal macrophages and a macrophage cell line, RAW264.7. Am J Physiol Cell Physiol. 2003;285:C592–8. doi:10.1152/ajpcell.00470.2002.
23. Gantz I, et al. Molecular cloning of a gene encoding the histamine H2 receptor. Proc Natl Acad Sci U S A. 1991;88:429–33.
24. Yamashita M, et al. Expression cloning of a cDNA encoding the bovine histamine H1 receptor. Proc Natl Acad Sci U S A. 1991;88:11515–9.
25. Lovenberg TW, et al. Cloning and functional expression of the human histamine H3 receptor. Mol Pharmacol. 1999;55:1101–7.
26. Nakamura T, Itadani H, Hidaka Y, Ohta M, Tanaka K. Molecular cloning and characterization of a new human histamine receptor, HH4R. Biochem Biophys Res Commun. 2000;279:615–20.
27. Talreja J, Kabir MH, B Filla M, Stechschulte DJ, Dileepan KN. Histamine induces Toll-like receptor 2 and 4 expression in endothelial cells and enhances sensitivity to Gram-positive and Gram-negative bacterial cell wall components. Immunology. 2004;113:224–33.
28. Oda T, Morikawa N, Saito Y, Masuho Y, Matsumoto S. Molecular cloning and characterization of a novel type of histamine receptor preferentially expressed in leukocytes. J Biol Chem. 2000;275:36781–6.
29. Masaki T, Yoshimatsu H. The hypothalamic H1 receptor: a novel therapeutic target for disrupting diurnal feeding rhythm and obesity. Trends Pharmacol Sci. 2006;27:279–84.
30. Inoue I, et al. Impaired locomotor activity and exploratory behavior in mice lacking histamine H1 receptors. Proc Natl Acad Sci U S A. 1996;93:13316–20.
31. Kobayashi T, et al. Abnormal functional and morphological regulation of the gastric mucosa in histamine H2 receptor-deficient mice. J Clin Invest. 2000;105:1741–9.
32. Weltman JK. Histamine as a regulator of allergic and asthmatic inflammation. Allergy Asthma Proc. 2003;24:227–9.
33. Betten A, Dahlgren C, Hermodsson S, Hellstrand K. Histamine inhibits neutrophil NADPH oxidase activity triggered by the lipoxin A4 receptor-specific peptide agonist Trp-Lys-Tyr-Met-Val-Met. Scand J Immunol. 2003;58:321–6.
34. Werner K, Neumann D, Seifert R. Analysis of the histamine H2-receptor in human monocytes. Biochem Pharmacol. 2014;92:369–79. doi:10.1016/j.bcp.2014.08.028.
35. Ferstl R, et al. Histamine receptor 2 is a key influence in immune responses to intestinal histamine-secreting microbes. J Allergy Clin Immunol. 2014;134:744–6.e743.
36. Frei R, et al. Histamine receptor 2 modifies dendritic cell responses to microbial ligands. J Allergy Clin Immunol. 2013;132:194–204.
37. Teuscher C, et al. Attenuation of Th1 effector cell responses and susceptibility to experimental allergic encephalomyelitis in histamine H2 receptor knockout mice is due to dysregulation of cytokine production by antigen-presenting cells. Am J Pathol. 2004;164:883–92.
38. Giovannini MG, et al. Improvement in fear memory by histamine-elicited ERK2 activation in hippocampal CA3 cells. J Neurosci. 2003;23:9016–23.
39. da Silva WC, Bonini JS, Bevilaqua LR, Izquierdo I, Cammarota M. Histamine enhances inhibitory avoidance memory consolidation through a H2 receptor-dependent mechanism. Neurobiol Learn Mem. 2006;86:100–6.
40. Vo TS, Kim SK. Down-regulation of histamine-induced endothelial cell activation as potential anti-atherosclerotic activity of peptides from Spirulina maxima. Eur J Pharm Sci. 2013;50:198–207.

41. Karlstedt K, Jin C, Panula P. Expression of histamine receptor genes Hrh3 and Hrh4 in rat brain endothelial cells. Br J Pharmacol. 2013;170:58–66.
42. Boertje SB, Le Beau D, Williams C. Blockade of histamine-stimulated alterations in cerebrovascular permeability by the H2-receptor antagonist cimetidine. Neuropharmacology. 1989;28:749–52.
43. Butt AM, Jones HC. Effect of histamine and antagonists on electrical resistance across the blood-brain barrier in rat brain-surface microvessels. Brain Res. 1992;569:100–5.
44. Arrang JM, Garbarg M, Schwartz JC. Auto-inhibition of brain histamine release mediated by a novel class (H3) of histamine receptor. Nature. 1983;302:832–7.
45. Passani MB, Blandina P. Histamine receptors in the CNS as targets for therapeutic intervention. Trends Pharmacol Sci. 2011;32:242–9.
46. Giannoni P, et al. Regional differential effects of the novel histamine H3 receptor antagonist 6-[(3-cyclobutyl-2,3,4,5-tetrahydro-1H-3-benzazepin-7-yl)oxy]-N-methyl-3-pyridine carboxamide hydrochloride (GSK189254) on histamine release in the central nervous system of freely moving rats. J Pharmacol Exp Ther. 2010;332:164–72.
47. Molina-Hernandez A, Nunez A, Sierra JJ, Arias-Montano JA. Histamine H3 receptor activation inhibits glutamate release from rat striatal synaptosomes. Neuropharmacology. 2001;41:928–34.
48. Yamamoto Y, Mochizuki T, Okakura-Mochizuki K, Uno A, Yamatodani A. Thioperamide, a histamine H3 receptor antagonist, increases GABA release from the rat hypothalamus. Methods Find Exp Clin Pharmacol. 1997;19:289–98.
49. Parmentier R, et al. The brain H3-receptor as a novel therapeutic target for vigilance and sleep-wake disorders. Biochem Pharmacol. 2007;73:1157–71.
50. Sander K, Kottke T, Stark H. Histamine H3 receptor antagonists go to clinics. Biol Pharm Bull. 2008;31:2163–81.
51. Onodera K, Miyazaki S, Imaizumi M, Stark H, Schunack W. Improvement by FUB 181, a novel histamine H3-receptor antagonist, of learning and memory in the elevated plus-maze test in mice. Naunyn-Schmiedebergs Arch Pharmacol. 1998;357:508–13.
52. Teuscher C, et al. Central histamine H3 receptor signaling negatively regulates susceptibility to autoimmune inflammatory disease of the CNS. Proc Natl Acad Sci U S A. 2007;104:10146–51.
53. Jansen FP, Mochizuki T, Yamamoto Y, Timmerman H, Yamatodani A. In vivo modulation of rat hypothalamic histamine release by the histamine H3 receptor ligands, immepip and clobenpropit. Effects of intrahypothalamic and peripheral application. Eur J Pharmacol. 1998;362:149–55.
54. Wiecek M, et al. N-Alkenyl and cycloalkyl carbamates as dual acting histamine H3 and H4 receptor ligands. Bioorg Med Chem. 2011;19:2850–8.
55. Damaj BB, Becerra CB, Esber HJ, Wen Y, Maghazachi AA. Functional expression of H4 histamine receptor in human natural killer cells, monocytes, and dendritic cells. J Immunol. 2007;179:7907–15.
56. Liu C, et al. Cloning and pharmacological characterization of a fourth histamine receptor (H(4)) expressed in bone marrow. Mol Pharmacol. 2001;59:420–6.
57. Leurs R, Chazot PL, Shenton FC, Lim HD, de Esch IJ. Molecular and biochemical pharmacology of the histamine H4 receptor. Br J Pharmacol. 2009;157:14–23.
58. Strakhova MI, et al. Localization of histamine H4 receptors in the central nervous system of human and rat. Brain Res. 2009;1250:41–8.
59. Connelly WM, et al. The histamine H4 receptor is functionally expressed on neurons in the mammalian CNS. Br J Pharmacol. 2009;157:55–63.
60. Ahmad SF, et al. Regulation of TNF-alpha and NF-kappaB activation through the JAK/STAT signaling pathway downstream of histamine 4 receptor in a rat model of LPS-induced joint inflammation. Immunobiology. 2015;220(7):889–98. doi:10.1016/j.imbio.2015.01.008.
61. Dunford PJ, et al. The histamine H4 receptor mediates allergic airway inflammation by regulating the activation of CD4+ T cells. J Immunol. 2006;176:7062–70.

62. Gschwandtner M, Mommert S, Kother B, Werfel T, Gutzmer R. The histamine H4 receptor is highly expressed on plasmacytoid dendritic cells in psoriasis and histamine regulates their cytokine production and migration. J Invest Dermatol. 2011;131:1668–76.
63. Hochrein H, O'Keeffe M, Wagner H. Human and mouse plasmacytoid dendritic cells. Hum Immunol. 2002;63:1103–10.
64. Vasicek O, Lojek A, Jancinova V, Nosal R, Ciz M. Role of histamine receptors in the effects of histamine on the production of reactive oxygen species by whole blood phagocytes. Life Sci. 2014;100:67–72.
65. Ballerini C, et al. Antagonism of histamine H4 receptors exacerbates clinical and pathological signs of experimental autoimmune encephalomyelitis. Br J Pharmacol. 2013;170:67–77.
66. del Rio R, et al. Histamine H4 receptor optimizes T regulatory cell frequency and facilitates anti-inflammatory responses within the central nervous system. J Immunol. 2012;188:541–7.
67. Hong ST, et al. Histamine and its receptors modulate temperature-preference behaviors in Drosophila. J Neurosci. 2006;26:7245–56.
68. Pantazis A, et al. Distinct roles for two histamine receptors (hclA and hclB) at the Drosophila photoreceptor synapse. J Neurosci. 2008;28:7250–9.
69. Hatton GI, Yang QZ. Ionotropic histamine receptors and H2 receptors modulate supraoptic oxytocin neuronal excitability and dye coupling. J Neurosci. 2001;21:2974–82.
70. Schilling L, Wahl M. Opening of the blood-brain barrier during cortical superfusion with histamine. Brain Res. 1994;653:289–96.
71. Bartels AL, et al. Decreased blood-brain barrier P-glycoprotein function in the progression of Parkinson's disease, PSP and MSA. J Neural Transm. 2008;115:1001–9.
72. Emmerling MR, et al. Traumatic brain injury elevates the Alzheimer's amyloid peptide A beta 42 in human CSF. A possible role for nerve cell injury. Ann N Y Acad Sci. 2000;903:118–22.
73. Minagar A, Alexander JS. Blood-brain barrier disruption in multiple sclerosis. Mult Scler. 2003;9:540–9.
74. Suzuki Y, Nagai N, Umemura K. Novel situations of endothelial injury in stroke—mechanisms of stroke and strategy of drug development: intracranial bleeding associated with the treatment of ischemic stroke: thrombolytic treatment of ischemia-affected endothelial cells with tissue-type plasminogen activator. J Pharmacol Sci. 2011;116:25–9.
75. Wang X, et al. Physical exercise training and neurovascular unit in ischemic stroke. Neuroscience. 2014;271:99–107. doi:10.1016/j.neuroscience.2014.04.030.
76. Abbott NJ, Patabendige AA, Dolman DE, Yusof SR, Begley DJ. Structure and function of the blood-brain barrier. Neurobiol Dis. 2010;37:13–25.
77. Pardridge WM. The blood-brain barrier: bottleneck in brain drug development. NeuroRx. 2005;2:3–14.
78. Khatri R, McKinney AM, Swenson B, Janardhan V. Blood-brain barrier, reperfusion injury, and hemorrhagic transformation in acute ischemic stroke. Neurology. 2012;79:S52–7.
79. Deli MA. Potential use of tight junction modulators to reversibly open membranous barriers and improve drug delivery. Biochim Biophys Acta. 2009;1788:892–910.
80. Tiligada E, Kyriakidis K, Chazot PL, Passani MB. Histamine pharmacology and new CNS drug targets. CNS Neurosci Ther. 2011;17:620–8.
81. Brown TP, Forouzan O, Shevkoplyas SS, Khismatullin DB. Histamine reduces GPIbalpha-mediated adhesion of platelets to TNF-alpha-activated vascular endothelium. Thromb Res. 2013;131:150–7.
82. Li H, et al. Histamine upregulates gene expression of endothelial nitric oxide synthase in human vascular endothelial cells. Circulation. 2003;107:2348–54.
83. Toda N. Mechanism of histamine actions in human coronary arteries. Circ Res. 1987;61:280–6.
84. Li H, Forstermann U. Nitric oxide in the pathogenesis of vascular disease. J Pathol. 2000;190:244–54.

85. Bae ON, Majid A. Role of histidine/histamine in carnosine-induced neuroprotection during ischemic brain damage. Brain Res. 2013;1527:246–54.
86. Hu WW, Chen Z. Role of histamine and its receptors in cerebral ischemia. ACS Chem Neurosci. 2012;3:238–47.
87. Tosaki A, Szerdahelyi P, Joo F. Treatment with ranitidine of ischemic brain edema. Eur J Pharmacol. 1994;264:455–8.
88. Adachi N, et al. Reduction of the infarct size by simultaneous administration of L-histidine and diphenhydramine in ischaemic rat brains. Resuscitation. 2011;82:219–21.
89. Bernardino L, et al. Histamine stimulates neurogenesis in the rodent subventricular zone. Stem Cells. 2012;30:773–84. doi:10.1002/stem.1042.
90. Molina-Hernandez A, Velasco I. Histamine induces neural stem cell proliferation and neuronal differentiation by activation of distinct histamine receptors. J Neurochem. 2008;106:706–17.
91. Yoshikawa T, et al. Molecular mechanism of histamine clearance by primary human astrocytes. Glia. 2013;61:905–16. doi:10.1002/glia.22484.
92. Lipnik-Stangelj M, Carman-Krzan M. Activation of histamine H1-receptor enhances neurotrophic factor secretion from cultured astrocytes. Inflamm Res. 2004;53:245–52.
93. Lipnik-Stangelj M, Carman-Krzan M. Histamine and IL-6 interaction in the stimulation of nerve growth factor secretion from cultured astrocytes. Inflamm Res. 2005;54 Suppl 1:S36–7.
94. Villoslada P, Genain CP. Role of nerve growth factor and other trophic factors in brain inflammation. Prog Brain Res. 2004;146:403–14.
95. Pannell M, Szulzewsky F, Matyash V, Wolf SA, Kettenmann H. The subpopulation of microglia sensitive to neurotransmitters/neurohormones is modulated by stimulation with LPS, interferon-gamma, and IL-4. Glia. 2014;62:667–79.
96. Zhu J, Qu C, Lu X, Zhang S. Activation of microglia by histamine and substance P. Cell Physiol Biochem. 2014;34:768–80.
97. Apolloni S, Fabbrizio P, Parisi C, Amadio S, Volonte C. Clemastine confers neuroprotection and induces an anti-inflammatory phenotype in SOD1 mouse model of amyotrophic lateral sclerosis. Mol Neurobiol. 2016;53(1):518–31.
98. Farokhi M, Etemadifar M, Rezaei A, Amani A, Jahanbani H. P021—Role of histamine and diamine oxidase enzyme in Multiple Sclerosis. Mult Scler Relat Disord. 2014;3:746.
99. Lu C, et al. Endothelial histamine H1 receptor signaling reduces blood-brain barrier permeability and susceptibility to autoimmune encephalomyelitis. Proc Natl Acad Sci U S A. 2010;107:18967–72.
100. Pedotti R, et al. Multiple elements of the allergic arm of the immune response modulate autoimmune demyelination. Proc Natl Acad Sci U S A. 2003;100:1867–72.
101. Emerson MR, Orentas DM, Lynch SG, LeVine SM. Activation of histamine H2 receptors ameliorates experimental allergic encephalomyelitis. Neuroreport. 2002;13:1407–10.
102. Lapilla M, et al. Histamine regulates autoreactive T cell activation and adhesiveness in inflamed brain microcirculation. J Leukoc Biol. 2011;89:259–67.
103. Saligrama N, Noubade R, Case LK, del Rio R, Teuscher C. Combinatorial roles for histamine H1-H2 and H3-H4 receptors in autoimmune inflammatory disease of the central nervous system. Eur J Immunol. 2012;42:1536–46.
104. Luo T, et al. Histamine H2 receptor activation exacerbates myocardial ischemia/reperfusion injury by disturbing mitochondrial and endothelial function. Basic Res Cardiol. 2013;108:342.
105. Rocha SM, Pires J, Esteves M, Graca B, Bernardino L. Histamine: a new immunomodulatory player in the neuron-glia crosstalk. Front Cell Neurosci. 2014;8:120.
106. ElAli A, Theriault P, Rivest S. The role of pericytes in neurovascular unit remodeling in brain disorders. Int J Mol Sci. 2014;15:6453–74.
107. Murphy DD, Wagner RC. Differential contractile response of cultured microvascular pericytes to vasoactive agents. Microcirculation. 1994;1:121–8.
108. Sims DE, Miller FN, Donald A, Perricone MA. Ultrastructure of pericytes in early stages of histamine-induced inflammation. J Morphol. 1990;206:333–42.
109. Iadecola C. Neurovascular regulation in the normal brain and in Alzheimer's disease. Nat Rev Neurosci. 2004;5:347–60.

110. Fang YI, et al. Marked increase in the histamine content of neointima after stent implantation of pig coronary artery and growth-promoting effects of histamine in cultured smooth muscle cells. Life Sci. 2005;77(3):241–51.
111. Satoh T, et al. Histamine as an activator of cell growth and extracellular matrix reconstruction for human vascular smooth muscle cells. Atherosclerosis. 1994;110:53–61.
112. Tanimoto A, et al. Histamine upregulates the expression of inducible nitric oxide synthase in human intimal smooth muscle cells via histamine H1 receptor and NF-kappaB signaling pathway. Arterioscler Thromb Vasc Biol. 2007;27:1556–61.
113. Nistri S, et al. Relaxin induces mast cell inhibition and reduces ventricular arrhythmias in a swine model of acute myocardial infarction. Pharmacol Res. 2008;57:43–8.
114. Chan SL, Sweet JG, Cipolla MJ. Treatment for cerebral small vessel disease: effect of relaxin on the function and structure of cerebral parenchymal arterioles during hypertension. FASEB J. 2013;27:3917–27.
115. Masini E, et al. Beneficial effects of a plant histaminase in a rat model of splanchnic artery occlusion and reperfusion. Shock. 2007;27:409–15.
116. Masini E, et al. Protective effects of a plant histaminase in myocardial ischaemia and reperfusion injury in vivo. Biochem Biophys Res Commun. 2003;309:432–9.
117. Masini E, et al. Pea seedling histaminase as a novel therapeutic approach to anaphylactic and inflammatory disorders. A plant histaminase in allergic asthma and ischemic shock. ScientificWorldJournal. 2007;7:888–902.
118. Hernandez F, Martinez AM, Piedra D, Catalan RE. Endothelin inhibits histamine-induced cyclic AMP accumulation in bovine brain vessels. Microvasc Res. 2000;60:49–54.
119. Vizuete ML, et al. Histamine infusion induces a selective dopaminergic neuronal death along with an inflammatory reaction in rat substantia nigra. J Neurochem. 2000;75:540–52.

# Chapter 9
# Histamine Function in Nervous Systems

**Olga A. Sergeeva and Helmut L. Haas**

**Abstract** Histamine is evolutionarily highly conserved as a signaling molecule, a neuromodulator, and a neurotransmitter from bacteria to mammals that fulfills basic demands of living like the organization of behavioral state. From mussels to mammals, it gates ion channels like classical ionotropic transmitters. These relatively neglected functions are treated for arthropods and mollusks. In vertebrate brains, histamine actions are mostly mediated by three of the four known G-protein-coupled receptors: H1R, H2R, and H3R. Histamine also modifies other receptor proteins through allosteric interactions. We describe and discuss the anatomical, biophysical, and physiological properties of histaminergic neurons as well as their projections and actions on target neurons.

**Keywords** Histaminergic neurons • Electrophysiology • Invertebrates • Mammals • Hypothalamus • Sleep and waking • Histamine receptors • Neural plasticity

## 9.1 Introduction

Histamine is not anymore an autacoid in search of a function. Its role as a neurotransmitter and neuromodulator is firmly established, with some delay after the fellow amine acetylcholine, the two catecholamines and serotonin. The identification of release from axons of histamine-containing neurons causing synaptic responses in follower neurons and behavioral consequences provided the first

---

O.A. Sergeeva, Ph.D.
Faculty of Medicine, Department of Neuro- and Sensory Physiology, Heinrich-Heine-University Düsseldorf, Düsseldorf 40205, Germany

Department of Clinical Neuroscience and Medical Psychology, Heinrich-Heine-University Düsseldorf, Düsseldorf 40205, Germany
e-mail: olga.sergeeva@uni-duesseldorf.de

H.L. Haas, M.D. (✉)
Faculty of Medicine, Department of Neuro- and Sensory Physiology, Heinrich-Heine-University Düsseldorf, Düsseldorf 40205, Germany
e-mail: haas@hhu.de



P. Blandina, M.B. Passani (eds.), *Histamine Receptors*, The Receptors 28,
DOI 10.1007/978-3-319-40308-3_9

unequivocal evidence for such a function in mollusks [1, 2]. With the progress in understanding more complex signaling pathways, it became clear that most of the actions of these four biogenic amines are not directly ionotropic but second messenger mediated. In vertebrates the central histaminergic neurons are located in the tuberomamillary nucleus at the posterior basis of the median eminence from where they innervate the whole central nervous system. This pattern is quite comparable to that of the other aminergic systems; it was, however, recognized much later, when the histochemical identification was presented by Panula et al. [3] and Watanabe et al. [4] in 1984. A further reason for the initial relative lack of interest in histamine and its receptors was the missing link to major diseases. However, an involvement in sleep disorders and in neurological diseases such as Alzheimer's, Down syndrome, multiple sclerosis, motion sickness, addiction, and hepatic encephalopathy has long been suggested. Tourette syndrome occurred in a family in association with a mutation in the histidine decarboxylase (HDC) encoding gene [5]. In this chapter we'll discuss the electrophysiological analysis of histaminergic neurons and the responses of their targets to histamine from mussels to man.

## 9.2 Histaminergic Neurons in Vertebrate Brains

In vertebrates the histaminergic neurons are localized exclusively in the tuberomammillary nucleus of the posterior hypothalamus [3, 4] from where they send widely branching unmyelinated predominantly ipsilateral projections to most regions of the brain and spinal cord. The density of innervation varies considerably with highest densities in the hypothalamus itself, the septum, the nucleus of the diagonal band of Broca, and the ventral tegmentum. The cerebral cortex, hippocampus, amygdala, thalamus, striatum, and substantia nigra also receive mostly moderate innervation.

In rodents the nucleus contains a few thousand neurons, subdivided in a large ventral group around the mammillary bodies close to the brain surface, a median group close to the mamillary recess of the third ventricle, and a diffuse part (Fig. 9.1).

The tuberomamillary histamine neurons have large somata (up to 30 μm) with 2–3 long subdividing dendrites overlapping with those from neighboring neurons and often laying close to the brain surface. The axons mostly arise from a dendrite. The cytoplasm contains many mitochondria and a well-developed Golgi apparatus. Most histaminergic neurons in the rat are stained also intensely with adenosine deaminase and contain a number of further transmitters. The GABA-synthesizing enzymes GAD 65/67 indicate a GABAergic phenotype, and GABA release has recently been suggested [7]. Subpopulations of the histaminergic neurons express enkephalins, galanin, substance P, and thyrotropin-releasing hormone (TRH).

The human brain contains about 64,000 large multipolar histaminergic neurons occupying a similar but comparatively larger proportion of the posterior hypothalamus [8]. There may be some functional distinction between different neuronal populations of the nucleus [9]. The tuberomamillary nucleus shows considerable degeneration in Alzheimer brains [10].

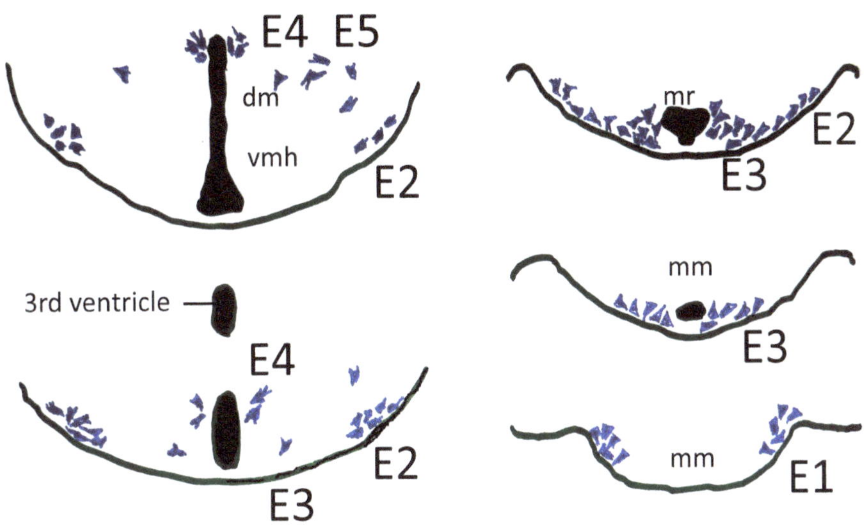


**Fig. 9.1** Frontal sections of the tuberomammillary nucleus in the posterior hypothalamus with histaminergic neurons, close to the inner (third ventricle, *MR* mammillary recess) and the outer surface of the brain. *MM* medial mammillary, *DM* dorsomedial, *VMH* ventromedial hypothalamic nuclei. The five subdivisions are described as E1–E5 [6]

Histaminergic cell bodies and axons are covered with H3R mediating a feedback inhibition of firing, release, and synthesis. A widespread heterosynaptic action blocks also the release of other transmitters, such as glutamate, acetylcholine, GABA, catecholamines, and probably others (Figs. 9.2, 9.3, and 9.4).

### 9.2.1 *Electrophysiology*

In vivo and in vitro, the histaminergic neurons fire regularly at about 1–4 Hz. In behaving rodents and cats, they are only active during the waking state but silent during all phases of sleep [11]. A non-inactivating sodium current is sufficient to drive spontaneous firing [12], and two opposing membrane currents, a hyperpolarization-activated inward and two outward currents of the A type, give a characteristic shape and keep the resting potential rather stable near −50 mV. A wide action potential near 2 ms at mid-amplitude at body temperature is followed by an afterhyperpolarization reaching about −75 mV where A-current inactivation is removed [13]. $Ca^{++}$-dependent prepotentials presumably arising in the dendrites are observed under tetrodotoxin (blocks sodium channels) or during artificial hyperpolarization [14]. Histaminergic neurons express five types of $Ca^{2+}$ channels [15] including N and P type which are blocked by histamine H3R antagonists (Fig. 9.5).

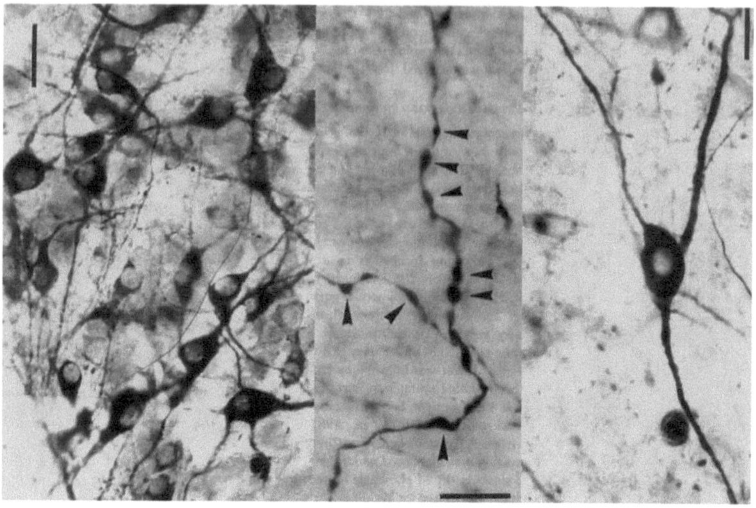

**Fig. 9.2** Histaminergic neurons from the median part of the rat tuberomammillary nucleus in organotypic culture stained for histidine decarboxylase. Large cell bodies with a diameter of typically 25 μm and two to three dendrites. In the *middle* axons with varicosities (*arrows*) in a cocultured hippocampus. *Scale bars* from *left* to *right*: 50, 20, and 10 μm (Modified from Diewald et al. [35])

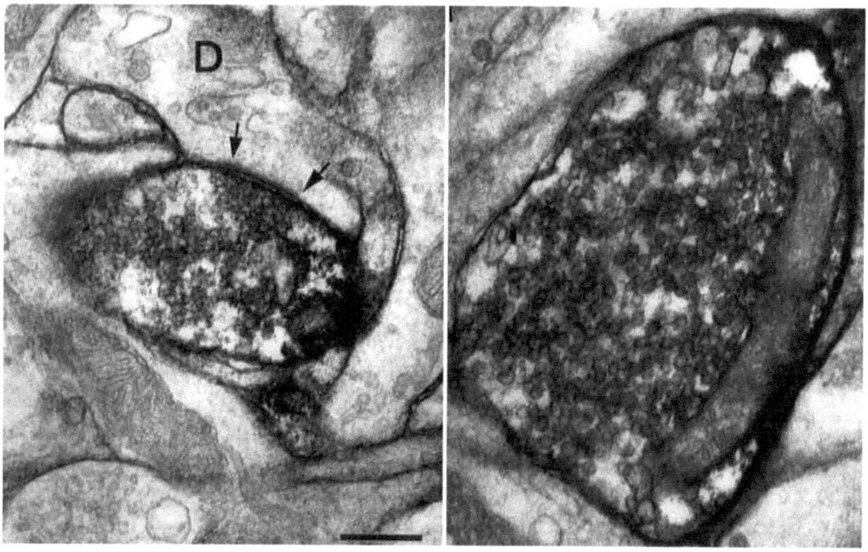


**Fig. 9.3** Electron microscopic pictures of histidine decarboxylase-immunoreactive axonal varicosities in a coculture of the posterior hypothalamus and hippocampus. The labeled varicosities are densely filled with synaptic vesicles. On the left the rare case of a classical synapse with postsynaptic density, forming a symmetric contact on a dendritic shaft of the cocultured hippocampus, on the right the more typical aminergic varicosity. *D* dendrite. Scale bar of 0.25 Fm (Modified from Diewald et al. [35])

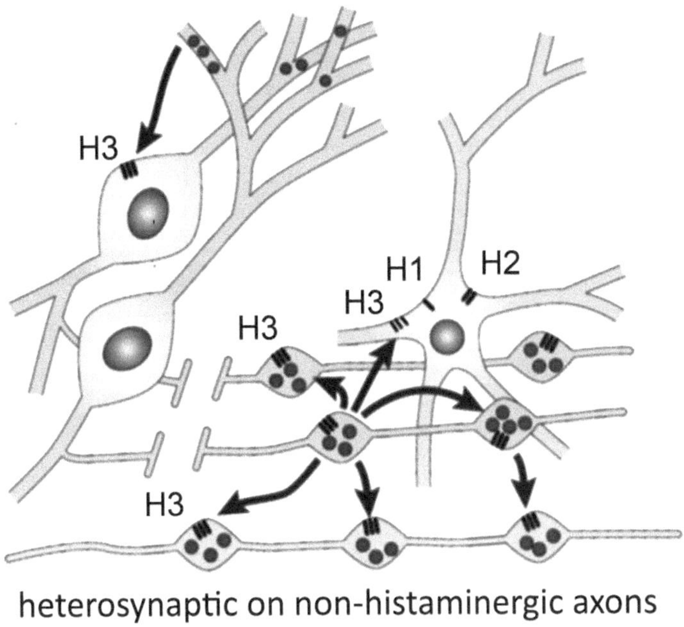


**Fig. 9.4** Histamine actions on autoreceptors and heteroreceptors (Modified from Nature Reviews Neuroscience 4:121–130, (2003), Haas HL and Panula P, The role of histamine and the tuberomamillary nucleus in the nervous system)

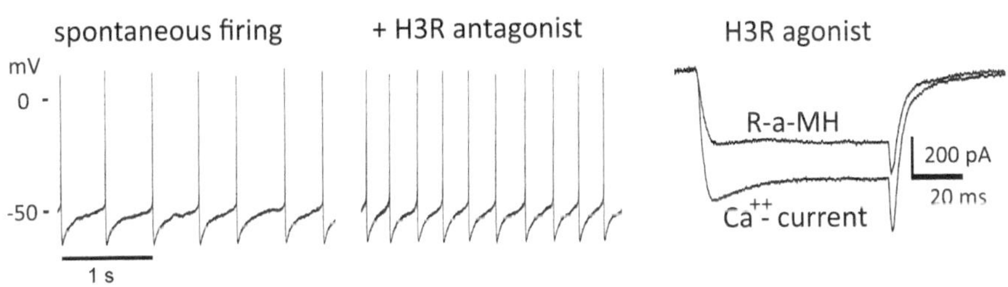


**Fig. 9.5** Intracellular recording from a spontaneously firing histaminergic neuron in a slice in vitro. The autoinhibition through H3R is demonstrated by the increased firing under perfusion with an H3R antagonist (thioperamide). On the right a voltage-clamp recording showing an inward $Ca^{2+}$ current, in response to a depolarizing command, reduced by perfusion with an H3R agonist (R-α-methyl histamine)

### 9.2.2 *Afferent Connections*

The activity of histamine neurons depends on behavioral state, synaptic, paracrine, and endocrine signals. Inputs come from hypothalamic and cortical areas, the subiculum and dorsal tegmentum.

#### 9.2.2.1 Amino Acids

Stimulation of hypothalamic neurons or fibers can evoke excitatory or inhibitory synaptic potentials in histamine neurons mediated by glutamate or GABA. Spontaneous postsynaptic potentials are only observed as inhibitory GABAergic ones. Both AMPA (GluR1,2,4, GluR3 missing) and NMDA (NR1, 2A, 2B) receptors are expressed.

GABAergic inputs from the preoptic area silence the histamine neurons during sleep. There are three different $GABA_AR$ groups with different sensitivities to GABA, depending on the expression of the γ-subunit [16]; α2 and β3 containing $GABA_AR$ are the most relevant for sleep. The GABAergic input to histamine neurons is regulated by a presynaptic feedback control through the $GABA_BR$. The osmolyte taurine can reach concentrations in the brain that gate $GABA_A$ and strychnine-sensitive glycine receptors. Its efficacy on $GABA_AR$ is independent of the receptor composition and thus inhibits a large range of neurons. Glycine inhibits only a subpopulation of histamine neurons, preferably the larger ones [17].

#### 9.2.2.2 Biogenic Amines

The histaminergic nucleus receives fibers from all amines including acetylcholine, which elicits fast desensitizing excitation through bungarotoxin-sensitive α7-receptors. These could be responsible for the waking action of nicotine. A muscarinic modulation of histamine release may occur on histaminergic varicosities [18]. Dopamine excites histaminergic neurons through D2R, and noradrenaline presynaptically controls GABAergic input through α2-receptors [19]. Serotonin excites histaminergic neurons by activating an electrogenic sodium/calcium exchanger that imports three $Na^+$ and exports one $Ca^{2+}$. Serotonin 2C receptors undergo posttranscriptional editing by three ADAR enzymes [20].

#### 9.2.2.3 Purines

Histaminergic neurons express mainly $P2X_2$ receptors through which ATP evokes fast desensitizing inward currents. A number of further purines excite histamine neurons through metabotropic $P2Y_1$ and $P2Y_4$ receptors. Adenosine, the dephosphorylation product of ATP, is a homoeostatic sleep factor and inhibits many neurons in the nervous system, not the wake-active histaminergic neurons. Adenosine deaminase is strongly expressed in most rat histaminergic neurons and adenosine has been suspected to be a cotransmitter.

#### 9.2.2.4 Peptides and Metabolic Signaling

Galanin is expressed in histaminergic neurons as well as in afferent fibers. Galanin inhibits histaminergic somata and release from axon varicosities and promotes sleep and food intake. Hypocretins (orexins), exclusively located in the close neighborhood, excite histaminergic neurons and maintain wakefulness. Their degeneration is the cause of narcolepsy: excessive daytime sleepiness and cataplexy [21, 22]. These neurons also express dynorphin which suppresses the GABAergic input to histamine neurons. Thyrotropin-releasing hormone (TRH) excites most histamine neurons and reduces food intake and sleep [23]; several further peptides involved in metabolic regulation are interacting. Estrogen receptor expression in the tuberomammillary nucleus varies with metabolic activity, sex, and age. Insulin-induced hypoglycemia, prostaglandins, and endocannabinoids activate histamine neurons, which are also involved in $CO_2$-mediated arousal.

### 9.2.3 *Histamine Targets*

Similar projection patterns are found in most species but differences exist in the innervation density. The histaminergic fibers are unmyelinated and vastly arborizing. They leave the tuberomammillary nucleus through two ascending and one descending bundle to contact somata, dendrites, and axons of many neurons all over the central nervous system. The fibers are extensively crossing and many neurons are branching to different pathways. Four metabotropic receptors have been cloned; H1R, H2R, and H3R occur in the central nervous system; there is so far no convincing evidence for the presence and function of the H4R, which occurs mainly in peripheral tissues. Table 9.1 gives a brief overview over histamine receptor G-protein-mediated signaling and actions in nervous systems.

## 9.3 Receptor-Channel Complexes Gated by Histamine

Histamine-gated chloride channels have been described in bacteria [24], mollusks [1], and arthropods [25, 26]. In the *Aplysia* cerebral ganglion, the two symmetric C2 neurons contain histamine and innervate about 15 follower neurons on each side.

**Table 9.1** Histamine receptor-mediated actions and signaling in nervous systems

| Receptor | G-protein enzyme | Second messenger | Effect | Channel transporter |
|---|---|---|---|---|
| H1R | Gq/11 | DAG | Depolarization | Cation |
| | PLC | IP3 | | $K^+$ block, NCX |
| | | – | Hyperpolarization | $K^+(Ca^{++})$ |
| | | NO | | |
| H2R | Gs | | AHP, accom. block | $K^+(Ca^{++})$ |
| | AC | cAMP | Depolarization (Ih) | HCN |
| | PKA | CREB | Synaptic plasticity | |
| H3R | $G_{i/o}$ | – | Inhibition | HVACC block |

*AC* adenylyl cyclase, *Accom* accommodation, *AHP* afterhyperpolarization, *CREB* cyclic AMP response element-binding protein, *DAG* diacylglycerol, *Ih* or *HCN*, hyperpolarization-activated inward current, *HVACC* high-voltage-activated $Ca^{2+}$ channel, *IP3* inositol trisphosphate, *NCX* sodium/calcium exchanger, *NO* nitric oxide, *PKA* protein kinase A, *PLC* phospholipase C

Pantazis et al. [27] identified distinct roles for two histamine receptors (*hclA and HclB*) at the *Drosophila* photoreceptor synapse.

### 9.3.1 Histamine-Gated Ion Channels in Invertebrates

The superfamily of proteins forming ligand-gated ion channels activated by all major neurotransmitters including histamine (ART-LGIC, Cys-loop receptors) is almost omnipresent even in bacteria. Both ligand-gating and channel-forming transmembrane domains have been described in prokaryotic organisms. An early transfer in phylogenesis to metazoan lineages has been suggested [24].

Histaminergic systems have been extensively investigated in mollusks, where a functional connection with feeding-related arousal exists, a function also attributed to the vertebrate histamine system [28]. In *Aplysia* the histaminergic mechanosensory C2 cell projects to several follower cells, where fast and slow excitatory and inhibitory responses are evoked. Different responses show pharmacological similarity to H1R- or H2R-mediated actions in vertebrates. A slow inhibitory postsynaptic potential (IPSP) but not a fast excitatory postsynaptic potential (EPSP) was induced by a $K^+$ conductance and blocked by the H2R antagonist cimetidine [1, 2, 29] (Fig. 9.6).

Paired recordings with follower cells revealed combinations of various depolarizing and hyperpolarizing postsynaptic responses. For instance, two interneurons in the buccal-cerebral ganglion involved in the coordination of feeding behavior (B17 and B18) receive presynaptic inhibition from C2 [30]. Histamine is the transmitter released from arthropod photoreceptors that gates chloride channels in the following postsynaptic large monopolar cells. In *Drosophila*, two genes, *hclA* (*ort*) and *hclB* (*hisCl1*), encode for HCLA and HCLB receptors that can build homomers and dimers. HCLA homomers are found in the large monopolar cells, whereas HCLB is restricted to glial cells [27].

### 9.3.2 Gating of Vertebrate Ion Channels by Histamine

GABAA receptors are heteropentameric proteins constructed from up to seven different (mostly α, β, and γ) subunits in variable composition. Homopentameric receptors composed of β subunits expressed in Xenopus oocytes or HEK293 cells are gated by histamine [31]. Such homomeric receptors most likely do not occur in the brain. Native receptors composed of αβγ-subunits are not gated but can be modulated by histamine acting as an inverse agonist at strychnine-sensitive α1β-glycine receptors in tuberomammillary (histaminergic) neurons [32–34].

There are indications of histamine-gated chloride channels in native neurons. This may occur at classical synapses with a close apposition of pre- and postsynaptic membranes, which are relatively rare in aminergic junctions [35] (see Fig. 9.3). Stimulation of afferent fibers from the tuberomamillary to the supraoptic nucleus elicited cimetidine-sensitive hyperpolarizing potentials [36]. Histaminergic neurons

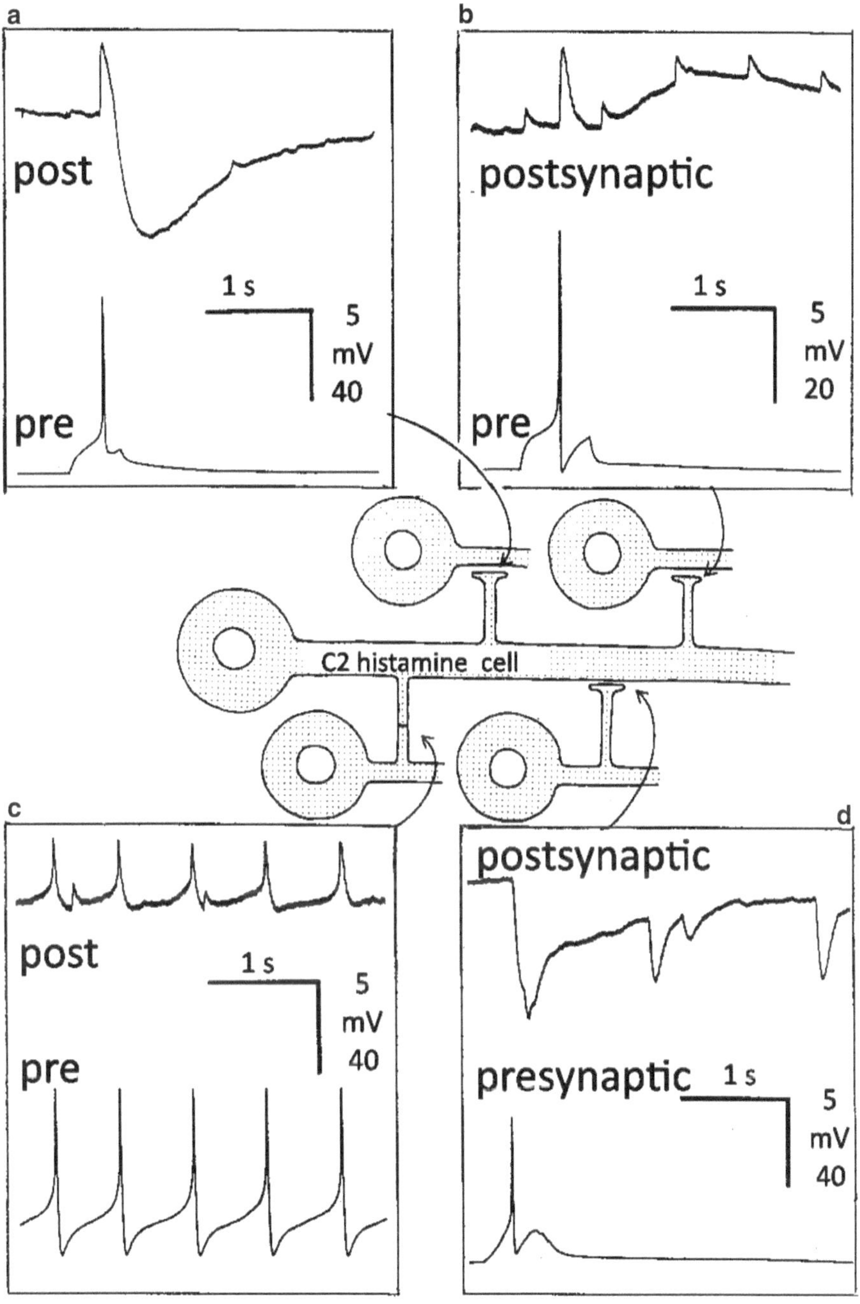


**Fig. 9.6** Schematic drawing of the histaminergic C2 neuron in the cerebral ganglion of *Aplysia* with follower neurons and results from combined recordings of pre- and postsynaptic structures. (**a**) One action potential evokes mainly a fast hyperpolarizing potential in the follower cell. (**b**) The postsynaptic response consists of a fast and a slow depolarizing potential. (**c**) Electrical transmission through a gap junction. (**d**) A purely hyperpolarizing response (From Weinreich, with permission)

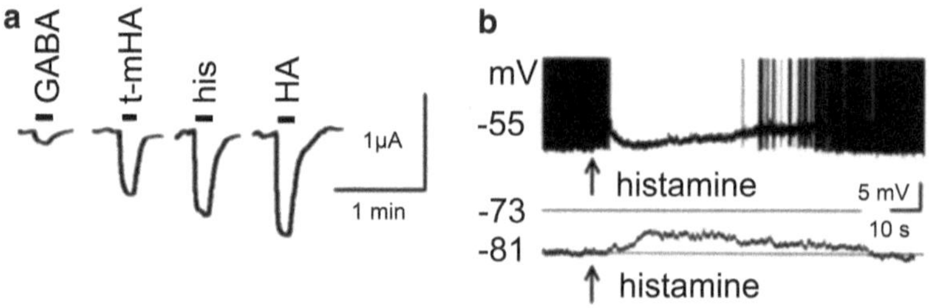


**Fig. 9.7** Histamine gating of chloride channels. (**a**) $GABA_AR$ β3 homomeric channels expressed in Xenopus oocytes are activated by (3 mM) GABA, tele-methylhistamine, histidine, and histamine (HA). (**b**) GABAergic neuron from a ferret thalamus is hyperpolarized and its firing blocked by histamine in a slice. The response is reversed at −73 mV, the chloride equilibrium potential, and displays an H2R pharmacology (Modified from Saras et al. [31] and Lee et al. [37], with permission)

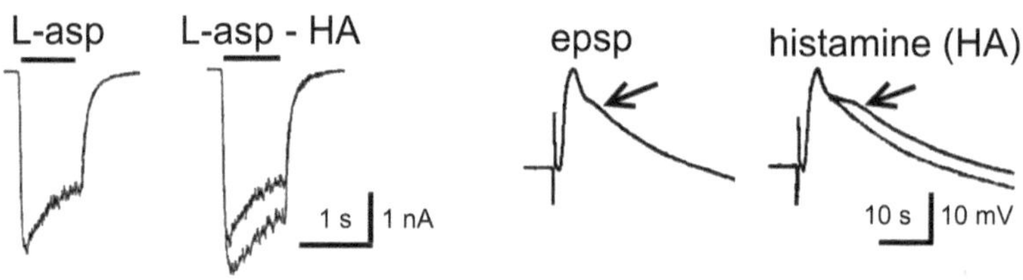


**Fig. 9.8** Potentiation of NMDA receptor-mediated responses in the hippocampus. *Left*: Brief perfusion of an isolated pyramidal cell with L-aspartate evokes an inward current that is potentiated in the presence of histamine. *Right*: Excitatory postsynaptic potential (EPSP) in CA1 pyramidal cell with NMDA component that is potentiated in the presence of histamine (*arrows*) (Modified from Vorobjev et al. [42] and Yanovsky et al. [45])

also express GABA but GABA release from their axons has been excluded in this case. More recently, the release of GABA from histaminergic endings has been proposed to provide coordinated extrasynaptic inhibition over large neocortical and striatal areas [7]. Lee et al. [37] measured inhibitory chloride conductances mediated by a H2R-related receptor in thalamic interneurons (see Fig. 9.8). Imidazole acetic acid, a histamine metabolite, has long been known to activate GABAAR [38], and mediation of the inhibitory action of histamine as well as cimetidine in raphe neurons through GABAAR was suggested [39].

### *9.3.3 Modulation of N-Methyl-D-aspartate (NMDA): Receptors by Histamine*

Modulation of ionotropic transmitter actions through G proteins is common: Payne and Neumann [40] described a H1R-mediated facilitation of NMDA receptors through protein kinase C and a reduction of the Mg2+ block. However there is also an allosteric modulation, i.e., a direct action of histamine on the NMDA receptor.

The reduction of NMDA-mediated membrane currents was detected by two groups: Bekkers [41] observed it on autapses in hippocampal cultures; Vorobjev et al. [42] in pyramidal cells isolated from hippocampal slices presented evidence for mediation through the polyamine modulatory site. (One reason for the late appearance of histamine in the transmitter scene was its cross-reaction with the polyamine spermidine that jeopardized histochemical demonstration through phthalaldehyde fluorescence that had worked well for the monoamine transmitters [43].) Saybasili et al. [44] and Yanovsky et al. [45] demonstrated the pH-sensitive effect also on the NMDA component of excitatory postsynaptic potentials (EPSPs) in the CA1 region of hippocampal slices: it was more prominent during slight acidification, such as occurs during intense and synchronous neuronal discharges. The effect is mimicked by the histamine metabolite 1-methylhistamine but not mediated by any of the known histamine receptors, and it is selective for the GluN2B subunit [46] which is important for synaptic plasticity and learning. Some H3R antagonists like clobenpropit are GluN2B selective NMDA antagonists [47]. Faucard et al. [48] have also described a potentiation of NMDA-mediated noradrenaline release from hippocampal synaptosomes through an allosteric site different from the polyamine site.

## 9.4 Histamine Receptor-Mediated Functions in Vertebrate Nervous Systems

Histamine from enterochromaffin cells, small intensely fluorescent (SIF) cells in ganglia, glomus cells, and mast cells, acts in a paracrine or endocrine fashion. Cutaneous itch is mediated by a specific type of C fibers expressing H1R and TRPV1 (the capsaicin receptor). Histamine is the major cause of itch in insect bites and urticaria [49]. The sensor cells for oxygen in the carotid bodies use both histamine and dopamine as transmitters to signal hypoxia to the respiratory centers [50].

### *9.4.1 Spinal Cord, Brain Stem, and Cerebellum*

The spinal cord receives histaminergic innervation from the tuberomamillary nucleus. Microionophoretic experiments have revealed histamine inhibitory actions on several spinal neurons. In lateral horn spinal sympathetic neurons are depolarized through H1R activation and block of a $K^+$ channel [51].

Many different neurons in the brain stem respond to histamine acting on H1R, H2R, and H3R on somatic and axonal sites. The periaqueductal gray is involved in waking regulation, pain control, and defense behavior. The trigeminal nuclei are reciprocally connected with the tuberomamillary nucleus with functional implication for finding and consuming food. The aminergic nuclei in the brain stem and midbrain are mutually connected also with the histaminergic neurons in the tuberomamillary nucleus, forming a functional network that organizes behavioral state.

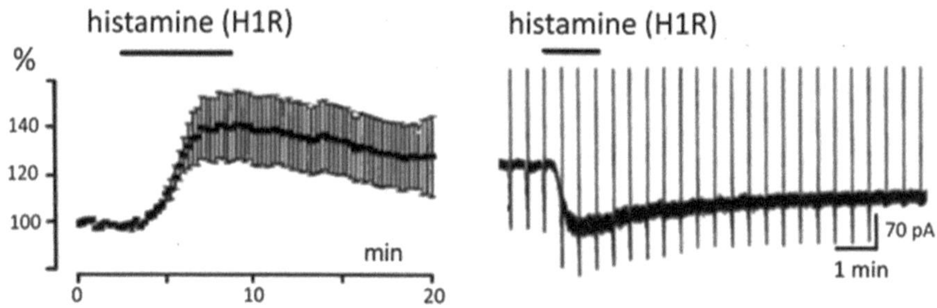


**Fig. 9.9** Examples of histamine excitations through H1R activation. *Left*: Averaged increase in firing rate of 15 GABAergic neurons in the substantia nigra. *Right*: Histamine-evoked inward current with increased channel noise in a serotonergic raphe neuron (Modified from Korotkova et al. [56]

The cholinergic neurons in the pons, the basal forebrain, and the septum [52] are a major target of histaminergic innervation and are all strongly excited by histamine through H1R activation. There is only a minimal reciprocal action of acetylcholine on histamine neurons through nicotinic receptors [53, 54]. A dense histaminergic innervation of the nucleus tractus solitarius represents the major influence on the vagal complex and thus the whole vegetative nervous system (Fig. 9.9).

The area postrema is a circumventricular organ in the rhombencephalon implicated with motion sickness that is treated with H1R antagonists. This therapeutic effect is likely located on vestibular nuclei too, where all three "neural" histamine receptors are present. Even vestibular hair cells express these three receptors, and their response is modulated by histamine. The cerebellum also receives histaminergic innervation in most species. Purkinje cells and interpositus neurons are excited through H2R and granule cells through both H1R and H2R.

The noradrenergic locus coeruleus is excited through H1R and H2R. Although the neurons express H3R, no respective action is seen at the cell body level. However, an inhibitory action mediated by H3R has been found on varicosities of noradrenergic axons. Histaminergic neurons are disinhibited presynaptically through α2R. The locus coeruleus also contains dopaminergic neurons projecting to the hippocampus [55].

Release of dopamine from fibers emanating from the substantia nigra and the ventral tegmental area (VTA) is under the control of H3R on varicosities. GABAergic but not dopaminergic neurons in the tegmental structures are excited by H1R and H2R activation; the dopaminergic neurons are thus indirectly inhibited [56, 57].

### 9.4.2 *Hypothalamus*

The hypothalamus is not only the seat of the histaminergic neurons; it is also a major target of histaminergic innervation. Neurons of the preoptic area involved in temperature and sleep regulation are directly or indirectly under histaminergic influence [58].

Suprachiasmatic neurons involved in the phase shift from an internal to the astronomic rhythm are excited through H1R and inhibited through H2R in rats but not in some other species [59–61].

The supraoptic and paraventricular nuclei releasing the important hormones vasopressin (AVP, adiuretin) and oxytocin from endings in the posterior lobe of the hypophysis are excited by histamine, and a number of further hypothalamic peptides (ACTH, endorphin, MSH, prolactin, TRH, and growth hormone) are thus under the indirect control of histamine. Stimulation with single pulses in the histaminergic nucleus elicits EPSPs in vasopressinergic neurons; multiple pulses blunt this excitation [62]. Oxytocin neurons display fast $Cl^-$-dependent IPSPs in response to stimulation of tuberomamillary neurons. A further histamine effect in this region is a block of gap junctional coupling which reduces neuronal synchronization and consequently pulsatile release of the peptide hormones [36]. These actions are particularly important during parturition and lactation [63, 64].

The large arcuate nucleus integrates metabolic signals and controls neuroendocrine and feeding rhythms. Histamine can excite and inhibit neurons in this nucleus through H1R and H2R, respectively [65, 66]. The histaminergic neurons cooperate in a complementary way with the hypocretinergic (orexinergic) neurons in the perifornical area in the organization of behavioral state, specifically waking, and the coordination of food seeking, feeding, and arousal.

### *9.4.3 Thalamus and Basal Ganglia*

The thalamus is the gate to the cortex, to consciousness; the relay neurons in the lateral geniculate nucleus are the gatekeepers that fire in two distinct modes for open and closed states of the gate: when they fire in alternating up (burst) and down states at around −60 mV, the door to consciousness is closed. Histaminergic, cholinergic, and other aminergic inputs depolarize them and shift them to a regular equal firing rate which enables information transfer to the cortex [67]. Both H1R and H2R contribute to this action through blocking a potassium current and enhancing a hyperpolarization-dependent cation current (Ih). GABAergic perigeniculate neurons are inhibited through an H2R-like ionotropic action [37]. This combined action could dampen thalamic oscillations during sleep-waking transitions.

The GABAergic medium spiny neurons in the striatum can be excited by H2R activation, but the overwhelming effect of histamine is an H3R-mediated inhibition on the excitatory inputs to these neurons from the thalamus and cortex. Histamine reduces glutamatergic transmission and plasticity. A pharmacological interference with this inhibition may be responsible for possible beneficial effects in basal ganglia diseases. The giant cholinergic interneurons in the striatum are excited by H1R activation. Histaminergic endings reach the same target regions as the dopaminergic ones, and there are dimeric H3R and D2R. As histaminergic neurons express dopa decarboxylase, they are able to synthesize dopamine and compensate for the loss of dopamine in Parkinson's disease, if ambient L-dopa is available. This is particularly the case under L-dopa therapy [19].

### 9.4.4 Hippocampus and Neuronal Plasticity

Histaminergic innervation reaches the hippocampus through two pathways, the fornix and a caudal route. The input from the entorhinal cortex to the dentate gyrus is blocked by H3R heteroreceptor action on the glutamate releasing varicosities in vivo by stimulation of the tuberomamillary nucleus [68] and in vitro [69–71]. The glutamatergic synapses at the CA1 and CA3 pyramidal neurons of the hippocampus are not suppressed in this way, but instead undergo an extraordinary long-lasting potentiation mediated through H2R, cyclic AMP, and PKA showing the properties of classical long-lasting (protein synthesis dependent) LTP [72]. When this long-term potentiation is elicited by repetitive stimulation [73], the effect is potentiated by histamine [70]. This is quite important in the light of the mechanisms of memory trace formation in the hippocampus: sensory input reaches a selected number of pyramidal neurons in the CA3 area, which exhibit a synchronous burst that enters CA1, where LTP is evoked on pyramidal dendrite glutamate receptors at specifically distributed locations [74, 75]. This is a more physiological trigger for LTP than high-frequency electrical stimulation. Along with this synaptic potentiation, a long-lasting H2R-mediated increase of firing rate is observed [76]. Similar plasticity occurs on entorhinal-dentate and CA1–CA3 pathways. A striking potentiation is also caused by histamine on the response to depolarization in principal dentate and pyramidal cells of the hippocampus: through blocking the accommodation of firing that depends on a $Ca^{2+}$-dependent $K^+$ conductance, a given depolarizing pulse will evoke an increased number of action potentials in the presence of histamine [77] (Figs. 9.10 and 9.11).

Histamine through cyclic AMP has also a striking effect on a long-lasting afterhyperpolarization and consequently the accommodation of action potential discharge in response to a depolarizing stimulus, which considerably outlasts the presence of histamine. This effect results from a block of the "small" $K^+$ channel and perfectly suits the role of histamine as a waking substance specifically for raising vigilance. Histamine also excites many interneurons in the cerebral cortex and the hippocampus [78], an action that may confer an overall inhibitory influence and is in line with proconvulsive effects of antihistamines [79]. The inhibition of principal neurons by excited interneurons (H1R) combined with potentiation (H2R) may provide an important shift in cortical network activity toward increased attention and vigilance. The H1R-mediated excitation of principal cortical neurons is the target of the sedative antihistamines and is considered the main mechanism for waking [11, 80]. As shown above for the cooperation of H1R with H2R in the promotion of plasticity, a combined action of the two receptor types may also be involved in the management of wakefulness. This likely relevant behavioral role of H2R warrants further study, in particular with respect to neuropsychiatric disorders.

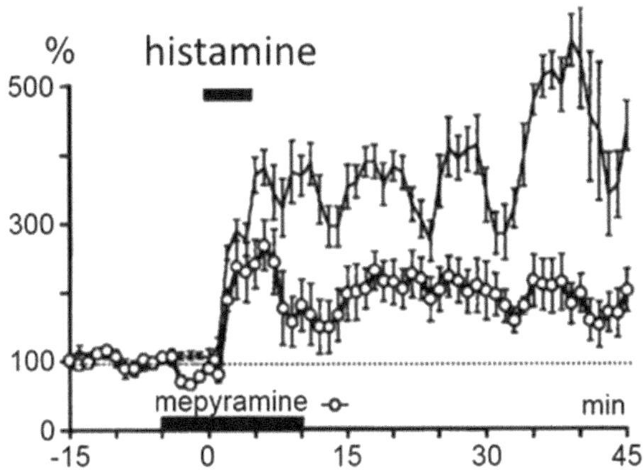


**Fig. 9.10** Histamine evokes long-lasting excitation of a hippocampal pyramidal neuron synaptically isolated in low $Ca^{2+}$/high $Mg^{2+}$ through a combined activation of H2R and H1R. A perfusion for 5 min enhances the firing severalfold for many hours (45 min is shown, average of four experiments). In a separate set of experiments, the H1R antagonist mepyramine was present during the histamine perfusion. In this case, the solely H2R-mediated actions consisted of an initial small inhibition followed by a long-lasting increase of firing. Thus the maximal histamine-induced potentiation is seen when both H1R and H2R are activated. The effects are cAMP-PKA dependent [28, 76]

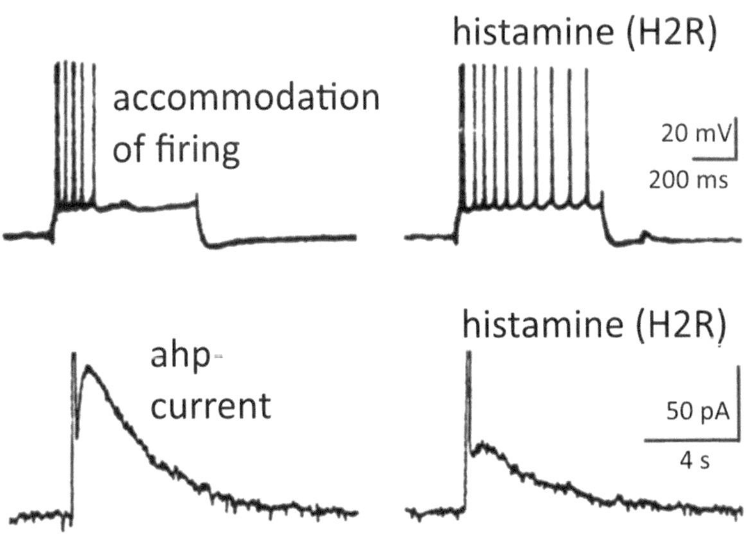


**Fig. 9.11** Histamine blocks accommodation of firing and the $I_{ahp}$ current which is also responsible for long-lasting afterhyperpolarizations in a hippocampal pyramidal cell

## 9.5 Conclusions

Release of histamine modulates neuronal activity in many ways. H1R activation causes mainly excitation, H2R mediate potentiation of excitation, and H3R are located on histaminergic neuron somata, dendrites, and axons to control firing, histamine synthesis, and release, acting as autoreceptors. As heteroreceptors they control transmitter release from a wide variety of non-histaminergic axons and in some cases directly inhibit the activity of non-histaminergic neurons. Therefore, H3R are a difficult but promising target for pharmaceutical development.

## References

1. McCaman RE, Weinreich D. On the nature of histamine-mediated slow hyperpolarizing synaptic potentials in identified molluscan neurones. J Physiol. 1982;328:485–506.
2. Carpenter DO, Gaubatz GL. H1 and H2 histamine receptors on Aplysia neurones. Nature. 1975;254(5498):343–4.
3. Panula P, Yang HY, Costa E. Histamine-containing neurons in the rat hypothalamus. Proc Natl Acad Sci U S A. 1984;81(8):2572–6.
4. Watanabe T, Maeyama K, Wada H. Histamine. Tanpakushitsu Kakusan Koso. 1984;29(12 Suppl):1443–58.
5. Ercan-Sencicek AG, Stillman AA, Ghosh AK, Bilguvar K, O'Roak BJ, Mason CE, et al. L-histidine decarboxylase and Tourette's syndrome. N Engl J Med. 2010;362(20):1901–8.
6. Ericson H, Watanabe T, Kohler C. Morphological analysis of the tuberomammillary nucleus in the rat brain: delineation of subgroups with antibody against L-histidine decarboxylase as a marker. J Comp Neurol. 1987;263(1):1–24.
7. Yu X, Ye Z, Houston CM, Zecharia AY, Ma Y, Zhang Z, et al. Wakefulness is governed by GABA and histamine cotransmission. Neuron. 2015;87(1):164–78.
8. Airaksinen MS, Paetau A, Paljarvi L, Reinikainen K, Riekkinen P, Suomalainen R, et al. Histamine neurons in human hypothalamus: anatomy in normal and Alzheimer diseased brains. Neuroscience. 1991;44(2):465–81.
9. Giannoni P, Medhurst AD, Passani MB, Giovannini MG, Ballini C, Corte LD, et al. Regional differential effects of the novel histamine H3 receptor antagonist 6-[(3-cyclobutyl-2,3,4,5 -tetrahydro-1H-3-benzazepin-7-yl)oxy]-N-methyl-3-pyridine carboxamide hydrochloride (GSK189254) on histamine release in the central nervous system of freely moving rats. J Pharmacol Exp Ther. 2010;332(1):164–72.
10. Airaksinen MS, Reinikainen K, Riekkinen P, Panula P. Neurofibrillary tangles and histamine-containing neurons in Alzheimer hypothalamus. Agents Actions. 1991;33(1–2):104–7.
11. Lin JS. Brain structures and mechanisms involved in the control of cortical activation and wakefulness, with emphasis on the posterior hypothalamus and histaminergic neurons. Sleep Med Rev. 2000;4(5):471–503.
12. Taddese A, Bean BP. Subthreshold sodium current from rapidly inactivating sodium channels drives spontaneous firing of tuberomammillary neurons. Neuron. 2002;33(4):587–600.
13. Greene RW, Haas HL, Reiner PB. Two transient outward currents in histamine neurones of the rat hypothalamus in vitro. J Physiol. 1990;420:149–63.
14. Stevens DR, Eriksson KS, Brown RE, Haas HL. The mechanism of spontaneous firing in histamine neurons. Behav Brain Res. 2001;124(2):105–12.
15. Takeshita Y, Watanabe T, Sakata T, Munakata M, Ishibashi H, Akaike N. Histamine modulates high-voltage-activated calcium channels in neurons dissociated from the rat tuberomammillary nucleus. Neuroscience. 1998;87(4):797–805.

16. Sergeeva OA, Eriksson KS, Sharonova IN, Vorobjev VS, Haas HL. GABA(A) receptor heterogeneity in histaminergic neurons. Eur J Neurosci. 2002;16(8):1472–82.
17. Sergeeva OA, Eriksson KS, Haas HL. Glycine receptor mediated responses in rat histaminergic neurons. Neurosci Lett. 2001;300(1):5–8.
18. Prast H, Prast M, Philippu A. H3 autoreceptors and muscarinic acetylcholine receptors modulate histamine release in the anterior hypothalamus of freely moving rats. Agents Actions. 1994;41(Spec No):C64–5.
19. Yanovsky Y, Li S, Klyuch BP, Yao Q, Blandina P, Passani MB, et al. L-Dopa activates histaminergic neurons. J Physiol. 2011;589(Pt 6):1349–66.
20. Sergeeva OA, Amberger BT, Haas HL. Editing of AMPA and serotonin 2C receptors in individual central neurons, controlling wakefulness. Cell Mol Neurobiol. 2007;27(5):669–80.
21. Siegel JM, Boehmer LN. Narcolepsy and the hypocretin system—where motion meets emotion. Nat Clin Pract Neurol. 2006;2(10):548–56.
22. Zeitzer JM, Nishino S, Mignot E. The neurobiology of hypocretins (orexins), narcolepsy and related therapeutic interventions. Trends Pharmacol Sci. 2006;27(7):368–74.
23. Parmentier R, Kolbaev S, Klyuch BP, Vandael D, Lin JS, Selbach O, et al. Excitation of histaminergic tuberomamillary neurons by thyrotropin-releasing hormone. J Neurosci. 2009;29(14):4471–83.
24. Tasneem A, Iyer LM, Jakobsson E, Aravind L. Identification of the prokaryotic ligand-gated ion channels and their implications for the mechanisms and origins of animal Cys-loop ion channels. Genome Biol. 2005;6(1):R4.
25. McClintock TS, Ache BW. Histamine directly gates a chloride channel in lobster olfactory receptor neurons. Proc Natl Acad Sci U S A. 1989;86(20):8137–41.
26. Hardie RC. A histamine-activated chloride channel involved in neurotransmission at a photoreceptor synapse. Nature. 1989;339(6227):704–6.
27. Pantazis A, Segaran A, Liu CH, Nikolaev A, Rister J, Thum AS, et al. Distinct roles for two histamine receptors (hclA and hclB) at the Drosophila photoreceptor synapse. J Neurosci. 2008;28(29):7250–9.
28. Haas HL, Sergeeva OA, Selbach O. Histamine in the nervous system. Physiol Rev. 2008;88(3):1183–241.
29. Chiel HJ, Weiss KR, Kupfermann I. Multiple roles of a histaminergic afferent neuron in the feeding behavior of Aplysia. Trends Neurosci. 1990;13(6):223–7.
30. Chiel HJ, Kupfermann I, Weiss KR. An identified histaminergic neuron can modulate the outputs of buccal-cerebral interneurons in Aplysia via presynaptic inhibition. J Neurosci. 1988;8(1):49–63.
31. Saras A, Gisselmann G, Vogt-Eisele AK, Erlkamp KS, Kletke O, Pusch H, et al. Histamine action on vertebrate GABAA receptors: direct channel gating and potentiation of GABA responses. J Biol Chem. 2008;283(16):10470–5.
32. Kletke O, Gisselmann G, May A, Hatt H, A Sergeeva O. Partial agonism of taurine at gamma-containing native and recombinant GABAA receptors. PLoS One. 2013;8(4), e61733.
33. Kletke O, Sergeeva OA, Lorenz P, Oberland S, Meier JC, Hatt H, et al. New insights in endogenous modulation of ligand-gated ion channels: histamine is an inverse agonist at strychnine sensitive glycine receptors. Eur J Pharmacol. 2013;710(1–3):59–66.
34. Yanovsky Y, Schubring SR, Yao Q, Zhao Y, Li S, May A, et al. Waking action of ursodeoxycholic acid (UDCA) involves histamine and GABAA receptor block. PLoS One. 2012;7(8), e42512.
35. Diewald L, Heimrich B, Busselberg D, Watanabe T, Haas HL. Histaminergic system in co-cultures of hippocampus and posterior hypothalamus: a morphological and electrophysiological study in the rat. Eur J Neurosci. 1997;9(11):2406–13.
36. Hatton GI, Yang QZ. Ionotropic histamine receptors and H2 receptors modulate supraoptic oxytocin neuronal excitability and dye coupling. J Neurosci. 2001;21(9):2974–82.
37. Lee KH, Broberger C, Kim U, McCormick DA. Histamine modulates thalamocortical activity by activating a chloride conductance in ferret perigeniculate neurons. Proc Natl Acad Sci U S A. 2004;101(17):6716–21.

38. Haas HL, Anderson EG, Hosli L. Histamine and metabolites: their effects and interactions with convulsants on brain stem neurons. Brain Res. 1973;51:269–78.
39. Lakoski JM, Aghajanian GK, Gallager DW. Interaction of histamine H2-receptor antagonists with GABA and benzodiazepine binding sites in the CNS. Eur J Pharmacol. 1983;88(2–3):241–5.
40. Payne GW, Neuman RS. Effects of hypomagnesemia on histamine H1 receptor-mediated facilitation of NMDA responses. Br J Pharmacol. 1997;121(2):199–204.
41. Bekkers JM, Stevens CF. NMDA receptors at excitatory synapses in the hippocampus: test of a theory of magnesium block. Neurosci Lett. 1993;156(1–2):73–7.
42. Vorobjev VS, Sharonova IN, Walsh IB, Haas HL. Histamine potentiates N-methyl-D-aspartate responses in acutely isolated hippocampal neurons. Neuron. 1993;11(5):837–44.
43. Green JP. Histamine and the nervous system. Fed Proc. 1964;23:1095–102.
44. Saybasili H, Stevens DR, Haas HL. pH-dependent modulation of N-methyl-D-aspartate receptor-mediated synaptic currents by histamine in rat hippocampus in vitro. Neurosci Lett. 1995;199(3):225–7.
45. Yanovsky Y, Reymann K, Haas HL. pH-dependent facilitation of synaptic transmission by histamine in the CA1 region of mouse hippocampus. Eur J Neurosci. 1995;7(10):2017–20.
46. Williams K. Subunit-specific potentiation of recombinant N-methyl-D-aspartate receptors by histamine. Mol Pharmacol. 1994;46(3):531–41.
47. Hansen KB, Mullasseril P, Dawit S, Kurtkaya NL, Yuan H, Vance KM, et al. Implementation of a fluorescence-based screening assay identifies histamine H3 receptor antagonists clobenpropit and iodophenpropit as subunit-selective N-methyl-D-aspartate receptor antagonists. J Pharmacol Exp Ther. 2010;333(3):650–62.
48. Faucard R, Armand V, Heron A, Cochois V, Schwartz JC, Arrang JM. N-methyl-D-aspartate receptor antagonists enhance histamine neuron activity in rodent brain. J Neurochem. 2006;98(5):1487–96.
49. Ikoma A, Steinhoff M, Stander S, Yosipovitch G, Schmelz M. The neurobiology of itch. Nat Rev Neurosci. 2006;7(7):535–47.
50. Koerner P, Hesslinger C, Schaefermeyer A, Prinz C, Gratzl M. Evidence for histamine as a transmitter in rat carotid body sensor cells. J Neurochem. 2004;91(2):493–500.
51. Whyment AD, Blanks AM, Lee K, Renaud LP, Spanswick D. Histamine excites neonatal rat sympathetic preganglionic neurons in vitro via activation of H1 receptors. J Neurophysiol. 2006;95(4):2492–500.
52. Gorelova N, Reiner PB. Histamine depolarizes cholinergic septal neurons. J Neurophysiol. 1996;75(2):707–14.
53. Uteshev VV, Knot HJ. Somatic Ca(2+) dynamics in response to choline-mediated excitation in histaminergic tuberomammillary neurons. Neuroscience. 2005;134(1):133–43.
54. Uteshev VV, Stevens DR, Haas HL. Alpha-bungarotoxin-sensitive nicotinic responses in rat tuberomammillary neurons. Pflugers Arch. 1996;432(4):607–13.
55. Smith CC, Greene RW. CNS dopamine transmission mediated by noradrenergic innervation. J Neurosci. 2012;32(18):6072–80.
56. Korotkova TM, Haas HL, Brown RE. Histamine excites GABAergic cells in the rat substantia nigra and ventral tegmental area in vitro. Neurosci Lett. 2002;320(3):133–6.
57. Zhou FW, Xu JJ, Zhao Y, LeDoux MS, Zhou FM. Opposite functions of histamine H1 and H2 receptors and H3 receptor in substantia nigra pars reticulata. J Neurophysiol. 2006;96(3):1581–91.
58. Tabarean IV. Histamine receptor signaling in energy homeostasis. Neuropharmacology. 2015.
59. Liou SY, Shibata S, Yamakawa K, Ueki S. Inhibitory and excitatory effects of histamine on suprachiasmatic neurons in rat hypothalamic slice preparation. Neurosci Lett. 1983;41(1–2):109–13.
60. Stehle J. Effects of histamine on spontaneous electrical activity of neurons in rat suprachiasmatic nucleus. Neurosci Lett. 1991;130(2):217–20.
61. Scott G, Piggins HD, Semba K, Rusak B. Actions of histamine in the suprachiasmatic nucleus of the Syrian hamster. Brain Res. 1998;783(1):1–9.
62. Li Z, Hatton GI. Histamine suppresses non-NMDA excitatory synaptic currents in rat supraoptic nucleus neurons. J Neurophysiol. 2000;83(5):2616–25.
63. Bealer SL, Crowley WR. Histaminergic control of oxytocin release in the paraventricular nucleus during lactation in rats. Exp Neurol. 2001;171(2):317–22.

64. Hatton GI, Li ZH. Neurophysiology of magnocellular neuroendocrine cells: recent advances. Prog Brain Res. 1998;119:77–99.
65. Haas HL. Histamine: action on single hypothalamic neurones. Brain Res. 1974;76(2):363–6.
66. Renaud LP. Histamine microiontophoresis on identified hypothalamic neurons: 3 patterns of response in the ventromedial nucleus of the rat. Brain Res. 1976;115(2):339–44.
67. McCormick DA, Williamson A. Modulation of neuronal firing mode in cat and guinea pig LGNd by histamine: possible cellular mechanisms of histaminergic control of arousal. J Neurosci. 1991;11(10):3188–99.
68. Weiler HT, Hasenohrl RU, van Landeghem AA, van Landeghem M, Brankack J, Huston JP, et al. Differential modulation of hippocampal signal transfer by tuberomammillary nucleus stimulation in freely moving rats dependent on behavioral state. Synapse. 1998;28(4):294–301.
69. Brown RE, Haas HL. On the mechanism of histaminergic inhibition of glutamate release in the rat dentate gyrus. J Physiol. 1999;515(Pt 3):777–86.
70. Brown RE, Reymann KG. Histamine H3 receptor-mediated depression of synaptic transmission in the dentate gyrus of the rat in vitro. J Physiol. 1996;496(Pt 1):175–84.
71. Manahan-Vaughan D, Reymann KG, Brown RE. In vivo electrophysiological investigations into the role of histamine in the dentate gyrus of the rat. Neuroscience. 1998;84(3):783–90.
72. Haas H, Panula P. The role of histamine and the tuberomamillary nucleus in the nervous system. Nat Rev Neurosci. 2003;4(2):121–30.
73. Bliss TV, Lomo T. Long-lasting potentiation of synaptic transmission in the dentate area of the anaesthetized rabbit following stimulation of the perforant path. J Physiol. 1973;232(2):331–56.
74. Yanovsky Y, Haas HL. Histamine increases the bursting activity of pyramidal cells in the CA3 region of mouse hippocampus. Neurosci Lett. 1998;240(2):110–2.
75. Buzsaki G, Draguhn A. Neuronal oscillations in cortical networks. Science. 2004;304 (5679):1926–9.
76. Selbach O, Brown RE, Haas HL. Long-term increase of hippocampal excitability by histamine and cyclic AMP. Neuropharmacology. 1997;36(11–12):1539–48.
77. Haas HL, Konnerth A. Histamine and noradrenaline decrease calcium-activated potassium conductance in hippocampal pyramidal cells. Nature. 1983;302(5907):432–4.
78. Chepkova AN, Sergeeva OA, Haas HL. Carbenoxolone impairs LTP and blocks NMDA receptors in murine hippocampus. Neuropharmacology. 2008;55(2):139–47.
79. Knoche A, Yokoyama H, Ponomarenko A, Frisch C, Huston J, Haas HL. High-frequency oscillation in the hippocampus of the behaving rat and its modulation by the histaminergic system. Hippocampus. 2003;13(2):273–80.
80. Lin JS, Anaclet C, Sergeeva OA, Haas HL. The waking brain: an update. Cell Mol Life Sci. 2011;68(15):2499–512.

# Chapter 10
# Heterogeneity of Histaminergic Neurons

**Patrizia Giannoni**

**Abstract** The central histaminergic system has a complex neuroanatomical and functional organisation. It originates in a small area of the posterior hypothalamus, the tuberomammillary nucleus (TMN). Despite the restricted location of cell bodies, anatomical studies showed that histamine neurons project to almost the entire brain. Indeed, neuronal histamine (HA) has been proven to modulate a plethora of body functions. The TMN was initially considered a single functional entity with neurons working in a coordinated and synchronous way. Recently though, several works are indicating that histaminergic neurons are organised in heterogeneous populations with distinct roles. Accumulating evidence based on multiple techniques suggests different properties among histamine TMN neurons. Although further studies are needed to fully characterise the organisation of the central histaminergic system and its activation following specific stimuli, interesting observations are emerging on the selective activity of clusters of histaminergic neurons according to the homeostatic or behavioural status. The heterogeneity of histamine neurons might represent the key for a fine-tuned modulation of specific functions regulated by neuronal HA. With the present chapter, we analyse the main findings and discuss future directions.

**Keywords** Neuronal histamine • Tuberomammillary nucleus (TMN) • Heterogeneity • Projection areas • Stressors • Stoichiometry

## 10.1 Introduction

In the brain, histamine (HA) is produced from neurons and mast cells. Yamatodani and colleagues in the early 1980s estimated that approximately 50 % of HA content in the brain derives from brain mast cells [1]. Although a recent study performed in mast cell-deficient mice (W/Wv) demonstrated that mast cell-derived HA is involved in sleep regulation, feeding behaviour, anxiety and depression [2], most studies conducted on histaminergic functions in the CNS focused on HA released from neurons. The source of central neuronal HA has been established simultaneously by

P. Giannoni, Ph. D. (✉)
Department of Neuroscience, Institute of Functional Genomics, CNRS-University of Montpellier (UMR 5203) 34094, 141 rue de la Cardonille, Montpellier, Cedex 5, France
E-Mail: patrizia.giannoni@igf.cnrs.fr



P. Blandina, M.B. Passani (eds.), *Histamine Receptors*, The Receptors 28,
DOI 10.1007/978-3-319-40308-3_10

two different laboratories more than 30 years ago. Thanks to the development of selective antibodies, the origin of the central histaminergic system was located in the tuberomammillary nucleus (TMN), a region of the posterior hypothalamus. In 1983, Watanabe and colleagues showed histamine-containing cells in the rat brain by indirect immunofluorescent histochemistry [3] with an antibody against histidine decarboxylase (HDC), the enzyme converting L-histidine to HA [4]. Numerous HDC immunoreactive neurons were shown in the posterior hypothalamic area; furthermore, HDC-immunopositive nerve fibres with a varicose appearance were found widely distributed in the brain. With a similar approach, using an antibody against HA, Panula and collaborators described in 1984 a comparable distribution of histaminergic neurons [5]. Histamine-immunoreactive neuronal cell bodies were found only in the hypothalamic and premamillary areas. The largest group of cells was seen in the caudal magnocellular nucleus and medially on the dorsal and ventral aspects of the ventral premamillary nucleus. Immunoreactive nerve fibres, but not cell bodies, were detected in other parts of the brain. Histamine-containing neurons appeared organised in clusters. In the rat brain, neuronal populations named E1, E2 and E3 have been described in the posterior hypothalamus, E4 in the ventromedial area and E5 in the more dorsal part [3, 5–9]. Recently, a similar organisation has been described in the mouse brain [10]. Slight differences between species, Wistar rats and C57/Bl6 mice, were mostly attributed to a different body mass. It is well documented that histaminergic neurons send projections to almost the entire brain. Several studies [6–8, 11] showed that retrograde tracers injected into different brain regions labelled histaminergic cell bodies scattered through the TMN. Thus, although divided in clusters, projection neurons do not seem to be organised in a highly topographic way, and individual TMN neurons give rise to both ascending and descending projections. Interestingly, histaminergic neurons originate from the TMN in all vertebrates studied so far [6, 12, 13]. In the rat brain a total of about 4600 histaminergic neurons is estimated [7], while an approximately 64,000 were estimated in the human brain [14]. The arborisation of histaminergic neurons present diffuse varicosities containing synaptic vesicles, and rarely form classical synapses, like other neurotransmitter systems [15]. Given these neuroanatomical features, HA can exert specific and selective actions through direct contacts, but it also can diffuse from the site of release and act as a local hormone [16]. The activation of histaminergic neurons was initially associated to the wake state and arousal [17, 18]. Recent elegant works from the groups of Wisden [19] and Arrigoni [20] have shown the complex action of HA in concert with co-released GABA in the control of the wake/sleep cycle and alertness.

Considering its widespread distribution in the central nervous system, it is not surprising that neuronal histamine is involved in the regulation of several brain functions ranging from homeostatic to cognitive ones (Fig. 10.1). Histamine controls thermoregulation [21] and vestibular function [22]; it controls appetite and satiety in concert with other neurohormones [23–25]; it affects learning and memory associated to adverse events [26, 27]. It is becoming clear that the diverse effects that follow the activation of the histaminergic system impose a selective recruitment of specific histaminergic pathways in a coordinated manner according to the

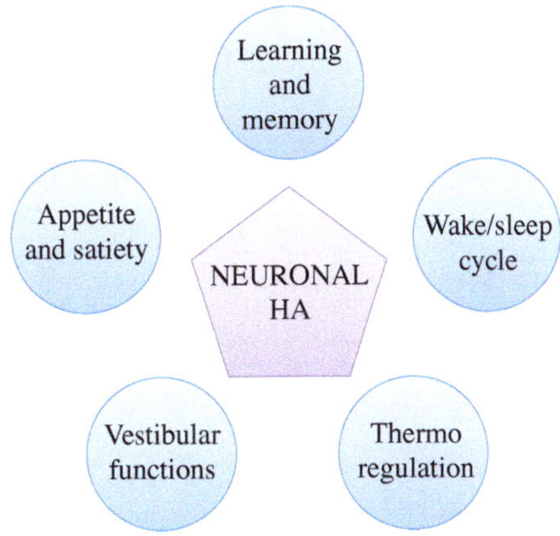


**Fig. 10.1** Body functions modulated by neuronal histamine. Research evidence has proven the implication of neuronal histamine in the modulation of numerous body functions. To note, histamine often works in concert with other neurotransmitters and/or neurohormones to exert this modulatory activity (see text for details)

environmental challenges. Indeed, despite the retrograde studies indicating the unitary nature of the TMN, there is now convincing evidence that histaminergic neurons are functionally and neuroanatomically heterogeneous. Here we review our current knowledge on the complexity of histaminergic neuron organisation.

## 10.2 Evidences of Histamine Neuron Heterogeneity

### *10.2.1 Histaminergic Neurons Differ in Sensitivities to Activity-Modulating Neurotransmitters*

The activity of histaminergic neurons is strictly dependent on behavioural state. They fire tonically during waking, little during slow-wave sleep and not at all during rapid eye movement (REM) sleep [28]. Inhibition of the histaminergic system is fundamental for sleep induction. Investigating the contribution of different systems to histaminergic inhibition, Sergeeva and colleagues found evidence of heterogeneity in histaminergic neurons [29]. Enzymatically isolated neurons were used to verify the contribution of glycine to histaminergic inhibition. In a whole-cell patch-clamp configuration, the majority of TMN cells (64 %) had pronounced glycine responses, whereas in 28 % of cells, the response to glycine was negligible. The remaining 8 % did not respond at all. Therefore, a different sensitivity to glycine stimulation suggested for the first time the presence of heterogeneous subpopulations of TMN neurons. Furthermore, glycine sensitivity correlated with soma size. Indeed, pronounced responses to glycine were associated with large cell bodies (25 μm), while smaller ones (19.5 μm) gave very small or no response at all. Two different subpopulations of TMN neurons were described as a function of glycine sensitivity and soma size. The physiological heterogeneity of TMN neurons was confirmed investigating GABAA receptor expression on TMN neurons [30]. It was known that GABAergic innervation from the ventrolateral preoptic area (VLPO)

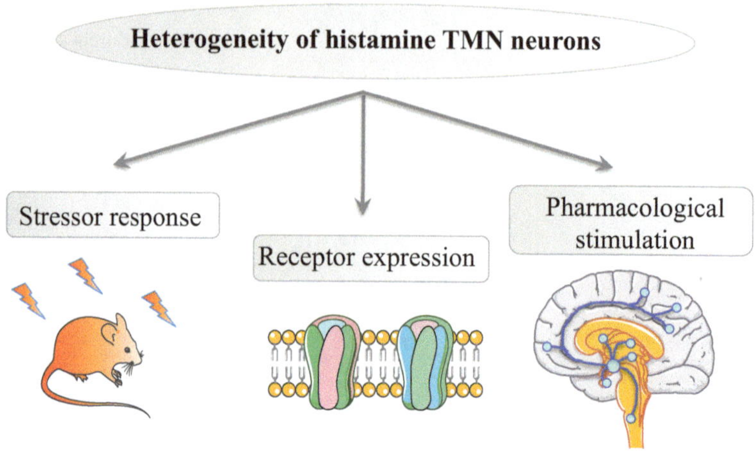


**Fig. 10.2** Evidence of histamine TMN neuron heterogeneity. The presence of distinct subpopulations among histamine TMN neurons is strongly suggested by recent findings. Main approaches applied to experimental investigations are schematically represented in the figure

regulates the firing of histaminergic neurons in the TMN [31]. Electrophysiology experiments coupled to biochemical studies investigated GABAA receptor subunit composition and stoichiometry in individual neurons in relation to GABA-evoked responses [28]. All 14 subunits previously identified were screened by PCR. Among the $\alpha$ subunits, $\alpha1$, $\alpha2$ and $\alpha5$ were detected, while $\beta1$, $\beta2$ and $\beta3$ and $\gamma1$ and $\gamma2$ were described for the $\beta$ and $\gamma$ subunits. In some cases, a $\varepsilon$ subunit was found.

Patch-clamp experiments revealed differences in sensitivity to GABA in the modulation of IPSC-decay times by zolpidem in those histaminergic neurons that express the $\gamma$ subunits at different levels.

Summarising, different stoichiometries were found in the population of TMN neurons analysed, as they presented one or two $\gamma$ subunits. It is suggested that, most likely, the majority of functional GABAA receptors in the TMN are assemblies either constituted by $\alpha2$, $\beta3$ and $\gamma2$ or $\alpha2$, $\beta3$ and $\gamma1$ subunits. The presence of one or two $\gamma$ subunits is correlated with GABA sensitivity. Therefore, different sensitivity to GABA appears to be associated with receptor subunit stoichiometry and to correlate with receptor function (Fig. 10.2). These seminal observations hint to the possibility of heterogeneous assembly of other types of receptors in different populations of TMN neurons.

### *10.2.2 Subpopulations of Histaminergic Neurons Are Selectively Activated by Different Stressors*

The implication of histamine in stress-related responses is well documented. Several observations indicate a crosstalk between HA, adrenocorticotropin (ACTH) and corticosterone secretion [32, 33]. Experiments following central administration of

HA or interference with histaminergic transmission with H3 receptor agonists and H1 or H2 receptor antagonists confirmed that HA is involved in responses to specific stressors that increase ACTH, β-endorphin and prolactin. In 2001, Haxhiu and colleagues demonstrated that, together with noradrenergic, dopaminergic and serotonergic neurons, histaminergic neurons were activated in response to hypercapnic stress [34] as assessed evaluating c-fos expression, a marker of neuronal activity. The authors suggested that the role of histaminergic neurons in response to hypercapnia might be related in part to the control of behavioural state. While in fact breathing is impaired by loss of wakefulness during sleep, histamine neuron stimulation might determine a respiratory response also regulated by the increased state of alertness. Interestingly, Miklós and Kovács in 2003 [35] demonstrated the presence of heterogeneous subpopulations of TMN neurons that specifically respond to different stressors (Fig. 10.2). Rats were exposed to a number of different stress stimuli (restraint, footshock, immobilisation, dehydration, hypoglycaemia, hypertonic salt, ether or LPS). Histaminergic neurons were identified by IHC and in situ hybridisation for HDC. Activation of specific subsets of histaminergic neurons was demonstrated by using c-fos immunostaining. The impact of the circadian rhythm was taken into account. No differences in neuronal activation were found between light and dark cycle. In control animals no c-fos staining was found in the E5, E2 and E1 subdivisions, while 0.2 % and 0.4 % histaminergic neurons were activated, respectively, in E3 and E4 subgroups. Noticeable differences in stress-induced c-fos activation emerged between subgroups, as restraint was the only stressor impacting on all histaminergic subdivisions (E1–E5). Footshock and insulin-induced hypoglycaemia preferentially activated the rostral subgroups (E3–E5), while immobilisation and dehydration acted mostly on E5 and E4. Hyperosmotic stimulation, ether stress and LPS treatment did not result in significant c-fos activation in any subgroup of histaminergic cells. Thus, histaminergic neurons do not respond generically to all stressors. The highest percentage of histaminergic neurons activated was determined by restraint, while intermediate responses were obtained with footshock, dehydration and immobilisation. The lack of significant activation following LPS stress was further confirmed by a recent study [36] demonstrating that LPS treatment reduced behavioural activity, in particular motivated behaviour such as exploration, play behaviour, social interaction and sweetened milk consumption. The behavioural responses were associated with reduction of c-fos expression in TMN neurons, an effect suggested to be mediated by the activation of the dorsal vagal complex (DVC). Hence, the histaminergic system represents an important component in the neuronal circuitry relevant for sickness behaviour. No clear anatomical segregation of histaminergic neurons according to the type of stressor was ever observed as histaminergic neurons activated by emotional stressors (e.g. restraint and footshock) intermingled with others activated by stressors classified as systemic challenge (e.g. hypoglycaemia). Nonetheless, these findings demonstrate the existence of functional subpopulations of histaminergic neurons.

Recently, Umehara and colleagues [37–39] identified a specific subpopulation of histaminergic neurons that are activated during food restriction by assessing suppression of c-fos expression with antihistamine pretreatment. The target region of these histaminergic neurons is the caudal part of the arcuate nucleus of the hypothalamus (cARC).

It is well known that histaminergic neurons convey satiety signalling by activating H1 receptors in the hypothalamic paraventricular (PVH) and the ventromedial (VMH) nuclei [40, 41]. Food deprivation under scheduled feeding induced c-fos expression in the cARC, while no differences were seen in the PVH and VMH [38]. Double immunofluorescent staining in the cARC showed that c-fos-positive cells also expressed H1 receptors. Thus, cARC neurons receive projections from the TMN and are activated by food deprivation through H1 receptors. Furthermore, while in normally fed rats c-fos expression in TMN was observed in the E1, E2 and E3 subdivisions of the TMN, food deprivation under scheduled feeding strongly and selectively activated the E3 group of TMN neurons, whereas the E1 and E2 subregions showed little or no activation [38]. These results suggested that the E3 subpopulation makes direct connection with the H1R-expressing cARC neurons, and the TMN-cARC circuit is selectively activated by food deprivation under restricted feeding [38]. Clearly, the activation of this histaminergic circuit is not related to the anorectic function of histaminergic neurons, but it may be related to a stress response induced by withdrawal of anticipated food under restricted feeding schedule [38].

### 10.2.3 Pharmacological Stimulation Demonstrates the Existence of Histamine Neuron Subpopulations that Project to Specific Brain Areas

Investigations on the function of histaminergic cells in response to drug treatments can be performed by microdialysis experiments. This technique enables continuous collection of interstitial fluid samples and monitoring of neurotransmitter levels. The effect of pharmacological stimulation can be verified in specific brain areas after systemic or local administration of a compound, and dual-probe experiments allow investigations in projection areas (Fig. 10.3). Studies performed with this technique suggested a possible selective pharmacological manipulation of specific subsets of histaminergic neurons. Among the first reports, investigations of the interaction between the endocannabinoid system and histaminergic neurons are found [42, 43]. Several functions such as cognition, locomotion, appetite and processing of emotional information are controlled by both systems [44–46], although evidence of a real interaction has been missing. CB1 receptor ligands were administered systemically as well as locally in the TMN. Histamine release was monitored directly in the TMN but also in the nucleus basalis magnocellularis (NBM) in the striatum and perirhinal cortex, all brain structures that receive histaminergic projections. The NBM is known to modulate the cortical activity [47, 48], and the stimulation of H1 receptors in the NBM increases acetylcholine release in the cortex [49]. The striatum presents a high concentration of H2 and H3 receptors and principally controls motor functions and stimulus–response habit formation [50],

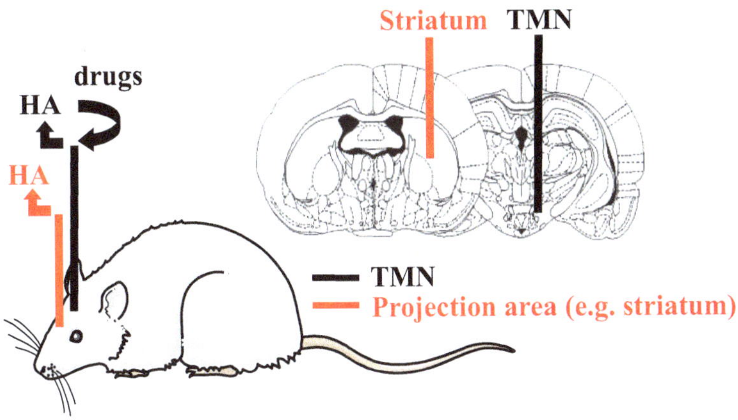


**Fig. 10.3** Schematic representation of a dual-probe microdialysis experiment. Rats were implanted two guide cannulae, one in the TMN and the second one in a histaminergic projection area (e.g. the striatum). Dialysates from both structures were analysed through HPLC to measure histamine content

whereas the perirhinal cortex is part of a neural circuit involved in recognition memory [51]. These studies demonstrated that systemic administration (i.p.) or intra-TMN perfusion of arachidonyl-2′ chloroethylamide/N-(2chloroethyl)-5Z, 8Z,11Z,14Z-eicosatetraenamide (ACEA) or R(+)-methanandamide (mAEA), two selective CB1 receptor agonists, significantly increased histamine release in the TMN, NBM and dorsal striatum, but not from the perirhinal cortex. Interestingly, the administration of an endocannabinoid membrane uptake blocker (AM404) determined an increase of HA release in the TMN, but not in the NBM nor in the striatum. Thus, exogenous and endogenous cannabinoids are suggested to exert different effects. When administered in the TMN, both ACEA and bicuculline significantly increased HA release from the TMN, whereas only the CB1 receptor agonist augmented HA release also from the NBM. Therefore, these observations indicate that excitation of histaminergic neurons might not necessarily produce a broad activation of all histaminergic projections and suggest the existence of subpopulations of histaminergic cells that respond differently to pharmacological manipulations and/or project to different brain regions.

Immunohistochemical analysis showed an overall low expression of CB1 receptors in the hypothalamus. To note, CB1 receptor immunostaining surrounded clusters of HDC-immunonegative cells. No colocalisation of CB1 receptor and HDC-positive cells was found [43].

Following these outcomes, other investigations were directly aimed at verifying whether histaminergic neurons are organised into functionally distinct circuits impinging on different brain regions [52, 53]. These studies used the double probe microdialysis technique with one probe in the proximity of the TMN, the other in one of the histaminergic projection areas, and HA release was measured in all these regions upon TMN perfusion with different compounds acting on different receptors known to be present on TMN neurons (H3 and GABAA receptors) (Fig. 10.2). The studies [52, 53] showed consistent effects on HA release of H3 receptor antagonists such as GSK189254 thioperamide, or ABT-239 administered into the TMN, where blockade of somatic and presynaptic H3 autoreceptors converge in augmenting HA levels in the synaptic cleft and increase histaminergic cell firing [54]. Acting as auto- and hetero-receptors, H3 receptors modulate also the release of numerous neurotransmitters, including acetylcholine (ACh), glutamate, noradrenaline and serotonin [55–58]. The results were compared to those obtained with intra-TMN infusion of the GABAA receptor antagonist bicuculline, that by blocking GABAA receptors on TMN neurons, local HA release and cell firing increase [52]. H3 antagonists were chosen as they have been proposed as drugs for the treatment of highly debilitating and socially devastating pathologies like obesity, sleep disorders, Alzheimer's and Parkinson's disease [59], although only one of them, pitolisant [60], was recently approved for the treatment of narcolepsy (http://www.ema.europa.eu)[61]. It was demonstrated that histaminergic neurons respond differently to the administered drugs despite the fact that both classes of ligands increase HA output within the posterior hypothalamus. While both bicuculline and H3 receptor antagonists always determined an increase of histamine release in the TMN, the effects in projection areas were strikingly different. Both classes of drugs increased HA release in the prefrontal cortex, whereas only H3 receptor antagonists increased HA output in the NBM; conversely, only bicuculline increased HA release in the nucleus accumbens (NAcc). None of the drugs augmented HA in the dorsal striatum. Single-probe experiments showed also that thioperamide infusion directly in the NBM or the prefrontal cortex increased HA release, but not from the dorsal striatum nor the NAcc [52].

To gather further insight into the mechanism of action of H3 receptor antagonists, the pattern of c-fos activation was examined in rat brain regions after perfusion of the TMN with ABT-239 [55]. In keeping with the microdialysis results, increased expression of c-fos with ABT-239 occurred in the prefrontal cortex and in the NBM, but neither in the NAcc nor in the striatum. Hence, despite neuroanatomical studies had shown that TMN histaminergic neurons are a rather homogeneous cell group with diffuse, overlapping projections throughout the neuraxis [8], the microdialysis studies support the hypothesis that subsets of histaminergic neurons form independent functional units modulated by selective mechanisms according to their respective origin and terminal projections.

Immunostaining performed with antibodies directed against HDC and H3 receptors revealed in the E2–E3 subdivision of TMN two histaminergic neuronal populations that differed significantly for H3R expression levels [62]. In fact, as confirmed by other authors, H3 receptor expression is a reliable marker for histaminergic neurons. The group of Sergeeva [63] recently verified the expression of H3 receptors in histaminergic neurons by single-cell RT-PCR and further characterised their response. Notably, even neurons projecting to the striatum, as shown by retrograde tracing, expressed H3 receptors. Thus, the lack of increase in HA release in this area after systemic or intra-TMN administration with an H3 antagonist [47, 48] cannot be explained on the basis of lack of expression of H3 receptors on striatum-projecting neurons. Sergeeva and co-authors [53] rightly speculate that GABA co-released with HA in the striatum may generate a tonic inhibitory effect counteracting histamine action [19]. Also, response to H3 receptors antagonists may be dampened within the striatum by activation of the TRPV1 channel [64], which is highly expressed in the dorsal striatum. Indeed, the striatum produces high levels of a "capsaicin-like" substance, N-arachidonoyl dopamine (NADA) that may decrease H3 receptor-mediated autoinhibition. Nonetheless, it is conceivable that the magnitude of neuronal responses to extracellular signals may depend on membrane receptor density with histaminergic neurons displaying low levels of the H3R, which are presumably those that innervate the NAcc or striatum.

Despite these seemingly unresolved questions, it is now clear that histaminergic neurons are not a homogenous neuronal population, and presumably functional differences of response relate to their heterogeneity with respect to projection fields, local environment (e.g. striatum) and co-release of other neurotransmitters.

As already mentioned, TMN neurons synthesise GABA [65] that is released from histaminergic neurons presumably in a paracrine manner similar to histamine in the cortex and striatum, as demonstrated by optogenetic studies [19]. The authors suggest that these TMN "GABA-histamine" neurons contribute to tonic inhibition of many neural circuits simultaneously. A different scenario was described in the posterior hypothalamus where the wake-active TMN and sleep-active ventrolateral preoptic nucleus (VLPO) are reciprocally connected [20], suggesting that each region can inhibit its counterpart when activated. Arrigoni and coworkers found that photostimulation evoked histamine release in the ventrolateral TMN (vlTMN) and the VLPO but found no evidence of GABA release. These results suggest that GABA is not released from histaminergic collaterals in the vlTMN.

Taken together, these observations clearly demonstrate the functional heterogeneity of histaminergic neurons impinging on different neuronal circuits to enhance wakefulness and alertness to shape motivation, cognition, locomotion and feeding.

## 10.3 Conclusions

The histaminergic nervous system has been the focus of extensive studies in the last three decades. Numerous advances have been done in order to clarify its complex structure and physiological organisation. Notably, central histamine regulates a plethora of body functions [62, 66]. If in the beginning TMN histamine neurons were suggested to act as a single functional entity, several findings based on functional and biochemical studies are now proving their organisation in heterogeneous subpopulations. Such heterogeneity is not based on topographical organisation, as evidenced by retrograde tracer investigations, but other characteristics as receptor expression or subunit stoichiometry have been proven. In some cases, experimental observations were not followed by univocal interpretations, and further studies are required to fully elucidate the underlying mechanisms. Experimental conditions and methodology limitations, as well as species differences, revealed to be key factors. As it emerges from this concise overview, additional investigations are needed, but recent findings open fascinating possibilities of a fine-tuned modulation of the histaminergic system. Results might help in the development of selective and safe drugs for the treatment of pathologies with a high social impact, ranging from obesity to cognitive disorders.

## References

1. Yamatodani A, Maeyama K, Watanabe T, Wada H, Kitamura Y. Tissue distribution of histamine in a mutant mouse deficient in mast cells: clear evidence for the presence of non-mast-cell histamine. Biochem Pharmacol. 1982;31(3):305–9.
2. Chikahisa S, Kodama T, Soya A, Sagawa Y, Ishimaru Y, Sei H, et al. Histamine from brain resident MAST cells promotes wakefulness and modulates behavioral states. PLoS One. 2013;8(10), e78434.
3. Watanabe T, Watanabe Y, Tani N, Miwa T, Watanabe K. Method for detection of antibody against low molecular weight antigen (hapten)—production and detection of antibody against histamine H2 receptor of antibody against histamine H2 receptor antagonist "Famotidine". Tokai J Exp Clin Med. 1983;8(4):307–15.
4. Green JP, Prell GD, Khandelwal JK, Blandina P. Aspects of histamine metabolism. Agents Actions. 1987;22(1–2):1–15.
5. Panula P, Yang HY, Costa E. Histamine-containing neurons in the rat hypothalamus. Proc Natl Acad Sci U S A. 1984;81(8):2572–6.
6. Inagaki N, Panula P, Yamatodani A, Wada H. Organization of the histaminergic system in the brain of the turtle Chinemys reevesii. J Comp Neurol. 1990;297(1):132–44.
7. Ericson H, Watanabe T, Kohler C. Morphological analysis of the tuberomammillary nucleus in the rat brain: delineation of subgroups with antibody against L-histidine decarboxylase as a marker. J Comp Neurol. 1987;263(1):1–24.
8. Kohler C, Swanson LW, Haglund L, Wu JY. The cytoarchitecture, histochemistry and projections of the tuberomammillary nucleus in the rat. Neuroscience. 1985;16(1):85–110.
9. Staines WA, Daddona PE, Nagy JI. The organization and hypothalamic projections of the tuberomammillary nucleus in the rat: an immunohistochemical study of adenosine deaminase-positive neurons and fibers. Neuroscience. 1987;23(2):571–96.
10. Moriwaki C, Chiba S, Wei H, Aosa T, Kitamura H, Ina K, et al. Distribution of histaminergic neuronal cluster in the rat and mouse hypothalamus. J Chem Neuroanat. 2015;68:1–13.
11. Lee SB, Chang BJ, Lee HS. Organization of histamine-immunoreactive, tuberomammillary neurons projecting to the dorsal tier of the substantia nigra compacta in the rat. Brain Res. 2008;1203:79–88.
12. Barroso CFM, Fasolo A, Panula P. Organization of histamine-containing neurons in the brain of the crested newt, Triturus carnifex. Cell Tissue Res. 1993;272(1):147–54.
13. Inagaki N, Panula P, Yamatodani A, Wada H. Organization of the histaminergic system in the brain of the teleost, Trachurus trachurus. J Comp Neurol. 1991;310(1):94–102.
14. Airaksinen MS, Paetau A, Paljarvi L, Reinikainen K, Riekkinen P, Suomalainen R, et al. Histamine neurons in human hypothalamus: anatomy in normal and Alzheimer diseased brains. Neuroscience. 1991;44(2):465–81.
15. Panula P, Karlstedt K, Sallmen T, Peitsaro N, Kaslin J, Michelsen KA, et al. The histaminergic system in the brain: structural characteristics and changes in hibernation. J Chem Neuroanat. 2000;18(1–2):65–74.

16. Wada H, Inagaki N, Yamatodani A, Watanabe T. Is the histaminergic neuron system a regulatory center for whole-brain activity? Trends Neurosci. 1991;14(9):415–8.
17. Lin JS, Anaclet C, Sergeeva OA, Haas HL. The waking brain: an update. Cell Mol Life Sci. 2011;68(15):2499–512.
18. Lin JS, Hou Y, Sakai K, Jouvet M. Histaminergic descending inputs to the mesopontine tegmentum and their role in the control of cortical activation and wakefulness in the cat. J Neurosci. 1996;16(4):1523–37.
19. Yu X, Ye Z, Houston CM, Zecharia AY, Ma Y, Zhang Z, et al. Wakefulness is governed by GABA and histamine cotransmission. Neuron. 2015;87(1):164–78.
20. Williams RH, Chee MJ, Kroeger D, Ferrari LL, Maratos-Flier E, Scammell TE, et al. Optogenetic-mediated release of histamine reveals distal and autoregulatory mechanisms for controlling arousal. J Neurosci. 2014;34(17):6023–9.
21. Sakata T, Yoshimatsu H, Kurokawa M. Hypothalamic neuronal histamine: implications of its homeostatic control of energy metabolism. Nutrition. 1997;13(5):403–11.
22. Tighilet B, Leonard J, Lacour M. Betahistine dihydrochloride treatment facilitates vestibular compensation in the cat. J Vestib Res. 1995;5(1):53–66.
23. Provensi G, Coccurello R, Umehara H, Munari L, Giacovazzo G, Galeotti N, et al. Satiety factor oleoylethanolamide recruits the brain histaminergic system to inhibit food intake. Proc Natl Acad Sci U S A. 2014;111(31):11527–32.
24. Provensi G, Blandina P, Passani MB. The histaminergic system as a target for the prevention of obesity and metabolic syndrome. Neuropharmacology. 2015.
25. Provensi G. Histamine receptors and appetite. In: editions S, editor. Histamine receptors. Springer; 2016.
26. Izquierdo F, de Carvalho. Histamine and histamine receptors in fear extinction. In: editions S, editor. Histamine receptors. 2016.
27. Benetti F, Furini CR, de Carvalho MJ, Provensi G, Passani MB, Baldi E, et al. Histamine in the basolateral amygdala promotes inhibitory avoidance learning independently of hippocampus. Proc Natl Acad Sci U S A. 2015;112(19):E2536–42.
28. Lin JS, Sakai K, Jouvet M. Evidence for histaminergic arousal mechanisms in the hypothalamus of cat. Neuropharmacology. 1988;27(2):111–22.
29. Sergeeva OA, Eriksson KS, Haas HL. Glycine receptor mediated responses in rat histaminergic neurons. Neurosci Lett. 2001;300(1):5–8.
30. Sergeeva OA, Eriksson KS, Sharonova IN, Vorobjev VS, Haas HL. GABA(A) receptor heterogeneity in histaminergic neurons. Eur J Neurosci. 2002;16(8):1472–82.
31. Sherin JE, Elmquist JK, Torrealba F, Saper CB. Innervation of histaminergic tuberomammillary neurons by GABAergic and galaninergic neurons in the ventrolateral preoptic nucleus of the rat. J Neurosci. 1998;18(12):4705–21.
32. Rudolph C, Richards GE, Kaplan S, Ganong WF. Effect of intraventricular histamine on hormone secretion in dogs. Neuroendocrinology. 1979;29(3):169–77.
33. Knigge U, Matzen S, Hannibal T, Jorgensen H, Warberg J. Involvement of histamine in the mediation of the stress-induced release of alpha-melanocyte-stimulating hormone in male rats. Neuroendocrinology. 1991;54(6):646–52.
34. Haxhiu MA, Tolentino-Silva F, Pete G, Kc P, Mack SO. Monoaminergic neurons, chemosensation and arousal. Respir Physiol. 2001;129(1–2):191–209.
35. Miklos IH, Kovacs KJ. Functional heterogeneity of the responses of histaminergic neuron subpopulations to various stress challenges. Eur J Neurosci. 2003;18(11):3069–79.
36. Gaykema RP, Park SM, McKibbin CR, Goehler LE. Lipopolysaccharide suppresses activation of the tuberomammillary histaminergic system concomitant with behavior: a novel target of immune-sensory pathways. Neuroscience. 2008;152(1):273–87.
37. Umehara H, Mizuguchi H, Mizukawa N, Matsumoto M, Takeda N, Senba E, et al. Innervation of histamine neurons in the caudal part of the arcuate nucleus of hypothalamus and their activation in response to food deprivation under scheduled feeding. Methods Find Exp Clin Pharmacol. 2010;32(10):733–6.

38. Umehara H, Mizuguchi H, Mizukawa N, Matsumoto M, Takeda N, Senba E, et al. Deprivation of anticipated food under scheduled feeding induces c-Fos expression in the caudal part of the arcuate nucleus of hypothalamus through histamine H(1) receptors in rats: potential involvement of E3 subgroup of histaminergic neurons in tuberomammillary nucleus. Brain Res. 2011;1387:61–70.
39. Umehara H, Mizuguchi H, Fukui H. Identification of a histaminergic circuit in the caudal hypothalamus: an evidence for functional heterogeneity of histaminergic neurons. Neurochem Int. 2012;61(6):942–7.
40. Masaki T, Chiba S, Yasuda T, Noguchi H, Kakuma T, Watanabe T, et al. Involvement of hypothalamic histamine H1 receptor in the regulation of feeding rhythm and obesity. Diabetes. 2004;53(9):2250–60.
41. Ookuma K, Sakata T, Fukagawa K, Yoshimatsu H, Kurokawa M, Machidori H, et al. Neuronal histamine in the hypothalamus suppresses food intake in rats. Brain Res. 1993;628(1–2):235–42.
42. Passani MB, Giannoni P, Bucherelli C, Baldi E, Blandina P. Histamine in the brain: beyond sleep and memory. Biochem Pharmacol. 2007;73(8):1113–22.
43. Cenni G, Blandina P, Mackie K, Nosi D, Formigli L, Giannoni P, et al. Differential effect of cannabinoid agonists and endocannabinoids on histamine release from distinct regions of the rat brain. Eur J Neurosci. 2006;24(6):1633–44.
44. Marsicano GWC, Azad SC, Bisogno T, Rammes G, Cascio MG, Hermann H, Tang J, Hofmann C, Zieglgänsberger W, Di Marzo V, Lutz B. The endogenous cannabinoid system controls extinction of aversive memories. Nature. 2002;418:530–4.
45. Kirkham TCWC, Fezza F, Di Marzo V. Endocannabinoid levels in rat limbic forebrain and hypothalamus in relation to fasting, feeding and satiation: stimulation of eating by 2-arachidonoyl glycerol. Br J Pharmacol. 2002;136(4):550–7.
46. Di Marzo VMI. Endocannabinoid control of food intake and energy balance. Nat Neurosci. 2005;8(5):585–9.
47. Knox D, Berntson GG. Cortical modulation by nucleus basalis magnocellularis corticopetal cholinergic neurons during anxiety-like states is reflected by decreases in delta. Brain Res. 2008;1227:142–52.
48. Wenk H, Bigl V, Meyer U. Cholinergic projections from magnocellular nuclei of the basal forebrain to cortical areas in rats. Brain Res. 1980;2(3):295–316.
49. Cecchi M, Passani MB, Bacciottini L, Mannaioni PF, Blandina P. Cortical acetylcholine release elicited by stimulation of histamine H1 receptors in the nucleus basalis magnocellularis: a dual-probe microdialysis study in the freely moving rat. Eur J Neurosci. 2001;13(1):68–78.
50. Gerdeman GL, Partridge JG, Lupica CR, Lovinger DM. It could be habit forming: drugs of abuse and striatal synaptic plasticity. Trends Neurosci. 2003;26(4):184–92.
51. Warburton EC, Brown MW. Neural circuitry for rat recognition memory. Behav Brain Res. 2015;285:131–9.
52. Giannoni P, Passani MB, Nosi D, Chazot PL, Shenton FC, Medhurst AD, et al. Heterogeneity of histaminergic neurons in the tuberomammillary nucleus of the rat. Eur J Neurosci. 2009;29(12):2363–74.
53. Giannoni P, Medhurst AD, Passani MB, Giovannini MG, Ballini C, Corte LD, et al. Regional differential effects of the novel histamine H3 receptor antagonist 6-[(3-cyclobutyl-2,3,4,5-tetrahydro-1H-3-benzazepin-7-yl)oxy]-N-methyl-3-pyridine carboxamide hydrochloride (GSK189254) on histamine release in the central nervous system of freely moving rats. J Pharmacol Exp Ther. 2010;332(1):164–72.
54. Haas H, Panula P. The role of histamine and the tuberomamillary nucleus in the nervous system. Nat Rev Neurosci. 2003;4(2):121–30.
55. Munari L, Provensi G, Passani MB, Blandina P. Selective brain region activation by histamine H(3) receptor antagonist/inverse agonist ABT-239 enhances acetylcholine and histamine release and increases c-Fos expression. Neuropharmacology. 2013;70:131–40.

56. Osorio-Espinoza A, Alatorre A, Ramos-Jimenez J, Garduno-Torres B, Garcia-Ramirez M, Querejeta E, et al. Pre-synaptic histamine H(3) receptors modulate glutamatergic transmission in rat globus pallidus. Neuroscience. 2011;176:20–31.
57. Aquino-Miranda G, Osorio-Espinoza A, Escamilla-Sanchez J, Gonzalez-Pantoja R, Ortiz J, Arias-Montano JA. Histamine H(3) receptors modulate depolarization-evoked [(3) H]-noradrenaline release from rat olfactory bulb slices. Neuropharmacology. 2012;62(2):1127–33.
58. Threlfell S, Cragg SJ, Kallo I, Turi GF, Coen CW, Greenfield SA. Histamine H3 receptors inhibit serotonin release in substantia nigra pars reticulata. J Neurosci. 2004;24(40):8704–10.
59. Passani MB, Blandina P. Histamine receptors in the CNS as targets for therapeutic intervention. Trends Pharmacol Sci. 2011;32(4):242–9.
60. Dauvilliers Y, Bassetti C, Lammers GJ, Arnulf I, Mayer G, Rodenbeck A, et al. Pitolisant versus placebo or modafinil in patients with narcolepsy: a double-blind, randomised trial. Lancet Neurol. 2013;12(11):1068–75.
61. http://www.ema.europa.eu. 2016.
62. Blandina P, Munari L, Provensi G, Passani MB. Histamine neurons in the tuberomamillary nucleus: a whole center or distinct subpopulations? Front Syst Neurosci. 2012;6:33.
63. De Luca R, Suvorava T, Yang D, Baumgartel W, Kojda G, Haas HL, et al. Identification of histaminergic neurons through histamine 3 receptor-mediated autoinhibition. Neuropharmacology. 2015.
64. Huang SM, Bisogno T, Trevisani M, Al-Hayani A, De Petrocellis L, Fezza F, et al. An endogenous capsaicin-like substance with high potency at recombinant and native vanilloid VR1 receptors. Proc Natl Acad Sci U S A. 2002;99(12):8400–5.
65. Kukko-Lukjanov TK, Panula P. Subcellular distribution of histamine, GABA and galanin in tuberomamillary neurons in vitro. J Chem Neuroanat. 2003;25(4):279–92.
66. Passani MB, Lin JS, Hancock A, Crochet S, Blandina P. The histamine H3 receptor as a novel therapeutic target for cognitive and sleep disorders. Trends Pharmacol Sci. 2004;25(12):618–25.

# Chapter 11
# Modulation by Histamine $H_3$ Receptors of Neurotransmitter Release in the Basal Ganglia

**Ricardo Márquez-Gómez, Ana-Maricela García-Gálvez, Guadalupe-Elide Morales-Figueroa, and José-Antonio Arias-Montaño**

**Abstract** Among the four G-protein coupled receptors ($H_1$-$H_4$ ) identified as the mediators of the biological effects of histamine, the $H_3$ receptor ($H_3R$) distinguishes for its almost exclusive expression in the nervous system and its dual function as auto- and hetero-receptor that enables $H_3Rs$ to modulate the histaminergic and other neurotransmitter systems. The basal ganglia are neuronal nuclei that form a subcortical circuitry responsible for integrating motor and sensorial information originated in the cerebral cortex and the thalamus. The abundant presence of $H_3Rs$ in the basal ganglia confers these receptors a preferential and strategic position to modulate both the incoming and the outgoing synaptic information. In this chapter we review the control by $H_3Rs$ of the release of the neurotransmitters involved in the basal ganglia circuitry.

**Keywords** Acetylcholine • Basal ganglia • Calcium channels • Cerebral cortex • Central nervous system • GABA • G protein-coupled receptors • Dopamine • Globus pallidus • Glutamate • Histamine • Histamine $H_3$ receptors • Hypothalamus • Neurotransmitter release • Neuromodulation • Noradrenaline • Serotonin (5-Hydroxytryptamine) • Striatum • Substantia nigra • Subthalamic nucleus • Thalamus

---

R. Márquez-Gómez, M.Sc. • A.-M. García-Gálvez, M.Sc. • G.-E. Morales-Figueroa, M.Sc.
J.-A. Arias-Montaño, M.D., Ph.D. (✉)
Departamento de Fisiología, Biofísica y Neurociencias, Centro de Investigación y de Estudios Avanzados del IPN, Av. IPN 2508, Zacatenco, Ciudad de México, D.F. 07360, Mexico
e-mail: ricardo.marquez3@gmail.com; anikcson_06@hotmail.com; mofi130@hotmail.com; jaarias@fisio.cinvestav.mx



P. Blandina, M.B. Passani (eds.), *Histamine Receptors*, The Receptors 28,
DOI 10.1007/978-3-319-40308-3_11

## Abbreviations

| | |
|---|---|
| 5-HT | 5-Hydroxytryptamine (serotonin) |
| $A_1R$ | Adenosine $A_1$ receptor |
| $A_{2A}R$ | Adenosine $A_{2A}$ receptor |
| BG | Basal ganglia |
| CNS | Central nervous system |
| $D_1R$ | Dopamine $D_1$-like receptor |
| $D_2R$ | Dopamine $D_2$-like receptor |
| EPSCs | Excitatory postsynaptic currents |
| FP | Field potential |
| GABA | γ-Aminobutyric acid |
| GIRK | G protein-gated inwardly rectifying $K^+$ channel |
| GP | Globus pallidus |
| GPCR | G protein-coupled receptor |
| $H_1R$ | Histamine $H_1$ receptor |
| $H_2R$ | Histamine $H_2$ receptor |
| $H_3R$ | Histamine $H_3$ receptor |
| $H_4R$ | Histamine $H_4$ receptor |
| HDC | Histidine decarboxylase |
| IPSCs | Inhibitory postsynaptic currents |
| MSN | Medium-sized spiny neuron |
| PKA | Protein kinase A |
| PKC | Protein kinase C |
| PLC | Phospholipase C |
| RN | Raphe nuclei |
| SN | Substantia nigra |
| SNc | Substantia nigra *pars compacta* |
| SNr | Substantia nigra *pars reticulata* |
| STN | Subthalamic nucleus |
| TH | Tyrosine hydroxylase |
| TMN | Tuberomammillary nucleus |

## 11.1 Introduction

Among the four G protein-coupled receptors ($H_1$, $H_2$, $H_3$, and $H_4$) identified as the mediators of the effects of histamine in mammals, the $H_3$ receptor ($H_3R$) stands out for its almost exclusive expression in the nervous system, the large variety of isoforms naturally found, and its dual function as auto- and hetero-receptor, capable of modulating the histaminergic and other neurotransmitter systems.

Movement is one of the central nervous system's (CNS) finest tuned processes, and for this purpose the CNS has evolved to form a subcortical circuitry, the basal ganglia (BG), which integrates motor and sensorial information originated in the cerebral cortex and the thalamus. The abundant presence of $H_3Rs$ in the BG confers these receptors a preferential and strategic position to modulate both the incoming and outgoing synaptic information. $H_3Rs$ have been implicated in the control of the BG function and therefore as a potential target for the treatment of neurological disorders such as addiction, depression, schizophrenia, and Parkinson's disease, among others, in which BG are affected.

One of the most experimentally tested consequences of $H_3R$ activation in the BG is its nearly ubiquitous inhibitory effect on neurotransmitter release, which will be reviewed in this chapter.

## 11.2 The Basal Ganglia (BG)

The BG comprise five nuclei [1]: the striatum (caudate-putamen), the globus pallidus (GP; external segment of the globus pallidus in primates), the entopeduncular nucleus (internal segment of the globus pallidus in primates), the subthalamic nucleus (STN), and the substantia nigra (*compacta* (SNc) and *reticulata* (SNr)).

The striatum is the main input nucleus of the BG, and the most abundant type of striatal neuronal cells, the GABAergic medium-sized spiny neurons (MSNs), originates the major BG synaptic pathways. MSNs can be divided almost equitably into two populations identifiable by their genetic profile [2]. One MSN population synthesizes enkephalins and expresses dopamine $D_2$-like receptors ($D_2Rs$) and adenosine $A_{2A}$ receptors ($A_{2A}Rs$), whereas a second MSN population synthesizes dynorphins and expresses dopamine $D_1$-like receptors ($D_1Rs$) and adenosine $A_1$ receptors ($A_1Rs$). $D_1Rs$ ($D_1$ and $D_5$ subtypes) couple primarily to $G\alpha_s$ proteins, and $D_2Rs$ ($D_2$, $D_3$, and $D_4$ subtypes) activate mainly $G\alpha_{i/o}$ proteins [3].

The division of MSNs provides the basis for BG functional connectivity with the two populations originating the direct ($D_1R$-expressing MSNs ($D_1$-MSNs)) and indirect ($D_2R$-expressing MSNs ($D_2$-MSNs)) synaptic pathways, and the functional balance between them provides a general model to explain how the BG participate in the control of motor behavior in both normal and pathological conditions (Fig. 11.1).

### *11.2.1 BG Synaptic Circuitry*

Whereas under resting conditions MSNs are silent and their activation is driven by glutamatergic afferents, all the neurons in the GP, STN, and SNr are autonomous pacemakers, that is, they generate action potentials on their own, without synaptic input [4].

Each MSN population has particular targets within the BG (Fig. 11.1). The tonic firing of GABAergic neurons in the SNr, the main output nucleus of the system, keeps thalamic neurons inhibited. $D_1$-MSNs project to the SNr, and upon activation by glutamatergic corticostriatal, afferents exert a phasic and robust inhibitory effect

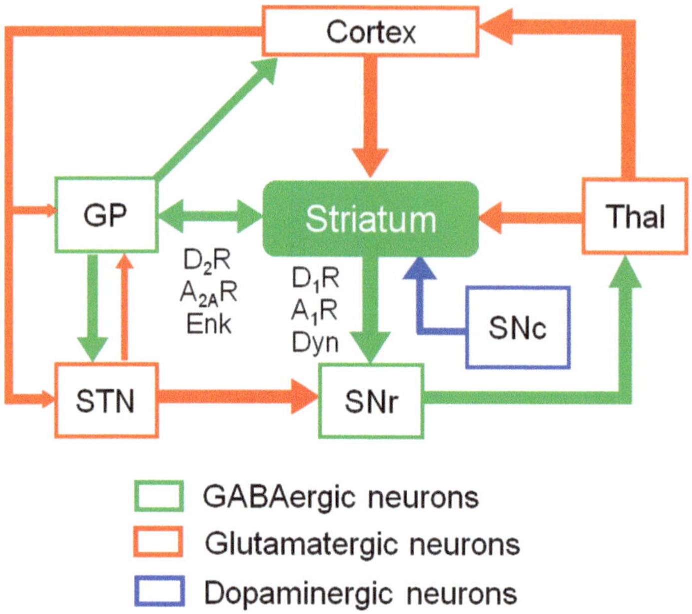


**Fig. 11.1** General scheme of the basal ganglia synaptic organization. Striatal GABAergic medium-sized spiny neurons (MSNs) originate the direct and indirect pathways. The direct pathway projects monosynaptically to the substantia nigra *pars reticulata* (SNr), and the indirect pathway follows the sequence striatum → globus pallidus (GP) → subthalamic nucleus (STN) → SNr. The striatum receives glutamatergic afferents from the cerebral cortex and the thalamus (Thal), and a dense dopaminergic innervation from the substantia nigra *pars compacta* (SNc). Cholinergic interneurons provide a significant innervation to the striatum, and histamine and 5-hydroxytryptamine (5-HT) are released by axons of neurons located in the tuberomammillary nucleus and raphe nuclei, respectively (not shown). *$A_1R$* adenosine $A_1$ receptor, *$A_{2A}R$* adenosine $A_{2A}$ receptor, *$D_1R$* dopamine $D_1$-like receptor, *$D_2R$* dopamine $D_2$-like receptor, *Dyn* dynorphins, *Enk* enkephalins

on the SNr neurons that leads to relief of the brake over the thalamus and thus to facilitation of thalamocortical glutamatergic transmission. In turn, $D_2$-MSNs project to the GP, a structure that exerts a tonic inhibition on the neurons of the STN, the only glutamatergic nucleus of the BG circuitry. The activation of $D_2$-MSNs inhibits GP neurons resulting in disinhibition of the STN neuronal cells and increased stimulation of SNr neurons, leading to inhibition of thalamocortical transmission [1, 5].

In this simplified view (the so-called rate model), activation of the BG direct pathway leads to facilitation of motor behaviors, whereas stimulation of the indirect pathway inhibits movement. Although this model has borne a great heuristic value for the understanding of BG physiology and pathophysiology, it is now recognized that the BG cannot be seen as a “go through” structure, with functional interactions occurring unidirectionally along the cortico-BG-thalamocortical circuits. For instance, the STN receives afferents from the cerebral cortex (the BG hyper-direct pathway) and thalamus (Fig. 11.1), and the STN and GP have emerged as key players in the regulation of the BG synaptic output through their reciprocal connections

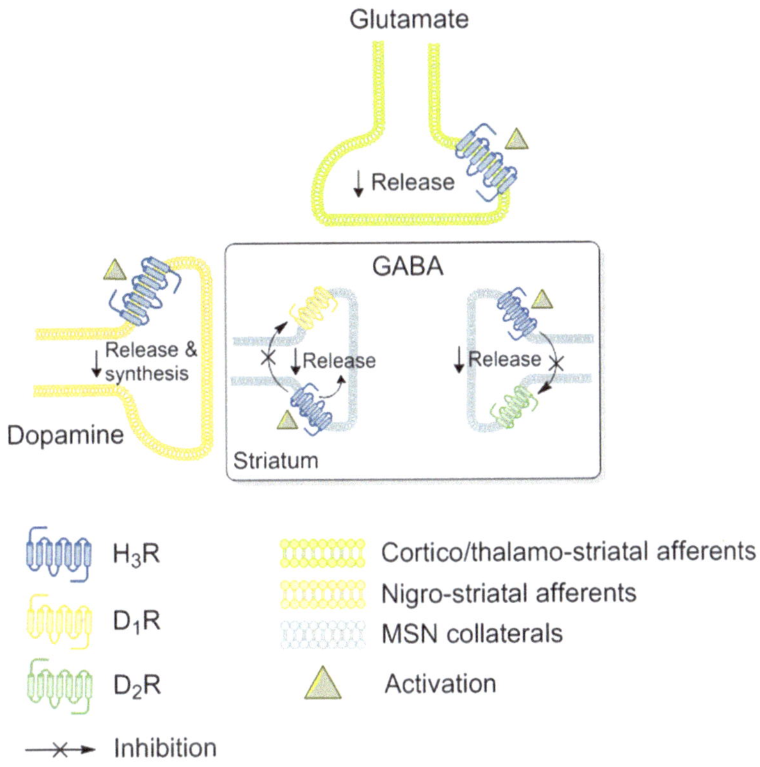


**Fig. 11.2** Summary of $H_3R$-mediated modulation of striatal afferents. Through $H_3R$ activation the histaminergic system exerts an inhibitory effect on the synaptic afferents originated in the cerebral cortex, the thalamus, and the substantia nigra *pars compacta* (SNc), as well as on the collaterals of the GABAergic medium-sized spiny neurons (MSNs). In the MSNs $H_3Rs$ keep a fine tune of GABA release by opposing the effects of dopamine $D_1$- and $D_2$-like receptors ($D_1Rs$ and $D_2Rs$)

and efferents to other BG nuclei [5, 6]. In addition, a direct GABAergic output from the GP to the prefrontal cortex was recently described [7].

### 11.2.2 Modulation of MSN Activity by Striatal Afferents

As previously mentioned, the striatum is the main input nucleus of the BG, and MSN activity is regulated by synaptic afferents from diverse brain areas (Fig. 11.2).

*Glutamatergic modulation.* The striatum receives afferents from numerous cortical areas and the thalamus, providing altogether a rich glutamatergic innervation that shifts the responsive state of MSNs. Glutamate exerts its synaptic effects by acting at metabotropic receptors (mGluRs), coupled to either $G\alpha_{i/o}$ proteins (mGluRs

2, 3, 4, 6, 7, and 8) or $G\alpha_{q/11}$ proteins (mGluRs 1 and 5) and ionotropic receptors (iGluRs) with three types, N-methyl-D-aspartate (NMDA), α-amino-3-hydroxy-5--methyl-4-isoxazolepropinate (AMPA), and kainate receptors [3, 8].

Under resting conditions, MSNs stay in a nonresponsive state ("downstate") with their membrane potential (Vm) hyperpolarized (−90 mV). Cortical and thalamic activities bring the MSNs to an "upstate" (Vm −60 mV or more depolarized), near their spike threshold, where they can exhibit spontaneous activity and burst firing. This response is regulated by several modulators, unable to directly change MSN excitability but capable to modifying the response to glutamate [4, 9].

*GABAergic modulation.* In the CNS, γ-aminobutyric acid (GABA) is the main inhibitory neurotransmitter, and this function is mediated by $Cl^-$-permeant ionotropic $GABA_A$ receptors and metabotropic $GABA_B$ receptors coupled to $G\alpha_{i/o}$ proteins [10]. Striatal GABAergic innervation arises from three known sources: "feedback" MSN collaterals, "feed-forward" GABAergic interneurons, and axons of GP neurons [11–15].

The intra-striatal MSN projections contribute to keep the balance between both MSN populations through reciprocal axon collaterals that negatively modulate the neighbor population. Interestingly, $D_2$-MSNs exert a stronger effect over $D_1$-MSNs due to their higher degree of collateralization and the larger expression of $GABA_A$ receptors by $D_1$-MSNs [16]. The striatum receives a GABAergic input originated in the GP, either by the collaterals of pallidal efferents to other BG nuclei (in particular the STN) or by the axons of GP neurons that exclusively innervate the striatum and target both types of MSNs and interneurons [17].

*Dopaminergic modulation.* The striatum is strongly innervated by axons of dopaminergic neurons located in the SNc that synapse onto both MSN populations, and dopamine modulates cortical and thalamic excitatory inputs through the activation of $D_1$Rs and $D_2$Rs that regulate both the transition to the upstate and the ion channels controlling neuronal spiking. In $D_1$-MSNs, dopamine facilitates the upstate transition and spiking through the cAMP/protein kinase A (PKA) pathway that increases L-type $Ca^{2+}$ channel currents, decreases $K^+$ currents, and has positive effects on AMPA and NMDA receptor function and trafficking. Conversely, in $D_2$-MSNs $D_2$R signaling hampers the upstate transition and diminishes spiking, by decreasing the activity of L-type $Ca^{2+}$ channels and $Na^+$ channels, reducing AMPA receptor currents and increasing $K^+$ currents. Further, presynaptic $D_2$Rs inhibit glutamate release from corticostriatal afferents [4, 9].

*Cholinergic modulation.* The striatum possesses one of the highest levels of cholinergic markers in the brain, and for a long time, the cholinergic interneurons were considered the only source of acetylcholine. However, recent information indicates that the pedunculopontine and laterodorsal tegmental nuclei also provide cholinergic afferents to the striatum [17, 18]. Cholinergic interneurons typically fire slowly and regularly, and by activating muscarinic $M_1$ receptors, acetylcholine increases the responsiveness of MSNs to cortical excitation through the closure of dendritic $K^+$ channels whose constitutive opening holds the membrane potential near the $K^+$ equilibrium potential [19]. In parallel, through the activation of nicotinic receptors, acetylcholine excites fast-spiking GABAergic interneurons that reduce MSN excitability [20].

## 11.3 Histaminergic Innervation of the Basal Ganglia

Histamine is produced in the brain by mast cells and neurons. The contribution of mast cell-derived histamine to the total levels is modest in the adult brain, but during early postnatal development is the principal source of the amine [21, 22].

In the adult mammalian brain, all histaminergic neurons are located in the hypothalamus tuberomammillary nucleus (TMN), and their axons innervate almost all brain regions including the BG [23–29]. Histaminergic fibers follow three major pathways: two ascending bundles which innervate the forebrain structures and one descending bundle reaching the spinal cord [23, 29].

Histamine is synthesized from the amino acid L-histidine by the enzyme histidine decarboxylase (HDC; [25]). The histaminergic neurons fire at a slow regular rate of $2.1 \pm 0.6$ Hz and exhibit a marked circadian rhythmicity, being more active during wakefulness [25, 30]. This activity is regulated by excitatory glutamatergic inputs from the cortex and the hypothalamus, via NMDA and AMPA receptors, and by inhibitory GABAergic afferents from the hypothalamic ventrolateral preoptic nucleus, through the activation of $GABA_A$ receptors. Further, orexins released by neighboring neuronal cells excite the histaminergic neurons [25, 31, 32].

Histamine-containing nerve fibers are characterized by prominent varicosities [29, 33], and most of them do not make typical synaptic contacts [34]. Therefore upon nerve terminal depolarization, histamine is released from several points along the fibers allowing the transmitter to act on a large number of cells [24, 35]. The synthesis and release of histamine are regulated by $H_3$ autoreceptors which inhibit both processes [36, 37], and the release is also inhibited by presynaptic $\alpha_2$-adrenoceptors and by $A_1$Rs [38–40].

There is no evidence for a high-affinity uptake system for histamine in the brain, although human astrocytes take up histamine with low affinity through the activity of the plasma membrane monoamine transporter (PMAT) and, to a lesser extent, the organic cation transporter 3 (OCT3) [41–43]. Once histamine is released into the extracellular space, two enzymes participate in its degradation, diamine oxidase and histamine N-methyltransferase, with the latter being the major responsible of the catabolism of histamine in the brain [44, 45].

Histamine content varies greatly among brain areas in keeping with the extent of histaminergic innervation. The highest levels occur in the hypothalamus and the lowest values are found in the cerebellum [46]. In regard to the BG in the rat striatum, the extracellular concentration of histamine estimated by microdialysis was 50 nM [24], but lower values have also been reported for conscious freely moving rats (6.2 nM, [47]) and anesthetized rats (1.2 nM; [28]). These concentrations are tentative, because dialysates might contain nonneuronal histamine. The total striatal histamine content was 0.22 and 0.44 pmol/mg for mouse and rat, respectively [27, 46], and for human BG the highest level was found in SNc (1.06 pmol/mg) and SNr (0.82 pmol/mg), in agreement with the density of histaminergic fibers, followed by the putamen (0.70 pmol/mg), the GP (0.66 and 0.61 pmol/mg for internal and external GP, respectively), and the caudate head (0.48 pmol/mg) [48].

Histamine levels are significantly increased in the SNc, GP, and striatum of Parkinson's disease patients [48]; conversely, decreased levels in the caudate nucleus were reported for Alzheimer's disease patients (−25 % of the control values; [49]). Although changes in the histaminergic system suggest its participation in neurodegenerative diseases, it is not yet known whether the changes are primary or secondary to the disease.

### *11.3.1 Histamine Receptors*

The four G protein-coupled histamine receptors ($H_1$–$H_4$) show differential distribution in the body and the CNS [32]. The molecular and pharmacological characteristics of these receptors are addressed in detail in other chapter of this book by R. Leurs and colleagues, and therefore only a brief description of these receptors with emphasis on the $H_3R$ is provided herein.

The $H_1$ receptor ($H_1R$) is expressed in peripheral tissues, including the gut, smooth muscle, adrenal medulla, bronchia, blood vessels, and lymphocytes [32, 50]. In the human and monkey brain, a high density of $H_1Rs$ is observed in the most internal areas (laminae V and VI) of the neocortex, claustrum, hippocampal formation, posterior hypothalamus, and thalamus [50, 51].

The $H_2$ receptor ($H_2R$) is also expressed in peripheral tissues and the CNS. It has been localized in the gastric parietal cells, smooth muscle, and immune cells and in the CNS in hippocampus, amygdala, raphe nuclei, BG, and cerebral cortex [25, 50, 52].

The $H_4$ receptor ($H_4R$) is mainly expressed in peripheral cells (mast cells, eosinophils, and T cells) and tissues such as the lung and gut, and its expression by cells of the immune system implies a role in immune responses and inflammation [32, 53–56].

In contrast to the other three histamine receptors, the $H_3$ receptor ($H_3R$) is almost exclusively expressed by neuronal cells of the CNS and the peripheral nervous system. The receptor has been found in the cerebral cortex, thalamus, BG, cerebellum, amygdala, hippocampus, hypothalamus, corpus callosum, raphe nuclei, locus coeruleus, and spinal cord [57–64].

In the BG $H_3Rs$ are heterogeneously distributed in keeping with the areas known to receive histaminergic projections [25]. A detailed mapping of the rat brain has been reported for $H_3R$ binding sites and mRNA expression, and for binding sites in the human and monkey brain [24, 36, 51, 65, 66]. In the human and rat brain, $H_3R$ binding levels follow the rank order SNr, nucleus accumbens > striatum > GP > SNc > STN, whereas mRNA signal intensity follows the rank order striatum, nucleus accumbens > STN > SN > GP ([51, 66, 67]; Table 11.1).

**Table 11.1** Density of histaminergic innervation and $H_3R$ binding sites in rodent basal ganglia nuclei

| | Fiber density | | $H_3R$ (rat) | |
|---|---|---|---|---|
| Nucleus | Rat | Guinea pig | Binding | mRNA |
| Striatum | Low-moderate | Low-moderate | 3–4+ | 3+ |
| GP | Low | Low | 3–4+ | 0–1+ |
| SNr | Moderate | Moderate | 4+ | 1+ |
| SNc | Moderate | Moderate-high | 1+ | 1+ |
| STN | High | High | 1+ | 2+ |

Scale for $H_3R$ binding sites and mRNA: 1+, very low; 2+, low; 3+, moderate; 4+, high. *GP* globus pallidus, *SNc* substantia nigra *pars compacta*, *SNr* substantia nigra *pars reticulata*, *STN* subthalamic nucleus

*References*: [23, 33, 66, 68]

## 11.3.2 Molecular and Functional Characteristics of the $H_3R$

The originally cloned human $H_3R$ is a protein of 445 amino acids whose gen is located in the long arm of the chromosome 20 (20q13.32–20q13.33; [69]). The architecture of the gen consists of three exons and two introns, explaining the expression in the brain of other receptor isoforms generated by alternative splicing [64, 70, 71]. The $H_3R$ receptor was first identified as an autoreceptor inhibiting the synthesis and release of histamine [36, 72, 73] and then a role as hetero-receptor was elicited, where the receptor inhibits the release of several neuroactive substances [74, 75]. There is also evidence for a postsynaptic location of the $H_3R$ in the striatum, cerebral cortex, hippocampus, and nucleus accumbens [32, 66, 75, 76].

The sensitivity of $H_3R$-stimulated [$^{35}$S]-GTP$\gamma$S binding to pertussis toxin [77] demonstrated the receptor coupling to $G\alpha_{i/o}$ proteins. The signaling pathways triggered by $H_3R$ activation are inhibition of adenylyl cyclase (AC) activity, activation of phospholipase $A_2$, inhibition of the N- and P-/Q- types of voltage-gated $Ca^{2+}$ channels, activation of the phosphatidylinositol 3-kinase (PI3K) pathway, activation of the mitogen-activated protein kinase (MAPK) pathway, mobilization of calcium from intracellular stores by activation of phospholipase C, activation of G protein-gated inwardly rectifying $K^+$ channels (GIRKs), and inhibition of $Na^+/H^+$ exchanger activity [32, 71, 78, 79].

In addition to the intracellular effects initiated by agonist-occupied receptors, $H_3Rs$ signal in an agonist-independent manner (constitutive signaling), implying that the receptor spontaneously stabilizes in an active conformational state leading to G protein activation and downstream signaling. Constitutive activity has been reported for rat, human, and mouse $H_3Rs$, and agonist-independent signaling has been evaluated in recombinant systems and in brain tissues. The constitutive activity of native and recombinant $H_3Rs$ was initially determined by the inhibition by the inverse agonists ciproxifan, thioperamide, and FUB-465 of high basal [$^{35}$S]-GTP$\gamma$S binding; $H_3R$ constitutive activity has also been evaluated through the inhibition of cAMP formation and agonist-evoked [$^3$H]-arachidonic acid release in heterologous expression systems and in cerebrocortical isolated nerve terminals (synaptosomes) by using [$^3$H]-histamine release assays [37, 80, 81].

## 11.4 Regulation by $H_3$Rs of Neurotransmitter Release in the Basal Ganglia

$H_3$Rs are mainly expressed on nerve terminals and control the release and synthesis of histamine as well as the release of several other neurotransmitters and neuromodulators, namely, noradrenaline, dopamine, glutamate, GABA, 5-hydroxytryptamine (5-HT; serotonin), acetylcholine, and neuropeptides [74, 75]. The wide and high distribution of $H_3R$ binding sites and mRNA in the BG suggested an important role in the modulation of the function of these nuclei and thus in the control of motor activity, and several reports indicate that $H_3$Rs control the release of transmitters and modulators that participate in the BG synaptic circuitry.

### *11.4.1 Mechanisms Involved in $H_3R$-Mediated Regulation of Neurotransmitter Release*

$H_3$Rs activate $G\alpha_{i/o}$ proteins, and their inhibitory effect on neurotransmitter release is thus likely to involve the reduction in depolarization-induced $Ca^{2+}$ entry (Fig. 11.3), via the action of $G\beta\gamma$ complexes at the pore-forming $\alpha_1$-subunit of N- and P-/Q-type voltage-gated $Ca^{2+}$ channels [82]. In this regard $H_3R$ activation reduces depolarization-induced $Ca^{2+}$ entry in dissociated hypothalamic histaminergic neurons [83], striatal synaptosomes [84], and transfected human neuroblastoma SH-SY-5Y cells [85]. Further, and as discussed below, the facilitation by $D_1$Rs and the inhibition by $H_3$Rs of depolarization-evoked [$^3$H]-GABA release from striatal nerve terminals appear to converge at P-/Q-type $Ca^{2+}$ channels [86, 87].

A complementary mechanism by which $H_3R$ can modulate neurotransmitter release is the inhibition of the cAMP/PKA pathway via the action of the $G\alpha_{i/o}$ subunits on adenylyl cyclases (Fig. 11.3). The $\alpha_1$ subunit of P-/Q-type voltage-gated $Ca^{2+}$ channels is a known substrate for PKA-mediated phosphorylation [88], and PKA also phosphorylates a number of exocytosis proteins such as snapin, synapsins, cysteine string protein (CSP), rabphilin 3A, αSNAP, SNAP-25, syntaxin, and RIM1α/RIM2α, leading to increased vesicle release probability and enhanced size of the readily releasable vesicle pool. Neurotransmitter release can also be modulated by cAMP in a PKA-independent manner, through the cAMP-binding protein guanine nucleotide exchange factor, GEF/Epac [89–93].

G protein-gated inwardly rectifying $K^+$ channels (GIRKs) have been recently shown to inhibit corticostriatal transmission [94]. Transfected $H_3$Rs activate GIRKs

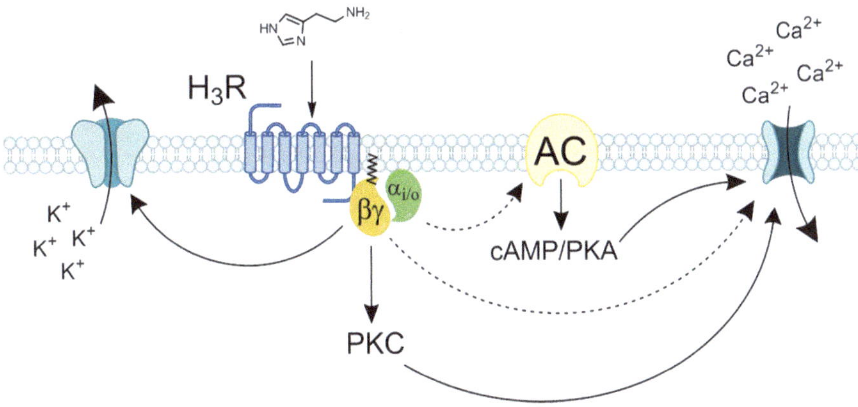


**Fig. 11.3** Mechanisms likely to underlie the $H_3$R-mediated modulation of neurotransmitter release. $H_3$Rs can modulate the opening of voltage-gated $Ca^{2+}$ channels by one or more of the following mechanisms: (**a**) direct binding of Gβγ complexes to the pore-forming $\alpha_1$ subunit of the N and P/Q types of $Ca^{2+}$ channels, (**b**) inhibition of adenylyl cyclases (*AC*) through $G\alpha_{i/o}$ subunits leading to inhibition of the cAMP/protein kinase A (*PKA*) pathway, and (**c**) activation by Gβγ complexes of $K^+$ channels (GIRKs) leading to membrane hyperpolarization and hampering of nerve terminal depolarization driven by action potentials. Through the Gβγ-mediated stimulation of phospholipase C, $H_3$Rs can activate protein kinase C (*PKC*), which phosphorylates the $\alpha_1$ subunit of $Ca^{2+}$ channels hindering the binding of Gβγ complexes

[95], and an additional mechanism for $H_3$R-mediated modulation of neurotransmitter release in the BG would be thus an action at presynaptic GIRKs.

In transfected cells Gβγ complexes released upon $H_3$R activation stimulate phospholipase C, PLC [57, 96], which produces diacylglycerol (DAG) and inositol-1,4,5-trisphosphate ($IP_3$). Alone or in combination with $Ca^{2+}$ ions, DAG stimulates the classical and novel isoforms of protein kinase C, PKC [97]. PKC can phosphorylate the I-II loop of the $\alpha_1$-subunit of voltage-gated $Ca^{2+}$ channels in a site near the Gβγ binding site hampering this process [98, 99]. Through this mechanism, the $H_3$R may be capable of self-limiting its effects on $Ca^{2+}$ channels.

## 11.4.2 Regulation of Neurotransmitter Release in the Striatum

In the striatum, histaminergic fibers are scattered and the density varies; moderate density is observed in the anterior parts of the dorsal striatum, while in all other areas the density is low to very low [23, 34, 78]. In mouse and rat striatum HDC mRNA is absent but the protein is found at high levels, comparable to the hypothalamus HDC expression [27].

The striatum expresses high levels of $H_3$Rs and the corresponding mRNA, indicating the presence of $H_3$Rs on both the intrinsic neurons and the afferents to the nucleus ([66]; Table 11.1). This ubiquitous distribution allows for the $H_3$R to simultaneously modulate the striatal incoming and outgoing information.

#### 11.4.2.1 Modulation of Glutamate Release

The main striatal glutamatergic afferents originate in the cerebral cortex, and most cortical areas and layers project to the striatum [100]. $H_3R$ mRNA is highly expressed in layer V of the primary motor cortex, layers III and V of the secondary motor cortex, layers V and VI of the somatosensory cortex, layers IV and V of the auditory cortex, and layer V of the visual cortex [66]. A high $H_3R$ density (358 fmol/mg protein) was found in striatal synaptosomes [101], and the vesicular glutamate transporter 1 (VGLUT1) colocalizes with $H_3Rs$ in the striatum indicating the presence of these receptors on corticostriatal afferents [76].

In the striatal synaptosomal preparation, $H_3R$ activation by different agonists, namely histamine, (*R*)-(−)-α-methyl-histamine (RAMH) and immepip, inhibited by 53 % the release of endogenous glutamate stimulated by depolarization with 4-aminopyridine [84] and in striatal slices reduced by 25 % the amplitude of field potentials (FPs) evoked by stimulation of cortical afferents [102]. The antagonist thioperamide prevented the effects of the agonists in both synaptosomes and slices, indicating that presynaptic $H_3Rs$ control glutamate release in the striatum. Thioperamide also acts as an inverse agonist at the $H_3R$ and increased on its own the FP amplitude and decreased the paired-pulse ratio, suggesting constitutive activity of the receptor [76]. $H_3R$-mediated inhibition of corticostriatal transmission was confirmed in striatal slices from transgenic $D_1R$-GFP and $D_2R$-GFP mice, with no significant difference in the inhibitory effect of histamine between $D_1$- and $D_2$-MSNs ($81.8 \pm 6.6$ % and $71.0 \pm 10.0$ % of control values, respectively; [103]).

The striatum also receives glutamatergic inputs from the thalamus [104], mainly from the center median/parafascicular complex [105] and in a minor extent from the rostral intralaminar, mediodorsal, pulvinar, lateral posterior, medial posterior nuclei, and the ventral motor nuclear group [106], and most of these nuclei express high to very high levels of $H_3R$ mRNA [66]. In optogenetic experiments $H_3R$-mediated modulation of thalamo-striatal glutamatergic transmission was also observed, with histamine reducing in both $D_1$- and $D_2$-MSNs the amplitude of the excitatory postsynaptic currents (EPSCs) evoked by thalamic stimulation. The inhibition was similar for both MSN populations (−56 % and −48 %, respectively) but higher than for corticostriatal transmission [103, 107].

The increased paired-pulse ratio confirmed the presynaptic location of $H_3Rs$ in corticostriatal and thalamo-striatal synapses. Further, high-frequency stimulation of corticostriatal synapses resulted in short-term facilitation followed by a constant depressed response, whereas thalamo-striatal synapses exhibited a constant depressed response. Interestingly, whereas $H_3R$ activation had no effect on the plasticity of corticostriatal synapses, the thalamo-striatal synapses became significantly facilitatory [103].

#### 11.4.2.2 Modulation of GABA Release

As previously mentioned, MSNs conform the majority of the neuronal striatal population, and at least 85 % of MSNs express $H_3Rs$ [76, 108]. Inhibitory inputs to MSNs are provided mainly by collaterals of neighbor neurons and axons of fast-spiking

GABAergic interneurons [11–15]. For a long time, MSN collaterals were thought to be nonfunctional connections, but as of the year 2000, several studies have shown not only a high degree of collateral arborization but also a significant role in the striatal synaptic microcircuitry [16, 109–111].

The first neurochemical evidence for the participation of $H_3Rs$ in regulating GABA release in the brain was obtained with rat SNr slices where [$^3$H]-GABA release induced by depolarization with 15 mM KCl was reduced by histamine and the $H_3R$ agonist immepip (−78 % and −80 %, respectively), and these effects were blocked by the $H_3R$ antagonist thioperamide [112]. $D_1$-MSNs that project to the SNr express high levels of $H_3Rs$ and the corresponding mRNA [66, 113], and in striatal slices $K^+$-evoked [$^3$H]-GABA release was also reduced by histamine and immepip (−78 % and −81 %, respectively), and these actions were blocked by the antagonists thioperamide and clobenpropit [86]. Depolarization-induced [$^3$H]-GABA release was largely dependent on the presence of dopamine, and in slices obtained from dopamine-depleted rats, elevated $K^+$ caused little release of [$^3$H]-GABA unless the $D_1R$ agonist SKF-38393 was also present. The $D_1R$-stimulated release was reduced by immepip to the level obtained in the absence of SKF-38393 indicating that $H_3Rs$ inhibit the same component of GABA release modulated by $D_1R$ stimulation.

The stimulatory action of $D_1Rs$ was mimicked by 8-bromo-cyclic AMP, prevented by PKA inhibition, and markedly reduced by ω-agatoxin TK, a blocker of P-/Q-type voltage-gated $Ca^{2+}$ channels, but not by ω-conotoxin MVIIA or nimodipine (blockers of N- or L-type $Ca^{2+}$ channels, respectively). Further, the effect of 8-bromo-cyclic AMP was practically abolished by $H_3R$ activation [87]. These observations indicate that $D_1R$-induced facilitation and $H_3R$-mediated inhibition of GABA release from $D_1$-MSN collaterals converge at P-/Q-type $Ca^{2+}$ channels. This conclusion is supported by data obtained with dissociated neurons from the hypothalamic ventromedial nucleus in which the $H_3R$-mediated inhibition of the frequency of GABAergic spontaneous inhibitory postsynaptic currents (IPSCs) was occluded by blocking P-/Q-type but not N- or L-type $Ca^{2+}$ channels [114].

The modulation by $H_3Rs$ of GABA release in the striatum has also been studied with electrophysiological methods. $H_3R$ activation reduced the amplitude of GABAergic IPSCs in both populations of MSNs to a similar extent (−40 %), although thioperamide completely blocked the histamine-mediated reduction in IPSC amplitude in $D_2$-MSNs, whereas for $D_1$-MSNs a partial inhibition was observed, and the $H_2$ receptor antagonist ranitidine blocked part of the effect of histamine. These data indicate that histamine strongly attenuates GABAergic transmission by acting at $H_3Rs$ for synapses onto $D_2$-MSNs and at both $H_3$ and $H_2$ receptors for synapses onto $D_1$-MSNs [103]. Importantly, this study also showed that $H_3Rs$ modulate GABAergic transmission between MSNs but not between fast-spiking GABAergic interneurons and MSNs.

#### 11.4.2.3 Modulation of Dopamine Release

Dopamine released from the nerve terminals of neurons located in the SNc modulates the responses of both MSN populations as well as the synaptic plasticity of the corticostriatal projections through the activation of $D_1$Rs and $D_2$Rs [4, 9, 115, 116]. $D_2$Rs function as autoreceptors and as such underlie the negative modulation of dopamine synthesis and release [117–119].

Although still a matter of controversy, $H_3$Rs appear to be expressed by dopaminergic neurons located in the SNc and ventral tegmental area [76, 113], and $H_3$R-mediated inhibition of [$^3$H]-dopamine release has been reported for mouse striatum [120] and rat SNr [112]. In mouse striatal slices, both histamine and the $H_3$R agonist RAMH reduced the electrically evoked release of [$^3$H]-dopamine by 18 % and 16 %, respectively, and the effect of histamine was significantly increased to 38 % by blocking $D_2$ autoreceptors [120], whereas in slices from rat SNr $K^+$-evoked [$^3$H]-dopamine release was reduced (−38 %) by the $H_3$R agonist immepip [112]. In contrast, $H_3$R activation had no effect on depolarization-evoked [$^3$H]-dopamine release from rat or rabbit striatal slices or synaptosomes ([120, 121] and unpublished results of our own) or rat nucleus accumbens slices [122], indicating interspecies and inter-region differences in $H_3$R expression and function. Furthermore, microdialysis studies showed that $H_3$R antagonists enhance dopamine release in rat prefrontal cortex but not in the striatum [123–125].

Synaptic transmission also depends on the neurotransmitter availability for release, and $H_3$R activation inhibits dopamine synthesis in rat striatum and nucleus accumbens [76, 101, 122]. In line with this information, the $H_3$R agonist imetit reduces the DOPA-induced increase in dopamine content in microdialysis samples from rat striatum [126].

As mentioned above, $H_3$R activation does not modulate [$^3$H]-dopamine release but inhibits dopamine synthesis in slices of rat striatum or nucleus accumbens, and in the latter nucleus, the effect appears to rely on the inhibition of the cAMP/PKA pathway [122]. Likewise for other GPCRs, the mechanism of $H_3$R-mediated actions is likely to involve the formation of complexes with signal transduction and adaptor molecules. The discrepancy between the effects on dopamine release and synthesis may therefore be due to the differential coupling of $H_3$ hetero-receptors to the inhibition of voltage-gated $Ca^{2+}$ channels via $G\beta\gamma$ dimers and to the $G\alpha_{i/o}$ subunit-mediated modulation of adenylyl cyclases responsible for cAMP formation.

Given the possible relevance of the histaminergic control of dopaminergic transmission, the expression of $H_3$Rs by SNc and ventral tegmental area neurons needs to be confirmed. Further studies employing techniques such as high-resolution high-sensitivity in situ hybridization [127] and single-cell PCR [128] would be useful to address this issue.

#### 11.4.2.4 Acetylcholine Release

$H_3$Rs are expressed by striatal cholinergic interneurons [76], but no studies on the effect of $H_3$Rs on striatal acetylcholine release have been reported. Several in vitro (slices) and in vivo (microdialysis) studies reported $H_3$R-mediated regulation of acetylcholine release in rat entorhinal cortex [129], frontoparietal cortex [130], hippocampus [131, 132], nucleus accumbens [133], and basolateral amygdala [134], but this effect appears to involve transsynaptic actions rather than a direct effect at receptors located on cholinergic nerve terminals.

### *11.4.3 Regulation of Neurotransmitter Release in the Globus Pallidus (GP)*

In spite of a modest histaminergic innervation [23, 33], the GP expresses a high density of $H_3$Rs, and the very low level of the corresponding mRNA [66] indicates that the vast majority, if not all, of such receptors are located on the nerve terminals of neurons projecting to the nucleus.

The main synaptic afferents to the GP are GABAergic fibers from the striatum [1], glutamatergic axons originated in the STN and, to a minor extent, in the cerebral cortex and thalamus [135, 136], and dopaminergic afferents from SNc [137]. $H_3$R mRNA is expressed by the neurons of all the areas or nuclei that project to the GP [66], and the density of $H_3$Rs found for rat GP synaptosomal membranes ($1327 \pm 79$ fmol/mg protein; [138]) is eightfold higher than that found for membranes obtained from the whole GP ($162 \pm 29$ fmol/mg protein; [139]), supporting that $H_3$Rs are concentrated on the GP synaptic inputs.

#### 11.4.3.1 Modulation of GABA Release

In rat GP slices, the $H_3$R agonist immepip had a modest and nonsignificant effect on depolarization-evoked [$^3$H]-GABA release ($93 \pm 4$% of controls; [139]). However, in striatonigral axons, the inhibitory action of $H_3$Rs on $K^+$-induced [$^3$H]-GABA release depends on the concomitant activation of $D_1$Rs and stimulation of the cAMP/PKA pathway. MSNs that project to the GP express $A_{2A}$Rs whose activation enhances GABA release via the cAMP/PKA pathway [140–142]. Collaterals of intrinsic GP neurons also release GABA, but these cells do not express $A_{2A}$R mRNA indicating that the $A_{2A}$R-mediated effect is selectively exerted on the striato-pallidal projections [143]. In rat GP synaptosomes, $K^+$-evoked [$^3$H]-GABA release was enhanced by $A_{2A}$R activation, and whereas neither $H_3$R agonists nor antagonists affected on their own the release, $H_3$R activation inhibited the stimulatory effect of $A_{2A}$Rs [138]. This situation resembles that observed for SNr and striatum where $H_3$Rs selectively modulate the component of [$^3$H]-GABA release depending on $D_1$R activation.

As discussed before, $D_1$R-mediated facilitation and $H_3$R-mediated inhibition of GABA release from striatal terminals appear to converge at P-/Q-type voltage-gated $Ca^{2+}$ channels. Because both $D_1$Rs and $A_{2A}$Rs stimulate adenylyl cyclase activity, one plausible explanation for the selective effect of $H_3$R activation on the $A_{2A}$R-mediated facilitation of [$^3$H]-GABA release is that the cAMP/PKA pathway modulates the opening of $Ca^{2+}$ channels present on the striato-pallidal axons and that are also controlled by $H_3$Rs.

In transfected cells and MSNs, $H_3Rs$ form heteromers with $D_1Rs$ or $D_2Rs$ [108, 144, 145], and in striatal membranes $H_3R$ activation decreased the affinity of $D_2R$ for the agonist quinpirole [144]. Further, in SK-N-MC cells expressing $H_3Rs$ and $D_1Rs$, the coupling of the latter shifted from $G\alpha_s$ to $G\alpha_{i/o}$ proteins, and their activation no longer resulted in cAMP formation but in inhibition of forskolin-induced cAMP accumulation [145]. Thus, another and interesting possibility is that $H_3R$-$A_{2A}R$ complexes underlie the functional interaction between $H_3Rs$ and $A_{2A}Rs$. In this regard, preliminary data show that in rat GP synaptosomal membranes the $H_3R$ agonist immepip reduces in a modest (twofold) but significant manner the affinity of the $A_{2A}R$ for its agonist CGS-21680, suggesting that dimerization between $H_3$ and $A_{2A}$ receptors does take place. $A_{2A}R$-mediated facilitation of GABA release depends on the cAMP/PKA pathway, and another explanation for the $H_3R$ selective effect is therefore a simultaneous action on adenylyl cyclases, because $H_3R$ activation reduced by 50–75 % $A_{2A}R$-mediated cAMP accumulation in GP slices [146]. $H_3Rs$ appear thus to modulate the GABAergic striato-pallidal transmission through direct and functional interactions with $A_{2A}Rs$.

#### 11.4.3.2 Modulation of Dopamine Release

Perfusion of GP slices with the $H_3R$ agonist immepip had a modest but nonsignificant effect on $K^+$-evoked [$^3H$]-dopamine release (94 ± 6 % of controls; [139]).

#### 11.4.3.3 Modulation of Glutamate Release

In rat GP slices, the $H_3R$ agonist immepip significantly inhibited $K^+$-evoked [$^3H$]-D-aspartate release (−64 %), an effect prevented by the non-imidazole $H_3R$ antagonist A-331440. In line with this effect, in anesthetized rats the spontaneous firing rate of GP neurons was reduced by ~70 % by the intra-pallidal infusion of immepip, and A-331440 reversed partially this effect [139].

Further, the injection of immepip into the GP followed by the intraperitoneal administration of apomorphine, an agonist at $D_1/D_2$ receptors, resulted in turning behavior in rats, ipsilateral to the injected GP and reduced by the $H_3R$ antagonist A-331440 [139]. The effect of the $H_3R$ agonist was similar and not additive to that induced by the intra-pallidal injection of the glutamate NMDA and AMPA receptor antagonists AP-5 and CNQX [147] and is therefore consistent with presynaptic $H_3R$-mediated inhibition of glutamate release which leads to diminished activity of GP neurons. Through their reciprocal connections and efferents to other BG nuclei, the GP and the STN play a key role in the regulation of the BG synaptic output [5], and the modulation of rat GP glutamatergic transmission by presynaptic $H_3Rs$ could therefore contribute to regulate the activity of GP neurons and thus BG function.

### *11.4.4 Regulation of Neurotransmitter Release in the Substantia Nigra Pars Reticulata (SNr)*

A dense network of histaminergic fibers is found in the substantia nigra [23, 33], and the number of these fibers is increased in a rat parkinsonian model and in post-mortem brains with Parkinson's disease [67, 148, 149].

As mentioned previously, the SNr GABAergic neurons originate the main BG output pathway and are thus critical for the control of movement. The SNr neurons receive GABAergic projections from the striatum and glutamatergic afferents from the STN through the BG direct and indirect pathways, respectively. In turn SNr neurons send synaptic information to the thalamus, superior colliculus, and brainstem motor structures [150, 151].

In the rat and human brain, a high density of $H_3R$ binding sites is observed in the SNr accompanied by a low mRNA expression [66], suggesting that most $H_3Rs$ are presynaptically located on striatonigral terminals and to a lesser extent on subthalamo-nigral projections, axons of neurons located in the raphe nuclei, and dendrites of SNc dopaminergic neurons [66, 70, 71, 152].

#### 11.4.4.1 Modulation of GABA Release

The intra-striatal administration of quinolinic acid results in a reduction in $H_3R$ binding sites that matches the loss of $D_1Rs$ in the ipsilateral striatum and SNr, indicating that both receptors are coexpressed by striatonigral neurons [113]. In rat SNr slices, depolarization-induced [$^3$H]-GABA release was inhibited (−70 %) by blocking $D_1Rs$, consistent with a large component of release being dependent on the activation of these receptors by endogenous dopamine. Both histamine and the $H_3R$ agonist immepip inhibited the evoked [$^3$H]-GABA release to a similar extent (78 % and 80 %, respectively), and the effect of histamine was antagonized by thioperamide, whereas [$^3$H]-GABA release remaining after $D_1R$ blockade was not affected by immepip. Further, in rats depleted of dopamine by pretreatment with reserpine, immepip had no effect, but in the presence of the $D_1R$ agonist SKF-38393, which produced a sevenfold stimulation of release, the $H_3R$ agonist reduced [$^3$H]-GABA efflux to levels not different from controls [112]. This evidence indicates that likewise for the striatum, in the SNr $H_3R$ activation inhibits the component of GABA release enhanced by $D_1R$ stimulation.

#### 11.4.4.2 Modulation of Dopamine Release

In the SNr histamine-immunoreactive varicosities are found in the vicinity of dendrites immunopositive for tyrosine hydroxylase (TH) [153], and in rat SNr slices, [$^3$H]-dopamine release induced by depolarization was reduced (−38 %) by the $H_3R$ agonist immepip [112]. $H_3R$ mRNA hybridization signals were observed at the cellular level in the SNc [66], and SNc neurons coexpress mRNAs coding for the $H_3R$ and TH as well as immunoreactivity for the corresponding proteins [76]. It therefore seems that some $H_3Rs$ are located on the dendrites of the SNc neurons where they regulate dopamine release.

#### 11.4.4.3 Modulation of 5-Hydroxytryptamine (5-HT) Release

A low to moderate $H_3R$ mRNA expression is observed in the raphe nuclei [66], mostly in the dorsal nucleus, which provides a dense serotonergic innervation to the SNr [154, 155], and varicosities immunoreactive for the 5-HT transporter (SERT) are found in the vicinity of TH-immunoreactive processes [153].

Fast-scan cyclic voltammetry in rat midbrain slices showed that the $H_3R$ agonists RAMH and immepip reduced the electrically evoked release of 5-HT by 51 % and 61 %, respectively, with the effect being prevented by the antagonist thioperamide. The inhibition of 5-HT release prevailed in the presence of antagonists of $GABA_A$, $GABA_B$, and glutamate (ionotropic and metabotropic) receptors, supporting the participation of $H_3Rs$ located on serotonergic nerve terminals [153].

#### 11.4.4.4 Glutamate Release

There is no information in regard to the effect of $H_3R$ activation on glutamate release in the SNr, but the firing rate of SNr neurons is controlled by excitatory projections from the STN, which possesses dense $H_3R$ mRNA labeling [66]. The possible regulation by $H_3Rs$ of glutamate release in the SNr deserves thus further consideration.

### *11.4.5 Regulation of Neurotransmitter Release in the Subthalamic Nucleus (STN)*

As mentioned above, the very low density of $H_3R$ binding sites and dense mRNA labeling in the STN [66] suggest that the STN excitatory projections to SNr and GP are endowed with $H_3Rs$. The STN presents a dense network of histaminergic fibers [23, 33], and the so-called BG hyper-direct pathway provides cortical information to the STN, making $H_3R$-mediated modulation of glutamate release in the STN likely.

### *11.4.6 Regulation of Neurotransmitter Release in the Thalamus*

Although the thalamus is not part of the BG, it links these nuclei to the cerebral cortex and serves as a relay transferring sensory information from the periphery and motor information originated in the cerebellum to the cerebral cortex. Histaminergic fibers innervate the periventricular areas and some of the ventral and lateral areas of the thalamus, and a dense innervation is observed in the dorsal thalamus [156]. The wide distribution of the $H_3R$ mRNA and the low to moderate density of the binding sites indicate that $H_3Rs$ expressed by thalamic relay neurons are primarily located on their axons projecting to areas of the cerebral cortex specialized in motor control [66, 70, 71, 152, 156, 157].

The thalamic function is mostly based on excitatory glutamatergic transmission and inhibitory GABAergic modulation. A highly heterogeneous distribution of $H_3R$ binding sites is observed in the thalamus, with the highest levels in the midline and intralaminar (paraventricular, intermedio dorsal, reuniens, and rhomboid nuclei), sensory (posterior group and medial geniculate nuclei), and association (lateral posterior and dorsal) nuclei. In the dorsal association nuclei, $H_3Rs$ may therefore be located on corticothalamic afferents originated in cortical layers V or VI, and their presence in the lateral dorsal and posterior nuclei suggests the involvement of $H_3Rs$ in visual systems.

$H_3Rs$ are present on thalamic synaptosomes at significant level ($141 \pm 12$ fmol/mg protein), and their activation reduced by 25–38 % the depolarization-induced release of endogenous glutamate from thalamic synaptosomes. In the thalamic ventrobasal complex, $H_3R$ activation reduced by 38 % the amplitude of field potentials (FPs) evoked by electrical stimulation of corticothalamic glutamatergic afferents and increased the FP2/FP1 ratio in the paired-pulse paradigm confirming a presynaptic inhibitory action [158].

The reticular nucleus provides GABAergic innervation to the thalamic relay neurons and expresses moderate to high levels of $H_3R$ mRNA [66]. However, $H_3R$ activation had no effect on depolarization-evoked [$^3$H]-GABA release from thalamic synaptosomes or slices [158].

## 11.5 Final Remarks

The highest levels of $H_3R$ expression in the brain are found in some of the BG nuclei, namely, the striatum, GP, and SNr. In these nuclei, $H_3R$ activation modulates presynaptically the release of the main neurotransmitters, glutamate, and GABA, as well as the release of the neuromodulators dopamine and 5-HT (Fig. 11.4 and Table 11.2). This function would allow $H_3Rs$ to regulate BG synaptic circuitry and thus the control of movement and the learning of motor programs.

$H_3Rs$ may also regulate synaptic transmission in striatonigral and striato-pallidal neurons by mechanisms different from the modulation of neurotransmitter release, such as heterodimerization with dopamine ($D_1$- and $D_2$-like) receptors [144, 145] and, presumably, with adenosine $A_{2A}$ receptors [146, 160].

The effects of histamine in the BG are not restricted to actions at the $H_3R$, and, for instance, histamine increases the firing rate of GP neurons and depolarizes both classes of striatal MSNs by acting at $H_2$ receptors [103, 147] and through $H_1$ receptor activation depolarizes striatal cholinergic interneurons [161]. In this regard, the

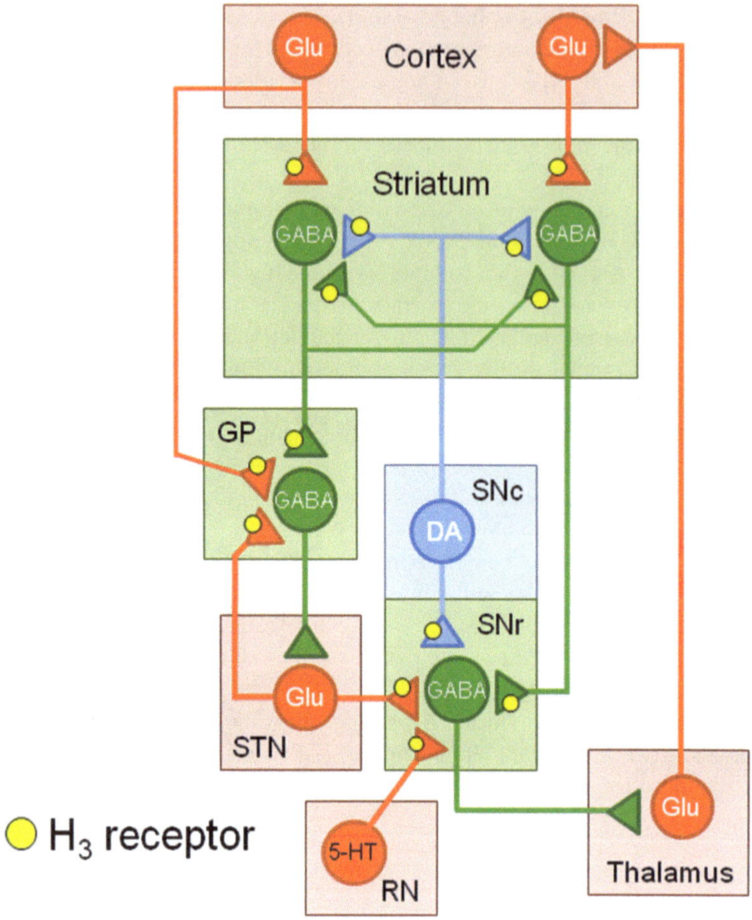


**Fig. 11.4** Location in the basal ganglia synaptic circuitry, of $H_3$Rs shown or likely to modulate neurotransmitter release. *5-HT* 5-hydroxytryptamine (serotonin), *GABA* γ-aminobutyric acid, *$H_3R$* histamine $H_3$ receptor, *Glu* glutamic acid, *GP* globus pallidus, *DA* dopamine, *RN* raphe nuclei, *SNc* substantia nigra *pars compacta*, *SNr* substantia nigra *pars reticulata*, *STN* subthalamic nucleus

differences in affinity for histamine of the three receptors expressed in the CNS (*K*i 2 μM, 12 μM, and 2.5 nM for the human $H_1$, $H_2$, and $H_3$ receptors, respectively; [70, 71, 162]) suggest that the modulation by histamine of neuronal activity in the BG may be related to the diurnal rhythm of histamine release, higher during wakefulness.

Our understanding of the BG function has significantly increased in the last years, and additional research is necessary to unveil in detail the participation of histamine and $H_3$Rs in normal and abnormal BG activity.

**Table 11.2** Summary of the effect of histamine $H_3$ receptor activation on neurotransmitter release in the basal ganglia

| Nucleus | Species | Preparation | Neurotransmitter/ neuromodulator | Effect | Reference |
|---|---|---|---|---|---|
| Striatum | Rat | Slices | Histamine | Inhibition | [159] |
| | Mouse | Slices | [$^3$H]-dopamine | Inhibition | [120] |
| | Rat | Slices | [$^3$H]-dopamine | No effect | [120] |
| | Rabbit | Slices | [$^3$H]-dopamine | No effect | [121] |
| | Rat | Slices | Glutamate | Inhibition | [102] |
| | Mouse | Slices | Glutamate | Inhibition | [102] |
| | Rat | Synaptosomes | Glutamate | Inhibition | [84] |
| | Rat | Slices | [$^3$H]-GABA | Inhibition[a] | [86, 87] |
| | Mouse | Slices | Glutamate | Inhibition | [103] |
| GP | Rat | Slices | [$^3$H]-D-aspartate | Inhibition | [139] |
| | Rat | Slices | [$^3$H]-GABA | No effect | [139] |
| | Rat | Synaptosomes | [$^3$H]-GABA | Inhibition[b] | [138] |
| SNr | Rat | Slices | [$^3$H]-dopamine | Inhibition | [112] |
| | Rat | Slices | [$^3$H]-GABA | Inhibition | [112] |
| | Rat | Slices | 5-HT | Inhibition | [153] |

*5-HT* 5-hydroxytryptamine (serotonin), *GP* globus pallidus, *SNr* substantia nigra *pars reticulata*
[a]Inhibition of the facilitatory action of dopamine $D_1$-like receptors
[b]Inhibition of the facilitatory action of adenosine $A_{2A}$ receptors

**Acknowledgement** Research in our laboratory is supported by Cinvestav and the Mexican Council for Science and Technology (Conacyt).

# References

1. Bolam JP, Hanley JJ, Booth PA, Bevan MD. Synaptic organisation of the basal ganglia. J Anat. 2000;196:527–42.
2. Fuxe K, Ferré S, Genedani S, Franco R, Agnati LF. Adenosine receptor-dopamine receptor interactions in the basal ganglia and their relevance for brain function. Physiol Behav. 2007;92:210–7.
3. Alexander SPH, Mathie A, Peters JA. Guide to receptors and channels (GRAC), 5th edition. Br J Pharmacol. 2011;164 Suppl 1:S32.
4. Gerfen CR, Surmeier DJ. Modulation of striatal projection systems by dopamine. Annu Rev Neurosci. 2011;34:441–66.
5. Obeso JA, Rodriguez-Oroz MC, Benitez-Temino B, Blesa FJ, Guridi J, Marin C, Rodriguez M. Functional organization of the basal ganglia: therapeutic implications for Parkinson's disease. Mov Disord. 2008;23 Suppl 3:S548–59.
6. Obeso JA, Lanciego JL. Past, present, and future of the pathophysiological model of the Basal Ganglia. Front Neuroanat. 2011;5:39.
7. Saunders A, Oldenburg IA, Berezovskii VK, Jhonson CA, Kingery ND, Elliot HL, Xie T, Gerfen CR, Sabatini BL. A direct GABAergic output from the basal ganglia to prefrontal cortex. Nature. 2015;521:85–9.

8. Harpsøe K, Isberg V, Tehan BG, Weiss D, Arsova A, Marshall FH, Bräuner-Osborne H, Gloriam DE. Selective negative allosteric modulation of metabotropic glutamate receptors—a structural perspective of ligands and mutants. Sci Rep. 2015;5:13869.
9. Surmeier DJ, Ding J, Day M, Wang Z, Shen W. $D_1$ and $D_2$ dopamine-receptor modulation of striatal glutamatergic signaling in striatal medium spiny neurons. Trends Neurosci. 2007;30:228–35.
10. Bormann J. The “ABC” of GABA receptor. Trends Pharmacol Sci. 2000;21:16–9.
11. Bolam JP, Somogyi P, Takagi H, Fodor I, Smith AD. Localization of substance P-like immunoreactivity in neurons and nerve terminals in the neostriatum of the rat: a correlated light and electron microscopic study. J Neurocytol. 1983;12:325–44.
12. Koós T, Tepper JM. Inhibitory control of neostriatal projection neurons by GABAergic interneurons. Nat Neurosci. 1999;2:467–72.
13. Mallet N, Micklem BR, Henny P, Brown MT, Williams C, Bolam JP, Nakamura KC, Magill PJ. Dichotomous organization of the external globus pallidus. Neuron. 2012;74:1075–86.
14. Somogyi P, Priestley JV, Cuello AC, Smith AD, Takagi H. Synaptic connections of enkephalin-immunoreactive nerve terminals in the neostriatum: a correlated light and electron microscopic study. J Neurocytol. 1982;11:779–807.
15. Wilson CJ, Groves PM. Fine structure and synaptic connections of the common spiny neuron of the rat neostriatum: a study employing intracellular inject of the horseradish peroxidase. J Comp Neurol. 1980;194:599–615.
16. Taverna S, Ilijic E, Surmeier DJ. Recurrent collateral connections of striatal medium spiny neurons are disrupted in models of Parkinson’s disease. J Neurosci. 2008;28:5504–12.
17. Silberberg G, Bolam JP. Local and afferent synaptic pathways in the striatal microcircuitry. Curr Opin Neurobiol. 2015;33:182–7.
18. Dautan D, Huerta-Ocampo I, Witten IB, Deisseroth K, Bolam JP, Gerdjikov T, Mena-Segovia J. A major external source of cholinergic innervations of the striatum and nucleus accumbens originates in the brainstem. J Neurosci. 2014;34:4509–18.
19. Goldberg JA, Ding JB, Surmeier DJ. Muscarinic modulation of striatal function and circuitry. Handb Exp Pharmacol. 2012;208:223–41.
20. Tepper JM, Bolam JP. Functional diversity and specificity of neostriatal interneurons. Curr Opin Neurobiol. 2004;14:685–92.
21. Molina-Hernández A, Díaz NF, Arias-Montaño JA. Histamine in brain development. J Neurochem. 2012;122:872–82.
22. Panula P, Sundvik M, Karlstedt K. Developmental roles of brain histamine. Trends Neurosci. 2014;37:159–68.
23. Airaksinen MS, Panula P. The histaminergic system in the guinea pig central nervous system: an immunocytochemical mapping study using an antiserum against histamine. J Comp Neurol. 1988;273:163–86.
24. Cumming P, Damsma G, Fibiger HC, Vincent SR. Characterization of extracellular histamine in the striatum and bed nucleus of the stria terminalis of the rat: an in vivo microdialysis study. J Neurochem. 1991;56:1797–803.
25. Haas HL, Sergeeva OA, Selbach O. Histamine in the nervous system. Physiol Rev. 2008;88:1183–241.
26. Inagaki N, Yamatodani A, Shinoda K, Panula P, Watanabe T, Shiotani Y, Wada H. Histaminergic nerve fibers in the median eminence and hypophysis of rats demonstrated immunocytochemically with antibodies against histidine decarboxylase and histamine. Brain Res. 1988;439:402–5.
27. Krusong K, Ercan-Sencicek AG, Xu M, Ohtsu H, Anderson GM, State MW, Pittenger C. High levels of histidine decarboxylase in the striatum of mice and rats. Neurosci Lett. 2011;495:110–4.
28. Russell WL, Henry DP, Phebus LA, Clemens JA. Release of histamine in rat hypothalamus and corpus striatum in vivo. Brain Res. 1990;512:95–101.
29. Watanabe T, Taguchi Y, Shiosaka S, Tanaka J, Kubota H, Terano Y, Tohyama M, Wada H. Distribution of the histaminergic neuron system in the central nervous system of rats; a

fluorescent immunohistochemical analysis with histidine decarboxylase as a marker. Brain Res. 1984;295:13–25.
30. Haas HL, Reiner PB. Membrane properties of histaminergic tuberomammillary neurones of the rat hypothalamus in vitro. J Physiol. 1988;399:633–46.
31. Ericson H, Blomqvist A, Köhler C. Origin of neuronal inputs to the region of the tuberomammillary nucleus of the rat brain. J Comp Neurol. 1991;311:45–64.
32. Panula P, Chazot PL, Cowart M, Gutzmer R, Leurs R, Liu WL, Stark H, Thurmond RL, Haas HL. International union of basic and clinical pharmacology. XCVIII. Histamine receptors. Pharmacol Rev. 2015;67:601–55.
33. Panula P, Pirvola U, Auvinen S, Airaksinen MS. Histamine-immunoreactive nerve fibers in the rat brain. Neuroscience. 1989;28:585–610.
34. Takagi H, Morishima Y, Matsuyama T, Hayashi H, Watanabe T, Wada H. Histaminergic axons in the neostriatum and cerebral cortex of the rat: a correlated light and electron microscopic immunocytochemical study using histidine decarboxylase as a marker. Brain Res. 1986;364:114–23.
35. Taylor KM, Snyder SH. The release of histamine from tissue slices of rat hypothalamus. J Neurochem. 1973;21:1215–23.
36. Arrang J-M, Garbarg M, Schwartz J-C. Autoregulation of histamine synthesis mediated by presynaptic $H_3$-receptors. Neuroscience. 1987;15:553–62.
37. Morisset S, Rouleau A, Ligneau X, Gbahou F, Tardivel-Lacombe J, Stark H, Schunack W, Ganellin CR, Schwartz JC, Arrang JM. High constitutive activity of native $H_3$ receptors regulates histamine neurons in brain. Nature. 2000;408:860–4.
38. Gulat-Marnay C, Lafitte A, Arrang JM, Schwartz JC. Modulation of histamine release and synthesis in the brain mediated by $\alpha_2$-adrenoceptors. J Neurochem. 1989;53:513–8.
39. Oishi Y, Huang ZL, Fredholm B, Urade Y, Hayaishi O. Adenosine in the tuberomammillary nucleus inhibits the histaminergic system via $A_1$ receptors and promotes non-rapid eye movement sleep. Proc Natl Acad Sci U S A. 2008;105:19992–7.
40. Prast H, Heistracher M, Philippu A. In vivo modulation of the histamine release in the hypothalamus by adrenoreceptor agonists and antagonists. Naunyn Schmiedebergs Arch Pharmacol. 1991;344:183–6.
41. Amphoux A, Vialou V, Drescher E, Brüss M, Mannoury La Cour C, Rochat C, Millan MJ, Giros B, Bönisch H, Gautron S. Differential pharmacological in vitro properties of organic cation transporters and regional distribution in rat brain. Neuropharmacology. 2006;50:941–52.
42. Engel K, Zhou M, Wang J. Identification and characterization of a novel monoamine transporter in the human brain. J Biol Chem. 2004;279:50042–9.
43. Yoshikawa T, Naganuma F, Iida T, Nakamura T, Harada R, Mohsen AS, Kasajima A, Sasano H, Yanai K. Molecular mechanism of histamine clearance by primary human astrocytes. Glia. 2013;61:905–16.
44. Barnes WG, Hough LB. Membrane bound histamine N-methyltransferase in mouse brain: possible role in the synaptic inactivation of neuronal histamine. J Neurochem. 2002;82:1262–71.
45. Maldonado M, Maeyama K. The metabolism of histamine in rat hypothalamus and cortex after reserpine treatment. Neurochem Int. 2015;85–86:31–9.
46. Taylor KM, Snyder SH. Histamine in rat brain: sensitive assay of endogenous levels, formation in vivo and lowering by inhibitors of histidine decarboxylase. J Pharmacol Exp Ther. 1971;179:619–33.
47. Chikai T, Oishi R, Saeki K. Microdialysis study of the effects of sedative drugs on extracellular histamine in the striatum of freely moving rats. J Pharmacol Exp Ther. 1993;266:1277–81.
48. Rinne JO, Anichtchik OV, Eriksson KS, Kaslin J, Tuomisto L, Kalimo H, Röyttä M, Panula P. Increased brain histamine levels in Parkinson's disease but not in multiple system atrophy. J Neurochem. 2002;81:954–60.

49. Mazurkiewicz-Kwilecki IM, Nsonwah S. Changes in the regional brain histamine and histidine levels in postmortem brains of Alzheimer patients. Can J Physiol Pharmacol. 1989;67:75–8.
50. Hill SJ, Ganellin CR, Timmerman H, Schwartz JC, Shankley NP, Young JM, Schunack W, Levi R, Haas HL. International Union of Pharmacology. XIII. Classification of histamine receptors. Pharmacol Rev. 1997;49:253–78.
51. Martinez-Mir MI, Pollard H, Moreau J, Arrang JM, Ruat M, Traiffort E, Schwartz JC, Palacios JM. Three histamine receptors (H1, H2 and H3) visualized in the brain of human and non-human primates. Brain Res. 1990;526:322–7.
52. Mobarakeh J, Takahashi K, Sakurada S, Kuramasu A, Yanai K. Enhanced antinociceptive effects of morphine in histamine H2 receptor gene knockout mice. Neuropharmacology. 2006;51:612–22.
53. Breunig E, Michel K, Zeller F, Seidl S, Weyhern CW, Schemann M. Histamine excites neurones in the human submucous plexus through activation of $H_1$, $H_2$, $H_3$ and $H_4$ receptors. J Physiol. 2007;583:731–42.
54. de Esch IJ, Thurmond RL, Jongejan A, Leurs R. The histamine $H_4$ receptor as a new therapeutic target for inflammation. Trends Pharmacol Sci. 2005;26:462–9.
55. Liu C, Ma X, Jiang X, Wilson SJ, Hofstra CL, Blevitt J, Pyati J, Li X, Chai W, Carruthers N, Lovenberg TW. Cloning and pharmacological characterization of a fourth histamine receptor ($H_4$) expressed in bone marrow. Mol Pharmacol. 2001;59:420–6.
56. Morgan RK, McAllister B, Cross L, Green DS, Kornfeld H, Center DM, Cruikshank WW. Histamine 4 receptor activation induces recruitment of $FoxP3^+$ T cells and inhibits allergic asthma in a murine model. J Immunol. 2007;178:8081–9.
57. Cogé F, Guénin SP, Audinot V, Renouard-Try A, Beauverger P, Macia C, Ouvry C, Nagel N, Rique H, Boutin JA, Galizzi JP. Genomic organization and characterization of splice variants of the human histamine $H_3$ receptor. Biochem J. 2001;355:279–88.
58. Drutel G, Peitsaro N, Karlstedt K, Wieland K, Smit MJ, Timmerman H, Panula P, Leurs R. Identification of rat $H_3$ receptor isoforms with different brain expression and signaling properties. Mol Pharmacol. 2001;59:1–8.
59. Esbenshade TA, Browman KE, Bitner RS, Strakhova M, Cowart MD, Brioni JD. The histamine $H_3$ receptor: an attractive target for the treatment of cognitive disorders. Br J Pharmacol. 2008;154:1166–81.
60. Morisset S, Sasse A, Gbahou F, Héron A, Ligneau X, Tardivel-Lacombe J, Schwartz JC, Arrang JM. The rat $H_3$ receptor: gene organization and multiple isoforms. Biochem Biophys Res Commun. 2001;280:75–80.
61. Strakhova MI, Fox GB, Carr TL, Witte DG, Vortherms TA, Manelli AM, Miller TR, Yao BB, Brioni JD, Esbenshade TA. Cloning and characterization of the monkey histamine $H_3$ receptor isoforms. Eur J Pharmacol. 2008;601:8–15.
62. Tardivel-Lacombe J, Morisset S, Gbahou F, Schwartz JC, Arrang JM. Chromosomal mapping and organization of the human histamine $H_3$ receptor gene. Neuroreport. 2001;12:321–4.
63. Wellendorph P, Goodman MW, Burstein ES, Nash NR, Brann MR, Weiner DM. Molecular cloning and pharmacology of functionally distinct isoforms of the human histamine $H_3$ receptor. Neuropharmacology. 2002;42:929–40.
64. Wiedemann P, Bönisch H, Oerters F, Brüss M. Structure of the human histamine $H_3$ receptor gene (HRH3) and identification of naturally occurring variations. J Neural Transm. 2002;109:443–53.
65. Cumming P, Laliberté C, Gjedde A. Distribution of histamine $H_3$ binding in forebrain of mouse and guinea pig. Brain Res. 1994;664:276–9.
66. Pillot C, Heron A, Cochois V, Tardivel-Lacombe J, Ligneau X, Schwartz JC, Arrang JM. A detailed mapping of the histamine $H_3$ receptor and its gene transcripts in rat brain. Neuroscience. 2002;114:173–93.
67. Anichtchik OV, Peitsaro N, Rinne JO, Kalimo H, Panula P. Distribution and modulation of histamine $H_3$ receptors in basal ganglia and frontal cortex of healthy controls and patients with Parkinson's disease. Neurobiol Dis. 2001;8:707–16.

68. Vizuete ML, Traiffort E, Bouthenet ML, Ruat M, Souil E, Tardivel-Lacombe J, Schwartz JC. Detailed mapping of the histamine $H_2$ receptor and its gene transcripts in guinea-pig brain. Neuroscience. 1997;80:321–43.
69. Lovenberg TW, Roland BL, Wilson SJ, Jiang X, Pyati J, Huvar A, Jackson MR, Erlander MG. Cloning and functional expression of the human histamine $H_3$ receptor. Mol Pharmacol. 1999;55:1101–7.
70. Bongers G, Krueger KM, Miller TR, Baranowski JL, Estvander BR, Witte DG, Strakhova MI, van Meer P, Bakker RA, Cowart MD, Hancock AA, Esbenshade TA, Leurs R. An 80-amino acid deletion in the third intracellular loop of a naturally occurring human histamine $H_3$ isoform confers pharmacological differences and constitutive activity. J Pharmacol Exp Ther. 2007;323:888–98.
71. Bongers G, Bakker R, Leurs R. Molecular aspects of the histamine $H_3$ receptor. Biochem Pharmacol. 2007;73:1195–204.
72. Arrang JM, Garbarg M, Schwartz JC. Auto-inhibition of brain histamine release mediated by a novel class ($H_3$) of histamine receptor. Nature. 1983;302:832–7.
73. Arrang JM, Garbarg M, Lancelot JC, Lecomte JM, Pollard H, Robba M, Schunack W, Schwartz JC. Highly potent and selective ligands for histamine $H_3$-receptors. Nature. 1987;327:117–23.
74. Feuerstein TJ. Presynaptic receptors for dopamine, histamine, and serotonin. Handb Exp Pharmacol. 2008;184:289–338.
75. Panula P, Nuutinen S. The histaminergic network in the brain: basic organization and role in disease. Nat Rev Neurosci. 2013;14:472–87.
76. González-Sepúlveda M, Rosell S, Hoffmann H, Castillo-Ruiz MM, Mignon V, Moreno-Delgado D, Vignes M, Díaz J, Sabriá J, Ortiz J. Cellular distribution of the histamine $H_3$ receptor in the basal ganglia: Functional modulation of dopamine and glutamate neurotransmission. Basal Ganglia. 2013;3:109–21.
77. Clark EA, Hill SJ. Sensitivity of histamine $H_3$ receptor agonist-stimulated [$^{35}$S]GTP $\gamma$[S] binding to pertussis toxin. Eur J Pharmacol. 1996;296:223–5.
78. Haas H, Panula P. The role of histamine and the tuberomammillary nucleus in the nervous system. Nat Rev Neurosci. 2003;4:121–30.
79. Leurs R, Bakker RA, Timmerman H, de Esch IJ. The histamine $H_3$ receptor: from gene cloning to $H_3$ receptor drugs. Nat Rev Drug Discov. 2005;4:107–20.
80. Rouleau A, Ligneau X, Tardivel-Lacombe J, Morisset S, Gbahou F, Schwartz JC, Arrang JM. Histamine $H_3$-receptor-mediated [$^{35}$S]GTP$\gamma$[S] binding: evidence for constitutive activity of the recombinant and native rat and human $H_3$ receptors. Br J Pharmacol. 2002;135:383–92.
81. Wieland K, Bongers G, Yamamoto Y, Hashimoto T, Yamatodani A, Menge WM, Timmerman H, Lovenberg TW, Leurs R. Constitutive activity of histamine $H_3$ receptors stably expressed in SK-N-MC cells: display of agonism and inverse agonism by $H_3$ antagonists. J Pharmacol Exp Ther. 2001;299:908–14.
82. Zamponi GW, Currie KP. Regulation of $Ca_V2$ calcium channels by G protein coupled receptors. Biochim Biophys Acta. 2013;1828:1629–43.
83. Takeshita Y, Watanabe T, Sakata T, Munakata M, Ishibashi H, Akaike N. Histamine modulates high-voltage-activated calcium channels in neurons dissociated from the rat tuberomammillary nucleus. Neuroscience. 1998;87:797–805.
84. Molina-Hernández A, Nuñez A, Sierra JJ, Arias-Montaño JA. Histamine $H_3$ receptor activation inhibits glutamate release from rat striatal synaptosomes. Neuropharmacology. 2001;41:928–34.
85. Silver RB, Poonwasi KS, Seyedi N, Wilson SJ, Lovenberg TW, Levi R. Decreased intracellular calcium mediates the histamine $H_3$-receptor-induced attenuation of norepinephrine exocytosis from cardiac sympathetic nerve endings. Proc Natl Acad Sci U S A. 2002;99:501–6.
86. Arias-Montaño JA, Floran B, Garcia M, Aceves J, Young JM. Histamine $H_3$ receptor-mediated inhibition of depolarization-induced, dopamine $D_1$ receptor-dependent release of [$^3$H]-$\gamma$-aminobutyric acid from rat striatal slices. Br J Pharmacol. 2001;133:165–71.

87. Arias-Montaño JA, Floran B, Floran L, Aceves J, Young JM. Dopamine $D_1$ receptor facilitation of depolarization-induced release of γ-amino-butyric acid in rat striatum is mediated by the cAMP/PKA pathway and involves P/Q-type calcium channels. Synapse. 2007;61:310–9.
88. Sakurai T, Hell JW, Woppmann A, Miljanich GP, Catterall WA. Immunochemical identification and differential phosphorylation of alternatively spliced forms of the $\alpha_{1A}$ subunit of brain calcium channels. J Biol Chem. 1995;270:21234–42.
89. Evans GJ, Morgan A. Regulation of the exocytotic machinery by cAMP-dependent protein kinase: implications for presynaptic plasticity. Biochem Soc Trans. 2003;31:824–7.
90. Leenders AG, Sheng ZH. Modulation of neurotransmitter release by the second messenger-activated protein kinases: implications for presynaptic plasticity. Pharmacol Ther. 2005;105:69–84.
91. Seino S, Shibasaki T. PKA-dependent and PKA-independent pathways for cAMP-regulated exocytosis. Physiol Rev. 2005;85:1303–42.
92. Südhof TC. The synaptic vesicle cycle. Annu Rev Neurosci. 2004;27:509–47.
93. Thakur P, Stevens DR, Sheng ZH, Rettig J. Effects of PKA-mediated phosphorylation of snapin on synaptic transmission in cultured hippocampal neurons. J Neurosci. 2004;24:6476–81.
94. Meneses D, Mateos V, Islas G, Barral J. G-protein-coupled inward rectifier potassium channels involved in corticostriatal presynaptic modulation. Synapse. 2015;69:446–52.
95. Sahlholm K, Nilsson J, Marcellino D, Fuxe K, Arhem P. Voltage sensitivities and deactivation kinetics of histamine $H_3$ and $H_4$ receptors. Biochim Biophys Acta. 2012;1818:3081–9.
96. Bongers GM. Signal transduction of the histamine $H_3$ receptor. 2008. PhD thesis, Vrije Universiteit Amsterdam, 2008. Accessed 4 Nov 2015.
97. Zeng L, Webster SV, Newton PM. The biology of protein kinase C. Adv Exp Med Biol. 2012;740:639–61.
98. Strock J, Diversé-Pierluissi MA. $Ca^{2+}$ channels as integrators of G protein-mediated signaling in neurons. Mol Pharmacol. 2004;66:1071–6.
99. Tedford HW, Zamponi GW. Direct G protein modulation of Cav2 calcium channels. Pharmacol Rev. 2006;58:837–62.
100. Kincaid AE, Wilson CJ. Corticostriatal innervations of the patch and matrix in the rat neostriatum. J Comp Neurol. 1996;374:578–92.
101. Molina-Hernández A, Nuñez A, Arias-Montaño JA. Histamine $H_3$-receptor activation inhibits dopamine synthesis in rat striatum. Neuroreport. 2000;11:163–6.
102. Doreulee N, Yanovsky Y, Flagmeyer I, Stevens DR, Haas HL, Brown RE. Histamine $H_3$ receptors depress synaptic transmission in the corticostriatal pathway. Neuropharmacology. 2001;40:106–13.
103. Ellender TJ, Huerta-Ocampo I, Deisseroth K, Capogna M, Bolam JP. Differential modulation of excitatory and inhibitory striatal synaptic transmission by histamine. J Neurosci. 2011;31:15340–51.
104. Smith Y, Galvan A, Ellender TJ, Doig N, Villalba RM, Huerta-Ocampo I, Wichmann T, Bolam JP. The thalamostriatal system in normal and diseased states. Front Syst Neurosci. 2014;8:5.
105. Lacey CJ, Bolam JP, Magill PJ. Novel and distinct operational principles of intralaminar thalamic neurons and their striatal projections. J Neurosci. 2007;27:4374–84.
106. Smith Y, Raju DV, Pare JF, Sidibe M. The thalamostriatal system: a highly specific network of the basal ganglia circuitry. Trends Neurosci. 2004;27:520–7.
107. Bolam JP, Ellender TJ. Histamine and the striatum. Neuropharmacology. 2015. doi:10.1016/j.neuropharm.2015.08.013.
108. Moreno E, Hoffmann H, Gonzalez-Sepúlveda M, Navarro G, Casadó V, Cortés A, Mallol J, Vignes M, McCormick PJ, Canela EI, Lluís C, Moratalla R, Ferré S, Ortiz J, Franco R. Dopamine $D_1$-histamine $H_3$ receptor heteromers provide a selective link to MAPK signaling in GABAergic neurons of the direct striatal pathway. J Biol Chem. 2011;286:5846–54.

109. Chuhma N, Tanaka KF, Hen R, Rayport S. Functional connectome of the striatal medium spiny neuron. J Neurosci. 2011;31:1183–92.
110. Planert H, Szydlowski SN, Hjorth JJ, Grillner S, Silberberg G. Dynamics of synaptic transmission between fast-spiking interneurons and striatal projection neurons of the direct and indirect pathways. J Neurosci. 2010;30:3499–507.
111. Tunstall MJ, Oorschot DE, Kean A, Wickens JR. Inhibitory interactions between spiny neurons in the rat striatum. J Neurophysiol. 2002;88:1263–9.
112. García M, Florán B, Arias-Montaño JA, Young JM, Aceves J. Histamine $H_3$ receptor activation selectively inhibits dopamine $D_1$ receptor-dependent [$^3$H]GABA release from depolarization-stimulated slices of rat substantia nigra pars reticulata. Neuroscience. 1997;80:241–9.
113. Ryu JH, Yanai K, Iwata R, Ido T, Watanabe T. Heterogeneous distribution of histamine $H_3$, and dopamine $D_1$ and $D_2$ receptors in rat brain. Neuroreport. 1994;5:621–4.
114. Jang I, Rhee JS, Watanabe T, Akaike N, Akaike N. Histaminergic modulation of GABAergic transmission in rat ventromedial hypothalamic neurons. J Physiol. 2001;534:791–803.
115. Calabresi P, Pisani A, Mercuri NB, Bernardi G. The corticostriatal projection: from synaptic plasticity to dysfunctions of the basal ganglia. Trends Neurosci. 1996;19:19–24.
116. Calabresi P, Galletti F, Saggese E, Ghiglieri V, Picconi B. Neuronal networks and synaptic plasticity in Parkinson's disease: beyond motor deficits. Parkinsonism Relat Disord. 2007;13 Suppl 3:259–62.
117. Benoti-Marand M, Borrelli E, Gonon F. Inhibition of dopamine release via presynaptic D2 receptors: time course and functional characteristics in vivo. J Neurosci. 2001;21:9134–41.
118. L'hirondel M, Chéramy A, Godeheu G, Artaud F, Saiardi A, Borrelli E, Glowinski J. Lack of autoreceptor-mediated inhibitory control of dopamine release in striatal synaptosomes of D2 receptor-deficient mice. Brain Res. 1998;792:253–62.
119. Lindgren N, Xu ZQ, Herrera-Marschitz M, Haycock J, Hökfelt T, Fisone G. Dopamine $D_2$ receptor regulate tyrosine hydroxylase activity and phosphorylation at Ser40 in rat striatum. Eur J Neurosci. 2001;13:773–80.
120. Schliker E, Fin K, Detzner M, Göther M. Histamine inhibits dopamine release in the mouse striatum via presynaptic $H_3$ receptors. J Neural Transm Gen Sect. 1993;93:1–10.
121. Smits RP, Mulder AH. Inhibitory effects of histamine on the release of serotonin and noradrenaline from rat brain slices. Neurochem Int. 1991;18:215–20.
122. Aquino-Miranda G, Escamilla-Sánchez J, González-Pantoja R, Bueno-Nava A, Arias-Montaño JA. Histamine $H_3$ receptors activation inhibits dopamine synthesis but not release or uptake in rat nucleus accumbens. Neuropharmacology. 2016;106:91–101.
123. Fox GB, Esbenshade TA, Pan JB, Radek RJ, Krueger KM, Yao BB, Browman KE, Buckley MJ, Ballard ME, Komater VA, Miner H, Zhang M, Faghih R, Rueter LE, Bitner RS, Drescher KU, Wetter J, Marsh K, Lemaire M, Porsolt RD, Bennani YL, Sullivan JP, Cowart MD, Decker MW, Hancock AA. Pharmacological properties of ABT-239 [4-(2-{2-[(2R)-2-Methylpyrrolidinyl]ethyl}-benzofuran-5-yl)benzonitrile]: II. Neurophysiological characterization and broad preclinical efficacy in cognition and schizophrenia of a potent and selective histamine $H_3$ receptor antagonist. J Pharmacol Exp Ther. 2005;313:176–90.
124. Ligneau X, Perrin D, Landais L, Camelin JC, Calmels TP, Berrebi-Bertrand I, Lecomte JM, Parmentier R, Anaclet C, Lin JS, Bertaina-Anglade V, la Rochelle CD, d'Aniello F, Rouleau A, Gbahou F, Arrang JM, Ganellin CR, Stark H, Schunack W, Schwartz JC. BF2.649 [1-{3-[3-(4-Chlorophenyl)propoxy]propyl}piperidine, hydrochloride], a nonimidazole inverse agonist/antagonist at the human histamine $H_3$ receptor: preclinical pharmacology. J Pharmacol Exp Ther. 2007;320:365–75.
125. Medhurst AD, Atkins AR, Beresford IJ, Brackenborough K, Briggs MA, Calver AR, Cilia J, Cluderay JE, Crook B, Davis JB, Davis RK, Davis RP, Dawson LA, Foley AG, Gartlon J, Gonzalez MI, Heslop T, Hirst WD, Jennings C, Jones DN, Lacroix LP, Martyn A, Ociepka S, Ray A, Regan CM, Roberts JC, Schogger J, Southam E, Stean TO, Trail BK, Upton N,

Wadsworth G, Wald JA, White T, Witherington J, Woolley ML, Worby A, Wilson DM. GSK189254, a novel $H_3$ receptor antagonist that binds to histamine $H_3$ receptors in Alzheimer's disease brain and improves cognitive performance in preclinical models. J Pharmacol Exp Ther. 2007;321:1032–45.
126. Nowak P, Bortel A, Dabrowska J, Biedka I, Slomian G, Roczniak W, Kostrzewa RM, Brus R. Histamine $H_3$ receptor ligands modulate L-dopa-evoked behavioral responses and L-dopa derived extracellular dopamine in dopamine-denervated rat striatum. Neurotox Res. 2008;13:231–40.
127. Swanger SA, Bassell GJ, Gross C. High-resolution fluorescence in situ hybridization to detect mRNAs in neuronal compartments in vitro and in vivo. Methods Mol Biol. 2011;714:103–23.
128. Yanovsky Y, Li S, Klyuch BP, Yao Q, Blandina P, Passani MB, Lin JS, Haas HL, Sergeeva OA. L-Dopa activates histaminergic neurons. J Physiol. 2011;589:1349–66.
129. Arrang JM, Drutel G, Schwartz JC. Characterization of histamine $H_3$ receptors regulating acetylcholine release in rat entorhinal cortex. Br J Pharmacol. 1995;114:1518–22.
130. Blandina P, Giorgetti M, Bartolini L, Cecchi M, Timmerman H, Leurs R, Pepeu G, Giovannini MG. Inhibition of cortical acetylcholine release and cognitive performance by histamine $H_3$ receptor activation in rats. Br J Pharmacol. 1996;119:1656–64.
131. Bacciottini L, Passani MB, Giovannelli L, Cangioli I, Mannaioni PF, Schunack W, Blandina P. Endogenous histamine in the medial septum-diagonal band complex increases the release of acetylcholine from the hippocampus: a dual-probe microdialysis study in the freely moving rat. Eur J Neurosci. 2002;15:1669–80.
132. Mochizuki T, Okakura-Mochizuki K, Horii A, Yamamoto Y, Yamatodani A. Histaminergic modulation of hippocampal acetylcholine release in vivo. J Neurochem. 1994;62:2275–82.
133. Prast H, Tran MH, Fischer H, Kraus M, Lamberti C, Grass K, Philippu A. Histaminergic neurons modulate acetylcholine release in the ventral striatum: role of $H_3$ histamine receptor. Naunyn Schmiedebergs Arch Pharmacol. 1999;360:558–64.
134. Passani MB, Cangioli I, Baldi E, Bucherelli C, Mannaioni PF, Blandina P. Histamine $H_3$ receptor-mediated impairment of contextual fear conditioning and *in-vivo* inhibition of cholinergic transmission in the rat basolateral amygdala. Eur J Neurosci. 2001;14:1522–32.
135. Bevan MD, Magill PJ, Terman D, Bolam JP, Wilson CJ. Move to the rhythm: oscillations in the subthalamic nucleus-external globus pallidus network. Trends Neurosci. 2002;25:525–31.
136. Naito A, Kita H. The cortico-pallidal projection in the rat: an anterograde tracing study with biotinylated dextran amine. Brain Res. 1994;653:251–7.
137. Anaya-Martinez V, Martinez-Marcos A, Martinez-Fong D, Aceves J, Erlij D. Substantia nigra compacta neurons that innervate the reticular thalamic nucleus in the rat also project to striatum or globus pallidus: implications for abnormal motor behavior. Neuroscience. 2006;143:477–86.
138. Morales-Figueroa GE, Márquez-Gómez R, González-Pantoja R, Escamilla-Sánchez J, Arias-Montaño JA. Histamine $H_3$ receptor activation counteracts adenosine $A_{2A}$ receptor-mediated enhancement of depolarization-evoked [$^3$H]-GABA release from rat globus pallidus synaptosomes. ACS Chem Neurosci. 2014;5:637–45.
139. Osorio-Espinoza A, Alatorre A, Ramos-Jiménez J, Garduño-Torres B, García-Ramírez M, Querejeta E, Arias-Montaño JA. Pre-synaptic histamine H3 receptors modulate glutamatergic transmission in rat globus pallidus. Neuroscience. 2011;176:20–31.
140. Mayfield R, Suzuki F, Zahniser NR. Adenosine $A_{2a}$ receptor modulation of electrically evoked endogenous GABA release from slices of rat globus pallidus. J Neurochem. 1993;60:2334–7.
141. Mayfield D, Larson G, Orona RA, Zahniser NR. Opposing actions of adenosine $A_{2a}$ and dopamine $D_2$ receptor activation on GABA release in the basal ganglia: evidence for an $A_{2a}$/$D_2$ receptor interaction in globus pallidus. Synapse. 1996;22:132–8.
142. Shindou T, Nonaka H, Richardson PJ, Mori A, Kase H, Ichimura M. Presynaptic adenosine $A_{2A}$ receptors enhance GABAergic synaptic transmission via a cyclic AMP dependent mechanism in the rat globus pallidus. Br J Pharmacol. 2002;136:296–302.

143. Shindou T, Nonaka H, Richardson PJ, Mori A, Kase H, Ichimura M. Adenosine modulates the striatal GABAergic inputs to the globus pallidus via adenosine $A_{2A}$ receptors in rats. Neurosci Lett. 2003;352:167–70.
144. Ferrada C, Ferré S, Casadó V, Cortés A, Justinova Z, Barnes C, Canela EI, Goldberg SR, Leurs R, Lluis C, Franco R. Interactions between histamine $H_3$ and dopamine $D_2$ receptors and the implications for striatal function. Neuropharmacology. 2008;55:190–7.
145. Ferrada C, Moreno E, Casadó V, Bongers G, Cortés A, Mallol J, Canela EI, Leurs R, Ferré S, Lluís C, Franco R. Marked changes in signal transduction upon heterodimerization of dopamine $D_1$ and histamine $H_3$ receptors. Br J Pharmacol. 2009;157:64–75.
146. Morales-Figueroa G-E, R. González-Pantoja R, J. Escamilla-Sánchez J, Arias-Montaño J-A. Functional interaction between histamine $H_3$ and adenosine $A_{2A}$ receptors in rat striato-pallidal nerve terminals. 2015; Program No. 119.04/A75. Neuroscience Meeting Planner. Chicago, IL: Society for Neuroscience. Online.
147. Chen L, Chan SC, Yung WH. Rotational behavior and electrophysiological effects induced by $GABA_B$ receptor activation in rat globus pallidus. Neuroscience. 2002;114:417–25.
148. Anichtchik OV, Huotari M, Peitsaro N, Haycock JW, Männistö PT, Panula P. Modulation of histamine $H_3$ receptors in the brain of 6-hydroxydopamine-lesioned rats. Eur J Neurosci. 2000;12:3823–32.
149. Anichtchik OV, Rinne JO, Kalimo H, Panula P. An altered histaminergic innervation of the substantia nigra in Parkinson's disease. Exp Neurol. 2000;163:20–30.
150. Da Cunha C, Boschen SL, Gómez-A A, Ross EK, Gibson WS, Min HK, Lee KH, Blaha CD. Toward sophisticated basal ganglia neuromodulation: Review on basal ganglia deep brain stimulation. Neurosci Biobehav Rev. 2015;58:186–210. doi:10.1016/j.neubiorev.2015.02.003.
151. Zhou FM, Lee CR. Intrinsic and integrative properties of substantia nigra pars reticulata neurons. Neuroscience. 2011;198:69–94.
152. Esbenshade TA, Strakhova MI, Carr TL, Sharma R, Witte DG, Yao BB, Miller TR, Hancock AA. Differential CNS expression and functional activity of multiple human $H_3$ receptor isoforms. Inflamm Res. 2006;55 Suppl 1:S38–9.
153. Threlfell S, Cragg SJ, Kalló I, Turi GF, Coen CW, Greenfield SA. Histamine $H_3$ receptors inhibit serotonin release in substantia nigra pars reticulata. J Neurosci. 2004;24:8704–10.
154. Corvaja N, Doucet G, Bolam JP. Ultrastructure and synaptic targets of the raphe-nigral projection in the rat. Neuroscience. 1993;55:417–27.
155. Moore RY, Halaris AE, Jones BE. Serotonin neurons of the midbrain raphe: ascending projections. J Comp Neurol. 2004;180:417–38.
156. Jin CY, Kalimo H, Panula P. The histaminergic system in human thalamus: correlation of innervation to receptor expression. Eur J Neurosci. 2002;15:1125–38.
157. Rouleau A, Héron A, Cochois V, Pillot C, Schwartz JC, Arrang JM. Cloning and expression of the mouse histamine $H_3$ receptor: evidence for multiple isoforms. J Neurochem. 2004;90:1331–8.
158. Garduño-Torres B, Treviño M, Gutiérrez R, Arias-Montaño JA. Pre-synaptic histamine $H_3$ receptors regulate glutamate, but not GABA release in rat thalamus. Neuropharmacology. 2007;52:527–35.
159. Arrang JM, Garbarg M, Schwartz JC. Autoregulation of histamine release in brain by presynaptic $H_3$-receptors. Neuroscience. 1985;15:553–62.
160. Márquez-Gómez R, Gutierrez-Rodelo C, Olivares-Reyes J-A, Arias-Montaño J-A. Adenosine $A_{2A}$ and histamine $H_3$ receptors gather to modulate intra-striatal GABAergic transmission. 2015; Program No. 119.05/A76. Neuroscience Meeting Planner. Chicago, IL: Society for Neuroscience. Online.
161. Bell MI, Richardson PJ, Lee K. Histamine depolarizes cholinergic interneurons in the rat striatum via a $H_1$-receptor mediated action. Br J Pharmacol. 2000;131:1135–42.
162. Leurs R, Traiffort E, Arrang JM, Tardivel-Lacombe J, Ruat M, Schwartz JC. Guinea pig histamine H1 receptor. II. Stable expression in Chinese hamster ovary cells reveals the interaction with three major signal transduction pathways. J Neurochem. 1994;62:519–27.

# Chapter 12
# Interaction of Brain Histaminergic and Dopaminergic Systems

**Saara Nuutinen and Outi Salminen**

**Abstract** Brain neurons containing the neurotransmitter dopamine have two well-characterized functions: they are key regulators of movements, and they mediate reward and motivation induced by natural rewards such as food and sex but also by drugs of abuse including nicotine, alcohol, and illegal drugs [1]. Dopamine neurons from the midbrain and histamine neurons from the posterolateral hypothalamus both send their axons to a brain area called the striatum which is the center for the control of movements, reward, and motivation. The striatum is divided into two subregions: the dorsal striatum which is classically referred to as the motor control region and the ventral striatum, including the nucleus accumbens, which regulates reward and motivation. Recent evidence shows however that the classical division of striatal functions is not that clear and, e.g., dorsal striatal areas have been shown to be involved in the regulation of reward too. The midbrain areas where dopaminergic neuron somas are located, substantia nigra and ventral tegmental area, also receive histaminergic projections [2, 3]. Of importance, the striatum expresses a high density of histamine H1–H3 receptors [4–6] suggesting that histamine can directly affect striatal function and basal ganglia output. The expression of the histamine H3 receptor in the striatum is exceptionally high [5–7]. H3 receptors are G protein-coupled receptors that regulate the release of histamine but also other neurotransmitter release (e.g., GABA, noradrenaline, acetylcholine, and possibly dopamine) [8]. Importantly, majority of the H3 receptors in the striatum are located postsynaptically at GABAergic medium spiny neurons [5, 6, 9], and there is evidence of a direct interaction between H3 and dopamine receptors in co-expressing culture systems [10–12] and in vivo [13].

**Keywords** Histamine • Histamine H1 receptor • Histamine H2 receptor • H3 receptor • Dopamine • Dopamine D1 receptor • Dopamine D2 receptor • Meso-limbic dopamine system • Nigro-striatal dopamine system • Reward • Motor functions • Receptor-receptor interaction • Heterodimerization

S. Nuutinen, Ph.D. (✉) • O. Salminen, Ph.D.
Division of Pharmacology and Pharmacotherapy, Faculty of Pharmacy, University of Helsinki, Viikinkaari 5E, P.O. Box 56, 00014 Helsinki, Finland
e-mail: saara.nuutinen@helsinki.fi; outi.salminen@helsinki.fi



P. Blandina, M.B. Passani (eds.), *Histamine Receptors*, The Receptors 28,
DOI 10.1007/978-3-319-40308-3_12

## 12.1 Introduction

Brain neurons containing the neurotransmitter dopamine have two well-characterized functions: they are key regulators of movements, and they mediate reward and motivation induced by natural rewards such as food and sex but also by drugs of abuse including nicotine, alcohol, and illegal drugs [1]. Dopamine neurons from the midbrain and histamine neurons from the posterolateral hypothalamus both send their axons to a brain area called the striatum which is the center for the control of movements, reward, and motivation. The striatum is divided into two subregions: the dorsal striatum which is classically referred to as the motor control region and the ventral striatum, including the nucleus accumbens, which regulates reward and motivation. Recent evidence shows however that the classical division of striatal functions is not that clear and, e.g., dorsal striatal areas have been shown to be involved in the regulation of reward too. The midbrain areas where dopaminergic neuron somas are located, substantia nigra and ventral tegmental area, also receive histaminergic projections [2, 3]. Of importance, the striatum expresses a high density of histamine H1–H3 receptors [4–6] suggesting that histamine can directly affect striatal function and basal ganglia output. The expression of the histamine H3 receptor in the striatum is exceptionally high [5–7]. H3 receptors are G protein-coupled receptors that regulate the release of histamine but also other neurotransmitter release (e.g., GABA, noradrenaline, acetylcholine, and possibly dopamine) [8]. Importantly, majority of the H3 receptors in the striatum are located postsynaptically at GABAergic medium spiny neurons [5, 6, 9], and there is evidence of a direct interaction between H3 and dopamine receptors in co-expressing culture systems [10–12] and in vivo [13].

We have demonstrated that alcohol-evoked reward, motivation, and relapse to alcohol use can all be inhibited by affecting histamine H3 receptors (H3Rs) in the brain [14–19]. Our studies suggest that the mechanism underlying these effects is an interaction of the brain histaminergic system with the dopaminergic neurons and/or their postsynaptic receptors. In addition to our observations, several other studies from behavioral to receptor level point to a role for histaminergic modulation of both nigrostriatal and mesolimbic dopamine pathways [20]. In the following chapters, the studies reporting evidence for this interaction are presented including possible mechanisms of the suggested histamine–dopamine interaction. The focus of this review is on the role of histaminergic system and its H3 receptor in the modulation of dopaminergic functions.

## 12.2 Behavioral Studies Suggesting Interaction with Mesolimbic and Nigrostriatal Dopamine Pathways

The first indications of histaminergic modulation of dopaminergic functions came from studies where intracerebroventricular histamine was found to induce alterations in locomotion of rodents [21–23]. The studies demonstrated a biphasic effect for histamine with an initial hypolocomotor response followed by hyperactivity. The use of histaminergic drugs supported the role of histaminergic regulation of the nigrostriatal dopamine movement pathway. H1 receptor antagonists were found to inhibit either both hypo- and hyperactivation [23] or only the hyperactivity induced by histamine [21, 22]. When injected into the hippocampus, an H1 antagonist inhibited the hypoactive response induced by histamine [24]. The H1 receptor antagonist itself did not produce any effect on motor activity suggesting that histamine had a modulatory rather than an essential role in the mediation of motor activity via H1 receptors. This view was later supported by the Watanabe group in Japan, who demonstrated that H1 receptor knockout mice displayed reduced explorative activity and rearing [25].

The role of the H2 receptor seems less important in the motor control, or at least a direct interaction with the nigrostriatal dopamine system is less likely than with H1 receptors. H2 receptor ligands do not inhibit the effects of histamine on motor activity [21, 23] via systemic administration. However, when injected into the hippocampus, the H2 receptor antagonist ranitidine blocked the hypolocomotor response and the effect on rearing evoked by histamine [26]. Chiavegatto and colleagues, on the other hand, found that intracerebroventricular ranitidine induced hyperactivity by itself [22]. There are to our knowledge no reports on the characterization of motor functions of H2 receptor (-/-) mice, but H1/H2 double (-/-) mice showed sensitized response to methamphetamine [27]. As the H1 receptor (-/-) mice displayed diminished locomotion [25], this suggests that the H2 receptor might have a role in the mediation of motor responses induced by methamphetamine. Further studies are needed to clarify this.

The evidence for H3 receptor interaction with the dopaminergic system is more abundant than for H1 and H2 receptors. Several studies have shown that thioperamide but also other H3R antagonists suppress hyperactivity induced by amphetamine or methamphetamine (Table 12.1). H3 receptor antagonists themselves do not seem to alter locomotion, supporting interaction with the dopaminergic system. Interestingly, Brabant and colleagues found that thioperamide enhanced cocaine-induced hyperactivity but suggested that this was probably due to a metabolic interaction of the drugs in the liver [28]. Thioperamide and other imidazole-based compounds are metabolized via liver CYP450 enzymes, and, when administered together with another CYP450-metabolized drug, the concentration of the other compound in the plasma can rise due to competition at the enzyme substrate site. This is important to keep in mind when interpreting in vivo studies conducted with thioperamide and, e.g., ciproxifan with other drugs. Also Zhang et al. reported a similar pharmacokinetic interaction for imidazole and haloperidol [29], indicating that the reported enhancement of haloperidol-induced catalepsy by thioperamide [30] results from a metabolic interaction.

Behavioral studies from Parkinson's disease animal models further support a role for histamine and histaminergic receptors in regulation of nigrostriatal dopaminergic motor functions but also in the regulation of dopamine cell survival (Table 12.1). In rats treated with the neurotoxin 6-hydroxydopamine (6-OHDA), drug-induced rotational behavior, reflecting the damage of dopaminergic neurons, was found to be greater when endogenous brain histamine levels were increased with histidine

**Table 12.1** Evidence of histaminergic modulation of motor functions suggesting interaction between the histaminergic system and the nigrostriatal dopamine pathway

| Histaminergic drug/condition | Dopaminergic drug | Assay | Result | References |
|---|---|---|---|---|
| *Thioperamide* 0.2–10 mg/kg, 0.3–20 μg i.c.v. | *Amphetamine* 1 mg/kg | Hyperactivity | ↓↓ | [64] |
| | *Apomorphine* 2 mg/kg | | ↓↓ | |
| | *Cocaine* 5 mg/kg | | ↓ | |
| *ABT-239* 0.1–1.0 mg/kg | *Methamphetamine* 1.0 mg/kg | Hyperactivity | ↓ No effect with ABT-239 alone | [65] |
| *Ciproxifan* 3 mg/kg | *Methamphetamine* 2 mg/kg | Hyperactivity | ↓ | [66] |
| *H3R (-/-) mice* | *Methamphetamine* 1 mg/kg | Hyperactivity and stereotypies in mice | ↓ | [67] |
| *Thioperamide* 2.5, 5, or 10 mg/kg | *Cocaine* 8 mg/kg | Hyperactivity in mice | ↑ due to a metabolic interaction | [28] |
| *A-331440* 2.5 and 20 mg/kg | *Cocaine* 8 mg/kg | | ±0 | |
| *Thioperamide* 3.75, 7.5, and 15 mg/kg | *Amphetamine* 2 mg/kg | Hyperactivity in mice | ↓ | [30] |
| | *Haloperidol* 2 mg/kg | Catalepsy | ↑ | |
| | *Apomorphine* 1.5 mg/kg | Climbing behavior | ↓ | |
| *GSK207040* 0.32–10.0 mg/kg | *Amphetamine* 0.4 mg/kg | Hyperactivity in rats | ±0 | [68] |
| *Pitolisant/BF2.649* 5 mg/kg | *Methamphetamine* 0.75 mg/kg | Hyperactivity | ↓ | [69] |
| | *Apomorphine* 1.5 mg/kg | Climbing behavior | ±0/↓ | |
| *Ciproxifan* 1.5 mg/kg | *Haloperidol* 0.1, 1 mg/kg | Hypoactivity and catalepsy in rats | ↑ | [5, 6] |

| | | | | |
|---|---|---|---|---|
| *JNJ-39220675*<br>10 mg/kg | *Quinpirole*<br>0.5 mg/kg | Hypoactivity in mice | ±0 in acute treatment<br>↓ after 5 days of admin<br>Mild ↓ by JNJ-39220675 alone acutely | [70] |
| *Histidine decarboxylase (-/-) mice* | *Amphetamine*<br>8.5 mg/kg | Stereotypies | ↑ | [52] |
| Parkinsonian models | | | | |
| *R-α-Methylhistamine*<br>2.5, 5, 10 mg/kg | *L-DOPA*<br>10 mg/kg | Contralateral turning in hemiparkinsonian rats | ↓ | [34] |
| *Thioperamide*<br>2.5, 5, 10 mg/kg | *L-DOPA*<br>10 mg/kg | | ±0/(↑) | |
| *Imetit*<br>5.0 mg/kg | *L-DOPA*<br>15 mg/kg | Rearings in dopamine-denervated rats | ↓ | [35] |
| *Immepip*<br>3.2 and 32 ng, intranigral | *Apomorphine*<br>0.5 mg/kg | Contralateral turning in hemiparkinsonian rats | ↓ | [33] |
| *L-Histidine*<br>200 or 500 mg/kg, 7–14 days | *Apomorphine*<br>0.5 mg/kg | Contralateral turning in hemiparkinsonian rats | ↑ | [31] |
| *α-Fluoromethylhistidine*<br>25 mg, i.c.v. | | | ↓ | |
| *L-Histidine*<br>500 mg/kg | *Apomorphine*<br>0.5 mg/kg | Contralateral turning in hemiparkinsonian rats | ↑ | [32] |
| *α-Fluoromethylhistidine*<br>25 μg, i.c.v. | *Apomorphine*<br>0.5 mg/kg | | ↓ | |
| *Pyrilamine*<br>50 μg, i.c.v. | *Apomorphine*<br>0.5 mg/kg | | ↓ | |
| *Cimetidine*<br>50 μg, i.c.v. | *Apomorphine*<br>0.5 mg/kg | | ↓ | |
| *Clobenprobit*<br>10 μg, i.c.v. | *Apomorphine*<br>0.5 mg/kg | | ±0 | |
| *Immepip*<br>10 μg, i.c.v | *Apomorphine*<br>0.5 mg/kg | | ↓ | |
| *R-α-Methylhistamine*<br>5 mg/kg | *Amphetamine*<br>2.5 mg/kg | Ipsilateral turning in hemiparkinsonian rats | ±0 | [34] |
| *Thioperamide*<br>5 mg/kg | *Amphetamine*<br>2.5 mg/kg | | ±0 | |

administration, suggesting that histamine might have enhanced the neuronal damage induced by 6-OHDA [31]. However, the effect was not dose dependent: only the smaller dose of histidine (200 mg/kg) but not the higher (500 mg/kg) increased apomorphine-induced rotations. A later study by the same group showed that by increasing the brain histamine levels after the lesion with 6-OHDA results also to increased rotational behavior by apomorphine now pointing to a more complex interaction of histamine with the dopaminergic system [32]. This was supported by findings that H1 and H2 receptor antagonists and an H3 receptor agonist all decreased apomorphine-induced rotations. Also other studies have shown that activation of the H3 receptor suppresses turning behavior induced by dopaminergic agonists L-DOPA and apomorphine in Parkinsonian rats [33–35]. Amphetamine-induced turning behavior was not affected by an H3 agonist [34], suggesting a post-synaptic interaction between the H3 receptor and dopamine receptors.

All the drugs were administered systemically via subcutaneous or intraperitoneal acute injections unless stated otherwise, intracerebroventricular (i.c.v.).

Further support for the role of histamine in regulating dopaminergic functions comes from studies concerning addictive behaviors (Table 12.2). Dopamine, released from dopaminergic axons innervating the nucleus accumbens in response to reward-related stimuli, and alterations in accumbal dopaminergic synaptic transmission have been implicated in the acquisition, maintenance, and relapse of addiction [36]. In the majority of the studies, histamine has been shown to suppress reward and reinforcement induced by addictive drugs (Table 12.2). One of the earliest findings suggesting that histamine inhibits reinforcement was that injection of histamine into the lateral ventricle caused an increase in reinforcement threshold, and injection of histamine into the lateral hypothalamus suppressed self-stimulation [37]. Later studies have shown, for example, that loading with L-histidine, the precursor of histamine, attenuates morphine-induced conditioned place preference, a paradigm that is commonly used to measure drug reward and reinforcement [38]. The inhibitory effect is likely to be mediated via H1 receptors since H1 receptor antagonists increase drug reward [39, 40], and methamphetamine induces stronger place preference in H1R (-/-) mice [41]. However, antihistamines have been found to increase the plasma levels of methamphetamine [42], suggesting that the H1 antagonist-evoked stronger reward is due to a pharmacokinetic interaction similar to what has been found for imidazole-structured H3 receptor antagonist and cocaine [43]. Also histamine H2 receptor antagonists can enhance reward in mice as was shown for zolantidine in combination with morphine [38].

Although alcohol is not only a "dopamine drug," several studies from our laboratory (Prof. Pertti Panula group) show that treatment with an H3R antagonist results in inhibition of alcohol-evoked reward, self-administration, and relapse to alcohol seeking [15–17, 19]. The inhibitory effect does not result from a metabolic interaction since both imidazole and non-imidazole-structured compounds elicit the same effect and that alcohol concentrations in the plasma remain unaltered. We have also shown that mice lacking endogenous histamine, histidine decarboxylase (-/-) mice, develop stronger alcohol-induced conditioned place preference supporting the inhibitory action of histamine in reward and reinforcement [14]. In contrast to our findings, no difference for cocaine reward was found in the conditioned place preference model between HDC (-/-) and wild-type mice [44].

**Table 12.2** Evidence for the involvement of histamine and histamine receptors in addictive behaviors suggesting interaction with mesolimbic dopamine system

| Histaminergic drug/ condition | Drug of abuse | Assay | Result | References |
|---|---|---|---|---|
| *L-Histidine*<br>*α-Fluoromethyl-histamine* | *Morphine*<br>*Morphine* | CPP | ↓<br>↑ | [38] |
| *Histidine decarboxylase (-/-) mice* | *Cocaine* | | Similar to wild types | [44] |
| *Histidine decarboxylase (-/-) mice* | *Alcohol* | CPP | ↑ | [14] |
| *Histidine decarboxylase (-/-) mice* | *Alcohol* | Self-administration | Similar to wild types | [19] |
| *Chlorpheniramine* | *Methamphetamine* | CPP in rats | ↑ Possibly due to a pharmacokinetic interaction | [39] |
| | *Cocaine* | | ↑ | |
| *Pyrilamine* | *Nicotine* | Self-administration in rats | ↓ | [71, 72] |
| *H1 receptor (-/-) mice* | *Methamphetamine* | CPP | ↑ | [41] |
| *Zolantidine* | *Morphine* | CPP in mice | ↑ | [38] |
| *Thioperamide* | *Methamphetamine* | Self-administration in rats | ↑ | [73] |
| *Clobenprobit* | *Methamphetamine* | | ↑ | |
| *JNJ-39220675* | *Amphetamine* | CPP in mice | ±0 | [70] |
| *Thioperamide* | *Cocaine* | CPP | ↑ | [28] |
| *Thioperamide* | *Alcohol* | Self-administration in alcohol-preferring rats | ↓ | [74] |
| *Clobenprobit* | *Alcohol* | | ↓ | |
| *Ciproxifan* | *Alcohol* | Self-administration in mice | ↓ | [15, 17] |
| *Immepip* | *Alcohol* | | ↑ | |
| *Ciproxifan* | *Alcohol* | CPP in mice | ↓ | [15, 17, 19] |
| *JNJ-10181457* | *Alcohol* | | ↓ | |
| *JNJ-39220675* | *Alcohol* | | ↓ | |
| *Ciproxifan* | *Alcohol* | Cue-induced reinstatement of self-administration | ↓ | [16] |
| *JNJ-39220675* | *Alcohol* | | ↓ | |

(continued)

**Table 12.2** (continued)

| Histaminergic drug/ condition | Drug of abuse | Assay | Result | References |
|---|---|---|---|---|
| *H3 receptor (-/-) mice* | *Methamphetamine* | CPP | Similar to wild types | [75] |
| | *MDMA* | | Similar to wild types | |
| *H3 receptor (-/-) mice* | *Alcohol* | Self-administration | ↓ | [15, 17] |
| | | CPP | ↓ | |

*CPP* conditioned place preference

## 12.3 Histaminergic Regulation of Dopamine Cell Development, Maintenance, and Survival

Lack of clear evidence on histamine-mediated dopamine release (see the paragraph below; [16, 45]) raises a question whether histamine H3 receptors expressed in the striatal and midbrain areas are localized on dopaminergic neurons or if they are, what is their functional role. Anichtchik and colleagues detected no or very low levels of H3 receptor mRNA in the substantia nigra of rats and humans [46–48]. Neither did Pillot et al. find high levels of H3R mRNA in substantia nigra or ventral tegmental area [5, 6]. In contrast, a recent study demonstrated that both H3 receptor mRNA and protein is expressed in dopaminergic cell bodies in the substantia nigra and in the ventral tegmental area [49]. These findings await confirmation from other groups, but in case the H3 receptor is localized on dopaminergic neurons, the observations supporting a role for histamine in dopamine cell development, maintenance, and survival are more easily explained. A few studies have reported that histamine inhibits forskolin or IBMX-stimulated DOPA accumulation via activation of H3 receptors in rat nucleus accumbens slice preparations [45, 50]. Dopamine synthesis is also inhibited by H3 receptor activation in rat striatal slices [51] and in dopamine-denervated rat striatum [35], suggesting that dopamine synthesis in both reward and motor controlling neurons is under H3 receptor-dependent regulation.

In support for the histamine-mediated control of dopamine synthesis, studies have shown that dopamine levels [52, 53] and dopamine turnover [54] are increased in histidine decarboxylase (-/-) mice. This long-lasting chronic effect could lead to alterations in the expression of dopamine receptors. This has been demonstrated with histidine decarboxylase (-/-) mice where the levels of D2/D3 receptor expression both in substantia nigra and pallidum are increased [52]. A mutation in the histidine decarboxylase gene was found in one family of Tourette's syndrome patients suggesting histaminergic dysfunction in the pathophysiology of the disease [55]. Animal studies suggest that reduced histamine production can produce symptoms of Tourette's syndrome through dysregulation of the dopaminergic modulation of the basal ganglia [52], further emphasizing the functional association between histaminergic and dopaminergic systems.

In addition to regulation of dopamine synthesis, a few studies suggest that histamine can affect dopamine neuron development and survival of mature neurons. Intrauterine administration of histamine was found to interfere with dopamine cell development, resulting in reduced tyrosine hydroxylase staining in rat embryos via an H1 receptor-dependent mechanism [56]. Histamine also diminished the expression of Lmx1a and Lmx1b, markers for midbrain dopaminergic precursors, and the expression of Pitx3, a protein important for the maintenance of dopaminergic identity. The effect of histamine was seen only at early developmental stages (E10 and E12) and was specific to dopaminergic progenitor cells only. However, it is important to note that the concentrations of histamine used were high and an H1 receptor antagonist alone did not affect dopamine cell development, indicating that endogenous histamine levels are not high enough to affect the development of dopamine cells.

Some work suggests that histamine can affect also the survival of mature dopaminergic neurons. Direct infusion of histamine to substantia nigra leads to a selective damage of dopaminergic neurons [57], and the neurotoxin 6-OHDA-induced dopamine cell loss was greater when endogenous histamine levels were increased, and this neurotoxic effect was mediated via H1 receptors [31]. Liu et al. concluded that endogenous histamine can accelerate the neurodegeneration of dopamine neurons. This view is supported by findings that the histamine levels and histaminergic innervation of substantia nigra were increased in Parkinson's disease patients' brains *post-mortem* [46, 48, 58].

## 12.4 Histaminergic Modulation of Firing of Dopamine Neurons

The findings described above suggest that histamine is a negative regulator of dopamine-dependent behaviors and dopamine cell development, maintenance, and survival. Thus, the next logical question to be asked is what is the mechanism for this functional antagonism? Histamine does not seem to directly affect the firing rate of dopamine neurons as was reported by Korotkova and colleagues in substantia nigra or ventral tegmental area of rat brain slices [59]. However, histamine was found to increase the firing of GABAergic neurons in both these areas, suggesting that the inhibition of dopaminergic functions is mediated via histamine-induced excitation of GABAergic neuronal activity. Zhou et al. [60] confirmed the same action of histamine on firing frequency of GABAergic neurons in mouse substantia nigra pars compacta and demonstrated that the effect was mediated via H1 and H2 receptors [60]. Activation of H3 receptors, on the other hand, hyperpolarized and inhibited the inhibitory GABAergic neurons of substantia nigra pars compacta. This suggests that the effects of H3 receptor ligands on dopamine-dependent motor functions could be mediated via alterations of the firing patterns of inhibitory GABAergic neurons that, in turn, regulate the activity of dopaminergic neurons in substantia nigra.

## 12.5 Histaminergic Modulation of Dopamine Release

Both histamine and the H3 receptor agonist α-methylhistamine are able to inhibit the electrically evoked release of 3H-dopamine from mouse striatal slices, suggesting that a fraction of H3 receptors is located on dopaminergic axon terminals [61]. However, later studies using in vivo microdialysis have not been able to show dopamine release in the striatum by H3 receptor ligands [16, 45]. Also the lack of clear evidence of H3 receptor expression in dopaminergic terminals suggests that H3 receptors do not directly alter dopamine release. However, indirect mechanisms for histamine to affect dopamine release in the striatum do exist. These include the ability of histamine to modulate the firing patterns of midbrain GABAergic neurons and the release of acetylcholine and glutamate within the striatum (for a review, see [7]).

## 12.6 Striatal Receptor–Receptor Interactions

Lesioning and co-localization studies with endogenous opioid peptides have shown that in the rat striatum, the major portion of the H3 receptors is located on GABAergic striatonigral projection neurons [5, 6, 9, 62], whereas in other brain areas, H3 receptors are mainly located on nerve terminals. There is ample evidence that the H3 receptor interacts with dopamine D1 and D2 receptors. The activation of the H3 receptor is able to inhibit D1 receptor agonist (SKF-81297)-induced cAMP accumulation in striatal slices [63]. Further, a study in a transfected cell line showed that H3 receptors and D1 receptors can heterodimerize, and signaling via one or the other of these two receptors was blocked not only by a selective antagonist but also by an antagonist of the partner receptor [11]. H3 receptor–D1 receptor heterodimers were found to exist in vivo in the mouse brain, and antagonists of either the H3 receptor or the D1 receptor were able to block MAPK signaling elicited by H3 or D1 agonists [13]. An H3 receptor agonist-induced activation of MAPK was absent in D1R knockout mice, further supporting the existence of functional postsynaptic H3R–D1R heterodimers. In a later study by the same group, an interesting observation was made: H3 receptor was confirmed to inhibit D1 receptor function via heterodimerization, and this inhibition was disturbed by cocaine binding to σ1 receptor that is able to modify the H3–D1 receptor complex [12]. This finding gives another aspect to the addiction research concerning the role of H3 receptor in the mediation of drug reward and suggests that the σ–H3–D1 receptor complex could be a novel target to treat cocaine addiction.

H3 receptors colocalize with proenkephalin A in rat striatopallidal neurons, which also express D2 receptors [5, 6]. Ciproxifan potentiates the increase of proenkephalin A mRNA expression that is induced by haloperidol in the rat striatum, suggesting that H3 receptors interact with D2 receptors as well. H3 receptors were found to form intramembrane interactions with D2 receptors, and H3 agonists decreased the affinity of D2 receptors for the D2 receptor agonist quinpirole in sheep striatal membranes [10]. Therefore, it could be postulated that postsynaptic H3 receptors negatively modulate dopaminergic transmission via interactions with D2 receptors.

Majority of all neurons (95 %) in the striatum are GABAergic medium spiny projection neurons (MSN), of which there are at least two classes: D1 dopamine receptor-expressing and D2 dopamine receptor-expressing MSNs. D1- and D2-expressing MSNs are the projection neurons of the striatum, and they comprise the direct and indirect pathways, respectively. Through balanced activity in these two pathways, the direct D1 pathway facilitates movement, and the indirect D2 pathway inhibits movement. The remaining 5 % of striatal neurons are composed of cholinergic and GABAergic interneurons [20]. Evidence presented above shows the proposed mechanism of histaminergic modulation of the dopaminergic system: there seem to be H3 receptor–D1 receptor interactions in postsynaptic GABA neurons comprising the direct pathway and/or H3 receptor–D2 receptor interactions in postsynaptic GABA neurons in the indirect pathway which regulate the signaling induced by dopamine released from striatal dopaminergic nerve terminals.

## 12.7 Summary

Several lines of evidence support the view that the brain histaminergic system is a regulator of dopaminergic functions (Fig. 12.1). Behavioral studies show that histamine and especially its H3 receptor modulate both motor functions and reward controlled by the nigrostriatal and mesolimbic dopamine neurons, respectively. However, studies using the old imidazole-based H3 receptor ligands must be interpreted with caution since these compounds are able to increase the blood concentrations of many drugs, including dopaminergic ligands, via metabolic interactions in the liver. Also some H1 receptor antagonists (antihistamines) can interrupt with

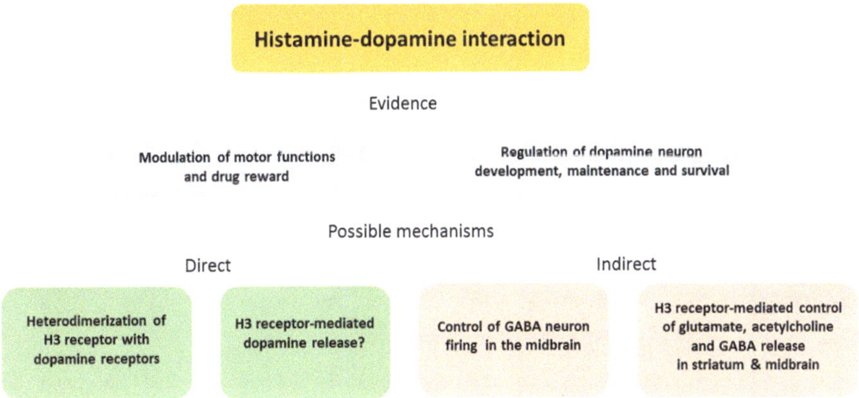


**Fig. 12.1** Summary of the evidence and possible direct and indirect mechanisms of the interaction of brain histamine and dopamine systems

the liver metabolism of dopaminergic drugs which is important to keep in mind when interpreting the early findings with these drugs. Rather convincing data exists for histamine as a regulator of the development, maintenance, and survival of dopaminergic neurons, suggesting an important role for the brain histaminergic system as a possible drug target in the treatment of disorders of the dopaminergic neurons. Mechanisms underlying the interaction of histaminergic and dopaminergic systems have been studied in great detail and include the ability of histamine via indirect mechanism to alter the firing frequency of GABAergic neurons and the histamine-induced GABA release in the midbrain and possibly via direct effect of histamine on dopamine release in the striatum. A direct mechanism of heterodimerization of histamine H3 receptors with striatal dopamine D1 and D2 receptors has also been suggested. More studies are still needed to clarify the role of histamine in dopaminergic functions by, e.g., repeating the behavioral studies with non-imidazole H3 receptor ligands, and to confirm whether the H3 receptor is expressed in dopaminergic neurons.

## References

1. Trudeau LE, Hnasko TS, Wallen-Mackenzie A, et al. The multilingual nature of dopamine neurons. Prog Brain Res. 2014;211:141–64. doi:10.1016/B978-0-444-63425-2.00006-4.
2. Airaksinen MS, Panula P. The histaminergic system in the guinea pig central nervous system: an immunocytochemical mapping study using an antiserum against histamine. J Comp Neurol. 1988;273(2):163–86. doi:10.1002/cne.902730204.
3. Takagi H, Morishima Y, Matsuyama T, et al. Histaminergic axons in the neostriatum and cerebral cortex of the rat: a correlated light and electron microscopic immunocytochemical study using histidine decarboxylase as a marker. Brain Res. 1986;364(1):114–23.
4. Martinez-Mir MI, Pollard H, Moreau J, et al. Three histamine receptors (H1, H2 and H3) visualized in the brain of human and non-human primates. Brain Res. 1990;526(2):322–7.
5. Pillot C, Heron A, Cochois V, et al. A detailed mapping of the histamine H(3) receptor and its gene transcripts in rat brain. Neuroscience. 2002;14(1):173–93.
6. Pillot C, Ortiz J, Heron A, et al. Ciproxifan, a histamine H3-receptor antagonist/inverse agonist, potentiates neurochemical and behavioral effects of haloperidol in the rat. J Neurosci. 2002;22(16):7272–80.
7. Panula P, Nuutinen S. The histaminergic network in the brain: basic organization and role in disease. Nat Rev Neurosci. 2013;14(7):472–87. doi:10.1038/nrn3526.
8. Ellenbroek BA. Histamine H(3) receptors, the complex interaction with dopamine and its implications for addiction. Br J Pharmacol. 2013;170(1):46–57. doi:10.1111/bph.12221.
9. Pillot C, Heron A, Schwartz JC, et al. Ciproxifan, a histamine H3-receptor antagonist/inverse agonist, modulates the effects of methamphetamine on neuropeptide mRNA expression in rat striatum. Eur J Neurosci. 2003;17(2):307–14.
10. Ferrada C, Ferre S, Casado V, et al. Interactions between histamine H3 and dopamine D2 receptors and the implications for striatal function. Neuropharmacology. 2008;55(2):190–7. doi:10.1016/j.neuropharm.2008.05.008.
11. Ferrada C, Moreno E, Casado V, et al. Marked changes in signal transduction upon heteromerization of dopamine D1 and histamine H3 receptors. Br J Pharmacol. 2009;157(1):64–75. doi:10.1111/j.1476-5381.2009.00152.x.
12. Moreno E, Moreno-Delgado D, Navarro G, et al. Cocaine disrupts histamine H3 receptor modulation of dopamine D1 receptor signaling: σ1-D1-H3 receptor complexes as key targets for reducing cocaine's effects. J Neurosci. 2014;34(10):3545–58. doi:10.1523/JNEUROSCI.4147-13.2014.

13. Moreno E, Hoffmann H, Gonzalez-Sepulveda M, et al. Dopamine D1-histamine H3 receptor heteromers provide a selective link to the map-kinase signalling in gabaergic neurons of the direct striatal pathway. J Biol Chem. 2011;286:5846–54. doi:10.1074/jbc.M110.161489.
14. Nuutinen S, Karlstedt K, Aitta-Aho T, et al. Histamine and H3 receptor-dependent mechanisms regulate ethanol stimulation and conditioned place preference in mice. Psychopharmacology (Berl). 2010;208(1):75–86. doi:10.1007/s00213-009-1710-5.
15. Nuutinen S, Lintunen M, Vanhanen J, et al. Evidence for the role of histamine H3 receptor in alcohol consumption and alcohol reward in mice. Neuropsychopharmacology. 2011;36(10):2030–40. doi:10.1038/npp.2011.90.
16. Nuutinen S, Maki T, Rozov S, et al. Histamine H3 receptor antagonist decreases cue-induced alcohol reinstatement in mice. Neuropharmacology. 2015;106:156–63. doi:10.1016/j.neuropharm.2015.06.006.
17. Nuutinen S, Vanhanen J, Pigni MC, et al. Effects of histamine H3 receptor ligands on the rewarding, stimulant and motor-impairing effects of ethanol in DBA/2J mice. Neuropharmacology. 2011;60(7–8):1193–9. doi:10.1016/j.neuropharm.2010.10.027.
18. Panula P, Nuutinen S. Histamine and H3 receptor in alcohol-related behaviors. J Pharmacol Exp Ther. 2011;336(1):9–16. doi:10.1124/jpet.110.170928.
19. Vanhanen J, Nuutinen S, Lintunen M, et al. Histamine is required for H3 receptor-mediated alcohol reward inhibition, but not for alcohol consumption or stimulation. Br J Pharmacol. 2013;170(1):177–87. doi:10.1111/bph.12170.
20. Bolam JP, Ellender TJ. Histamine and the striatum. Neuropharmacology. 2015;106:74–84. doi:10.1016/j.neuropharm.2015.08.013.
21. Bristow LJ, Bennett GW. Biphasic effects of intra-accumbens histamine administration on spontaneous motor activity in the rat; a role for central histamine receptors. Br J Pharmacol. 1988;95(4):1292–302.
22. Chiavegatto S, Nasello AG, Bernardi MM. Histamine and spontaneous motor activity: biphasic changes, receptors involved and participation of the striatal dopamine system. Life Sci. 1998;62(20):1875–88.
23. Kalivas PW. Histamine-induced arousal in the conscious and pentobarbital-pretreated rat. J Pharmacol Exp Ther. 1982;222(1):37–42.
24. Alvarez EO. Banzan AM Further evidence that histamine in hippocampus affects the exploratory behavior in the rat. Physiol Behav. 1985;34(5):661–4.
25. Inoue I, Yanai K, Kitamura D, et al. Impaired locomotor activity and exploratory behavior in mice lacking histamine H1 receptors. Proc Natl Acad Sci U S A. 1996;93(23):13316–20.
26. Alvarez EO, Banzan AM. Histamine in dorsal and ventral hippocampus. II. Effects of H1 and H2 histamine antagonists on exploratory behavior in male rats. Physiol Behav. 1986;37(1):39–45.
27. Iwabuchi K, Kubota Y, Ito C, et al. Methamphetamine and brain histamine: a study using histamine-related gene knockout mice. Ann N Y Acad Sci. 2004;1025:129–34. doi:10.1196/annals.1316.016.
28. Brabant C, Charlier Y, Quertemont E, et al. The H3 antagonist thioperamide reveals conditioned preference for a context associated with an inactive small dose of cocaine in C57BL/6J mice. Behav Brain Res. 2005;160(1):161–8. doi:10.1016/j.bbr.2004.11.029.
29. Zhang M, Ballard ME, Pan L, et al. Lack of cataleptogenic potentiation with non-imidazole H3 receptor antagonists reveals potential drug-drug interactions between imidazole-based H3 receptor antagonists and antipsychotic drugs. Brain Res. 2005;1045(1–2):142–9.
30. Akhtar M, Uma Devi P, Ali A, et al. Antipsychotic-like profile of thioperamide, a selective H3-receptor antagonist in mice. Fundam Clin Pharmacol. 2006;20(4):373–8.
31. Liu CQ, Chen Z, Liu FX, et al. Involvement of brain endogenous histamine in the degeneration of dopaminergic neurons in 6-hydroxydopamine-lesioned rats. Neuropharmacology. 2007;53(7):832–41.
32. Liu CQ, Hu DN, Liu FX, et al. Apomorphine-induced turning behavior in 6-hydroxydopamine lesioned rats is increased by histidine and decreased by histidine decarboxylase, histamine H1 and H2 receptor antagonists, and an H3 receptor agonist. Pharmacol Biochem Behav. 2008;90(3):325–30. doi:10.1016/j.pbb.2008.03.010.

33. Garcia-Ramirez M, Aceves J, Arias-Montano JA. Intranigral injection of the H3 agonist immepip and systemic apomorphine elicit ipsilateral turning behaviour in naive rats, but reduce contralateral turning in hemiparkinsonian rats. Behav Brain Res. 2004;54(2):409–15. doi:10.1016/j.bbr.2004.03.007.
34. Huotari M, Kukkonen K, Liikka N, et al. Effects of histamine H(3)-ligands on the levodopa-induced turning behavior of hemiparkinsonian rats. Parkinsonism Relat Disord. 2000;6(3):159–64.
35. Nowak P, Bortel A, Dabrowska J, et al. Histamine H(3) receptor ligands modulate L-dopa-evoked behavioral responses and L-dopa derived extracellular dopamine in dopamine-denervated rat striatum. Neurotox Res. 2008;13(3–4):231–40.
36. Koob GF, Volkow ND. Neurocircuitry of addiction. Neuropsychopharmacology. 2010;35(1):217–38. doi:10.1038/npp.2009.110.
37. Cohn CK, Ball GG, Hirsch J. Histamine: effect on self-stimulation. Science. 1973;180(87):757–8.
38. Suzuki T, Takamori K, Misawa M, et al. Effects of the histaminergic system on the morphine-induced conditioned place preference in mice. Brain Res. 1995;675(1–2):195–202.
39. Masukawa Y, Suzuki T, Misawa M. Differential modification of the rewarding effects of methamphetamine and cocaine by opioids and antihistamines. Psychopharmacology (Berl). 1993;111(2):139–43.
40. Suzuki T, Mori T, Tsuji M, et al. Generalization of D-, L- and DL-chlorpheniramine and zolantidine to the discriminative stimulus effects of cocaine and methamphetamine. Behav Pharmacol. 1997;8(8):718–24.
41. Takino N, Sakurai E, Kuramasu A, et al. Roles of the histaminergic neurotransmission on methamphetamine-induced locomotor sensitization and reward: a study of receptors gene knockout mice. Int Rev Neurobiol. 2009;85:109–16. doi:10.1016/S0074-7742(09)85008-3.
42. Okuda T, Ito Y, Nakagawa N, et al. Drug interaction between methamphetamine and antihistamines: behavioral changes and tissue concentrations of methamphetamine in rats. Eur J Pharmacol. 2004;505(1–3):135–44. doi:10.1016/j.ejphar.2004.10.022.
43. Brabant C, Alleva L, Grisar T, et al. Effects of the H3 receptor inverse agonist thioperamide on cocaine-induced locomotion in mice: role of the histaminergic system and potential pharmacokinetic interactions. Psychopharmacology (Berl). 2009;202(4):673–87. doi:10.1007/s00213-008-1345-y.
44. Brabant C, Quertemont E, Anaclet C, et al. The psychostimulant and rewarding effects of cocaine in histidine decarboxylase knockout mice do not support the hypothesis of an inhibitory function of histamine on reward. Psychopharmacology (Berl). 2007;190(2):251–63. doi:10.1007/s00213-006-0603-0.
45. Aquino-Miranda G, Escamilla-Sanchez J, Gonzalez-Pantoja R, et al. Histamine H receptor activation inhibits dopamine synthesis but not release or uptake in rat nucleus accumbens. Neuropharmacology. 2015;106:91–101. doi:10.1016/j.neuropharm.2015.07.006.
46. Anichtchik OV, Huotari M, Peitsaro N, et al. Modulation of histamine H3 receptors in the brain of 6-hydroxydopamine-lesioned rats. Eur J Neurosci. 2000;12(11):3823–32.
47. Anichtchik OV, Peitsaro N, Rinne JO, et al. Distribution and modulation of histamine H(3) receptors in basal ganglia and frontal cortex of healthy controls and patients with Parkinson's disease. Neurobiol Dis. 2001;8(4):707–16. doi:10.1006/nbdi.2001.0413.
48. Anichtchik OV, Rinne JO, Kalimo H, et al. An altered histaminergic innervation of the substantia nigra in Parkinson's disease. Exp Neurol. 2000;163(1):20–30. doi:10.1006/exnr.2000.7362.
49. Gonzalez-Sepulveda M, Rosell S, Hoffmann HM, et al. Cellular distribution of the histamine H3 receptor in the basal ganglia: functional modulation of dopamine and glutamate neurotransmission. Basal Ganglia. 2013;3:109–21.
50. Onali P, Olianas MC. Involvement of adenylate cyclase inhibition in dopamine autoreceptor regulation of tyrosine hydroxylase in rat nucleus accumbens. Neurosci Lett. 1989;102(1):91–6.
51. Molina-Hernandez A, Nunez A, Arias-Montano JA. Histamine H3-receptor activation inhibits dopamine synthesis in rat striatum. Neuroreport. 2000;11(1):163–6.
52. Castellan Baldan L, Williams KA, Gallezot JD, et al. Histidine decarboxylase deficiency causes tourette syndrome: parallel findings in humans and mice. Neuron. 2014;81(1):77–90. doi:10.1016/j.neuron.2013.10.052.

53. Rapanelli M, Pittenger C. Histamine and histamine receptors in Tourette syndrome and other neuropsychiatric conditions. Neuropharmacology. 2015;106:85–90. doi:10.1016/j.neuropharm.2015.08.019.
54. Dere E, De Souza-Silva MA, Topic B, et al. Histidine-decarboxylase knockout mice show deficient nonreinforced episodic object memory, improved negatively reinforced water-maze performance, and increased neo- and ventro-striatal dopamine turnover. Learn Mem. 2003;10(6):510–9. doi:10.1101/lm.67603.
55. Ercan-Sencicek AG, Stillman AA, Ghosh AK, et al. L-histidine decarboxylase and Tourette's syndrome. N Engl J Med. 2010;20;362(20):1901–8. doi:10.1056/NEJMoa0907006.
56. Escobedo-Avila I, Vargas-Romero F, Molina-Hernandez A, et al. Histamine impairs midbrain dopaminergic development in vivo by activating histamine type 1 receptors. Mol Brain. 2014;7:58. doi:10.1186/s13041-014-0058-x.
57. Vizuete ML, Merino M, Venero JL, et al. Histamine infusion induces a selective dopaminergic neuronal death along with an inflammatory reaction in rat substantia nigra. J Neurochem. 2000;75(2):540–52.
58. Rinne JO, Anichtchik OV, Eriksson KS, et al. Increased brain histamine levels in Parkinson's disease but not in multiple system atrophy. J Neurochem. 2002;81(5):954–60.
59. Korotkova TM, Haas HL, Brown RE. Histamine excites GABAergic cells in the rat substantia nigra and ventral tegmental area in vitro. Neurosci Lett. 2002;320(3):133–6.
60. Zhou FW, Xu JJ, Zhao Y, et al. Opposite functions of histamine H1 and H2 receptors and H3 receptor in substantia nigra pars reticulata. J Neurophysiol. 2006;96(3):1581–91.
61. Schlicker E, Fink K, Detzner M, et al. Histamine inhibits dopamine release in the mouse striatum via presynaptic H3 receptors. J Neural Transm Gen Sect. 1993;93(1):1–10.
62. Pollard H, Moreau J, Arrang JM, et al. A detailed autoradiographic mapping of histamine H3 receptors in rat brain areas. Neuroscience. 1993;52(1):169–89.
63. Sanchez-Lemus E, Arias-Montano JA. Histamine H3 receptor activation inhibits dopamine D1 receptor-induced cAMP accumulation in rat striatal slices. Neurosci Lett. 2004;364(3):179–84. doi:10.1016/j.neulet.2004.04.045.
64. Clapham J, Kilpatrick GJ. Thioperamide, the selective histamine H3 receptor antagonist, attenuates stimulant-induced locomotor activity in the mouse. Eur J Pharmacol. 1994;259(2):107–14.
65. Fox GB, Esbenshade TA, Pan JB, et al. Pharmacological properties of ABT-239 [4-(2-{2-[(2R)-2-Methylpyrrolidinyl]ethyl}-benzofuran-5-yl)benzonitrile]: II. Neurophysiological characterization and broad preclinical efficacy in cognition and schizophrenia of a potent and selective histamine H3 receptor antagonist. J Pharmacol Exp Ther. 2005;313(1):176–90. doi:10.1124/jpet.104.078402.
66. Morisset S, Pilon C, Tardivel-Lacombe J, et al. Acute and chronic effects of methamphetamine on tele-methylhistamine levels in mouse brain: selective involvement of the D(2) and not D(3) receptor. J Pharmacol Exp Ther. 2002;300(2):621–8.
67. Toyota H, Dugovic C, Koehl M, et al. Behavioral characterization of mice lacking histamine H(3) receptors. Mol Pharmacol. 2002;62(2):389–97.
68. Southam E, Cilia J, Gartlon JE, et al. Preclinical investigations into the antipsychotic potential of the novel histamine H3 receptor antagonist GSK207040. Psychopharmacology (Berl). 2009;201(4):483–94. doi:10.1007/s00213-008-1310-9.
69. Ligneau X, Perrin D, Landais L, et al. BF2.649 [1-{3-[3-(4-Chlorophenyl)propoxy]propyl} piperidine, hydrochloride], a nonimidazole inverse agonist/antagonist at the human histamine H3 receptor: preclinical pharmacology. J Pharmacol Exp Ther. 2007;320(1):365–75. doi:10.1124/jpet.106.111039.
70. Vanhanen J, Kinnunen M, Nuutinen S, et al. Histamine H3 receptor antagonist JNJ-39220675 modulates locomotor responses but not place conditioning by dopaminergic drugs. Psychopharmacology (Berl). 2015;232(6):1143–53. doi:10.1007/s00213-014-3751-7.
71. Cousins V, Rose JE, Levin ED. IV nicotine self-administration in rats using a consummatory operant licking response: sensitivity to serotonergic, glutaminergic and histaminergic drugs. Prog Neuropsychopharmacol Biol Psychiatry. 2014;54:200–5. doi:10.1016/j.pnpbp.2014.06.004.

72. Levin ED, Slade S, Wells C, et al. Histamine H(1) antagonist treatment with pyrilamine reduces nicotine self-administration in rats. Eur J Pharmacol. 2011;650(1):256–60. doi:10.1016/j.ejphar.2010.10.013.
73. Munzar P, Tanda G, Justinova Z, et al. Histamine h3 receptor antagonists potentiate methamphetamine self-administration and methamphetamine-induced accumbal dopamine release. Neuropsychopharmacology. 2004;29(4):705–17. doi:10.1038/sj.npp.1300380.
74. Lintunen M, Hyytia P, Sallmen T, et al. Increased brain histamine in an alcohol-preferring rat line and modulation of ethanol consumption by H(3) receptor mechanisms. FASEB J. 2001;15(6):1074–6.
75. Okuda T, Zhang D, Shao H, et al. Methamphetamine- and 3,4-methylenedioxymethamphetamine-induced behavioral changes in histamine H3-receptor knockout mice. J Pharmacol Sci. 2009;111(2):167–74.

# Chapter 13
# Histamine $H_1$ Receptor Occupancy in the Human Brain Measured by Positron Emission Tomography

**Kazuhiko Yanai, Kotaro Hiraoka, Anikó Kárpáti, Fumito Naganuma, Nobuyuki Okamura, Manabu Tashiro, Tadaho Nakamura, and Takeo Yoshikawa**

**Abstract** Since Sir Henry Dale, a Nobel laureate in Physiology or Medicine, discovered the activities of histamine in 1910, a large number of studies have been conducted on histamine's physiological and pathological actions. While histamine was considered an allergy-causing "bad guy," recent studies have indicated that histamine also has beneficial physiological activities. The guideline for allergic diseases, such as pollenosis and atopic dermatitis, recommends nonsedating antihistamines with low central nervous system (CNS) penetration to avoid suppressing histamine's CNS actions. Positron emission tomography (PET) is often used to evaluate the efficacy of CNS drugs by determining the drugs' neuronal receptor occupancy rate. While the blood concentration levels of a drug are frequently

K. Yanai, M.D., Ph.D. (✉)
Department of Pharmacology, Tohoku University School of Medicine, 2-1 Seiryocho, Aoba-ku, Sendai 980-8575, Japan

Cyclotron and Radioisotope Center (CYRIC), Tohoku University, Sendai, Japan
e-mail: yanai@med.tohoku.ac.jp

K. Hiraoka • M. Tashiro
Cyclotron and Radioisotope Center (CYRIC), Tohoku University, Sendai, Japan

A. Kárpáti • T. Yoshikawa
Department of Pharmacology, Tohoku University School of Medicine, 2-1 Seiryocho, Aoba-ku, Sendai 980-8575, Japan

F. Naganuma • T. Nakamura
Department of Pharmacology, Tohoku University School of Medicine, 2-1 Seiryocho, Aoba-ku, Sendai 980-8575, Japan

Department of Pharmacology, Tohoku Medical and Pharmaceutical University, Sendai, Japan

N. Okamura
Department of Pharmacology, Tohoku University School of Medicine, 2-1 Seiryocho, Aoba-ku, Sendai 980-8575, Japan

Cyclotron and Radioisotope Center (CYRIC), Tohoku University, Sendai, Japan

Department of Pharmacology, Tohoku Medical and Pharmaceutical University, Sendai, Japan



P. Blandina, M.B. Passani (eds.), *Histamine Receptors*, The Receptors 28,
DOI 10.1007/978-3-319-40308-3_13

determined clinically, i.e., via therapeutic drug monitoring (TDM), CNS drugs do not necessarily show a correlation between their blood concentration levels and effect. Previously, we have reported on brain $H_1$ receptor occupancy measurements of antihistamines, antidepressants, and antipsychotics. In the present review, the results of our previous studies on the significance of brain histamine $H_1$ receptor occupancy of histamine $H_1$ blockers are summarized from the perspective of histamine function in the CNS.

**Keywords** Brain histamine H1 receptor occuapncy • Positron emission tomography (PET) • Non-sedating antihistamines • [$^{11}$C]doxepin • Topically applied antihistamines • Learning and Memory • Histamine degradation • Gliotransmitters • Carnosine-histidine-histamine pathway

## 13.1 Recent Overview of Histamine as a "Good Guy"

Histamine is a biogenic amine synthesized from the amino acid L-histidine by histidine decarboxylase (HDC). The main histamine-producing cell types are histaminergic neurons that are located in the hypothalamic tuberomammillary nucleus, gastric enterochromaffin-like (ECL) cells, mast cells, and basocytes. Gastric ECL cells release histamine upon stimulation by gastrin and acetylcholine. Histamine then mediates gastric acid secretion by parietal cells [1, 2]. The intracellular granules of mast cells and basocytes form a histamine reservoir and degranulate upon stimulation/sensitization by antigens. Further, histamine present in food is an important source of histamine.

The content of the essential amino acid histidine in fish meat is higher than that of any other essential amino acid. As shown in Fig. 13.1, the metabolism and

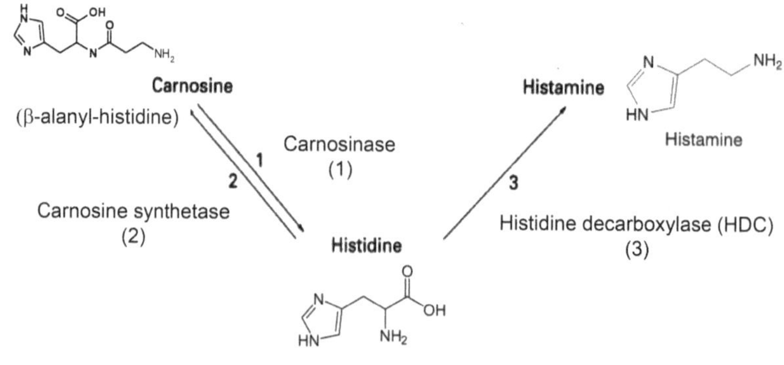


**Fig. 13.1** Carnosine-histidine-histamine pathway. Not only does histidine serve as a source of histamine in the body, but histidine is also converted to carnosine, mainly in the muscle. In the brain, carnosine is converted to histamine via histidine. Carnosine and histidine, both containing an imidazole skeleton, exhibit (1) antioxidant, (2) intracellular and extracellular metal-chelating ($Ca^{2+}$, $Zn^{2+}$, and $Cu^{2+}$), (3) pH-buffering, and (4) histaminergic neuron-activating activities

catabolism pathways of carnosine, histidine, and histamine are closely interrelated [3]. Carnosine has recently drawn attention as an imidazole dipeptide in anti-stress diet supplements. In addition to the synthesis of histamine from L-histidine by HDC, histaminergic neurons in particular contain carnosinase that degrades carnosine and, thus, can efficiently synthesize histamine from carnosine. One of the potential mechanisms underlying the enhancement of cognitive function by exercise is the activation of histaminergic neurons by carnosine produced in muscle [4]. A large number of epidemiological studies have demonstrated that exercise is the most effective way to prevent dementia, even though the underlying molecular mechanisms are still unknown. Brain function and cognitive function may be enhanced, and the hallmark protein in Alzheimer's disease, amyloid β (Aβ), and cellular toxicity may be reduced through the effects of the carnosine-histidine-histamine pathway, suggestive of a potential anti-Alzheimer's disease effect [5]. From this perspective, suppressing histamine's action in the brain should be avoided. Indeed, according to a recent report, the use of anticholinergic sedative antihistamines was related to the development of Alzheimer's disease [6].

## 13.2 Histamine Degradation and Release of Gliotransmitters by Astrocytes

A histamine-specific transporter has not been discovered yet. Upon administration of a drug that suppresses the function of astrocytes (glial cells), the brain histamine concentration increases substantially. Accordingly, astrocytes are important for clearance of histamine [7]. In the human brain, the number of glial cells, including astrocytes, is estimated to be 50 times the number of neurons. We propose that a nonspecific transporter expressed by astrocytes forms an important molecular mechanism underlying the clearance of histamine in the brain (Fig. 13.2). Our previous analysis of histamine transport kinetics and the results of drug-inhibition experiments and knockdown experiments have shown that both the organic cation transporter 3 (OCT3) and plasma membrane monoamine transporter (PMAT) are important for transport of histamine [8, 9]. Two types of histamine-metabolizing enzymes, namely, diamine oxidase and histamine *N*-methyltransferase (HNMT), are known. However, only HNMT is expressed in astrocytes and is localized solely in the cytoplasm. Therefore, histamine released into the synaptic cleft is taken up by astrocytes via OCT3 and PMAT and is subsequently metabolized and, hence, inactivated by HNMT in the cytoplasm.

Astrocytes play a key role in regulating brain homeostasis and neuronal activity, which is facilitated through various mechanisms, including sensing of neurotransmitters from neighboring neurons and releasing gliotransmitters such as ATP and glutamate [10]. Histamine in the synaptic cleft binds to histamine $H_1$ receptors expressed on the astrocyte membrane, thus activating signaling cascades that cause an increase in the intracellular $Ca^{2+}$ concentration ($[Ca^{2+}]_i$) (Fig. 13.3). An increase in $[Ca^{2+}]_i$ is essential for triggering gliotransmitter release, which may contribute to the physiological functions of neuronal histamine, such as in the sleep-wake cycle and in learning and memory [11–13].

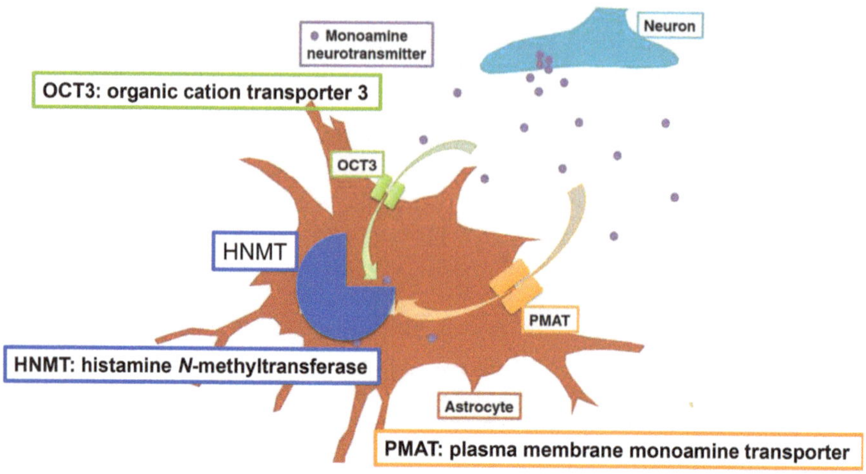


**Fig. 13.2** Histamine clearance mechanism. No histamine-specific transporter has been discovered yet. Histamine in the synaptic cleft is taken up by astrocytes via OCT3 and PMAT and is subsequently metabolized by HNMT. HNMT, histamine *N*-methyltransferase; OCT3, organic cation transporter 3; PMAT, plasma membrane monoamine transporter

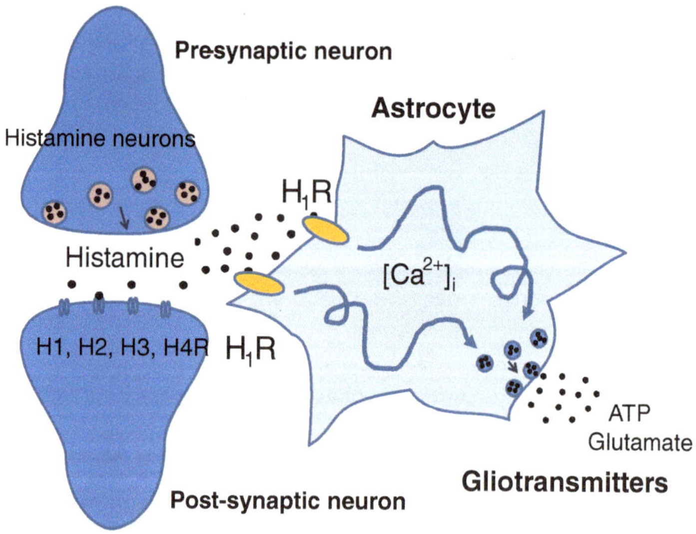


**Fig. 13.3** Schematic overview of gliotransmitter release by astrocytes

## 13.3 The Role of the Histamine $H_1$ Receptor in Learning and Memory

Although it is well known that sedative antihistamines induce cognitive decline in humans through blockage of the $H_1$ receptor [14, 15], both facilitatory and inhibitory effects of neuronal histamine on learning and memory have been described in animal behavioral studies [16]. We propose that the effects of histamine $H_1$ receptor blockage are mental state dependent. Figure 13.4 illustrates the conceptual hypothesis of the roles of histamine $H_1$ receptors in cognition on the basis of results from studies using $H_1$ receptor gene knockout mice. Under normal conditions, such as socially reared normal states, blocking $H_1$ receptors impairs cognition. It is well known that sedative antihistamines induce cognitive decline in humans. Blocking of the H3 receptor can stimulate the release of histamine from neurons and can also improve cognition [17, 18]. However, in psychiatric disorders, which are characterized by stressful conditions such as isolation-reared states, cognition is impaired by stress itself. In such conditions, the blockage of $H_1$ receptors attenuates the stress-induced impaired cognition. This finding might be the reason why both facilitatory and inhibitory effects of neuronal histamine on learning and memory have been described in animal behavioral studies [19]. The new-generation antipsychotics including clozapine, olanzapine, and quetiapine are potent $H_1$ receptor antagonists as well as tricyclic antidepressants and the newer generation of the antidepressant mirtazapine. As indicated by animal behavioral studies, the $H_1$ receptor antagonistic effects of these drugs might have some therapeutic efficacy in addition to their sedative properties [20].

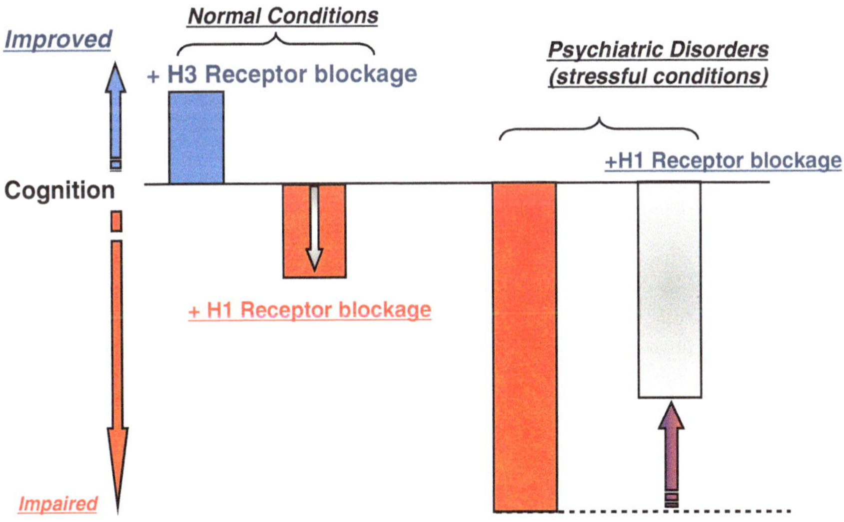


**Fig. 13.4** The effects of histamine $H_1$ receptor blockage on cognition are mental state dependent: a hypothesis

## 13.4 Imaging Studies of Histamine $H_1$ Receptors Using Positron Emission Tomography (PET)

Radiopharmaceuticals used for PET neuronal receptor studies should have high affinity and selectivity for the respective receptors as well as high blood–brain barrier (BBB) penetration. Fluorine-18 (half-life: ~110 min) and carbon-11 (half-life: ~20 min) are often used to radioactively label drugs. Carbon-11 is preferred because of its shorter half-life. PET is frequently used for the measurement of receptor occupancy to evaluate the efficacy of central nervous system (CNS) drugs. In clinical settings, therapeutic drug monitoring (TDM) is used, even though the blood concentration and drug effect of CNS drugs are not necessarily correlated. The neuronal receptor occupancy of CNS drugs has been used to determine occupancy of the dopamine D1 and 2 receptors, dopamine transporter, opiate receptor, benzodiazepine receptor, muscarinic acetylcholine receptor, serotonin 5-HT2 receptor, and NK1 receptor. Using PET to determine D2 receptor occupancy of antipsychotics, it was found that the therapeutic effect of classical antipsychotics became apparent at 65–80 % occupancy of D2 receptors [21].

We developed [$^{11}$C]pyrilamine and [$^{11}$C]doxepin for imaging histamine $H_1$ receptor in the human brain (Fig. 13.5). Pyrilamine is a standard histamine $H_1$ receptor antagonist, and doxepin is a tricyclic antidepressant with high histamine $H_1$ receptor affinity. Doxepin's high binding specificity and affinity for the histamine $H_1$ receptor were confirmed in histamine $H_1$ receptor knockout mice [22]. Comparison of the binding levels of [$^{11}$C]pyrilamine and [$^{11}$C]doxepin in the human brain as detected by PET showed that the distribution patterns of these two drugs were similar. However, [$^{11}$C]doxepin had a higher signal-to-noise ratio than that of [$^{11}$C]pyrilamine. Accordingly, we used [$^{11}$C]doxepin for the human PET study. As shown in Fig. 13.6, histamine $H_1$ receptor binding was highest in the cingulated cortex. The prefrontal cortex and temporal cortex also showed relatively high binding ratios, whereas in the cerebellum, binding was very weak and was rarely specific. Therefore, the binding signal in the cerebellum was used as a reference/control signal. The crystalline structure of the human $H_1$ receptor complex with doxepin revealed a specific interaction with the $H_1$ receptor protein and doxepin [23].

We applied [$^{11}$C]doxepin-PET to examine the changes in the brain $H_1$ receptor density in normal aging and in several neuropsychiatric disorders, including complex partial seizure, Alzheimer's disease, depression, schizophrenia, gender differentiation, and anorexia nervosa. We previously published a review on the changes in the brain histamine $H_1$ receptor density in these diseases as detected by human PET studies [22]. This chapter will focus primarily on brain histamine $H_1$ receptor occupancy of antihistamines, antidepressants, and antipsychotic drugs in young, healthy volunteers. Figure 13.7 shows the time course of the histamine $H_1$ receptor occupancy after the administration of the sedating antihistamines diphenhydramine and ketotifen. The brain $H_1$ receptor binding was markedly decreased at 3 h after drug administration and gradually returned to approximately the half-normal level

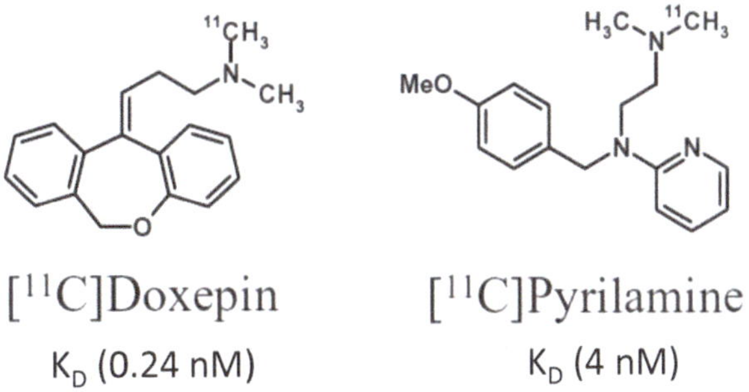


**Fig. 13.5** The chemical structure of the histamine $H_1$ receptor imaging probes [$^{11}$C]doxepin and [$^{11}$C] pyrilamine. The affinity for $H_1$ receptor was determined in the autopsied human brain using [$^{3}$H]doxepin and [$^{3}$H]pyrilamine

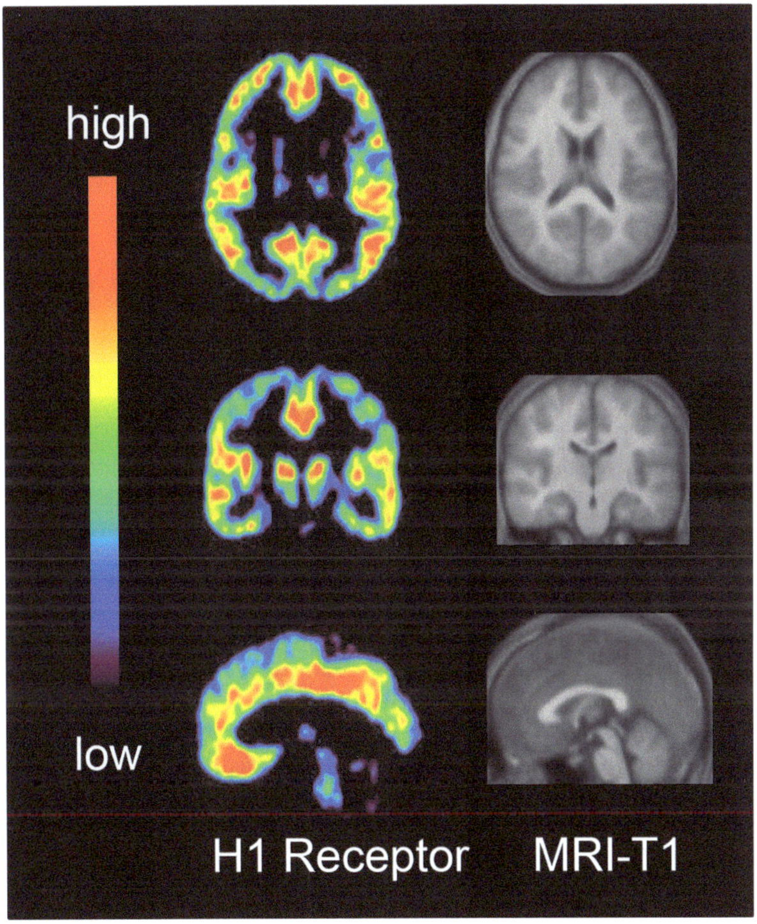


**Fig. 13.6** The distribution of the histamine $H_1$ receptor in the brain of healthy, young male volunteers. The distribution of the histamine $H_1$ receptor in the brain was measured in young, healthy male subjects using PET and [$^{11}$C]doxepin. $H_1$ receptor density was relatively high in the cingulate cortex, frontotemporal cortex, amygdala, and hippocampus. The lowest density was observed in the cerebellum

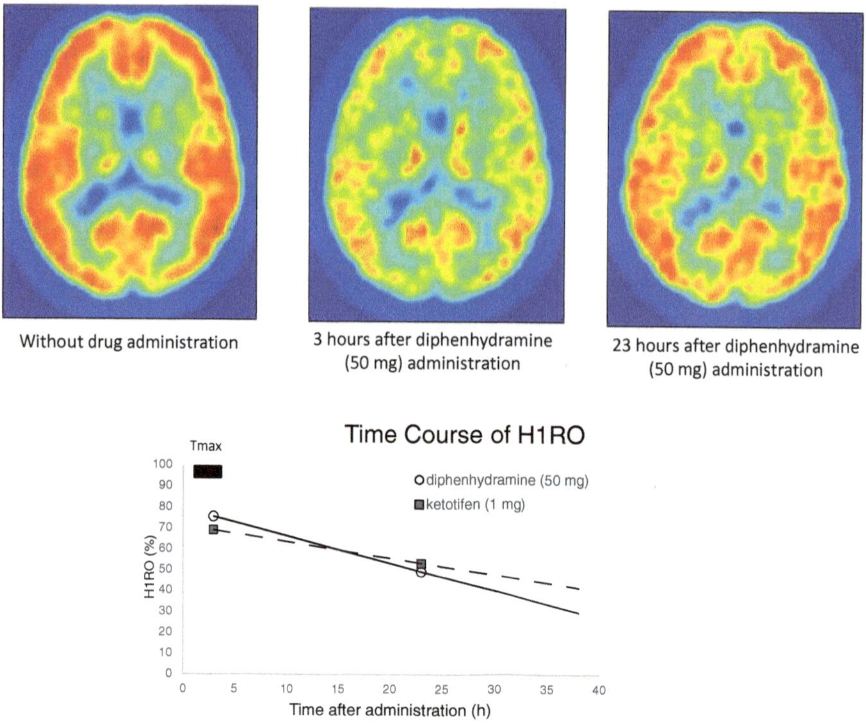


**Fig. 13.7** The time course of histamine $H_1$ receptor occupancy. Brain histamine $H_1$ receptors were measured using PET and [$^{11}$C]doxepin at 3 h and 23 h consecutively after diphenhydramine (50 mg) or ketotifen (1 mg) intake in healthy, young male subjects. The half-life of diphenhydramine and ketotifen in the brain was estimated to be ~30 h and 45 h, respectively

at 23 h after drug administration. We previously reported the next-day residual sedative effect after nighttime administration of diphenhydramine (50 mg) by direct PET measurement of histamine $H_1$ receptor occupancy. At the next day after intake of diphenhydramine, cortical $H_1$ receptor occupancy was approximately 50 %, explaining the next-day residual sedative effect of diphenhydramine.

Sedative $H_1$ antagonists that readily penetrate the BBB occupy a large proportion of histamine $H_1$ receptors in the brain; variations in brain $H_1$ receptor occupancy are due to differences in BBB permeability between the sedative $H_1$ antagonists. [$^{11}$C] Doxepin is usually administered at the time of the plasma peak concentration (Tmax) of test drugs. The histamine $H_1$ receptor occupancy at Tmax is a reliable method for evaluating the sedative properties of antihistamines in humans. We demonstrated a significant correlation between cognitive decline and brain histamine $H_1$ receptor occupancy at the Tmax of histamine $H_1$ antagonists. In addition, we showed that both cognitive decline and brain histamine $H_1$ receptor occupancy at Tmax as measured using PET significantly correlated with the plasma concentration of sedating $H_1$ blockers [24].

## 13.5 The Development of Nonsedating Antihistamines

Antihistamines are used as antiallergy drugs since the 1940s. The first antihistamine was synthesized by the Italian pharmacologist Daniel Bovet and his research assistant Anne-Marie Staub in 1937. The first compound tested was toxic in humans. Pyrilamine was discovered in 1944 because of screening of a large number of compounds and was used in humans. Even today, pyrilamine is used as a standard histamine $H_1$ antagonist in fundamental research The first-generation antihistamines, such as pyrilamine and promethazine, became the prototype for many CNS drugs, including antipsychotics and antidepressants. In fact, some antidepressants and antipsychotics have the highest histamine $H_1$ antagonistic activities compared to classical antihistamines [25]. Bovet was awarded the Nobel Prize for his and his colleagues' contribution to the advancement of psychoneuropharmacology.

While the classic first-generation antihistamines are recognized as effective agents to treat some allergy symptoms, their main side effect is the strong sedative effect due to their BBB penetration. Further, their selectivity for histamine $H_1$ receptors is weak, and the prevalence of anticholinergic adverse reactions, such as dry mouth, anuresis, and tachycardia, is high. Therefore, second-generation antihistamines with high histamine $H_1$ receptor selectivity and, hence, less severe CNS side effects (e.g., sedation) were developed. They are characterized by a relatively low BBB penetration and long plasma half-life, the former chiefly the result of the introduction of hydrophilic functional groups (–COOH and $-NH_2$) (Fig. 13.8). Epinastine, desloratadine, and mequitazine, which contain an amino group ($-NH_2$), are not selective for histamine $H_1$ receptors and have certain anticholinergic activities [26]. Nonsedating antihistamines containing a carboxy group (–COOH; zwitterionic compounds) include fexofenadine, levocetirizine, cetirizine, and acrivastine, all of which have high $H_1$ receptor selectivity [27].

## 13.6 Classification of the Sedative Properties of $H_1$ Antagonists by $H_1$ Receptor Occupancy

The current classification of antihistamines as "first generation" and "second generation" is based on the nonsedating properties. The second-generation antihistamines may be sold as nonsedative drugs, with the claim that no significant differences between the drug and the placebo exist; however, the results on which these claims are based are derived from cognitive function tests with insufficient sensitivity or which tested only sensations of sleepiness. To classify antihistamines objectively according to their sedative effects, we proposed the use of $H_1$ receptor occupancy at Tmax (Fig. 13.9) and, therefore, classified antihistamines into three groups: (1) sedative antihistamines (antihistamines with $H_1$ receptor occupancy higher than

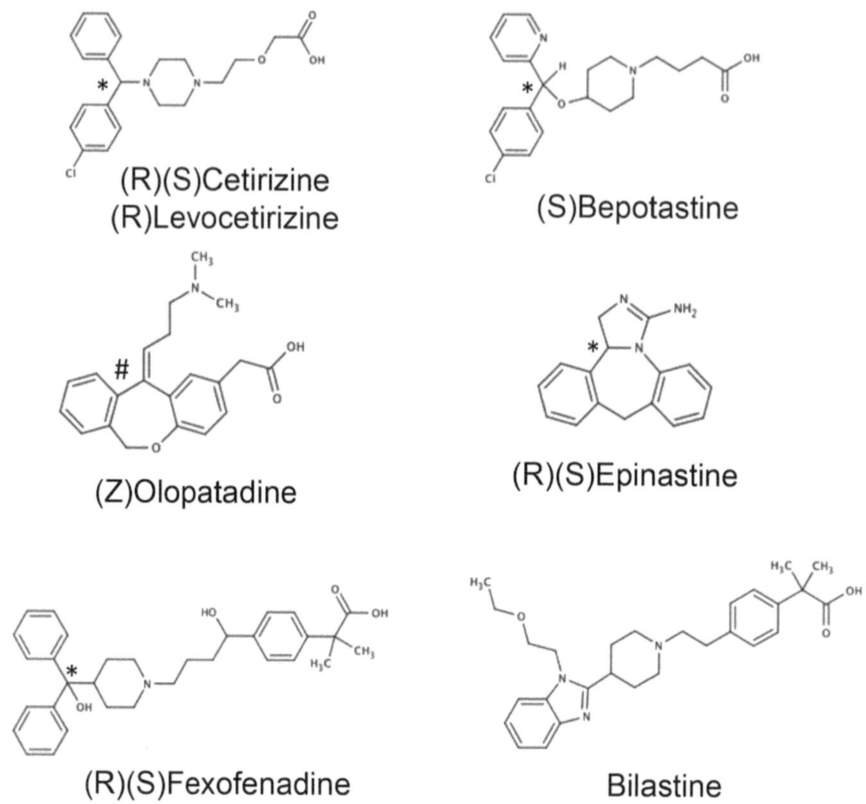


**Fig. 13.8** Nonsedating antihistamines. There are two types of nonsedating antihistamines: the carboxy group (zwitterionic compounds) and amino group type. (R)(S)epinastine is a representative amino group-type antihistamine, while (R)(S)cetirizine, (R)levocetirizine, (S)bepotastine, (Z) olopatadine, (R)(S)fexofenadine, and bilastine are carboxy group-type antihistamines. The asymmetrical carbon atom that is related to optical isomerism is marked with *asterisk*; *hash* denotes the double bond that is related to the structurally different geometric isomer (*cis-trans* isomer) without optical isomerism

50 %), (2) less-sedative antihistamines (antihistamines with $H_1$ receptor occupancy between 20 and 50 %), and (3) nonsedative antihistamines (antihistamines with $H_1$ receptor occupancy of 20 % or lower) [28].

Shamsi and Hindmarch [15] analyzed all published reports of randomized, placebo-controlled, double-blind trials and calculated the incidence rates of the sedative effects of various antihistamines in terms of subjective and objective aspects. The incident rates of sedation obtained were closely correlated with the results of $H_1$ receptor occupancy as measured by PET as previously reported [27]. The first-generation antihistamines are profoundly sedative and impair CNS function at virtually all doses tested because of their relatively high BBB penetration.

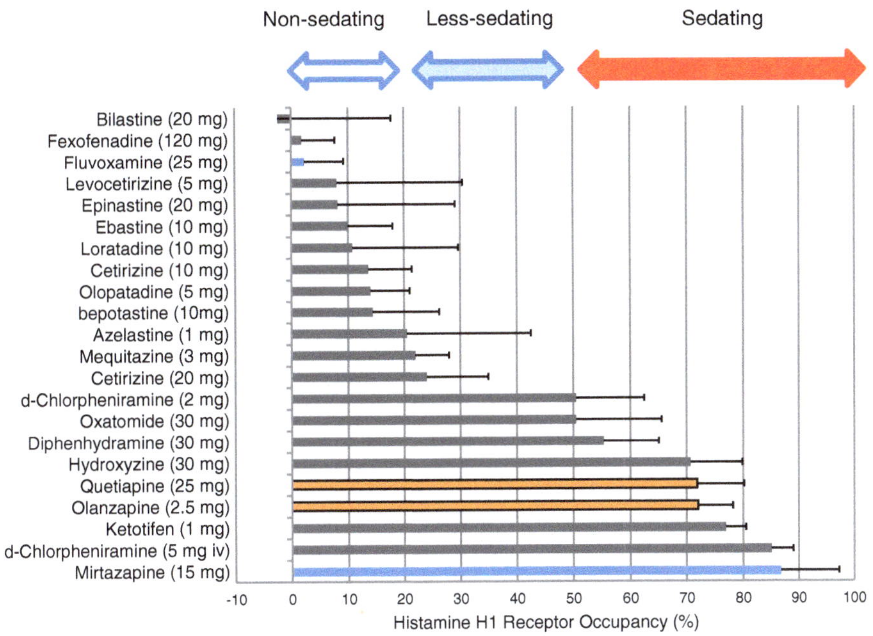


**Fig. 13.9** Histamine $H_1$ receptor occupancy of several drugs in the human brain. Histamine $H_1$ receptor occupancy of various drugs in the human brain. The values of the histamine $H_1$ receptor occupancy obtained from published and unpublished data ($N$=3–12) are summarized for the frontal cortex (mean +/- SD). Sedative, less-sedative, and nonsedative drugs are classified according to their histamine $H_1$ receptor occupancy of 50–100 %, 20–50 %, and 0–20 %, respectively. The *gray*, *blue*, and *orange columns* represent antihistamines, antidepressants, and antipsychotics, respectively

Several second-generation antihistamines also penetrate the BBB dose dependently, but their effects on cognitive performance are considerably lower than those of the first-generation antihistamines [29]. In fact, some of the second-generation antihistamines, such as cetirizine, bepotastine, and olopatadine, may show dose-dependent impairment of CNS function as their CNS side effects increase with increasing doses. Thus, when any impairment of CNS function is found at higher doses of a nonsedating antihistamine, the respective drug should be classified as a "relatively nonsedative" or "less-sedative" drug, but not a "nonsedative" drug. It should be noted that several second-generation antihistamines, such as fexofenadine and bilastine, do not penetrate the BBB at all and, consequently, have an $H_1$ receptor occupancy of almost zero [27, 30, 31]. Therefore, fexofenadine and bilastine may be classified as "non-brain-penetrating antihistamines," that is, antihistamines lacking CNS distribution and, thus, lacking significant effects on CNS function in daily life.

Brain-penetrating $H_1$ antagonists have hypnotic, appetite-promoting, and sedative side effects. Most second-generation antipsychotics and antidepressants show potent histamine $H_1$ receptor antagonism. Although several of these antipsychotics and antidepressants are clinically efficacious for the palliative treatment of psychiatric disorders, histamine $H_1$ antagonism is considered only to the hypnotic, appetite-promoting, and sedative side effects. The therapeutic actions of antipsychotics and antidepressants through their effect on histaminergic neurons remain incompletely understood. To date, the affinities of various antipsychotics for histamine receptors have been determined mainly in vitro. Previously, we also examined the histamine $H_1$ receptor occupancy of the new-generation antipsychotics and antidepressants at their minimum-permitted prescribed doses using [$^{11}$C]doxepin and PET [32, 33]. Our findings clearly demonstrated that more than 60 % of the histamine $H_1$ receptors in the frontal cortex were occupied by olanzapine, quetiapine, and mirtazapine at the minimum-permitted prescribed dose of 2.5, 25, and 15 mg, respectively. In contrast, fluvoxamine, a selective serotonin reuptake inhibitor (SSRI), shows negligible histamine $H_1$ receptor occupancy in the human brain.

## 13.7 Histamine $H_1$ Receptor Occupancy of Topically Applied Antihistamines

Topical antihistamines are one of the best treatment options for various ocular allergies, thanks to their safety, convenience of use, and rapid mechanism of action. Nevertheless, clinical sedative side effects of antihistamine eye drops are sometimes reported. On the basis of the results of a previous study, we propose that antihistamines contained in eye drops distribute from the eye to the brain, thereby blocking the histamine $H_1$ receptors in the brain resulting in sedation. We conducted this previous study to assess the potential influence of two different eye drops containing a different antihistamine, i.e., 0.05 % ketotifen or 0.1 % olopatadine, on CNS function by measuring the brain histamine $H_1$ receptor occupancy using PET. Ketotifen had a histamine $H_1$ receptor occupancy of approximately 50 %, whereas that of olopatadine was near 0 % (Fig. 13.10). These results were the first to suggest that eye drops containing the first-generation antihistamine ketotifen might potentially induce CNS sedation at higher doses—even though such a CNS side effect has rarely been reported for topical eye drops—while olopatadine did not have measurable CNS effects [34]. The considerable amount of a sedating antihistamine that penetrates the BBB may be partly attributed to the lack of a hepatic first-pass effect and the additional absorbance via the throat and nasal mucosa. Therefore, the use of nonsedating olopatadine, which exhibits no or little BBB penetration and, thus, sedative effects, may have advantages over the use of first-generation antihistamines, even if the latter are applied topically.

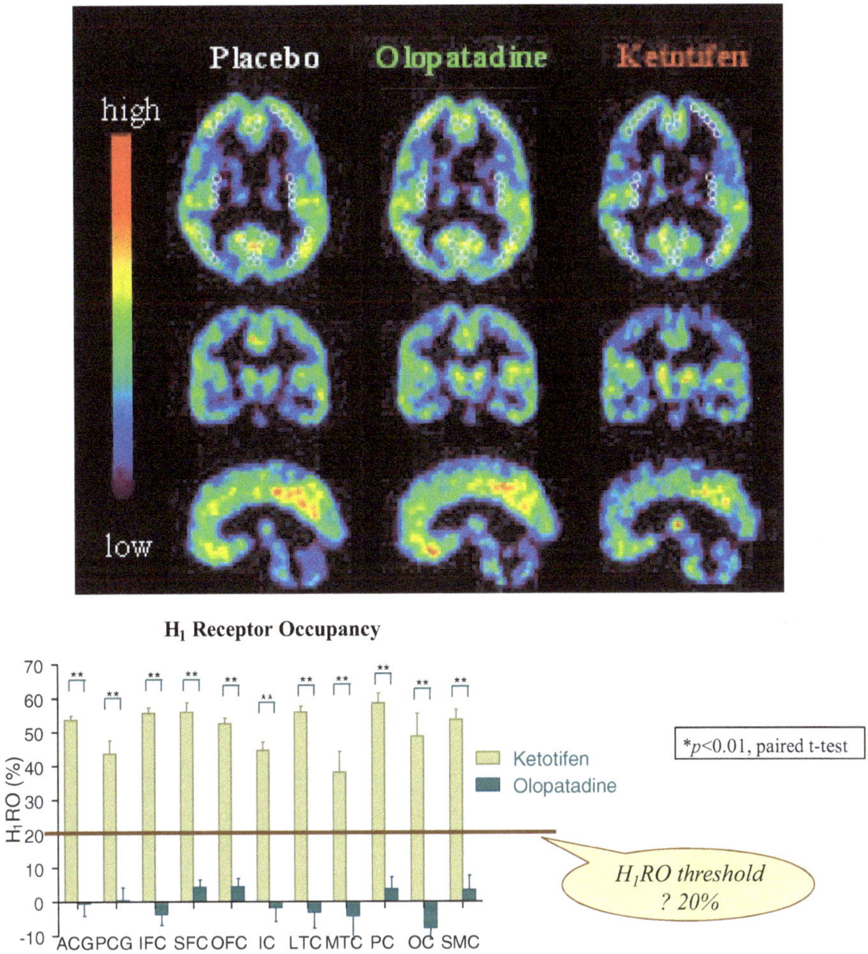


**Fig. 13.10** The sedative effect of antihistamine-containing eye drops. The histamine $H_1$ receptor occupancy was near 50 % (mean 45.7 %, $n=5$) and 0 % (mean −1.83 %, $n=7$) following ketotifen and olopatadine instillation, respectively. The lack of a hepatic first-pass effect of the antihistamines in the eye drops may cause the relatively high $H_1$ receptor occupancy of the sedating antihistamines. The data are expressed as the mean +/- SEM

## 13.8 Conclusion

Histamine is considered a "bad guy" from an allergy point of view. On the other hand, histamine is functioning as a "good guy" in several other physiological aspects. In particular, histaminergic neurons form one of the most important neuronal systems that maintain and stimulate wakefulness. Histamine also functions as a bioprotector against various noxious and undesirable stimuli, for example, convulsion, nociception, drug sensitization, ischemic lesions, and stress. It may be suggested that the activation of histaminergic neurons is important to maintain mental health. The clinically most important contribution of PET studies on histamine $H_1$ occupancy in the brain of antihistamine-treated allergic patients is probably to the identification of the narrow therapeutic window of less than 20 % of systemic and topical treatments that permit optimal treatment without causing significant sedative side effects. Thus, these PET results may be useful for the development of optimal antihistamine-dosing strategies. However, although this finding can be applied consistently to the sedation caused by antidepressants and antipsychotics, it cannot be applied to the treatment efficacy of psychological symptoms. Further studies are warranted to elucidate the relationship between histamine $H_1$ receptor occupancy in the brain and the main effects of antidepressants and antipsychotics in patients.

## References


1. Falus A, Grossman N, Darvas ZS. Histamine: biology and medical aspects. Switzerland: S. KARGER AG; 2004. ISBN ISBN 3-8055-7715-X.
2. Watanabe T, Wada H, editors. Histaminergic neurons: morphology and function. Boca Raton: CRC Press; 1991. ISBN ISBN 0-8493-6425-6.
3. Boldyrev AA, Aldini G, Derave W. Physiology and pathophysiology of carnosine. Physiol Rev. 2013;93(4):1803–45.
4. Babizhayev MA. Biochemical, biomedical and metabolic aspects of imidazole-containing dipeptides with the inherent complexity to neurodegenerative diseases and various states of mental well-being: a challenging correction and neurotherapeutic pharmaceutical biotechnology for treating cognitive deficits, depression and intellectual disabilities. Curr Pharm Biotechnol. 2014;15(8):738–78.
5. Corona C, Frazzini V, Silvestri E, Lattanzio R, La Sorda R, Piantelli M, et al. Effects of dietary supplementation of carnosine on mitochondrial dysfunction, amyloid pathology, and cognitive deficits in 3xTg-AD mice. PLoS One. 2011;6(3):e17971.
6. Gray SL, Anderson ML, Dublin S, Hanlon JT, Hubbard R, Walker R, et al. Cumulative use of strong anticholinergics and incident dementia: a prospective cohort study. JAMA Intern Med. 2015;175(3):401–7.
7. Huszti Z. Histamine uptake into non-neuronal brain cells. Inflamm Res. 2003;52 Suppl 1:S3–6.
8. Naganuma F, Yoshikawa T, Nakamura T, Iida T, Harada R, Mohsen AS, et al. Predominant role of plasma membrane monoamine transporters in monoamine transport in 1321N1, a human astrocytoma-derived cell line. J Neurochem. 2014;129(4):591–601.
9. Yoshikawa T, Naganuma F, Iida T, Nakamura T, Harada R, Mohsen AS, et al. Molecular mechanism of histamine clearance by primary human astrocytes. Glia. 2013;61:905–16.
10. Araque A, Carmignoto G, Haydon PG, Oliet SH, Robitaille R, Volterra A. Gliotransmitters travel in time and space. Neuron. 2014;81(4):728–39.
11. Haas HL, Sergeeva OA, Selbach O. Histamine in the nervous system. Physiol Rev. 2008;88:1183–41.
12. Watanabe T, Yanai K. Studies on functional roles of the histaminergic neuron system by using pharmacological agents, knockout mice and positron emission tomography. Tohoku J Exp Med. 2001;195:197–217.
13. Haas HL, Panula P. The role of histamine and the tuberomamillary nucleus in the nervous system. Nat Rev Neurosci. 2003;4:121–30.
14. Church MK, Maurer M, Simons FE, Bindslev-Jensen C, van Cauwenberge P, Bousquet J, et al. Risk of first-generation $H_1$-antihistamines: a GA(2)LEN position paper. Allergy. 2010;65:459–66.

15. Shamsi Z, Hindmarch I. Sedation and antihistamines: a review of inter-drug differences using proportional impairment ratios. Hum Psychopharmacol. 2000;15:S3–30.
16. Huston JP, Wagner U, Hasenohrl RU. The tuberomammillary nucleus projections in the control of learning, memory and reinforcement processes: evidence for an inhibitory role. Behav Brain Res. 1997;83:97–105.
17. Brioni JD, Esbenshade TA, Garrison TR, Bitner SR, Cowart MD. Discovery of histamine H3 antagonists for the treatment of cognitive disorders and Alzheimer's disease. J Pharmacol Exp Ther. 2011;336(1):38–46.
18. Passani MB, Blandina P. Histamine receptors in the CNS as targets for therapeutic intervention. Trends Pharmacol Sci. 2011;32(4):242–9.
19. Yanai K, Dai H, Sakurai E, Watanabe T. The roles of histamine $H_1$ receptors on cognition. Inflamm Res. 2008;57 Suppl 1:S39–40.
20. Jia F, Mobarakeh JI, Dai H, Kato M, Xu A, Okuda T, et al. Blocking histamine $H_1$ improves learning and mnemonic dysfunction in mice with social isolation plus repeated methamphetamine injection. J Pharmacol Sci. 2008;107(2):167–74.
21. Nord M, Farde L. Antipsychotic occupancy of dopamine receptors in schizophrenia. CNS Neurosci Ther. 2011;17(2):97–103.
22. Yanai K, Tashiro M. The physiological and pathophysiological roles of neuronal histamine: an insight from human positron emission tomography studies. Pharmacol Ther. 2007;113(1):1–15.
23. Shimamura T, Shiroishi M, Weyand S, Tsujimoto H, Winter G, Katritch V, et al. Structure of the human histamine $H_1$ receptor complex with doxepin. Nature. 2011;475:65–70.
24. Okamura N, Yanai K, Higuchi M, Sakai J, Iwata R, Ido T, et al. Functional neuroimaging of cognition impaired by a classical antihistamine, d-chlorpheniramine. Br J Pharmacol. 2000;129:115–23.
25. Kanba S, Richelson E. Interactions with psychotropic drugs. In: Watanabe T, Wada H, editors. Histaminergic neurons: morphology and function. Boca Raton: CRC Press; 1991. p. 271–82.
26. Yanai K. Anticholinergic activity of antihistamines. Clin Neurophysiol. 2012;123(4):633–4.
27. Yanai K, Zhang D, Tashiro M, Yoshikawa T, Naganuma F, Harada R, et al. Positron emission tomography evaluation of sedative properties of antihistamines. Expert Opin Drug Saf. 2011;10(4):613–22.
28. Holgate ST, Canonica GW, Simons FE, Taglialatela M, Tharp M, Timmerman H, Consensus Group on New-Generation Antihistamines, et al. Consensus Group on New-Generation Antihistamines (CONGA): present status and recommendations. Clin Exp Allergy. 2003;33(9):1305–24.
29. Tashiro M, Kato M, Miyake M, Watanuki S, Funaki Y, Ishikawa Y, Iwata R, Yanai K. Dose dependency of brain histamine $H_1$ receptor occupancy following oral administration of cetirizine hydrochloride measured using PET with [$^{11}$C]doxepin. Hum Psychopharmacol. 2009;24(7):540–8.
30. Farré M, Pérez-Mañá C, Papaseit E, Menoyo E, Pérez M, Martin S, et al. Bilastine vs. hydroxyzine: occupation of brain histamine $H_1$ receptors evaluated by positron emission tomography in healthy volunteers. Br J Clin Pharmacol. 2014;78:970–80.
31. Jáuregui I, Ramaekers JG, Yanai K, Farré M, Redondo E, Valiente R, et al. Bilastine: a new antihistamine with an optimal benefit-to-risk ratio for safety during driving. Expert Opin Drug Saf. 2016;15:89–98.
32. Sato H, Ito C, Hiraoka K, Tashiro M, Shibuya K, Funaki Y, et al. Histamine $H_1$ receptor occupancy by the new-generation antipsychotics olanzapine and quetiapine: a positron emission tomography study in healthy volunteers. Psychopharmacology (Berl). 2015;232(19):3497–505.
33. Sato H, Ito C, Tashiro M, Hiraoka K, Shibuya K, Funaki Y, et al. Histamine $H_1$ receptor occupancy by the new-generation antidepressants fluvoxamine and mirtazapine: a positron emission tomography study in healthy volunteers. Psychopharmacology (Berl). 2013;230(2):227–34.
34. Zhang D, Shibuya K, Tashiro M, Harada R, Okamura N, Yoshikawa T, et al. Potential central sedative effect of antihistamine eye-drops: histamine $H_1$ occupancy measured by positron emission tomography. J Pharmacol Sci. 2011;115 Suppl 1:221P.

# Chapter 14
# Modulation of Memory Consolidation, Retrieval and Extinction by Brain Histamine

**Jociane de Carvalho Myskiw, Cristiane R.G. Furini, and Ivan Izquierdo**

**Abstract** The brain histaminergic system, whose cell bodies are in the tuberomammilary nucleus, regulates various memory types. The best studied is inhibitory avoidance, which depends on histamine H2 receptors in hippocampus and basolateral amygdala, contextual fear conditioning, which depends on histamine H3 receptors in hippocampus, and the extinction of these two tasks which relies on histamine H2 receptors in ventromedial prefrontal cortex, hippocampus and the basolateral amygdala. In addition, histamine can promote fear extinction through H1 receptors and inhibit it through H2 receptors, both in hippocampus.

**Keywords** Memory consolidation • Memory retrieval • H1 receptors • H2 receptors • Hippocampus • Basolateral amygdala

The multiple consolidation of memory [1]: different parallel brain systems are in charge of memory consolidation [2].

For at least a century, perhaps even from before Pavlov [3] proposed that "temporary connections" are "closed" at specific sites "in the cerebral cortex," it was more or less taken for granted that different memories are made and stored at restricted sites in the brain. One memory, one site. This may have resulted from the remnants of Franz Joseph Gall's old phrenology, in which head bumps predicted character and brain function, or those of its also infelicitous successor, the "new neuropharmacological phrenology" (a name coined in irony by Steve [4]), which caught neuroscientists at the time when modern work on brain neurotransmitters was just beginning and led many to believe that each of the molecules, no matter where it was released, specialized in some function (acetylcholine in memory; dopamine in pleasure, norepinephrine in excitement). A giant leaps backward from Paul Ehrlich's concept of receptors. Still today, some science writers adhere to those strange ideas.

Growing evidence began to accumulate showing that many areas of the brain were involved in memory, and questions were raised as to where and how they are involved. It took the power of modern biochemistry and pharmacology [1, 2, 5–10]

J. de Carvalho Myskiw, Ph.D. • C.R.G. Furini, Ph.D. • I. Izquierdo, M.D., Ph.D. (✉)
Memory Center, Brain Institute of Rio Grande do Sul, Pontifical Catholic University of Rio Grande do Sul, Av. Ipiranga 6690, 90610-000 Porto Alegre, RS, Brazil
e-mail: jociane_carvalho@hotmail.com; cristianefurini@hotmail.com; izquier@terra.com.br



P. Blandina, M.B. Passani (eds.), *Histamine Receptors*, The Receptors 28,
DOI 10.1007/978-3-319-40308-3_14

and more recently and more spectacularly that of the optogenetic revolution [11–15] to accept the now older but finally undisputable fact that memories are made and modulated in several brain regions concomitantly and often simultaneously by different sets of neurons (see [16–21]). The making of memories and their modulation in many parts of the brain became a major research field, and the variety of modulatory systems guaranteed the diversity of the different memory engrams formed.

Over 20 years ago, Jorge Brioni called attention to "the multiple consolidation of memory" in an article commenting on the widely disseminated role of γ-aminobutyric acid (GABA) in memories formed in many places of the brain [1]. Our own group reported at the same time that the microinfusion of norepinephrine, timolol, glutamate, oxotremorine, scopolamine, picrotoxin, and muscimol or several combinations thereof affected the memory consolidation of inhibitory avoidance (IA) similarly when infused immediately posttraining into the CA1 region of the hippocampus, the basolateral amygdala (BLA), or the medial septum [2]. Likewise, glutamate receptor antagonists also affected IA memory consolidation similarly when microinfused into hippocampus CA1, BLA [22], or medial septum/diagonal band nucleus [23].

Clearly, both the data from Brioni [1] and our own set of data [2, 22, 23] in the early 1990s could be taken as reasonable suggestions that memories were made and modulated at several brain sites simultaneously or in close succession and that there is in fact parallel memory processing in several such places at the time of consolidation. But as will be seen, belief in parallel memory processing had to await the new methods of optogenetics to become accepted by one-memory/one-site diehards (see [24]). Very interestingly, as has been recently shown [25] and will be discussed in the last section of this chapter, in the case of histamine, **the same** neural pathway mediates both the enhancement and the inhibition of retrieval, the difference being only at the level of the receptor in the last neuron of the pathway: H1 for enhancing and H2 for inhibition. The switch that selects between the two must be located in the tuberomammillary nucleus itself.

Obviously, the larger the number of available concurrent or consecutive sites of consolidation, the bigger the chances for neurotransmitter systems with multiple projections to modulate memory [1], and the histaminergic system is clearly highly promiscuous as to the brain sites it innervates [16, 26–30, [69]. It is certainly safer to store information in more than one place of the brain in order to circumvent failures or injuries in one or in a few of them, exactly for the same reasons that it is safer to store important files in more than one electronic device. Parallel processing implies great advantages for the survival of important brain functions, such as memory. If one neuron or one subsystem fails, another one can pick up the flag and go; witness to this are the millions of mostly asymptomatic people and animals that daily survive strokes.

Then, as a consequence of this greatly renewed way of thinking, the "new neuropharmacologic phrenology" envisioned by Zornetzer [4] experienced a rebirth that began to make sense. Different traces of different duration and modulated by a diversity of systems clearly led to more interesting, colorful, and complex views of the organization of memory than the much more rigid idea of single punctual storage sites. For example, the neurons that have been best studied so far as memory makers are probably the pyramidal cells of the hippocampus (e.g., [31–34]). They are innervated by a large number of modulatory fibers: noradrenergic, cholinergic, serotoninergic, histaminergic, and various peptide-releasing fibers, as can be seen in any of the major recent reviews available on its function [35, 36]; in addition, these cells rely on modulation by neurotrophic factors [37] and neurogenesis [38] among other factors. How far we have come now from the days of one memory, one site.

## 14.1 Histamine and Memory: The Early Work

All the neurons that produce and contain histamine are restricted to one single nucleus (the tuberomammillary nucleus, [39]), with quite scattered projections reaching many areas of the brain, including some well known to be involved in memory consolidation processes [28, 29]. So it was only a question of time that, as had been the case with other major neurotransmitters (acetylcholine, serotonin, and the catecholamines), a role for brain histamine in memory modulation would be investigated.

The first to study this were de Almeida and Izquierdo [40], who showed that the immediate posttraining intracerebroventricular (i.c.v.) infusion of histamine facilitated memory of the IA in rats. In a subsequent experiment, 48/80, a well-known polymer then popularly produced by the condensation of *N*-methyl-*p*-methoxyphenethylamine with formaldehyde that releases histamine from mast cells, had no effect on memory when given i.c.v., which pointed to a role for neuronal and not mast cell histamine in memory modulation [41]. At that time, the presence of histamine in the brain [42], as well as that of histidine decarboxylase was determined [43] and that of mast cells containing histamine in brain [44] had already been well ascertained [42]. Brain neurons [44] and fibers [39] containing histamine were being discovered. The tuberomammillary nucleus was soon found to be involved in many aspects of brain function, including prominently regulating the waking state [46–50] and food intake [51], both of which can be affected by stress [51]. Stress, in turn, as is known, strongly affects memory, an interaction that involves endocannabinoids [24, 52–56].

Back in 1986, the effects of histamine were immediately correlated with the well-known depressant action and sleepiness induced by the popular over-the-counter "antihistamine" compounds available at that time, most of which acted on peripheral receptors of the H1 type. Its putative relation to memory processing, particularly fear-related memories, was suggested later in part by the parallel description of connections to the tuberomammillary nucleus from the septum/diagonal band nucleus [57], a nucleus that has a strong connection to the hippocampus [58] and through it plays a major role in fear motivated and many other types of learning [2, 23].

Histamine was soon discovered to have also other receptors: H2, H3, and H4, the former two in the brain and the latter only in the periphery. The histaminergic system in the brain was in the process of becoming known, so soon this system and its H1, H2, and H3 receptor subtypes began to be intensely studied as memory modulators, as they are now [16, 28, 30]. Much of these works are carried out using fear conditioning models and localized infusion of histamine or its mimetic and antagonists into brain areas known to regulate memory consolidation and by enhancing the action of endogenous histamine. The results are coincident in that histamine, acting on different receptors depending both on the brain site into which it is infused and on the task studied, enhances different forms of fear conditioning and fear extinction.

The tuberomammillary nucleus is composed of different subpopulations of histaminergic cells which innervate multiple brain areas each with different histamine receptor types [28, 29]. This suggests that histaminergic neurons are "heterogeneous, organized into functionally distinct circuits, impinging on different brain regions, and displaying selective control mechanisms. This could imply independent functions of subsets of histamine neurons according to their respective origin and terminal projections," to put it in the authors' own words.

Many effects of histamine on memory have been attributed to H1, H2, or H1 plus H2 receptors [40, 59–63] but as will be seen below, clearly other effects in areas critical for memory formation are mediated by H3 receptors [26].

Histamine given into several brain regions modulates memory consolidation of various learning tasks, including mainly fear-motivated tasks. As mentioned, the first report of memory modulation by histamine was an enhancement of the consolidation of the IA with posttraining i.c.v. administration [40]. In that paper, histamine was effective at low doses (1 or 10 ng/rat) and was blocked by the H1 receptor antagonist, promethazine, and by the H2 antagonist, cimetidine, given together, but not by either drug alone. Since then, the effects on memory of histamine, histamine releasers, enhancers, and antagonists given into various structures of the brain were studied in different forms of memory. Some reports have concluded that histamine facilitates consolidation and others that depresses it by actions on different receptors in different brain sites [26, 39, 59–61, 64–68]. It appears that at some receptors and in some brain areas, histamine enhances memory consolidation of certain tasks, and at other receptors and in other areas or tasks, it may have different effects. For example, memory facilitation of IA has been described on one hand with histamine given into i.c.v. [40] or into BLA [30, 69] (Fig. 14.1) and on the other with pharmacological inhibition of the tuberomammillary nucleus [70].

Here we will concentrate on mainstream studies on histamine modulation of well-known tasks, whose mechanisms have been well studied and on which there is wide agreement [24, 54, 56]: contextual fear conditioning (CFC) and extinction [27, 62, 71, 72], IA [7, 8, 16, 73], and object recognition [74–77].

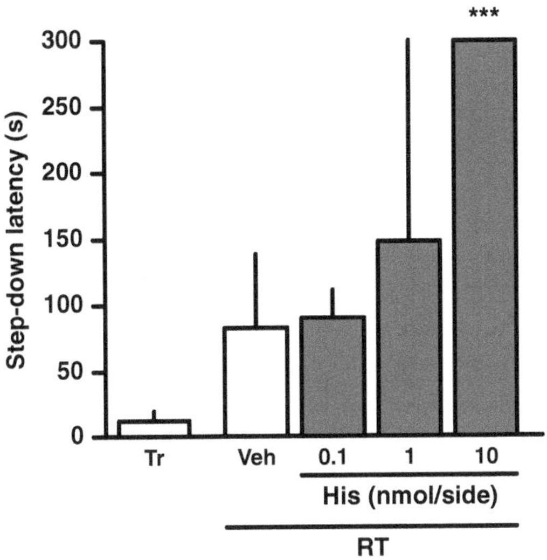


**Fig. 14.1** Infusion of histamine into the basolateral nuclear complex of the amygdala (BLA) enhances aversive memory. Rats with infusion cannulas implanted in the BLA were trained (Tr) in inhibitory avoidance and immediately after that received infusions of vehicle (Veh) or histamine (His). Retention was evaluated 24 h after (RT). (Figure modified from [69]; histamine infused into basolateral amygdala enhances memory consolidation of inhibitory avoidance. International Journal of Neuropsychopharmacology, 16:1539–1545)

## 14.2 Effects of Histaminergic Modulation of IA

IA has an edge of advantage over the other tasks: it is, by far, the behavioral model in which the molecular mechanisms of consolidation have been best and most extensively studied [7, 8, 16, 73, 78, 78]. Especially concerning the memory facilitation by histamine. Its effect depends on the phosphorylation of the constitutive transcription factor c-AMP response element-binding protein (CREB) both in the hippocampus and in the BLA [5, 6, 16], and on the activation of the extracellularly regulated kinase system (ERK) coupled with cholinergic stimulation [68].

The effect of the histamine *N*-methyltransferase inhibitor, SKF9188 that prolongs the half-life of histamine action [62, 69], and of a variety of histamine H1 and H2 receptor antagonists in BLA and hippocampus have been well studied, and the results are very consistent.

Endogenous histamine production can be blocked by α-fluoro-methylhistidine, a suicide inhibitor of histidine decarboxylase infused into the brain ventricles, which reaches the tuberomammillary nucleus and suppresses histamine production in the projections of that nucleus [16, 26, 27]. By this procedure, a full blockade of brain histamine-mediated transmission can be obtained [29]. This inhibits the consolidation of, for example, IA, which is known to be sustained by both the BLA and the hippocampus (see [7, 8, 80–83]). The deleterious effect of α-fluoro-methylhistidine on IA consolidation can be overcome by the infusion of histamine into either BLA or hippocampus; i.e., histamine can act on one of these structures while the other one is depleted and impaired. This observation suggests parallel processing of the modulation of memory consolidation by BLA and hippocampus (see [2, 16, 22, 84]). Therefore, histamine modulation of IA consolidation occurs independently in BLA and in hippocampus [30], which strongly suggests that in physiological conditions, it should be rather synchronous in both [7]. This adds to the literature on an independent modulation of fear memory consolidation by these two brain structures [82, 83, 85].

The modulatory effect of histamine fibers on IA consolidation is exerted both at the BLA and at the CA1. Interestingly, when one of these two histaminergic connection sites fails, the other one takes over [16].

Experiments with dimaprit, ranitidine, and thioperamide have shown that H1, H2, and H3 histamine receptors in BLA, hippocampus and ventromedial prefrontal cortex (vmPFC), facilitate memory consolidation of IA and CFC, and their specific antagonists have an opposite effect in the consolidation of different tasks [26, 62, 69]. The cognitive deficit in a Morris water maze, in an object recognition task [77] and in IA learning [30, 86, 87] visible in adult life of rats submitted to brief daily maternal deprivation during the first 10 days of life, is in part due to a histamine deficit, and it can be corrected by histamine given into the BLA or by the histamine *N*-methyltransferase inhibitor and histamine enhancer, SKF91488 [69]. This deficit can also be reversed by physical exercise [86, 87] or by pro-cholinergic drugs, such as galantamine and donepezil [88], so it is possible that this cognitive deficit is due to a failure of various transmitter systems. Interestingly, the H3 receptor antagonist, thioperamide, has been reported to enhance consolidation in some tasks or brain regions [89–91], and also to antagonize the enhancing action of histamine in others. H3 receptors act through fostering the release both of acetylcholine and of histamine itself from axon terminals [27, 92].

Histamine is a major modulator of fear extinction (Fig. 14.2). The histamine *N*-methyltransferase inhibitor SKF9188 enhances, and the H2 histamine receptor antagonist ranitidine inhibits extinction of both CFC and IA learning regardless of whether they were administered into the vmPFC, the dorsal hippocampus, or the BLA, which are the three main regions for consolidation of the extinction of both tasks. Only the indirect stimulant of the NMDA receptor, D-serine, has such a generalized effect on extinction among many drugs tested, including noradrenergic and dopaminergic agonists and antagonists ([62]; see also [61]).

## 14.3 Brain Histamine and the Consolidation of Contextual Fear Conditioning (CFC)

There have been several important hints in the literature on a role of histaminergic processes in the consolidation of CFC in the hippocampus. These hints point each in a different direction, so the picture so far is not very clear.

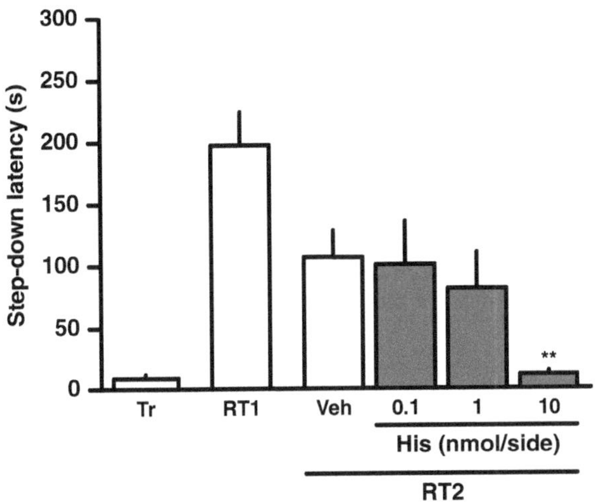


**Fig. 14.2** Histamine into the CA1 facilitates the extinction of inhibitory avoidance memory. Rats with infusion cannulas implanted in the CA1 region of hippocampus were trained (Tr) in inhibitory avoidance and 24 h later submitted to a non-reinforced test session (RT1). Immediately thereafter, animals received infusions of vehicle (Veh) or histamine (His). Retention was evaluated 24 h after RT1 (RT2). (Figure modified from [61]; histamine facilitates consolidation of fear extinction. International Journal of Neuropsychopharmacology, 14:1209–1217)

Liu et al. [93] reported improved learning and memory of CFC and hippocampal CA1 long-term potentiation (LTP) in histidine decarboxylase knockout mice. Brabant et al. [94] described a clear enhancing effect of the H3 inverse agonist, pitolisant, on CFC when given systemically to female mice. Pitolisant, 1-{3-[3-(4-Chlorophenyl) propoxy]propyl}piperidine hydrochloride), is the first H3 receptor inverse agonist that has been tested in human trials and is well tolerated.

The most complex but perhaps more definite or definable evidence for a role of histamine H3 receptors points in a different direction from those of the two previously cited data. Benetti et al. [27] infused histaminergic receptor ligands into the nucleus basalis magnocellularis (NBM) of rats right after CFC and observed increased CFC freezing behavior 72 h after training, which ensures that the infused material did not influence acquisition or retrieval but just consolidation. They found that posttraining blockade of H3 receptors with the antagonist/inverse agonist thioperamide or activation of those receptors with immepip in the NBM potentiates or decreases, respectively, freezing response at retrieval. Thioperamide-induced memory enhancement seemed to depend on H2 but not H1 receptor activation, as the H2 receptor antagonist zolantidine blocked the effect of thioperamide, whereas the H1 receptor antagonist pyrilamine was ineffective. The H2 agonist amphetamine improved fear memory expression independently of the H3 agonist effect. Their findings indicate that activation of postsynaptic H2 receptors within the NBM by endogenous histamine is responsible for the potentiated expression of fear responses.

## 14.4 Histamine Modulation of IA and CFC Fear Extinction

Until 2012 there had been a relative neglect of the modulatory mechanisms of fear extinction, despite the fact that it is involved in the delicate and often fragile treatment of the post-traumatic stress disorder (PTSD) by the so-called exposure therapy [95]. We launched a short but effective program to study the influence of well-known neuromodulatory substances (norepinephrine acting via β receptors, dopamine acting via D1/D5 receptors, histamine, and D-serine) infused into the three main sites of regulation of the memory consolidation of both CFC and IA extinction: the CA1 region of the hippocampus, the BLA, and the ventromedial prefrontal cortex [72]. We came out with two very solid modulators: D-serine, which acts at an allosteric site within glutamate *N*-methyl-aspartate (NMDA) receptors and would have been a prime suspect anyway, and histamine acting on H2 ranitidine-sensitive receptors [62], which may offer a translational possibility worthy of further study. Both D-serine and histamine enhanced extinction of the two fear tasks in all the three brain sites studied, and their effects were antagonized, respectively, by AP5 and by ranitidine [56, 62].

## 14.5 Histamine Modulation of Object Recognition

Ennaceur and Delacour [77] introduced the so-called object recognition task as a very welcome non-aversive learning procedure in which rodents express their preference for the remaining close to and investigating a novel object rather than a previously known object. Part of the considerable interest aroused by this task is that recognition memory fails quite specifically early in Alzheimer's disease [96].

The neural basis for object recognition involves the hippocampus [96], the entorhinal [97], and to a great extent the perirhinal cortex [74, 75]. It relies on CA1 LTP [99].

Brain histamine has a very ample and diversified modulatory influence on object recognition memory. When infused in the CA1 region immediately, 30, 120, or 360 min posttraining, the H1 receptor antagonist pyrilamine, the H2 receptor antagonist ranitidine, and the H3 receptor agonist imetit blocked the long-term memory retention in a time-dependent manner (30–120 min) without affecting general exploratory behavior, anxiety state, or hippocampal function [100]. Our data indicate that histaminergic system modulates consolidation of object recognition memory through H1, H2, and H3 receptors.

## 14.6 Histamine and Memory: The Latest Work. The Histamine Retrieval Switch

After this chapter was submitted, Fabbri, Furini, and coworkers showed that the retrieval of inhibitory avoidance was prevented by previous depletion of brain histamine by α-fluoro-methylhistidine, and it was reinstated by the microinfusion of histamine into the CA1 region of the hippocampus but not into the BLA or the vmPFC in the histamine-depleted animals; this effect of histamine was antagonized by the H1 receptor antagonist, pyrilamine [25]. This finding was in contrast to the previously shown inhibitory effect of histamine on retrieval mediated by H2 receptors in CA1, BLA, and vmPFC described above [62], an effect indeed that underlies the influence of histamine in the three structures on normal extinction of both contextual fear conditioning and inhibitory avoidance (see [25, 62]).

The fact that histamine can support retrieval by an action in CA1 mediated by H1 receptors and inhibit retrieval by an action on CA1, the BLA, and the vmPFC mediated by H1receptors clearly points to a **histamine switch** located at the tuberomammillary nucleus by which this structure must decide whether to activate the neurons that innervate CA1 H1 receptors and thereby stimulate retrieval or those that innervate CA1, BLA, and vmPFC H2 receptors and thereby inhibit extinction. It is not yet known whether this **histamine switch** presumably in the tuberomammillary nucleus acts only on the retrieval of fear-motivated memories or in that of other memories as well. This point deserves further investigation.

Meanwhile, evidence points to the necessary location of this **switch** that controls the decision of whether the animal should perform or inhibit retrieval in the tuberomammillary nucleus. The existence of cell subpopulations in that nucleus has been convincingly argued for by Blandina et al. [28]: there are no other histaminergic nuclei except that one, there are no intermediate stations between the tuberomammillary nucleus and the structures it projects to, and the decision of whether that nucleus will send this or the other contingent of neurons to release histamine on H1 or H2 receptors here or there must be taken ahead of the structures that contain those receptors (in this case, CA1, basolateral amygdala and vmPFC).

## 14.7 To Summarize

Recent evidence indicates a major role of the brain histaminergic system, whose cell bodies are in the tuberomammillary nucleus, in the regulation of various different types of memory. The best studied of these are IA, which depends on histamine H2 receptors in the hippocampus and in the BLA; CFC, where regulation depends on histamine H3 receptors in the hippocampus; and the extinction of IA and CFC, whose consolidation relies on histamine H2 receptors in the ventromedial prefrontal cortex, in the CA1 region of the hippocampus, and in the BLA. In all cases, histamine regulation should be viewed as integrated into modulatory networks related to pathways involving other neurotransmitters.

**Acknowledgments** Work supported by the National Research council of Brazil (CNPq) and the Rio Grande do Sul State Agency for the Support of Research (FAPERGS).

## References


1. Brioni JD. Role of GABA during the multiple consolidation of memory. Drug Dev Res. 1993;28:3–27.
2. Izquierdo I, Medina JH, Jeriisalinsky D, Da Cunha C. Post-training memory processing in amygdala, septum and hippocampus: role of benzodiazepine/GABAA receptors, and their interaction with other neurotransmitter systems. Rev Neurosci. 1992;3:11–24.
3. Pavlov IP, Anrep GV. Conditioned reflexes : an investigation of the physiological activity of the cerebral cortex. [S.l.]. Oxford: Oxford University Press; 1927.
4. Zornetzer SF. Neurotransmitter modulation and memory: a new neuropharmacological phrenology? In: Lipton MA, DiMascio A, Killam KF, editors. Psychopharmacology: a generation of progress. New York: Raven; 1978.
5. Bernabeu R, Bevilaqua L, Ardenghi P, Bromberg E, Schmitz P, Bianchin M, Izquierdo I, Medina JH. Involvement of hippocampal D1/D5 receptor - cAMP signaling pathways in a late memory consolidation phase of an aversively-motivated task in rats. Proc Natl Acad Sci U S A. 1997;94:7041–6.
6. Bernabeu R, Cammarota M, Izquierdo I, Medina JH. Involvement of hippocampal AMPA glutamate receptor changes and the cAMP/protein kinase A/CREB-P signalling pathway in memory consolidation of an avoidance task in rats. Braz J Med Biol Res. 1997;30:961–5.
7. Izquierdo I, Medina JH. Memory formation: the sequence of biochemical events in the hippocampus and its connection to activity in other brain structures. Neurobiol Learn Mem. 1997;68:285–316.
8. Izquierdo I, Bevilaqua LR, Rossato JI, Bonini JS, Medina JH, Cammarota M. Different molecular cascades in different sites of the brain control memory consolidation. Trends Neurosci. 2006;29:496–505.
9. Kandel ER, Dudai Y, Mayford MR. The molecular and systems biology of memory. Cell. 2014;157:163–86.
10. Matthies H. In search of cellular mechanisms of memory. Prog Neurobiol. 1989;32:277–349.
11. Boyden ES, Zhang F, Bamberg E, Nagel G, Deisseroth K. Millisecond-timescale, genetically targeted optical control of neural activity. Nat Neurosci. 2005;8:1263–8.
12. Deisseroth K. Optogenetics. Nat Methods. 2011;8:26–9.
13. Deisseroth K. Optogenetics: 10 years of microbial opsins in neuroscience. Nat Neurosci. 2015;18:1213–25.
14. Josselyn SA, Köhler S, Frankland PW. Finding the engram. Nat Neurosci. 2015;16:521–34.
15. Reiner A, Isacoff EY. The Brain Prize 2013: the optogenetics revolution. Trends Neurosci. 2013;36:557–60.
16. Benetti F, Furini CRG, de Carvalho MJ, Provensi G, Passani MB, Baldi E, Bucherelli C, Munari L, Izquierdo I, Blandina P. Histamine in the basolateral amygdala promotes inhibitory avoidance learning independently of hippocampus. Proc Natl Acad Sci U S A. 2015;112:E2536–42.
17. Cowansage KK, Shuman T, Dillingham BC, Chang A, Golshani P, Mayford M. Direct reactivation of a coherent neocortical memory of context. Neuron. 2014;84:432–41.
18. Kheirbek MA, Drew LJ, Burghardt NS, Costantini DO, Tannenholz L, Ahmari SE, Zeng H, Fenton AA, Hen R. Differential control of learning and anxiety along the dorsoventral axis of the dentate gyrus. Neuron. 2013;77:955–68.
19. Redondo RL, Kim J, Arons AL, Ramirez S, Liu X, Tonegawa S. Bidirectional switch of the valence associated with a hippocampal contextual memory engram. Nature. 2014;513:426–30.
20. Tanaka KZ, Pevzner A, Hamidi AB, Nakazawa Y, Graham J, Wiltgen BJ. Cortical representations are reinstated by the hippocampus during memory retrieval. Neuron. 2014;84:347–54.
21. Wittgen BJ, Tanaka KZ. Systems consolidation and the content of memory. Neurobiol Learn Mem. 2013;106:365–71.
22. Jerusalinsky D, Ferreira MB, Walz R, Da Silva RC, Bianchin M, Ruschel AC, Zanatta MS, Medina JH, Izquierdo I. Amnesia by post-training infusion of glutamate receptor antagonists into the amygdala, hippocampus, and entorhinal cortex. Behav Neural Biol. 1992;58:76–80.

23. Walz R, da Silva RC, Bueno e Silva M, Medina JH, Izquierdo I. Post-training infusion of glutamate receptor antagonists into the medial septum of rats causes amnesia. Ciência e Cultura. 1992;44:339–41.
24. Izquierdo I, Furini CRG, Myskiw JC. Fear memory. Physiol Rev. 2016;96:695–750.
25. Fabbri R, Furini CRF, Passani MB, Provensi G, Baldi E, Bucherelli C, Izquierdo I, Myskiw JC, Blandina PG. Memory retrieval of inhibitory avoidance requires histamine H1 receptors in the hippocampus. Proc Natl Acad Sci U S A. 2016;113:E2714–20.
26. Benetti F, da Silveira CKB, da Silva WC, Cammarota M, Izquierdo I. Histamine reverses a memory deficit induced in rats by early postnatal maternal deprivation. Neurobiol Learn Mem. 2012;97:54–8.
27. Benetti F, Baldi E, Bucherelli C, Blandina P, Passani MB. Histaminergic ligands injected into the nucleus basalis magnocellularis differentially affect fear conditioning consolidation. Int J Neuropsychopharmacol. 2013;16:575–82.
28. Blandina P, Munari L, Provensi G, Passani MB. Histamine neurons in the tuberomammillary nucleus: a whole center or distinct subpopulations? Front Syst Neurosci. 2012;6:33.
29. Passani MB, Blandina P. Histamine receptors in the CNS as targets for therapeutic intervention. Trends Pharmacol Sci. 2011;32:242–9.
30. Benetti F, da Silveira CK, Rosa J, Izquierdo I. Histamine acting on the basolateral amygdala reverts the impairment of aversive memory of rats submitted to neonatal maternal deprivation. Behav Brain Res. 2015;278:83–9.
31. De Carvalho MJ, Benetti F, Izquierdo I. Behavioral tagging of extinction learning. Proc Natl Acad Sci U S A. 2013;110:1071–6.
32. De Carvalho MJ, Furini CRG, Benetti F, Izquierdo I. Hippocampal molecular mechanisms involved in the enhancement of fear extinction caused by exposure to novelty. Proc Natl Acad Sci U S A. 2014;111:4572–7.
33. Frey U, Morris RG. Synaptic tagging and long-term potentiation. Nature. 1997;385:533–6.
34. Frey U, Morris RG. Synaptic tagging: implications for late maintenance of hippocampal long-term potentiation. Trends Neurosci. 1998;21:181–8.
35. Izquierdo I. The art of forgetting. New York: Springer; 2015.
36. Schiller D, Eichenbaum H, Buffalo EA, Davachi L, Foster DJ, Leutgeb S, Ranganath C. Memory and space: towards an understanding of the cognitive map. J Neurosci. 2015;35:13904–11.
37. Bekinschtein P, Kent BA, Oomen CA, Clemenson GD, Gage FH, Saksida LM, Bussey TJ. Brain-derived neurotrophic factor interacts with adult-born immature cells in the dentate gyrus during consolidation of overlapping memories. Hippocampus. 2014;24:905–11.
38. Aimone JB, Deng W, Gage FH. Resolving new memories: a critical look at the dentate gyrus, adult neurogenesis, and pattern separation. Neuron. 2011;70:589–96.
39. Panula P, Revuelta AV, Cheney DL, Wu JY, Costa E. An immunohistochemical study on the location of GABAergic neurons in rat septum. J Comp Neurol. 1984;222:69–80.
40. de Almeida MA, Izquierdo I. Memory facilitation by histamine. Arch Int Pharmacodyn Ther. 1986;283:193–8.
41. de Almeida MA, Izquierdo I. Intracerebroventricular histamine, but not 48/80, causes post-training memory facilitation in the rat. Arch Int Pharmacodyn Ther. 1987;291:202–7.
42. Picatoste F, Palacios JM, Blanco I. Subcellular localization of histamine in neonatal rat brain [proceedings]. Agents Actions. 1977;7:120.
43. Baudry M, Martres MP, Schwartz JC. The subcellular localization of histidine decarboxylase in various regions of rat brain. J Neurochem. 1973;21:1301–9.
44. Rönnberg AL, Edvinsson L, Larsson LI, Nielsen KC, Owman C. Regional variation in the presence of mast cells in the mammalian brain. Agents Actions. 1973;3:191.
45. Lin JS, Luppi PH, Salvert D, Sakai K, Jouvet M. Histamine-immunoreactive neurons in the hypothalamus of cats. C R Acad Sci III. 1986;303:371–6.
46. Monti JM. Involvement of histamine in the control of the waking state. Life Sci. 1993;53(17):1331–8.
47. Monti JM, Jantos H, Leschke C, Elz S, Schunack W. The selective histamine H1-receptor agonist 2-(3-trifluoromethylphenyl)histamine increases waking in the rat. Eur Neuropsychopharmacol. 1994;4:459–62.

48. Monti JM, Jantos H, Ponzoni A, Monti D. Sleep and waking during acute histamine H3 agonist BP 2.94 or H3 antagonist carboperamide (MR 16155) administration in rats. Neuropsychopharmacology. 1996;15:31–5.
49. Provensi G, Coccurello R, Umehara H, Munari L, Giacovazzo G, Galeotti N, Nosi D, Gaetani S, Romano A, Moles A, Blandina P, Passani MB. Satiety factor oleoylethanolamide recruits the brain histaminergic system to inhibit food intake. Proc Natl Acad Sci U S A. 2014;111:11527–32.
50. Passani MB, Giannoni P, Bucherelli C, Baldi E, Blandina P. Histamine in the brain: beyond sleep and memory. Biochem Pharmacol. 2007;73:1113–22.
51. Carlini VP, Schiöth HB, de Barioglio SR. Obestatin improves memory performance and causes anxiolytic effects in rats. Biochem Biophys Res Commun. 2007;352:907–12.
52. Atsak P, Hauer D, Campolongo P, Schelling G, Fornari RV, Roozendaal B. Endocannabinoid signaling within the basolateral amygdala integrates multiple stress hormone effects on memory consolidation. Neuropsychopharmacology. 2015;40:1485–94.
53. Atsak P, Hauer D, Campolongo P, Schelling G, Fornari RV, Roozendaal B. Endocannabinoid signaling integrates multiple stress hormone effects on memory consolidation. Psychoneuroendocrinology. 2015;61:5.
54. Furini CRG, Myskiw JC, Izquierdo I. The learning of fear extinction. Neurosci Biobehav Rev. 2014;47:670–83.
55. Morena M, Roozendaal B, Trezza V, Ratano P, Peloso A, Hauer D, Atsak P, Trabace L, Cuomo V, McGaugh JL, Schelling G, Campolongo P. Endogenous cannabinoid release within prefrontal-limbic pathways affects memory consolidation of emotional training. Proc Natl Acad Sci U S A. 2014;111:18333–8.
56. Myskiw JC, Izquierdo I, Furini CRG. Modulation of the extinction of fear learning. Brain Res Bull. 2014;105:61–9.
57. Wouterlood FG, Gaykema RP, Steinbusch HW, Watanabe T, Wada H. The connections between the septum-diagonal band complex and histaminergic neurons in the posterior hypothalamus of the rat. Anterograde tracing with Phaseolus vulgaris-leucoagglutinin combined with immunocytochemistry of histidine decarboxylase. Neuroscience. 1988;26:827–45.
58. Green JD. The hippocampus. Physiol Rev. 1964;44:561–608.
59. Alvarez EO, Ruarte MB. Glutamic acid and histamine-sensitive neurons in the ventral hippocampus and the basolateral amygdala of the rat: functional interaction on memory and learning processes. Behav Brain Res. 2004;152:209–19.
60. Alvarez EO, Banzan AM. The activation of histamine-sensitive sites of the ventral hippocampus modulates the consolidation of a learned active avoidance response in rats. Behav Brain Res. 2008;189:92–9.
61. Bonini JS, Da Silva WC, Da Silveira CKB, Köhler CA, Izquierdo I, Cammarota M. Histamine facilitates consolidation of fear extinction. Int J Neuropsychopharmacol. 2011;14:1209–17.
62. Fiorenza NG, Rosa J, Izquierdo I, Myskiw JC. Modulation of the extinction of two different fear-motivated tasks in three distinct brain areas. Behav Brain Res. 2012;232:210–6.
63. da Silva WC, Bonini JS, Bevilaqua LR, Izquierdo I, Cammarota M. Histamine enhances inhibitory avoidance memory consolidation through a H2 receptor-dependent mechanism. Neurobiol Learn Mem. 2006;86:100–6.
64. Baldi E, Bucherelli C, Schunack W, Cenni G, Blandina P, Passani MB. The H3 receptor protean agonist proxyfan enhances the expression of fear memory in the rat. Neuropharmacology. 2005;48:246–51.
65. Cangioli I, Baldi E, Mannaioni PF, Bucherelli C, Blandina P, Passani MB. Activation of histaminergic H3 receptors in the rat basolateral amygdala improves expression of fear memory and enhances acetylcholine release. Eur J Neurosci. 2002;16:521–8.
66. Passani MB, Cangioli I, Baldi E, Bucherelli C, Mannaioni PF, Blandina P. Histamine H3 receptor-mediated impairment of contextual fear conditioning and in-vivo inhibition of cholinergic transmission in the rat basolateral amygdala. Eur J Neurosci. 2001;14:1522–32.
67. Spieler RE, Nelson CA, Huston JP, Mattioli R. Post-trial administration of H1 histamine receptor blocker improves appetitive reversal learning and memory in goldfish, Carassius auratus. Neurosci Lett. 1999;277:5–8.

68. Giovannini MG, Efoudebe M, Passani MB, Baldi E, Bucherelli C, Giachi F, Corradetti R, Blandina P. Improvement in fear memory by histamine-elicited ERK2 activation in hippocampal CA3 cells. J Neurosci. 2003;23:9016–23.
69. Benetti F, Izquierdo I. Histamine infused into basolateral amygdala enhances memory consolidation of inhibitory avoidance. Int J Neuropsychopharmacol. 2013;16:1539–45.
70. Frisch A, Postilnick D, Rockah R, Michaelovsky E, Postilnick S, Birman E, Laor N, Rauchverger B, Kreinin A, Poyurovsky M, Schneidman M, Modai I, Weizman R. Association of unipolar major depressive disorder with genes of the serotonergic and dopaminergic pathways. Mol Psychiatry. 1999;4:389–92.
71. Orsini CA, Maren S. Neural and cellular mechanisms of fear and extinction memory formation. Neurosci Biobehav Rev. 2012;236:1773–802.
72. Quirk GJ, Mueller D. Neural mechanisms of extinction learning and retrieval. Neuropsychopharmacology. 2008;33:56–72.
73. Whitlock JR, Heynen AJ, Shuler MG, Bear MF. Learning induces long-term potentiation in the hippocampus. Science. 2006;313:1093–7.
74. Balderas I, Rodriguez-Ortiz CJ, Bermudez-Rattoni F. Retrieval and reconsolidation of object recognition memory are independent processes in the perirhinal cortex. Neuroscience. 2013;253:398–405.
75. Balderas I, Rodriguez-Ortiz CJ, Bermudez-Rattoni F. Consolidation and reconsolidation of object recognition memory. Behav Brain Res. 2015;285:213–22.
76. Broadbent NJ, Gaskin S, Squire LR, Clark RE. Object recognition memory and the rodent hippocampus. Learn Mem. 2010;17:5–11.
77. Ennaceur A, Delacour J. A new one-trial test for neurobiological studies of memory in rats. 1: behavioral data. Behav Brain Res. 1988;31:47–59.
78. Giovannini MG, Lana D, Pepeu G. The integrated role of ACh, ERK and mTOR in the mechanisms of hippocampal inhibitory avoidance memory. Neurobiol Learn Mem. 2015;119:18–33.
79. Walz R, Roesler R, Quevedo J, Rockenbach IC, Amaral OB, Vianna MR, Lenz G, Medina JH, Izquierdo I. Dosedependent impairment of inhibitory avoidance retention in rats by immediate post-training infusion of a mitogenactivated protein kinase kinase inhibitor into cortical structures. Behav Brain Res. 1999;105:219–23.
80. McGaugh JL. Time-dependent processes in memory storage. Science. 1966;153:1351–8.
81. McGaugh JL. Memory--a century of consolidation. Science. 2000;287:248–51.
82. McGaugh JL. Making lasting memories: remembering the significant. Proc Natl Acad Sci U S A. 2013;2:10402–7.
83. McGaugh JL. Consolidating memories. Annu Rev Psychol. 2015;66:1–24.
84. Cammarota M, Bevilaqua LR, Rossato JI, Lima RH, Medina JH, Izquierdo I. Parallel memory processing by the CA1 region of the dorsal hippocampus and the basolateral amygdala. Proc Natl Acad Sci U S A. 2008;105:10279–84.
85. Feng T, Yang S, Wen D, Sun Q, Li Y, Ma C, Cong B. Stress-induced enhancement of fear conditioning activates the amygdalar cholecystokinin system in a rat model of post-traumatic stress disorder. Neuroreport. 2014;25:1085–90.
86. Mello PB, Benetti F, Cammarota M, Izquierdo I. Effects of acute and chronic exercise and stress on different types of memory in rats. An Acad Bras Cienc. 2008;80:361–9.
87. Mello PB, Benetti F, Cammarota M, Izquierdo I. Physical exercise can reverse the deficit in fear memory induced by maternal deprivation. Neurobiol Learn Mem. 2009;92:364–9.
88. Benetti F, Mello PB, Bonini JS, Monteiro S, Cammarota M, Izquierdo I. Early postnatal maternal deprivation in rats induces memory deficits in adult life that can be reversed by donepezil and galantamine. Int J Dev Neurosci. 2009;27:59–64.
89. Bernaerts P, Lamberty Y, Tirelli E. Histamine H3 antagonist thioperamide dose-dependently enhances memory consolidation and reverses amnesia induced by dizocilpine or scopolamine in a one-trial inhibitory avoidance task in mice. Behav Brain Res. 2004;154:211–9.
90. Ghi P, di Carlo G, Molinengo L. Effects of thioperamide on locomotor activity and on memory processes. Prog Neuropsychopharmacol Biol Psychiatry. 1998;22:387–95.

91. Orsetti M, Ferretti C, Gamalero R, Ghi P. Histamine H3-receptor blockade in the rat nucleus basalis magnocellularis improves place recognition memory. Psychopharmacology (Berl). 2002;159:133–72.
92. Munari L, Provensi G, Passani MB, Blandina P. Selective brain region activation by histamine $H_3$ receptor antagonist/inverse agonist ABT-239 enhances acetylcholine and histamine release and increases c-Fos expression. Neuropharmacology. 2013;70:131–40.
93. Liu L, Zhang S, Zhu Y, Fu Q, Zhu Y, Gong Y, Ohtsu H, Luo J, Wei E, Chen Z. Improved learning and memory of contextual fear conditioning and hippocampal CA1 long-term potentiation in histidine decarboxylase knock-out mice. Hippocampus. 2007;17:634–41.
94. Brabant C, Charlier Y, Tirelli E. The histamine $H_3$-receptor inverse agonist pitolisant improves fear memory in mice. Behav Brain Res. 2013;243:199–204.
95. Singewald N, Schmuckermair C, Whittle N, Holmes A, Ressler KJ. Pharmacology of cognitive enhancers for exposure-based therapy of fear, anxiety and trauma-related disorders. Pharmacol Ther. 2015;149:150–90.
96. Coupé P, Fonov VS, Bernard C, Zandifar A, Eskildsen SF, Helmer C, Manjón JV, Amieva H, Dartigues JF, Allard M, Catheline G, Collins DL, Alzheimer's Disease Neuroimaging Initiative. Detection of Alzheimer's disease signature in MR images seven years before conversion to dementia: toward an early individual prognosis. Hum Brain Mapp. 2015;36:4758–70.
97. Grossberg S. From brain synapses to systems for learning and memory: object recognition, spatial navigation, timed conditioning, and movement control. Brain Res. 2015;1621:270–93.
98. Broadbent NJ, Gaskin S, Squire LR, Clark RE. Object recognition memory and the rodent hippocampus. Learn Mem. 2010;17:5–11.
99. Clarke JR, Cammarota M, Gruart A, Izquierdo I, Delgado-García JM. Plastic modifications induced by object recognition memory processing. Proc Natl Acad Sci U S A. 2010;107:2652–7.
100. da Silveira CK, Furini CR, Benetti F, Monteiro Sda C, Izquierdo I. The role of histamine receptors in the consolidation of object recognition memory. Neurobiol Learn Mem. 2013;103:64–71.

# Chapter 15
# Histamine and Appetite

**Gustavo Provensi, Patrizio Blandina, and Maria Beatrice Passani**

**Abstract** Our survival relies on the ability to search for food to attend immediate metabolic needs and to store excess energy in the form of fat to meet metabolic demands during fasting. Hunger and satiety are key factors driving eating behavior and are under control of a complex interplay of several central and peripheral neuroendocrine systems. Interest in the control of feeding has increased as a result of the obesity epidemic and rising incidence of metabolic diseases. The first evidence of the involvement of brain histaminergic system in the regulation of feeding dates back to the 1970s. Since then, many studies ensued, and up-to-date evidence suggests an inverse relationship between neuronal histamine and food intake. Preclinical studies demonstrated that brain histamine is released during both the appetitive and consummatory phases of feeding behavior and is also involved in the control of peripheral mechanisms regulating energy expenditure. Hypothalamic $H_1$ and $H_3$ receptors are crucial for the regulation of the diurnal rhythm of food consumption; furthermore, these receptors have been specifically recognized as mediators of energy intake and expenditure. All these features point for the histaminergic system as an attractive target for the development of new anti-obesity drugs. Unfortunately, so far, no selective brain-penetrating $H_1$ receptor agonists have been identified, and clinical trials of the potential $H_3$ receptor antagonists-induced weight loss did not meet up to the expectations or were interrupted. Not all is lost tough, recent clinical trials demonstrated the potential of betahistine (an $H_1$ agonist/$H_3$ antagonist) in opposing metabolic side effects associated with chronic antipsychotic treatment.

**Keywords** Histamine • Food consumption • Energy homeostasis • Body weight • Neuropeptides • H1 receptor • H3 receptor • Betahistine • Antipsychotics • Clinical trials

G. Provensi, Ph.D. (✉) P. Blandina, M.D.
Department of Neuroscience, Psychology, Drug Research and Child Health, University of Florence, Florence, Italy
e-mail: gustavo.provensi@unifi.it

M.B. Passani, Ph.D.
Department of Health Science, University of Florence, Florence, Italy



P. Blandina, M.B. Passani (eds.), *Histamine Receptors*, The Receptors 28,
DOI 10.1007/978-3-319-40308-3_15

## 15.1 Introduction

Body weight is tightly regulated by complex homeostatic mechanisms controlling the balance between food intake and energy expenditure; even subtle mismatches (less than 0.5 %) in this balance are sufficient to cause weight gain [1]. Thus, obesity can be defined as a state in which energy intake chronically exceeds energy expenditure and is widely recognized as a pervasive and fast-growing public health problem in many countries. In 2010, the World Health Organization estimated more than 700 million people worldwide are obese and nearly 2 billion people are overweight; thus, the neologism "globesity" was created to define the growing global epidemic of overweight and obesity [2]. The impact of morbidity and mortality associated with obesity on healthcare cost is also expected to rise with the increased incidence of obesity. Although prevention through education and changes in the lifestyle associated with psychological therapies are the first-line choices, this is not effective in all patients. A complementary strategy is the pharmacological therapy, but unfortunately, available anti-obesity drugs are scarce, and some were hastily withdrawn from market owing to their unacceptable side-effect burden [3].

Brain histamine plays a fundamental role in eating behavior. Animal studies have shown that brain histamine is released during the appetitive phase to provide a high level of arousal preparatory to feeding, but it also mediates satiety. Moreover, histamine regulates energy expenditure and regulates peripheral metabolic processes (Fig. 15.1). This chapter will provide an overview of the role of histamine and histaminergic receptors in feeding behavior and maintenance of body weight summarizing preclinical and clinical research and discuss the emerging clinical trials evaluating the potential utility of histaminergic compounds for the treatment of metabolic side effects associated with chronic antipsychotic treatment.

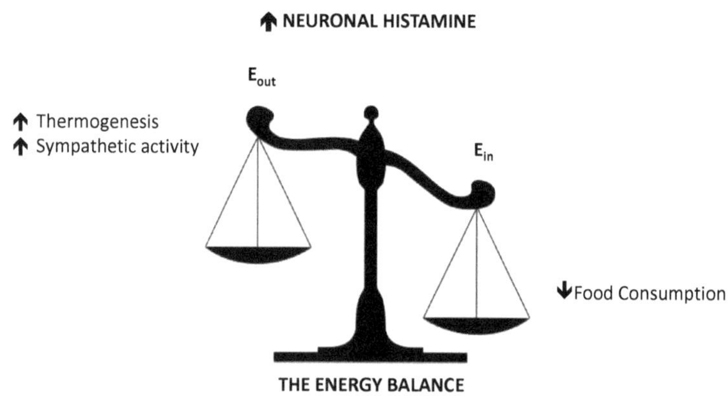


**Fig. 15.1** Brain histamine affects both sides of the energy balance, by reducing energy intake ($E_{in}$) and increasing energy expenditure ($E_{out}$)

## 15.2 Role of Histamine in Feeding and Weight Regulation

The first evidence of the inverse relationship between brain histaminergic activity and appetite dates back to 1973 when Clineschmidt and Lotti administered histamine into the lateral ventricle of cats and observed a long-term suppression of food intake [4]. Years later, a reduction on food intake was also observed when histamine was continuously infused into the suprachiasmatic nucleus of the hypothalamus [5] or acutely injected into the lateral ventricles [6] of rats. In analogy, systemic administration of the histamine precursor, L-histidine, [7–11] or the inhibitor of histamine catabolism, metoprine [12], also inhibited food consumption.

Conversely, restriction of neuronal histamine synthesis due to injections of the histidine decarboxylase (HDC) inhibitor alpha-fluoromethylhistidine (α-FMH) into the rat lateral ventricle increased significantly food consumption [13–15]. Bilateral microinfusion of α-FMH into the ventromedial hypothalamus (VMH) and the paraventricular nuclei (PVN) mimicked this effect, while injections into the lateral hypothalamus (LH), the dorsomedial hypothalamic (DMH), or the preoptic anterior hypothalamus (POAH) nuclei had no effect on food intake [14]. Histamine plays also an important role in regulating aspects of meal size and duration: depletion of histamine in the mesencephalic trigeminal sensory nucleus decreased the speed of meal consumption, whereas histamine depletion of VMH increased the amount of food they eat and the duration of feeding [16].

The generation of histidine decarboxylase gene knock-out (HDC-KO) mice, lacking functional HDC enzyme, and therefore unable to synthesize histamine [17] represented a very interesting tool to investigate the functions of endogenous histamine in the brain, including feeding behavior. Based on previous results, one could expect an obese phenotype for these animals, but surprisingly, no differences in caloric intake and body weight were observed between normal and histamine-deficient mice up to 11 weeks of age. Increases of 13 and 20 % in body weight were found in HDC-KO mice with respect to wild-type controls just at 16 and 30 weeks of age, respectively [18]. HDC-KO mice showed also an increased tendency to develop high-fat diet-induced obesity (DIO) when compared with wild-type littermates [19].

Neuronal histamine affects not only food intake but also regulates feeding circadian rhythms. Sustained infusion of α-FMH into the lateral ventricle disrupted light–dark cycles of feeding, drinking, and ambulatory activity in rats fed ad libitum [20]. It is known that food availability is a powerful circadian signal; thus, when food availability is restricted to a predictable time of the day, animals show increased motor activity and food searching behaviors before the anticipated daily meal, reflecting a state of increased arousal, related to an enhanced feeding motivation [21]. The involvement of the histaminergic system in feeding-induced arousal was demonstrated by a cluster-specific activation of neurons in the E3 subdivision of the histaminergic tuberomammillary nucleus (TMN) immediately before meal time in food-restricted rats under a scheduled feeding [22–24]. Furthermore, hypothalamic histamine increased when hungry rats were challenged to open a mesh container filled with palatable food [25]. On the contrary, rats fed ad libitum were not interested in the food, remained in quiet wake or sleeping during enticing, and showed no changes in histamine concentration [25]. Hence, in these experimental settings, increased activity of histaminergic neurons parallels a situation of arousal induced either by the expectation of food or the attempt to open the box with the food during enticing; thus, it is linked to the appetitive phase of feeding behavior.

There is also consistent evidence indicating that histamine regulates the consummatory phase of feeding behavior, as demonstrated by the transient but significant increase of histamine release in the hypothalamus when 24-h fasted rats were fed [26, 27]. When an animal eats, the oral cavity receives a variety of sensory information from food, such as taste and texture. Elegant studies demonstrated that the gustatory information can modulate the histaminergic activity by two mechanisms: the physiological excitation of the chorda tympani nerve, one of the taste nerves [28] and emotions elicited by taste perception, i.e., taste palatability [29]. Treesukosol and coworkers showed that adverse and hazardous taste stimuli like NaCl, HCl, or quinine caused significant increases in histamine release in the hypothalamus. On the other hand, histamine release was decreased by sucrose and saccharin solutions [29]. In rodents, chemicals that are described by humans as "bitter" or "nauseous" are rejected, while those described as "sweet" or "pleasurable" by humans are avidly accepted [30]. Therefore, it was postulated that histamine increase is related to aversive taste stimuli, but not to palatable tastes. Moreover, these findings suggest the possibility that palatable food blunts histamine release which results in overeating it [27]. Not only taste perception but also sensory information related to food texture can influence histaminergic activity. As an example, Ishizuka and colleagues [31] observed increased histamine release from the amygdala when rats were fed hard pellets, while no significant differences were observed when the animals ate soft pellets.

## 15.3 Role of Histaminergic Receptors in Feeding and Weight Regulation

Histamine exerts its actions through the activation of distinct subtypes of G-protein-coupled receptors. To date, four histamine receptor subtypes were identified, namely: $H_1R$, $H_2R$, $H_3R$, and $H_4R$ [32]. The involvement of the different histaminergic receptors on feeding regulation was extensively studied and, so far, a major role for $H_1$ and $H_3$ receptors acting at hypothalamic nuclei was identified.

In an interesting study, Haq and coworkers investigated the effects of dietary composition (quantity and quality of proteins and energetic level) on voluntary food intake and $H_1$ receptor expression in the whole rat brain [33]. High concentration of $H_1$ receptors, as assessed with binding assays on tissue homogenates, was associated with decreased food intake of rats fed with a low-protein diet. On the other hand, rats that received a low-energy diet displayed reduced $H_1$ receptor concentration that was associated with increased food intake [33]. Subsequent pharmacological studies demonstrated that centrally administered $H_1$ receptor agonists suppressed, whereas injection of $H_1$ receptor antagonist elicited food intake in rats [20, 34–41]. The effects of $H_1$ receptor antagonists seem to be site-specific, since microinfusions of these compounds locally into the VMH or PVN, but not into the LH or DMH, elicited feeding responses and increased both meal size and duration [40–42]. In keeping with the pharmacological manipulations of the $H_1$ receptor, genetically modified $H_1$R-KO mice exhibit an increase in daily food consumption and visceral adiposity [43], increased hepatic steatosis, hyperglycemia, and insulin resistance when fed with a high-fat diet [44].

Most experimental observations in rodents seem to agree that blockade of brain $H_3$ receptor is beneficial in decreasing energy intake, body weight, and plasma triglycerides [45–47]. Indeed, there are evidences demonstrating that $H_3$ receptor antagonists increase histamine release from the hypothalamus and reduce energy intake in normal and leptin-resistant mice with diet-induced obesity (reviewed in [45, 46]). In addition, administration of $H_1$ receptor antagonists attenuates the feeding suppression induced by $H_3$ antagonists [48]. Moreover, $H_3$ antagonists attenuate the orexigenic effect of neuropeptide Y (NPY) [49, 50] and enhance cholecystokinin (CCK)-induced satiety [51]. On the other hand, experiments with $H_3$ receptor agonists have provided mixed results. Activation of $H_3$ receptors increased food intake when administered intraperitoneally to mice [19], but failed to induce such an effect when centrally infused in rats [52]. However, despite the lack of effects per se, $H_3$ receptor agonists *R*-α-methylhistamine and imetit reduced satiety induced by amylin [53], bombesin [54] or CCK [51].

Not all data support an appetite-suppressant effect induced by $H_3$ receptor blockade. In diet-induced obese mice, Yoshimoto and colleagues reported that $H_3$ receptor activation reduced, whereas $H_3$ receptor antagonism increased food intake and body weight but presumably with a mechanism independent of histamine release [55]. In another study, Sindelar and coworkers observed that in spite of the administration route (oral or intraperitoneal) that produces similar pharmacokinetic profile, $H_3$ receptor occupancy and histamine turnover, thioperamide reduced food intake and conditioned place aversion in i.p.-treated animals only. The authors claim that blockade of central $H_3$ receptors does not play a direct role in decreasing food intake or altering energy homeostasis [56]. However, these data are hard to reconcile with studies in which i.c.v. administration was effective in reducing food consumption in a pathway clearly dependent on an intact histaminergic system [35].

Findings with $H_3$R-KO mice are also controversial. Toyota and colleagues described parallel growth curves for $H_3$R-KO and wild-type littermates, with the $H_3$R-KO animals displaying a slightly lower, but not significantly different, average body weight [57]. Conversely, Takahashi and coworkers demonstrated that $H_3$R-KO mice have a disrupted regulation of body weight, energy expenditure, and food intake resulting in obese hyperphagic mice with reduced energy expenditure that resembles the phenotype of $H_1$R-KO [58]. Such phenotype appears a paradox because one could expect that without presynaptic histamine receptors, inhibition of histamine synthesis and release would escape tonic control resulting in overstimulation of postsynaptic $H_1$ receptor and concomitant reduction of food intake. However, the authors found increased histamine concentration in the hypothalamus of $H_3$R-KO mice and proposed that elevated histamine content could desensitize $H_1$R resulting in hyperphagia [58]. Considering these data, it is clear that the effects of $H_3$R modulators on food consumption and metabolism are more complex and not only mediated by histamine release, but they are regulated through a variety of receptors and neurotransmitters and may be responsible for the discrepancies described above.

Nonhibernating seasonal mammals have adapted to temporal changes in food availability through behavioral and physiological mechanisms to store food and energy during times of predictable plenty and conserve energy during predicted shortage. Siberian hamsters (*Phodopus sungorus*) are seasonal animals that survive a winter climate by making adaptations in physiology and behavior, like reduced food intake and increased catabolism of fat reserves resulting in a natural loss of body weight [59]. Barrett and colleagues showed that $H_3$ receptor mRNA expression in the posterior hypothalamus is significantly decreased when animals are in a lean state during the short-day photoperiod of winter. After switching from an inhibitory short-day photoperiod to a stimulatory long-day photoperiod, increased expression of $H_3$ receptor occurs relatively rapidly along with body weight recovery [60]. Further studies demonstrated that administration of the $H_3$ receptor agonist imetit increased, whereas treatment with the $H_3$ receptor antagonists clobenpropit and thioperamide decreased food intake in hamsters in the lean state [61]. Differences in $H_3$ receptor mRNA expression were also described in golden hamster (*Mesocricetus auratus*) with a strong expression in the cortex and hippocampus of pubertals and in amygdalar areas of hibernating adult hamsters. Interestingly, thioperamide induced significant reduction of food intake in adults, but not in pubertals [62].

In contrast with $H_1$ receptor and $H_3$ receptor, there are few works using $H_2$ receptor ligands. Most studies demonstrate that treatment with either $H_2$ receptor agonist or antagonists had no effects in food consumption [38–41, 52]; therefore, $H_2$ receptor seems not to be involved in the regulation of feeding behavior.

## 15.4 Role of Histamine in Energy Homeostasis

Brain histamine affects both sides of the energy balance: by decreasing food intake and increasing expenditure [63] (Fig. 15.1). Maintenance of core temperature represents a major energy expenditure of a homeothermic organism and uncoupling proteins like UCP1 plays a central role in regulating energy expenditure and thermogenesis in rodents and neonates of larger mammalian species, including humans [64]. Infusion of histamine in the lateral ventricle or in the preoptic area, but not in the lateral hypothalamus or the ventromedial hypothalamic nucleus, caused upregulation of UCP1 mRNA expression in brown adipose tissue (BAT) and increased electrophysiological activity of sympathetic nerves that innervate it [43, 65] suggesting that the preoptic area is the principal locus of histaminergic modulation of thermogenesis. Interestingly, histamine-deficient animals have an impaired ability to express UCP1 in BAT [18], further suggesting a role of histamine signaling in the control of energy expenditure. Central administration of histamine or the $H_3$ receptor antagonist thioperamide increased the lipolytic response in white adipose tissue, whereas pretreatment, with a propranolol, beta-receptor antagonist, blocked the thioperamide-induced response, suggesting that the effect is mediated by sympathetic nerves that innervate white adipose tissue [66]. Furthermore, Kimura and coworkers recently proposed that central histamine downregulates hepatic gluconeogenic gene expression by activating $H_1$ receptors [67].

Histamine released in peripheral organs presumably is involved in metabolic and homeostatic processes related to food intake, but evidence is circumstantial. Intestinal mucosal mast cells are activated and degranulated to release histamine and other mediators to the circulation during fat absorption [68]. $H_1$ receptor signaling in the central nervous system (CNS), as well as in the pancreatic tissue regulates glucose metabolism, whereas $H_2$ receptor activation appears to be related to a peripheral action in the liver and skeletal muscles via the adiponectin system that regulates both lipid and glucose metabolisms [44].

## 15.5 Interactions Between Brain Histamine and Hormones that Control Feeding Behavior

The gastrointestinal tract and adipose tissue release more than 20 different hormones that regulate diverse physiological processes. In addition to local paracrine actions and peripheral endocrine effects, these hormones play a pivotal role relaying information on nutritional status to important appetite controlling centers within the CNS, such as hypothalamus and the brain stem, which integrates this peripheral information with brain signals (e.g., reward and mood) and contribute to regulate feelings of hunger and satiety [69, 70]. Thus, a very complex network of central and peripheral stimuli interacts to regulate feeding behavior, and brain histamine seems to act as a relay station integrating peripheral signals and central functions.

Leptin is predominantly secreted by adipocytes with circulating levels proportional to fat mass [71]. Both central and systemic administrations of leptin significantly increase central histamine availability [72], and serum leptin levels are highly elevated in HDC-KO mice [73]. Conversely, levels of hypothalamic histamine are reduced in high-fat diet-induced obesity (DIO) and diabetic mice due to leptin receptor point mutation (*db/db*) [74]. Accordingly, leptin-induced suppression of food intake was significantly attenuated in α-FMH-treated [75, 76] and $H_1$R-KO mice [72, 77]. Leptin-induced increase in UCP1 and UCP3 expressions of brown (BAT) and white (WAT) adipose tissues, respectively, were attenuated in $H_1$R-KO mice [66]. Increase in food intake and body weight observed in DIO and *db/db* mice was reversed by chronic i.c.v. infusion of histamine [77]. Histamine effect was attenuated when $H_1$ receptor expression was additionally disrupted in DIO and *db/db* mice [77].

Leptin also stimulates pro-opiomelanocortin (POMC) neurons, and POMC activates melanocortin-4 receptors (MC4Rs). Agouti yellow ($A^y$/a) mice develop obesity because they overexpress agouti-related protein, a physiological MC4R antagonist. Interestingly, administration of histamine (i.c.v.) to obese ($A^y$/a) mice reduced food intake and body weight and increased UCP1 expression in BAT. All these effects were attenuated in $H_1$R-deficient ($A^y$/a) obese mice [78].

The hypophagic effect of Glucagon-like peptide-1 (GLP-1) seems to be mediated, at least in part, by the neuronal histaminergic system. Central infusions with GLP-1 augmented the histamine turnover in the hypothalamus and induced hypophagic effect, which was partially attenuated in histamine-deprived rats [79].

Thyrotropin-releasing hormone (TRH) is secreted by neurons in the hypothalamic PVN. It suppresses food intake, activates the TMN neurons [80], and increases histamine turnover in the hypothalamus [81]. In food-deprived $H_1R$-KO mice and acute histamine-depleted rats, TRH-induced suppression of feeding is significantly attenuated [81]. Similar observations were reported for the anorectic effect of i.c.v. neurotensin [82] and nesfatin-1 [83]. The anorexic effect of the neuropeptide nesfatin-1 was partially attenuated in rats administered with α-FMH and in $H_1R$-KO mice. Nesfatin-1 central injection increased histamine turnover, vice versa histamine centrally injected increased nesfatin-1 expression in the hypothalamus. Moreover, immunohistochemical analysis revealed $H_1R$ expression on nesfatin-1 neurons in the PVN, and nesfatin-1 expression was significantly reduced in the hypothalamus of $H_1R$-KO mice as compared to wild-type littermates [83].

Estrogen inhibits food intake in mice, and consequently, ovariectomy results in hyperphagia and weight gain. This mechanism is also involved in the increased incidence of obesity in postmenopausal women. Interestingly, estrogen receptor α (Esα) is expressed on histaminergic neurons, and the anorexic effect of estrogen is attenuated in $H_1R$-KO mice. Moreover, estrogen supplementation completely reversed the effect of ovariectomy on weight gain and food intake in the wild type, but this response was attenuated in $H_1R$-KO mice [84].

Histamine and orexin neurons exert different, but complementary, controls on wakefulness: the former being more important for aspects of consciousness and cognitive functions, whereas the latter is involved primarily in behavioral arousal, including muscle tone, locomotion, and emotional reactions [85]. There is a close and reciprocal anatomical connection between histaminergic and orexinergic neurons. In vitro, orexin strongly excites TMN neurons [86]. Perfusion of orexin A into TMN increased histamine release from both the medial preoptic area and the frontal cortex and promoted wakefulness [87] in rats. When injected into the lateral ventricles, orexin A produces a significant increase in wakefulness [87], stimulates food intake, and upregulates mRNA expression of the orexigenic neuropeptide Y (NPY) in the wild type, but not in $H_1R$-KO mice [19]. Indeed, NPY mRNA expression was fourfold upregulated in $H_1R$-KO mice as compared with wild-type controls [19]. A delayed and short-lasting histamine release in rats and increased food intake to a much greater extent in $H_1R$-KO mice than in wild-type controls were observed after NPY i.c.v. injection suggesting that histamine may act on NPY system in a negative feedback loop downregulating NPY-stimulated food intake [88].

We recently reported a functional interaction between brain histamine and the endogenous lipid messenger oleoylethanolamide (OEA) [89]. OEA mediates fat-induced satiety by engaging sensory fibers of the vagus nerve that project centrally to the nucleus tractus solitarius (NTS). It was recently shown that noradrenergic NTS–PVN projections are involved in the activation of the hypothalamic oxytocin system [90, 91], and pharmacological blockade of oxytocin receptors in the brain prevents OEA anorexic effects [92]. We observed that in histamine-deficient mice,

OEA-induced hypophagia was significantly attenuated; thus, our hypothesis is that OEA induces anorexia indirectly stimulating also histamine neurons. We speculated that the nucleus of the solitary tract (NTS) adrenergic fibers projecting to the TMN disinhibit histaminergic neurons through $\alpha_2$A adrenoceptor-mediated mechanism (Fig. 15.2). OEA also increased c-Fos expression in a subgroup of TMN neurons and increased histamine release from the cortex of hungry mice [89]. As histamine neurons send broad projections within the CNS that are organized in functionally distinct circuits impinging on different brain regions, it is conceivable that OEA indirectly increases histamine release in the PVN where activation of $H_1$ receptors stimulate oxytocin release [93]. Accordingly, we observed that OEA-induced activation of oxytocic neurons in the PVN was blunted in histamine-deficient mice, an observation that could account for the inefficacy of OEA in these animals.

All together, these observations further prove the complexity of the histaminergic system as a regulator of food intake and energy metabolism, as both orexigenic and anorexigenic effects of endogenous molecules appear to require the integrity of the central histaminergic system.

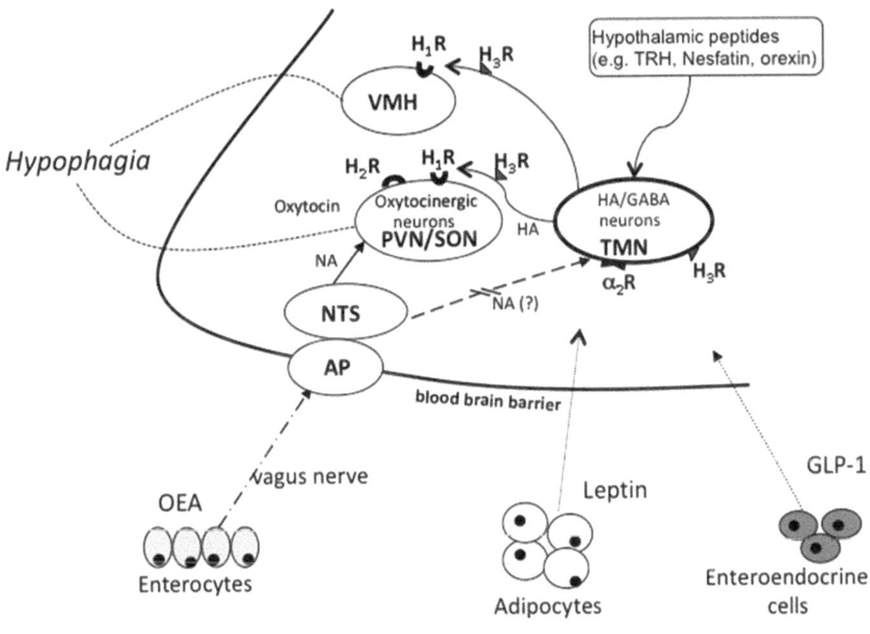


**Fig. 15.2** Schematic drawing illustrating the putative interactions between hypophagic hormones and the central histaminergic system. Histaminergic neurons are exclusively localized in the tuberomammillary nucleus (TMN) of the posterior hypothalamus. The *broken lines* designate presumed noradrenergic excitatory projections from the nucleus of the solitary tract (NTS) to the TMN. Several hormones directly or indirectly activate TMN histaminergic neurons and result in increase of histamine release in the TMN itself and also in the hypothalamic projection areas, where histamine by activating H1 and H2 receptors on feeding-related neurons mediates suppression of food intake

## 15.6 Brain Histamine and Eating Disorders: Clinical Studies

Recently, human studies are beginning to provide evidence that the pharmacological manipulation of the histaminergic system affects weight gain and body mass index. Ratliff and colleagues, using data available from the 2005 to 2006 National Health Examination Survey, described a relationship between the use of $H_1$ antihistamines (cetirizine, fexofenadine, and desloratadine) and an increased risk of obesity in US adults as compared with age- and gender-matched, healthy controls. Furthermore, $H_1$ antihistamine use was associated with higher plasma insulin concentrations [94]. In this regard, it was recently demonstrated that chronic administration of cetirizine or fexofenadine worsened progression of hepatosteatosis in mice that had been fed a high-fat diet; this effect was associated with significantly increased levels of glucose and hepatic bile acids [95]. These drugs have low affinity for muscarinic receptors that contribute to glucose metabolism via activation of vagal efferents. Hence, the homeostatic and metabolic effects of $H_1$ antihistamines are presumably due to their affinity for $H_1$ receptors both in the brain, despite poor brain penetration of these drugs, and in the periphery.

Results from preclinical studies suggest the $H_1$ receptor as a useful target for the development of new anti-obesity drugs, but from a therapeutic standpoint, though, no brain-penetrating $H_1$ receptor agonists have been identified devoid of intolerable peripheral side effects involving the cardiovascular, respiratory, or gastrointestinal systems. Therefore, the use of compounds that enhance histamine release from nerve terminals, such as $H_3$ receptor antagonists/inverse agonists, afforded an alternative strategy. Despite the encouraging preclinical results, though, clinical trials with $H_3$ receptor antagonists were disappointing. Patients enrolled in clinical trials to test the efficacy of other $H_3R$ antagonists (pitolisant or MK-0249) in narcolepsy [96], attention-deficit/hyperactivity disorder [97], schizophrenia [98], or epilepsy [99] did not report significant weight changes. Nonetheless, these compounds may turn out to be effective in tests evaluating specifically eating disorders. For instance, a multicenter, randomized, placebo-controlled phase II clinical trial that evaluated the efficacy of the $H_3$ receptor antagonist SCH 497079 on weight loss in obese and overweight subjects was recently completed, but the results were not disclosed (www.clinicatrials.gov). Given the substantial differences of the preclinical outcome and the discrepancies in clinical trials, considerable experimental effort remains necessary to prove the so far unclear concept of $H_3$ receptor antagonists in the treatment of obesity and weight gain [46].

Not only in obesity, alterations in the central histaminergic system were found also in other eating disorders. Positron emission tomography revealed in female anorexia nervosa patients an increase of [$^{11}$C]doxepin binding potential in the amygdala and lentiform nucleus when compared to healthy female controls [100]. This is a very interesting result because the amygdala certainly plays an important role in emotional responses [101], and histamine facilitates anxiety via $H_1R$ in the rat amygdala [102], but further studies, particularly in patients who have recovered from anorexia nervosa, are needed to clarify if and how higher binding potential of [$^{11}$C]doxepin are involved in anorexia nervosa.

## 15.7 Betahistine: A New Strategy to Prevent Antipsychotic-Induced Weight Gain

The inconsistent results obtained with $H_3R$ antagonists prompted researchers to change strategy and evaluate the effect of betahistine, a structural analog of histamine that combines $H_1$ receptor agonist and $H_3$ receptor antagonist properties [103] in weight control. An early study showed that acute treatment of pigmy goats with betahistine inhibited food intake and increased satiety [104]. In humans, betahistine is used in the symptomatic treatment of vestibular disorders with a remarkable safety profile that indicates that it does not cause cardiovascular, respiratory, or gastrointestinal side effects. Betahistine is orally available and readily penetrates the central nervous system [105]. These properties encouraged clinicians to examine the effects of acute and chronic regime in obese patients.

The acute effects of various doses of betahistine (48, 96, or 144 mg) on food intake and appetite were examined in a proof-of-concept, randomized, placebo-controlled study in obese, otherwise healthy women (BMI of 30–39.99 $kg/m^2$). Contrary to preclinical results, no significant effects of betahistine were observed in this cohort of obese women [106]. Another study evaluated weight loss and other parameters (e.g., blood pressure) during a 12-week treatment period, in an obese multiethnic population. The study reported no significant weight loss with betahistine; however, a post hoc subgroup analysis revealed a significant effect on body weight with minimal adverse effects only in women below 50 years of age [107].

Within the last 20 years, there has been a striking increase in the incidence of obesity and metabolic disorder in schizophrenic patients [108] associated with some first- and second-generation antipsychotic agents [109] that account for patients' non-compliance with these medications and increase the risk of obesity-related complications [110]. The pathophysiological mechanisms underlying antipsychotic-induced weight gain are yet to be elucidated, but the histaminergic neurotransmitter system has certainly a key role [111, 112]. For instance, olanzapine and clozapine that exhibit the highest binding affinities for the histamine $H_1$ and muscarinic receptors (reviewed in [113]) are associated with the greatest weight gain and metabolic impairments, including increased fasting glucose, insulin, and triglycerides. Thus, despite the disappointing results in obese patients, clinical research in schizophrenics treated with atypical antipsychotics with propensity to induce weight gain is continuing [114]. Poyurovsky and colleagues were the first to report the beneficial effects of betahistine treatment in three patients hospitalized for a first episode of schizophrenic disorder. Betahistine at the dosage used to treat vertigo was coadministered with olanzapine for 6 weeks. Although the lack of placebo controls precludes definitive conclusions, all patients after an initial weight gain during the first 2 weeks had no additional increments, suggesting a stabilizing effect of betahistine [115]. More recently, the same authors used a combination treatment with reboxetine, a selective norepinephrine reuptake inhibitor and betahistine to evaluate the olanzapine-induced weight gain in a small cohort of schizophrenic patients [116]. Compared to olanzapine/placebo-treated controls, patients in the combination therapy gained significantly less weight. It remains to be established if the combination reboxetine/betahistine offers a therapeutic advantage over betahistine alone. In this regard, the administration of betahistine in an animal model of olanzapine-induced weight gain was associated with decreased food intake and curbed weight gain [117]. These results open the possibility that betahistine might exert weight-mitigating effects also in patients affected by other pathologies associated with obesity (e.g., diabetes mellitus) and reduce metabolic parameters relevant to weight gain.

The promising effect of betahistine in preventing the metabolic side effects induced by atypical antipsychotic, but not in healthy obese patients may have to do with changes of the histaminergic system in the brain of people affected by schizophrenia, or plausibly obesity. The level of tele-methylhistamine, the histamine metabolite that mirrors histamine release, is increased in the cerebrospinal fluid of individuals with schizophrenia [118], although the relevance of this observation needs to be determined.

A significant association between genetic variants of $H_1$ receptors (rs346074–rs346070) and BMI/obesity has been identified in non-affective, psychotic disorder patients treated with high $H_1$ receptor affinity antipsychotic olanzapine, clozapine, and quetiapine [119]. Postmortem studies found reduced $H_1$ receptor binding in the frontal and prefrontal cortex and in the cingulate gyrus of individuals with schizophrenia [120], whereas $H_3$ receptor binding, as measured by receptor radioligand binding autoradiography, was increased in the dorsolateral prefrontal cortex but unchanged in the temporal cortex of patients with schizophrenia compared with the same brain regions in healthy control subjects [121]. These differences, though, may reflect structural abnormalities of the cortical network and change in cellular composition that underlies the functional impairments in this disorder. However, they may merely represent cytological adaptations in response to pharmacological treatment. It remains to be established if the morphological features of the histaminergic system in the brain of schizophrenic patients are responsible for the suggested beneficial effects of betahistine. Regarding obese patients, though, to our knowledge, there are no published data that correlate modifications of the histaminergic system with weight gain or dysmetabolic pathologies.

In light of these observations, several parameters were studied to understand the effect of betahistine in antipsychotic-induced metabolic disorders. Recently, it was shown in rats that both subchronic (2 weeks [122]) and chronic (up to 4 weeks [123]) betahistine co-treatment prevented olanzapine-induced weight gain and fat mass and regulated feeding efficiency. In addition, co-treatment with betahistine reverted or prevented olanzapine-induced cellular changes, such as increased expression of hypothalamic $H_1$ receptor, of pAMPK that senses cellular energy status [124]. Betahistine also prevented chronic olanzapine-induced decreased in the expression of UCP1 and PGC-1a [125], two biomarkers of thermogenesis in the BAT. These last observations suggest that betahistine may reduce olanzapine-induced weight gain and metabolic changes by modulating the hypothalamic $H_1$R-AMPK/BAT-UCP1-PGC-1a pathway. One possible mechanism of action is that histaminergic neurons stimulate the PVN or other hypothalamic nuclei to release peptides that in turn signal to the BAT by activating sympathetic nerves [126].

Therefore, during treatment with second-generation antipsychotics, hypothalamic $H_1$ receptor antagonism not only increases appetite but also reduces thermogenesis, presumably by inhibiting sympathetic outflow to the brainstem rostral raphe pallidus and rostral ventrolateral medulla. In addition to central effects, blocking peripheral $H_1$ receptors may contribute to fat accumulation by decreasing lipolysis and increasing lipogenesis in white adipose tissue. Also, $H_1$ receptor blockade in the liver and pancreatic tissue will contribute to the onset of metabolic disorders (see [111] for a review). We may attempt to provide a mechanistic explanation for the effects of betahistine. It is a weak $H_1$ agonist and a more potent $H_3$ antagonist that enhance histamine neuron activity [127] and histamine synthesis within the TMN [128]. Antagonists of the $H_3$ receptor decrease food intake in several mammalian species. Therefore, the pharmacodynamic profile of betahistine may be responsible for preventing antipsychotic-induced metabolic side effects. As a weak $H_1$ receptor agonist, betahistine would compete with antipsychotics for binding to this same receptor, both in the CNS and in peripheral organs, whereas antagonism at the $H_3$ receptor would increase brain histamine release to curb appetite.

## Concluding Remarks

More than 30 years have elapsed from the major discoveries that convinced the scientific community of the role of histamine as a neurotransmitter. Since then, many studies ensued, and it is now clear that brain histamine affects a variety brain functions: wakefulness, arousal, circadian rhythms, motor behavior, emotionality, and cognition [129, 130]. Brain histamine plays a central role in body weight maintenance by modulating both sides of the energy balance: decreasing food intake and increasing energy expenditure [63]. The paraventricular and ventromedial hypothalamic nuclei seem to be the brain sites where histamine, through mechanism involving $H_1$ and $H_3$ receptors, regulates food consumption. Moreover, histamine plays a major role in higher integrative brain functions, as arousal and cognition [130, 131]. Novelty-induced attention and arousal are of major importance for adaptation to changing environments by comparing new information with the recollection of past events. This has a major impact on feeding behavior, because histamine supposedly drives food intake by increasing the arousal state of the animal, and secondary to arousing the animal, it coordinates satiety and the consolidation of temporal information associated with food consumption [46]. Encouraging results are emerging from clinical trials using betahistine, a mixed $H_1$ receptor agonist and $H_3R$ antagonist, in the prevention of antipsychotic-induced weight gain. Therefore, we believe that understanding the actions of neuronal histamine especially at the hypothalamic circuits that control food intake and energy spending may be an important step toward the development of new pharmacotherapeutic approaches to the treatment of eating-related disorders.

## References


1. Rosenbaum M, Leibel RL, Hirsch J. Obesity. N Engl J Med. 1997;337(6):396–407.
2. WHO. Global status report on noncommunicable diseases 2010. Geneva: WHO; 2010.
3. Rodgers RJ, Tschop MH, Wilding JP. Anti-obesity drugs: past, present and future. Dis Model Mech. 2012;5(5):621–6.
4. Clineschmidt BV, Lotti VJ. Histamine: intraventricular injection suppresses ingestive behavior of the cat. Arch Int Pharmacodyn Ther. 1973;206(2):288–98.
5. Itowi N, Nagai K, Nakagawa H, Watanabe T, Wada H. Changes in the feeding behavior of rats elicited by histamine infusion. Physiol Behav. 1988;44(2):221–6.
6. Lecklin A, Tuomisto L. The blockade of H1 receptors attenuates the suppression of feeding and diuresis induced by inhibition of histamine catabolism. Pharmacol Biochem Behav. 1998;59(3):753–8.
7. Sheiner JB, Morris P, Anderson GH. Food intake suppression by histidine. Pharmacol Biochem Behav. 1985;23(5):721–6.
8. Vaziri P, Dang K, Anderson GH. Evidence for histamine involvement in the effect of histidine loads on food and water intake in rats. J Nutr. 1997;127(8):1519–26.
9. Kasaoka S, Tsuboyama-Kasaoka N, Kawahara Y, Inoue S, Tsuji M, Ezaki O, et al. Histidine supplementation suppresses food intake and fat accumulation in rats. Nutrition. 2004;20(11–12):991–6.
10. Orthen-Gambill N. Antihistaminic drugs increase feeding, while histidine suppresses feeding in rats. Pharmacol Biochem Behav. 1988;31(1):81–6.
11. Yoshimatsu H, Chiba S, Tajima D, Akehi Y, Sakata T. Histidine suppresses food intake through its conversion into neuronal histamine. Exp Biol Med (Maywood). 2002;227(1):63–8.
12. Lecklin A, Tuomisto L, MacDonald E. Metoprine, an inhibitor of histamine N-methyltransferase but not catechol-O-methyltransferase, suppresses feeding in sated and in food deprived rats. Methods Find Exp Clin Pharmacol. 1995;17(1):47–52.
13. Orthen-Gambill N, Salomon M. FMH-induced decrease in central histamine levels produces increased feeding and body weight in rats. Physiol Behav. 1992;51(4):891–3.
14. Ookuma K, Sakata T, Fukagawa K, Yoshimatsu H, Kurokawa M, Machidori H, et al. Neuronal histamine in the hypothalamus suppresses food intake in rats. Brain Res. 1993;628(1–2):235–42.
15. Tuomisto L, Yamatodani A, Jolkkonen J, Sainio EL, Airaksinen MM. Inhibition of brain histamine synthesis increases food intake and attenuates vasopressin response to salt loading in rats. Methods Find Exp Clin Pharmacol. 1994;16(5):355–9.
16. Sakata T, Yoshimatsu H, Masaki T, Tsuda K. Anti-obesity actions of mastication driven by histamine neurons in rats. Exp Biol Med (Maywood). 2003;228(10):1106–10.
17. Ohtsu H, Tanaka S, Terui T, Hori Y, Makabe-Kobayashi Y, Pejler G, et al. Mice lacking histidine decarboxylase exhibit abnormal mast cells. FEBS Lett. 2001;502(1–2):53–6.
18. Fulop AK, Foldes A, Buzas E, Hegyi K, Miklos IH, Romics L, et al. Hyperleptinemia, visceral adiposity, and decreased glucose tolerance in mice with a targeted disruption of the histidine decarboxylase gene. Endocrinology. 2003;144(10):4306–14.
19. Jorgensen EA, Knigge U, Watanabe T, Warberg J, Kjaer A. Histaminergic neurons are involved in the orexigenic effect of orexin-A. Neuroendocrinology. 2005;82(2):70–7.
20. Doi T, Sakata T, Yoshimatsu H, Machidori H, Kurokawa M, Jayasekara LA, et al. Hypothalamic neuronal histamine regulates feeding circadian rhythm in rats. Brain Res. 1994;641(2):311–8.
21. Mistlberger RE. Circadian food-anticipatory activity: formal models and physiological mechanisms. Neurosci Biobehav Rev. 1994;18(2):171–95.
22. Inzunza O, Seron-Ferre MJ, Bravo H, Torrealba F. Tuberomammillary nucleus activation anticipates feeding under a restricted schedule in rats. Neurosci Lett. 2000;293(2):139–42.
23. Meynard MM, Valdes JL, Recabarren M, Seron-Ferre M, Torrealba F. Specific activation of histaminergic neurons during daily feeding anticipatory behavior in rats. Behav Brain Res. 2005;158(2):311–9.

24. Umehara H, Mizuguchi H, Mizukawa N, Matsumoto M, Takeda N, Senba E, et al. Deprivation of anticipated food under scheduled feeding induces c-Fos expression in the caudal part of the arcuate nucleus of hypothalamus through histamine H(1) receptors in rats: potential involvement of E3 subgroup of histaminergic neurons in tuberomammillary nucleus. Brain Res. 2011;1387:61–70.
25. Valdes JL, Sanchez C, Riveros ME, Blandina P, Contreras M, Farias P, et al. The histaminergic tuberomammillary nucleus is critical for motivated arousal. Eur J Neurosci. 2010;31(11):2073–85.
26. Itoh Y, Oishi R, Saeki K. Feeding-induced increase in the extracellular concentration of histamine in rat hypothalamus as measured by in vivo microdialysis. Neurosci Lett. 1991;125(2):235–7.
27. Ishizuka T, Yamatodani A. Integrative role of the histaminergic system in feeding and taste perception. Front Syst Neurosci. 2012;6:44.
28. Smith KR, Treesukosol Y, Paedae AB, Contreras RJ, Spector AC. Contribution of the TRPV1 channel to salt taste quality in mice as assessed by conditioned taste aversion generalization and chorda tympani nerve responses. Am J Physiol Regul Integr Comp Physiol. 2012;303(11):R1195–205.
29. Treesukosol Y, Ishizuka T, Yamamoto C, Senda K, Tsutsumi S, Yamatodani A, et al. Hypothalamic histamine release by taste stimuli in freely moving rats: possible implication of palatability. Behav Brain Res. 2005;164(1):67–72.
30. Scott TR, Verhagen JV. Taste as a factor in the management of nutrition. Nutrition. 2000;16(10):874–85.
31. Ishizuka T, Sako N, Murotani T, Morimoto A, Yamatodani A, Ohura K. The effect of hardness of food on amygdalar histamine release in rats. Brain Res. 2010;1313:97–102.
32. Panula P, Chazot PL, Cowart M, Gutzmer R, Leurs R, Liu WL, et al. International union of basic and clinical pharmacology. XCVIII Histamine receptors. Pharmacol Rev. 2015;67(3):601–55.
33. Haq AU, Bundrant HM, Mercer LP. Food intake is inversely correlated with central nervous system histamine receptor (H1) concentrations in male Sprague-Dawley rats fed normal, low protein, low energy or poor quality protein diets. J Nutr. 1996;126(12):3083–9.
34. Machidori H, Sakata T, Yoshimatsu H, Ookuma K, Fujimoto K, Kurokawa M, et al. Zucker obese rats: defect in brain histamine control of feeding. Brain Res. 1992;590(1–2):180–6.
35. Sakata T. Hypothalamic neuronal histamine modulates energy balance in rats. Agents Actions Suppl. 1991;33:91–6.
36. Sakata T, Ookuma K, Fujimoto K, Fukagawa K, Yoshimatsu H. Histaminergic control of energy balance in rats. Brain Res Bull. 1991;27(3–4):371–5.
37. Yoshimatsu H, Machidori H, Doi T, Kurokawa M, Ookuma K, Kang M, et al. Abnormalities in obese Zuckers: defective control of histaminergic functions. Physiol Behav. 1993;54(3):487–91.
38. Fukagawa K, Sakata T, Shiraishi T, Yoshimatsu H, Fujimoto K, Ookuma K, et al. Neuronal histamine modulates feeding behavior through H1-receptor in rat hypothalamus. Am J Physiol. 1989;256(3 Pt 2):R605–11.
39. Sakata T, Fukagawa K, Fujimoto K, Yoshimatsu H, Shiraishi T, Wada H. Feeding induced by blockade of histamine H1-receptor in rat brain. Experientia. 1988;44(3):216–8.
40. Sakata T, Fukagawa K, Ookuma K, Fujimoto K, Yoshimatsu H, Yamatodani A, et al. Modulation of neuronal histamine in control of food intake. Physiol Behav. 1988;44(4–5):539–43.
41. Sakata T, Ookuma K, Fukagawa K, Fujimoto K, Yoshimatsu H, Shiraishi T, et al. Blockade of the histamine H1-receptor in the rat ventromedial hypothalamus and feeding elicitation. Brain Res. 1988;441(1–2):403–7.
42. Ookuma K, Yoshimatsu H, Sakata T, Fujimoto K, Fukagawa F. Hypothalamic sites of neuronal histamine action on food intake by rats. Brain Res. 1989;490(2):268–75.
43. Masaki T, Chiba S, Yasuda T, Noguchi H, Kakuma T, Watanabe T, et al. Involvement of hypothalamic histamine H1 receptor in the regulation of feeding rhythm and obesity. Diabetes. 2004;53(9):2250–60.

44. Wang KY, Tanimoto A, Yamada S, Guo X, Ding Y, Watanabe T, et al. Histamine regulation in glucose and lipid metabolism via histamine receptors: model for nonalcoholic steatohepatitis in mice. Am J Pathol. 2010;177(2):713–23.
45. Provensi G, Blandina P, Passani MB. The histaminergic system as a target for the prevention of obesity and metabolic syndrome. Neuropharmacology. 2016;106:3–12.
46. Passani MB, Blandina P, Torrealba F. The histamine H3 receptor and eating behavior. J Pharmacol Exp Ther. 2011;336(1):24–9.
47. Hancock AA, Brune ME. Assessment of pharmacology and potential anti-obesity properties of H3 receptor antagonists/inverse agonists. Expert Opin Investig Drugs. 2005;14(3):223–41.
48. Hancock AA, Diehl MS, Fey TA, Bush EN, Faghih R, Miller TR, et al. Antiobesity evaluation of histamine H3 receptor (H3R) antagonist analogs of A-331440 with improved safety and efficacy. Inflamm Res. 2005;54 Suppl 1:S27–9.
49. Itoh E, Fujimiya M, Inui A. Thioperamide, a histamine H3 receptor antagonist, suppresses NPY-but not dynorphin A-induced feeding in rats. Regul Pept. 1998;75–76:373–6.
50. Itoh E, Fujimiya M, Inui A. Thioperamide, a histamine H3 receptor antagonist, powerfully suppresses peptide YY-induced food intake in rats. Biol Psychiatry. 1999;45(4):475–81.
51. Attoub S, Moizo L, Sobhani I, Laigneau JP, Lewin MJ, Bado A. The H3 receptor is involved in cholecystokinin inhibition of food intake in rats. Life Sci. 2001;69(4):469–78.
52. Lecklin A, Etu-Seppala P, Stark H, Tuomisto L. Effects of intracerebroventricularly infused histamine and selective H1, H2 and H3 agonists on food and water intake and urine flow in Wistar rats. Brain Res. 1998;793(1–2):279–88.
53. Lutz TA, Del Prete E, Walzer B, Scharrer E. The histaminergic, but not the serotoninergic, system mediates amylin's anorectic effect. Peptides. 1996;17(8):1317–22.
54. Kent P, Plamondon H, Merali Z. Pharmaco-ontogeny of bombesin's suppression of food intake and its attenuation by histamine H3 receptor agonists. Brain Res Dev Brain Res. 1997;102(1):87–95.
55. Yoshimoto R, Miyamoto Y, Shimamura K, Ishihara A, Takahashi K, Kotani H, et al. Therapeutic potential of histamine H3 receptor agonist for the treatment of obesity and diabetes mellitus. Proc Natl Acad Sci U S A. 2006;103(37):13866–71.
56. Sindelar DK, Shepperd ML, Pickard RT, Alexander-Chacko J, Dill MJ, Cramer JW, et al. Central H3R activation by thioperamide does not affect energy balance. Pharmacol Biochem Behav. 2004;78(2):275–83.
57. Toyota H, Dugovic C, Koehl M, Laposky AD, Weber C, Ngo K, et al. Behavioral characterization of mice lacking histamine H(3) receptors. Mol Pharmacol. 2002;62(2):389–97.
58. Takahashi K, Suwa H, Ishikawa T, Kotani H. Targeted disruption of H3 receptors results in changes in brain histamine tone leading to an obese phenotype. J Clin Invest. 2002;110(12):1791–9.
59. Ebling FJ, Barrett P. The regulation of seasonal changes in food intake and body weight. J Neuroendocrinol. 2008;20(6):827–33.
60. Barrett P, Ross AW, Balik A, Littlewood PA, Mercer JG, Moar KM, et al. Photoperiodic regulation of histamine H3 receptor and VGF messenger ribonucleic acid in the arcuate nucleus of the Siberian hamster. Endocrinology. 2005;146(4):1930–9.
61. Jethwa PH, Barrett P, Turnbull Y, Enright RA, Warner A, Murphy M, et al. The role of histamine 3 receptors in the control of food intake in a seasonal model of obesity: the Siberian hamster. Behav Pharmacol. 2009;20(2):155–65.
62. Madeo M, Granata T, Facciolo RM, Tripepi S, Canonaco M. Feeding differences in pubertal and aged golden hamsters (Mesocricetus auratus) are related to specific cerebral expression pattern of histamine subtype 3 receptor. Behav Neurosci. 2006;120(6):1235–41.
63. Tabarean IV. Histamine receptor signaling in energy homeostasis. Neuropharmacology. 2016;106:13–9.
64. Nicholls DG, Locke RM. Thermogenic mechanisms in brown fat. Physiol Rev. 1984;64(1):1–64.
65. Yasuda T, Masaki T, Sakata T, Yoshimatsu H. Hypothalamic neuronal histamine regulates sympathetic nerve activity and expression of uncoupling protein 1 mRNA in brown adipose tissue in rats. Neuroscience. 2004;125(3):535–40.

66. Masaki T, Yoshimatsu H, Chiba S, Watanabe T, Sakata T. Central infusion of histamine reduces fat accumulation and upregulates UCP family in leptin-resistant obese mice. Diabetes. 2001;50(2):376–84.
67. Kimura K, Nakamura Y, Inaba Y, Matsumoto M, Kido Y, Asahara S, et al. Histidine augments the suppression of hepatic glucose production by central insulin action. Diabetes. 2013;62(7):2266–77.
68. Ji Y, Sakata Y, Yang Q, Li X, Xu M, Yoder S, et al. Activation of rat intestinal mucosal mast cells by fat absorption. Am J Physiol Gastrointest Liver Physiol. 2012;302(11):G1292–300.
69. Sam AH, Troke RC, Tan TM, Bewick GA. The role of the gut/brain axis in modulating food intake. Neuropharmacology. 2012;63(1):46–56.
70. Suzuki K, Jayasena CN, Bloom SR. Obesity and appetite control. Exp Diabetes Res. 2012;2012:824305.
71. Caro JF, Sinha MK, Kolaczynski JW, Zhang PL, Considine RV. Leptin: the tale of an obesity gene. Diabetes. 1996;45(11):1455–62.
72. Morimoto T, Yamamoto Y, Mobarakeh JI, Yanai K, Watanabe T, Yamatodani A. Involvement of the histaminergic system in leptin-induced suppression of food intake. Physiol Behav. 1999;67(5):679–83.
73. Hegyi K, Fulop KA, Kovacs KJ, Falus A, Toth S. High leptin level is accompanied with decreased long leptin receptor transcript in histamine deficient transgenic mice. Immunol Lett. 2004;92(1–2):193–7.
74. Itateyama E, Chiba S, Sakata T, Yoshimatsu H. Hypothalamic neuronal histamine in genetically obese animals: its implication of leptin action in the brain. Exp Biol Med (Maywood). 2003;228(10):1132–7.
75. Yoshimatsu H, Itateyama E, Kondou S, Tajima D, Himeno K, Hidaka S, et al. Hypothalamic neuronal histamine as a target of leptin in feeding behavior. Diabetes. 1999;48(12):2286–91.
76. Toftegaard CL, Knigge U, Kjaer A, Warberg J. The role of hypothalamic histamine in leptin-induced suppression of short-term food intake in fasted rats. Regul Pept. 2003;111(1–3):83–90.
77. Masaki T, Yoshimatsu H, Chiba S, Watanabe T, Sakata T. Targeted disruption of histamine H1-receptor attenuates regulatory effects of leptin on feeding, adiposity, and UCP family in mice. Diabetes. 2001;50(2):385–91.
78. Masaki T, Chiba S, Yoshimichi G, Yasuda T, Noguchi H, Kakuma T, et al. Neuronal histamine regulates food intake, adiposity, and uncoupling protein expression in agouti yellow (A(y)/a) obese mice. Endocrinology. 2003;144(6):2741–8.
79. Gotoh K, Fukagawa K, Fukagawa T, Noguchi H, Kakuma T, Sakata T, et al. Glucagon-like peptide-1, corticotropin-releasing hormone, and hypothalamic neuronal histamine interact in the leptin-signaling pathway to regulate feeding behavior. FASEB J. 2005;19(9):1131–3.
80. Parmentier R, Kolbaev S, Klyuch BP, Vandael D, Lin JS, Selbach O, et al. Excitation of histaminergic tuberomamillary neurons by thyrotropin-releasing hormone. J Neurosci. 2009;29 (14):4471–83.
81. Gotoh K, Fukagawa K, Fukagawa T, Noguchi H, Kakuma T, Sakata T, et al. Hypothalamic neuronal histamine mediates the thyrotropin-releasing hormone-induced suppression of food intake. J Neurochem. 2007;103(3):1102–10.
82. Ohinata K, Shimano T, Yamauchi R, Sakurada S, Yanai K, Yoshikawa M. The anorectic effect of neurotensin is mediated via a histamine H1 receptor in mice. Peptides. 2004;25(12):2135–8.
83. Gotoh K, Masaki T, Chiba S, Ando H, Shimasaki T, Mitsutomi K, et al. Nesfatin-1, corticotropin-releasing hormone, thyrotropin-releasing hormone, and neuronal histamine interact in the hypothalamus to regulate feeding behavior. J Neurochem. 2013;124(1):90–9.
84. Gotoh K, Masaki T, Chiba S, Higuchi K, Kakuma T, Shimizu H, et al. Hypothalamic neuronal histamine signaling in the estrogen deficiency-induced obesity. J Neurochem. 2009;110(6):1796–805.
85. Anaclet C, Parmentier R, Ouk K, Guidon G, Buda C, Sastre JP, et al. Orexin/hypocretin and histamine: distinct roles in the control of wakefulness demonstrated using knock-out mouse models. J Neurosci. 2009;29(46):14423–38.

86. Eriksson KS, Sergeeva OA, Haas HL, Selbach O. Orexins/hypocretins and aminergic systems. Acta Physiol (Oxf). 2010;198(3):263–75.
87. Huang ZL, Qu WM, Li WD, Mochizuki T, Eguchi N, Watanabe T, et al. Arousal effect of orexin A depends on activation of the histaminergic system. Proc Natl Acad Sci U S A. 2001;98(17):9965–70.
88. Ishizuka T, Nomura S, Hosoda H, Kangawa K, Watanabe T, Yamatodani A. A role of the histaminergic system for the control of feeding by orexigenic peptides. Physiol Behav. 2006;89(3):295–300.
89. Provensi G, Coccurello R, Umehara H, Munari L, Giacovazzo G, Galeotti N, et al. Satiety factor oleoylethanolamide recruits the brain histaminergic system to inhibit food intake. Proc Natl Acad Sci U S A. 2014;111(31):11527–32.
90. Romano A, Cassano T, Tempesta B, Cianci S, Dipasquale P, Coccurello R, et al. The satiety signal oleoylethanolamide stimulates oxytocin neurosecretion from rat hypothalamic neurons. Peptides. 2013;49:21–6.
91. Romano A, Tempesta B, Provensi G, Passani MB, Gaetani S. Central mechanisms mediating the hypophagic effects of oleoylethanolamide and N-acylphosphatidylethanolamines: different lipid signals? Front Pharmacol. 2015;6:137.
92. Gaetani S, Fu J, Cassano T, Dipasquale P, Romano A, Righetti L, et al. The fat-induced satiety factor oleoylethanolamide suppresses feeding through central release of oxytocin. J Neurosci. 2010;30(24):8096–101.
93. Bealer SL, Crowley WR. Histaminergic control of oxytocin release in the paraventricular nucleus during lactation in rats. Exp Neurol. 2001;171(2):317–22.
94. Ratliff JC, Barber JA, Palmese LB, Reutenauer EL, Tek C. Association of prescription H1 antihistamine use with obesity: results from the National Health and Nutrition Examination Survey. Obesity (Silver Spring). 2010;18(12):2398–400.
95. Raveendran VV, Kassel KM, Smith DD, Luyendyk JP, Williams KJ, Cherian R, et al. H1-antihistamines exacerbate high-fat diet-induced hepatic steatosis in wild-type but not in apolipoprotein E knockout mice. Am J Physiol Gastrointest Liver Physiol. 2014;307(2):G219–28.
96. Dauvilliers Y, Bassetti C, Lammers GJ, Arnulf I, Mayer G, Rodenbeck A, et al. Pitolisant versus placebo or modafinil in patients with narcolepsy: a double-blind, randomised trial. Lancet Neurol. 2013;12(11):1068–75.
97. Herring WJ, Wilens TE, Adler LA, Baranak C, Liu K, Snavely DB, et al. Randomized controlled study of the histamine H3 inverse agonist MK-0249 in adult attention-deficit/hyperactivity disorder. J Clin Psychiatry. 2012;73(7):e891–8.
98. Egan MF, Zhao X, Gottwald R, Harper-Mozley L, Zhang Y, Snavely D, et al. Randomized crossover study of the histamine H3 inverse agonist MK-0249 for the treatment of cognitive impairment in patients with schizophrenia. Schizophr Res. 2013;146(1–3):224–30.
99. Kasteleijn-Nolst Trenite D, Parain D, Genton P, Masnou P, Schwartz JC, Hirsch E. Efficacy of the histamine 3 receptor (H3R) antagonist pitolisant (formerly known as tiprolisant; BF2.649) in epilepsy: dose-dependent effects in the human photosensitivity model. Epilepsy Behav. 2013;28(1):66–70.
100. Yoshizawa M, Tashiro M, Fukudo S, Yanai K, Utsumi A, Kano M, et al. Increased brain histamine H1 receptor binding in patients with anorexia nervosa. Biol Psychiatry. 2009;65(4):329–35.
101. Hamann SB, Ely TD, Grafton ST, Kilts CD. Amygdala activity related to enhanced memory for pleasant and aversive stimuli. Nat Neurosci. 1999;2(3):289–93.
102. Zarrindast MR, Moghadam AH, Rostami P, Roohbakhsh A. The effects of histaminergic agents in the central amygdala of rats in the elevated plus-maze test of anxiety. Behav Pharmacol. 2005;16(8):643–9.
103. Fossati A, Barone D, Benvenuti C. Binding affinity profile of betahistine and its metabolites for central histamine receptors of rodents. Pharmacol Res. 2001;43(4):389–92.
104. Rossi R, Del Prete E, Scharrer E. Effect of the H1-histamine receptor agonist betahistine on drinking and eating behavior in pygmy goats. Physiol Behav. 1999;66(3):517–21.

105. Jeck-Thole S, Wagner W. Betahistine: a retrospective synopsis of safety data. Drug Saf. 2006;29(11):1049–59.
106. Ali AH, Yanoff LB, Stern EA, Akomeah A, Courville A, Kozlosky M, et al. Acute effects of betahistine hydrochloride on food intake and appetite in obese women: a randomized, placebo-controlled trial. Am J Clin Nutr. 2010;92(6):1290–7.
107. Barak N, Greenway FL, Fujioka K, Aronne LJ, Kushner RF. Effect of histaminergic manipulation on weight in obese adults: a randomized placebo controlled trial. Int J Obes (Lond). 2008;32(10):1559–65.
108. Baptista T, Zarate J, Joober R, Colasante C, Beaulieu S, Paez X, et al. Drug induced weight gain, an impediment to successful pharmacotherapy: focus on antipsychotics. Curr Drug Targets. 2004;5(3):279–99.
109. Arnoldy R, Curtis J, Samaras K. The effects of antipsychotic switching on diabetes in chronic schizophrenia. Diabet Med. 2014;31(3):e16–9.
110. Gohlke JM, Dhurandhar EJ, Correll CU, Morrato EH, Newcomer JW, Remington G, et al. Recent advances in understanding and mitigating adipogenic and metabolic effects of antipsychotic drugs. Front Psychiatry. 2012;3:62.
111. He M, Deng C, Huang XF. The role of hypothalamic H1 receptor antagonism in antipsychotic-induced weight gain. CNS Drugs. 2013;27(6):423–34.
112. Kim SF, Huang AS, Snowman AM, Teuscher C, Snyder SH. From the cover: antipsychotic drug-induced weight gain mediated by histamine H1 receptor-linked activation of hypothalamic AMP-kinase. Proc Natl Acad Sci U S A. 2007;104(9):3456–9.
113. Teff KL, Kim SF. Atypical antipsychotics and the neural regulation of food intake and peripheral metabolism. Physiol Behav. 2011;104(4):590–8.
114. Barak N. Betahistine: what's new on the agenda? Expert Opin Investig Drugs. 2008;17(5):795–804.
115. Poyurovsky M, Pashinian A, Levi A, Weizman R, Weizman A. The effect of betahistine, a histamine H1 receptor agonist/H3 antagonist, on olanzapine-induced weight gain in first-episode schizophrenia patients. Int Clin Psychopharmacol. 2005;20(2):101–3.
116. Poyurovsky M, Fuchs C, Pashinian A, Levi A, Weizman R, Weizman A. Reducing antipsychotic-induced weight gain in schizophrenia: a double-blind placebo-controlled study of reboxetine-betahistine combination. Psychopharmacology (Berl). 2013;226(3):615–22.
117. Deng C, Lian J, Pai N, Huang XF. Reducing olanzapine-induced weight gain side effect by using betahistine: a study in the rat model. J Psychopharmacol. 2012;26(9):1271–9.
118. Prell GD, Green JP, Kaufmann CA, Khandelwal JK, Morrishow AM, Kirch DG, et al. Histamine metabolites in cerebrospinal fluid of patients with chronic schizophrenia: their relationships to levels of other aminergic transmitters and ratings of symptoms. Schizophr Res. 1995;14(2):93–104.
119. Vehof J, Risselada AJ, Al Hadithy AF, Burger H, Snieder H, Wilffert B, et al. Association of genetic variants of the histamine H1 and muscarinic M3 receptors with BMI and HbA1c values in patients on antipsychotic medication. Psychopharmacology (Berl). 2011;216(2):257–65.
120. Iwabuchi K, Ito C, Tashiro M, Kato M, Kano M, Itoh M, et al. Histamine H1 receptors in schizophrenic patients measured by positron emission tomography. Eur Neuropsychopharmacol. 2005;15(2):185–91.
121. Jin CY, Anichtchik O, Panula P. Altered histamine H3 receptor radioligand binding in post-mortem brain samples from subjects with psychiatric diseases. Br J Pharmacol. 2009;157(1):118–29.
122. Lian J, Huang XF, Pai N, Deng C. Preventing olanzapine-induced weight gain using betahistine: a study in a rat model with chronic olanzapine treatment. PLoS One. 2014;9(8):e104160.
123. Lian J, Huang XF, Pai N, Deng C. Betahistine ameliorates olanzapine-induced weight gain through modulation of histaminergic, NPY and AMPK pathways. Psychoneuroendocrinology. 2014;48:77–86.
124. Minokoshi Y, Alquier T, Furukawa N, Kim YB, Lee A, Xue B, et al. AMP-kinase regulates food intake by responding to hormonal and nutrient signals in the hypothalamus. Nature. 2004;428(6982):569–74.

125. Stefanidis A, Verty AN, Allen AM, Owens NC, Cowley MA, Oldfield BJ. The role of thermogenesis in antipsychotic drug-induced weight gain. Obesity (Silver Spring). 2009;17(1):16–24.
126. Yasuda T, Masaki T, Chiba S, Kakuma T, Sakata T, Yoshimatsu H. L-histidine stimulates sympathetic nerve activity to brown adipose tissue in rats. Neurosci Lett. 2004;362(2):71–4.
127. Gbahou F, Davenas E, Morisset S, Arrang JM. Effects of betahistine at histamine H3 receptors: mixed inverse agonism/agonism in vitro and partial inverse agonism in vivo. J Pharmacol Exp Ther. 2010;334(3):945–54.
128. Lacour M, Sterkers O. Histamine and betahistine in the treatment of vertigo: elucidation of mechanisms of action. CNS Drugs. 2001;15(11):853–70.
129. Haas HL, Sergeeva OA, Selbach O. Histamine in the nervous system. Physiol Rev. 2008;88(3):1183–241.
130. Panula P, Nuutinen S. The histaminergic network in the brain: basic organization and role in disease. Nat Rev Neurosci. 2013;14(7):472–87.
131. Torrealba F, Riveros ME, Contreras M, Valdes JL. Histamine and motivation. Front Syst Neurosci. 2012;6:51.

# Index





P. Blandina, M.B. Passani (eds.), *Histamine Receptors*, The Receptors 28,
DOI 10.1007/978-3-319-40308-3

**N**

Kolibri 360 GmbH
Ferdinand-Jühlke-Straße 7
99095 Erfurt, Deutschland
produktsicherheit@kolibri360.de